CATALYSIS IN
ELECTROCHEMISTRY

Wiley Series on Electrocatalysis and Electrochemistry

Andrzej Wieckowski, Series Editor

Fuel Cell Catalysis: A Surface Science Approach, Marc T. M. Koper

Electrochemistry of Functional Supramolecular Systems, Margherita Venturi, Paola Ceroni, and Alberto Credi

Fuel Cell Science: Theory, Fundamentals, and Biocatalysis, Andrzej Wieckowski and Jens Nørskov

Catalysis in Electrochemistry: From Fundamentals to Strategies for Fuel Cell Development, Elizabeth Santos and Wolfgang Schmickler

CATALYSIS IN ELECTROCHEMISTRY

FROM FUNDAMENTALS TO STRATEGIES FOR FUEL CELL DEVELOPMENT

Edited by

Elizabeth Santos
Wolfgang Schmickler

WILEY

A JOHN WILEY & SONS, INC., PUBLICATION

Published by John Wiley & Sons, Inc., Hoboken, New Jersey
Published simultaneously in Canada

For general information on our other products and services or for technical support, please contact our
Customer Care Department within the United States at (800) 762-2974, outside the United States at (317)
572-3993 or fax (317) 572-4002.

Wiley also publishes its books in a variety of electronic formats. Some content that appears in print may
not be available in electronic formats. For more information about Wiley products, visit our web site at
www.wiley.com.

Library of Congress Cataloging-in-Publication Data:

Santos, Elizabeth
    Catalysis in electrochemistry : from fundamentals to strategies for fuel cell development / edited by
Elizabeth Santos and Wolfgang Schmickler
        p. cm.
    Includes bibliographical references and index.
    ISBN 978-0-470-40690-8 (cloth)

Printed in Singapore

eBook ISBN: 978-0-470-92941-4
oBook ISBN: 978-0-470-92942-1
ePub ISBN: 978-0-470-93473-9

10 9 8 7 6 5 4 3 2 1

To Cristina Giordano and Wolf Vielstich,
Our Mentors and Parents in Science

E.S. and W.S.

CONTENTS

By many, Professor Wolf Vielstich is considered to be the father of modern fuel cell research. His originally German textbook *Brennstoffelemente*, which appeared in 1965, was quickly translated into Russian, English (*Fuel Cells*, Wiley, 1970), and Spanish and is a classic. Even today, it is an excellent source for the scientific background behind electrochemical energy conversion. In total, Professor Vielstich has been active in fuel cell research for more than 50 years, and only a few years ago he

began to edit, together with A. Lamm , H. Yokokawa, and H. Gasteiger, the *Handbook of Fuel Cells—Fundamentals, Technology, Applications*, whose last volumes have just been published. We interviewed him electronically at his home in Salta, Argentina.

INTERVIEW WITH WOLF VIELSTICH

What made you become interested in fuel cells as early as the 1950s?
During the early 1950s, when I was still a student, I worked at the Ruhrchemie-Oberhausen during vacations. There, in cooperation with Professor Justi, Braunschweig, experiments on high-temperature FCs were in progress, while I worked on alkaline H_2/O_2 cells at low temperatures. Then, in the 1960s, it was a good idea to choose H_2/O_2 cells, with liquid reactants, for electric power and water supply (from the cell reaction) for the NASA Apollo programm. The Wernher von Braun organization asked my advice about introducing a 1- to 2-kW FC unit, and I suggested Dr. Jose Giner from our Electrochemistry Department at Bonn University to do the job. It was a success, and NASA is still using this alkaline FC today.

Since that time there has been an ebb and flow in fuel cell activities. What are the perspectives at present?
During the last 10 years, the automobile industry, including General Motors and Daimler, has been optimistic about developing an acid-type H_2–air unit for application in electric cars. But up to now, costs of production are still too high, and in addition the supply of hydrogen in the form of gas or liquid is an unsolved problem.

What is the most promising fuel for fuel cells: hydrogen, methanol, something else?

While power density and energy capacity are satisfactory in the case of H_2/O_2 cells, as has been demonstrated by their recent use in motorgliders (DLR-Motorsegler Antares), this is not at all the case for methanol or ethanol.

At present, there is much activity in fuel cell research. What are the challenges for fundamental science in this area today?

Nowadays, much fundamental fuel cell research is focused on the catalysis of methanol and ethanol oxidation at temperatures below 80°C. Theoretically, with oxidation to CO_2, methanol can deliver six e^- per molecule; at present, commercial cells deliver 250 mA cm^{-2} and a cell voltage of 400 mV at 60°C, and the six e^- per molecule is almost attained. But for application in electric cars a factor of at least 4 to 5 in power density would be necessary. Using ethanol, present-day catalysts show an even lower energy density than with methanol and offer no more than 2–3 electrons per molecule, while the complete oxidation to CO_2 involves 12 electrons. A catalyst breaking the C–C bond has still to be found.

This year, the first commercial all-electric cars, heavily subsidized, have been marketed in Japan. Are they just a marketing hype or are they already a viable alternative to conventional cars?

All-electric cars, using Li ion batteries, still have a problem. Due to the high costs, only small cars with limited power and energy capacity are being built. Hybrid systems, as produced by Toyota today, use only small battery sets; this makes sense in this particular application.

What role will electric cars play in 20 years time? Which technologies will they use: fuel cells, batteries, or a combination of both?

Without a marked change in the availability of gas and/or oil, even in 20 years time all-electric cars, using fuel cells or batteries, will make only a small contribution, mainly due to the high production costs.

E. Santos

W. Schmickler

■■■■■ PREFACE TO THE WILEY SERIES ON ELECTROCATALYSIS AND ELECTROCHEMISTRY

The Wiley Series on Electrocatalysis and Electrochemistry covers recent advances in electrocatalysis and electrochemistry and depicts prospects for their contribution to the industrial world. The series illustrates the transition of electrochemical sciences from its beginnings in physical electrochemistry (covering mainly electron transfer reactions, concepts of electrode potentials, and structure of the electrical double layer) to a filed in which electrochemical reactivity is shown as a unique aspect of heterogeneous catalysis, is supported by high-level theory, connects to other areas of science, and focuses on electrode surface structure, reaction environments, and interfacial spectroscopy.

The scope of this series ranges from electrocatalysis (practice, theory, relevance to fuel cell science and technology) to electrochemical charge transfer reactions, biocatalysis and photoelectrochemistry. While individual volumes may appear quite diverse, the series promises up-to-date and synergistic reports on insights to further the understanding of properties of electrified solid/liquid systems. Readers of the series will also find strong refernce to theoretical approaches for predicting electrocatalytic reactivity by high-level theories such as DFT. Beyond the theoretical perspective, further vehicles for growth are provided by the sound experimental background and demonstration of the significance of such topics as energy storage, syntheses of catalytic materials via rational design, nanometer-scale technologies, prospects in electrosynthesis, new research instrumentation, surface modifications in basic research on charge transfer and related interfacial reactivity. In this context, one might notice that new methods that are being developed for one specific field are readily adapted for application in another.

Electrochemistry has benefited from numerous monographs and review articles due to its applicability in the practical world. Electrocatalysis has also been the subject of individual reviews and compilations. The Wiley Series on Electrocatalysis and Electrochemistry hopes to address the current activity in both of these complementary fields by containing volumes that individually focus on topics of current and potential interest and application. At the same time, the chapters intend to demonstrate the connections of electrochemistry to areas in addition to chemistry and physics, such as chemical engineering, quantum mechanics, chemical physics, surface science, biochemistry, and biology, and thereby bring together a vast range of literature that covers each topic. While the title of each volume informs of the specific concentration chosen by the volume editors and chapter authors, the integral outcome of the series aims is to offer a broad-based analysis of the total development of the field.

The progress of the series will provide a global definition of what electrocatalysis and electrochemistry are concerned with now and how these fields will evolve overtime. The purpose is twofold; to provide a modern reference for graduate instruction and for active researchers in the two disciplines, and to document that electrocatalysis and electrochemistry are dynamic fields that are ever-expanding and ever-changing in their scientific profiles.

Creation of each volume required the editor involvement, vision, enthusiasm and time. The Series Editor thanks all the individual volume editors who graciously accepted the invitations. Special thanks are for Ms. Anita Lekhwani, the Series Acquisitions Editor, who extended the invitation to the Series Editor and is a wonderful help in the assembling process of the series.

ANDRZEJ WIECKOWSKI
Series Editor

CONTRIBUTORS

Antolini, Ermete Scuola Scienza Materiali, Via 25 Aprile 22, 16016 Cogoleto (Genova), Italy

Arvia, Alejandro Jorge Instituto de Investigaciones Fisicoquímicas Teóricas y Aplicadas (INIFTA), Universidad Nacional de La Plata (UNLP), Consejo Nacional de Investigaciones Científicas y Técnológicas (CONICET), Diagonal 113 y Calle 64, CC16, Suc. 4, 1900 La Plata, Argentina

Baltruschat, Helmut Abteilung Elektrochemie, Universität Bonn, Römerstr. 164, D-53117 Bonn, Germany

Beltramo, Guillermo Jülich Forschungzentrum, Institute of Complex Systems, D-52425 Jülich, Germany

Bogolowski, Nicky Abteilung Elektrochemie, Universität Bonn, Römerstr. 164, D-53117 Bonn, Germany

Bolzán, Agustín Eduardo Instituto de Investigaciones Fisicoquímicas Teóricas y Aplicadas (INIFTA), Universidad Nacional de La Plata (UNLP), Consejo Nacional de Investigaciones Científicas y Técnológicas (CONICET), Diagonal 113 y Calle 64, CC16, Suc. 4, 1900 La Plata, Argentina

Calatayud, Mónica UPMC Univ Paris 06, UMR 7616, Laboratoire de Chimie Théorique, F-75005, Paris, France; CNRS, UMR 7616, Laboratoire de Chimie Théorique, F-75005, Paris, France

Climent, Víctor Instituto de Electroquímica, Universidad de Alicante, Apdo. 99, E-03080 Alicante, Spain

Cuesta, Ángel Instituto de Química Física "Rocasolano," CSIC, C. Serrano, 119, E-28006 Madrid, Spain

Ernst, Siegfried Abteilung Elektrochemie, Universität Bonn, Römerstr. 164, D-53117 Bonn, Germany

Feliú, Juan M. Instituto de Electroquímica, Universidad de Alicante, Apdo. 99, E-03080 Alicante, Spain

Giesen, Margret Jülich Forschungzentrum, Institute of Complex Systems, D-52425 Jülich, Germany

Girault, Hubert H. Laboratoire d'Electrochimie Physique et Analytique, Ecole Polytechnique Fédérale de Lausanne, Station 6, CH-1015 Lausanne, Switzerland

González, Ernesto R. Instituto de Química de São Carlos, Universidade de São Paulo, 13560-970 São Carlos, Brasil

Gross, Axel Institute of Theoretical Chemistry, Ulm University, Albert-Einstein-Allee 11, D-89069 Ulm, Germany

Gutiérrez, Claudio Instituto de Química Física "Rocasolano," CSIC, C. Serrano, 119, E-28006 Madrid, Spain

Herrero, Enrique Instituto de Electroquímica, Universidad de Alicante, Apdo. 99, E-03080 Alicante, Spain

Koper, Marc T. M. Leiden Institute of Chemistry, Leiden University, PO Box 9502, 2300 RA Leiden, The Netherlands

Parsons, Roger 16 Thornhill Road, Bassett, Southampton 5016 TAT, United Kingdom

Pasquale, Miguel Ángel Instituto de Investigaciones Fisicoquímicas Teóricas y Aplicadas (INIFTA), Universidad Nacional de La Plata (UNLP), Consejo Nacional de Investigaciones Científicas y Técnológicas (CONICET), Diagonal 113 y Calle 64, CC16, Suc. 4, 1900 La Plata, Argentina

Samec, Zdeněk J. Heyrovsky Institute of Physical Chemistry, Academy of Sciences of Czech Republic, Dolejskova 3, 182 23, Prague 8, Czech Republic

Santos, Elizabeth Instituto de Física Enrique Gaviola (IFEG-CONICET), Facultad de Matemática, Astronomía y Física Universidad Nacional de Córdoba, Córdoba, Argentina and Institute of Theoretical Chemistry, Ulm University, Albert-Einstein-Allee 11, D-89069 Ulm, Germany

Schmickler, Wolfgang Institute of Theoretical Chemistry, Ulm University, Albert-Einstein-Allee 11, D-89069 Ulm, Germany

Schnur, Sebastian Institute of Theoretical Chemistry, Ulm University, Albert-Einstein-Allee 11, D-89069 Ulm, Germany

Su, Bin Laboratoire d'Electrochimie Physique et Analytique, Ecole Polytechnique Fédérale de Lausanne, Station 6, CH-1015 Lausanne, Switzerland

Ticianelli, Edson A. Instituto de Química de São Carlos, Universidade de São Paulo, 13560-970 São Carlos, Brasil

Tielens, Frederik UPMC Univ Paris 06, UMR 7609, Laboratoire de Réactivité de Surface, F-75005, Paris, France; CNRS, UMR 7609, Laboratoire de Réactivité de Surface, F-75005, Paris, France

CHAPTER 1

Volcano Curves in Electrochemistry

ROGER PARSONS

Bassett, Southhampton, 5016 TAT, United Kingdom

1.1 INTRODUCTION

The effect of bonding of a reactant with the catalyst is qualitatively described by Sabatier [1], who pointed out that some bonding is necessary for the reaction to be catalyzed but that strong bonding with the catalyst would block the surface and slow the reaction. He did not consider the implication that the rate as a function of the strength of the bond between the intermediate and the catalyst would form a curve with a maximum. This conclusion was reached by Balandin [2], who plotted "volcano" curves [3] from a thermochemical point of view. However, these were only a small part of his theory of catalysis, which he called "multiplet theory," a theory mainly concerned with the steric relation between reactant and the surface structure of the catalyst.

The origin of the volcano curve for a catalytic reaction can be demonstrated using a very simplified model. To do this it is necessary to invoke the Brønsted relation between rate constants and equilibrium constants [4]. This was based on a study of the relation between the catalytic constant (k) for homogeneous acid-catalyzed reactions and the dissociation constant (K) of the acid concerned:

$$k = GK^\alpha \tag{1.1}$$

where G and α are constants characteristic of the reaction and α is often close to $\frac{1}{2}$. The well-known relations between K and the Gibbs energy of reaction ΔG and between k and the Gibbs energy of activation $\Delta G^{\#}$ leads directly to

$$\Delta G^{\#} = \alpha \, \Delta G + \text{const} \tag{1.2}$$

This is an example of the linear Gibbs energy relationship, one of the best known being that of Hammett [5]. Frumkin [6] pointed out the common significance of

Catalysis in Electrochemistry: From Fundamentals to Strategies for Fuel Cell Development,
First Edition. Edited by Elizabeth Santos and Wolfgang Schmickler.
© 2011 John Wiley & Sons, Inc. Published 2011 by John Wiley & Sons, Inc.

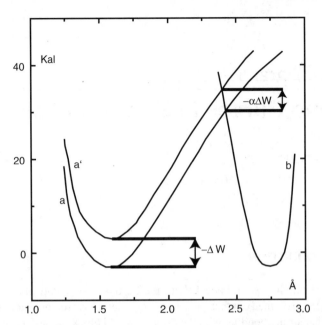

FIGURE 1.1   Relation between adsorption energy of hydrogen atom and activation energy. (Plot adapted from Horiuti and Polanyi [8].)

Brønsted's α and the transfer coefficient introduced by Erdey-Gruz and Volmer [7] to explain the slope of the Tafel plot obtained for electrochemical hydrogen evolution [7]. The meaning of Eq. (1.2) was explained by Horiuti and Polanyi [8] for a wide variety of proton transfer reactions when they represented the reaction path by a scheme of two approximately parabolic curves for the initial and final states, their intersection being the transition state (Fig. 1.1).

The relative slopes of the two curves at this intersection gave the value of α. They pointed out that this model could be applied to the proton transfer from a hydroxonium ion to a metal surface in the electrochemical process of hydrogen evolution. Although they discussed the effect of the metal–hydrogen bond strength on this process, this did not lead them to derive a volcano curve because they did not consider the effect of the coverage of the surface by hydrogen atoms.

Dogonadze, et al. [9] criticized the work of Horiuti and Polanyi because they used a quasi-classical approach. This cannot be valid at room temperature because the energy levels in each state are too widely spaced and a quantal approach is essential.

1.2   VOLCANO CURVES IN HETEROGENEOUS CATALYSIS

That the consideration of the surface coverage leads to a volcano curve may be demonstrated by using the simplest type of catalytic reaction, that known as the Eley–Rideal mechanism [10]. The reaction between two gases,

$$A(g) + B(g) \rightarrow AB(g) \tag{1.3}$$

is assumed to proceed by a two-step mechanism with only species A adsorbed:

$$A(g) \rightarrow A(ads) \tag{1.4}$$

$$B(g) + A(ads) \rightarrow AB(g) \tag{1.5}$$

The catalyst surface is assumed uniform so that the adsorption of A follows the Langmuir isotherm, so that, if reaction (1.4) were in equilibrium, the surface coverage θ would be given by

$$\theta = \frac{(p_A/p_{A,1/2})}{1 + p_A/p_{A,1/2}} \tag{1.6}$$

where p_A is the pressure of A is the gas phase and $p_{A,1/2}$ is the pressure at equilibrium with the half-covered surface. This may be related to the enthalpy of adsorption of A, ΔH_{ads},

$$p_{A,1/2} = K_{ads} \exp\left(\frac{\Delta H_{ads}}{RT}\right) \tag{1.7}$$

where K_{ads} is a constant.

The rate of adsorption of A may be written as

$$\vec{v}_1 = \vec{k}_1 (1 - \theta)\, p_A \tag{1.8}$$

and that of the reverse reaction as

$$\overleftarrow{v}_1 = \overleftarrow{k}_1 \theta \tag{1.9}$$

where $\vec{k}_1$ and $\overleftarrow{k}_1$ are rate constants. If (1.8) and (1.9) are equated, of course the equilibrium isotherm is obtained and it follows that

$$p_{A,1/2} = \frac{\overleftarrow{k}_1}{\vec{k}_1} \tag{1.10}$$

and $p_{A,1/2}$ is the reciprocal of the equilibrium constant of reaction (1.4). Brønsted's relation (1.1) together with the equilibrium constant (1.10) may be used to express the dependence of the rate constants on the enthalpy of adsorption:

$$\vec{k}_1 = \vec{k}_1^0 \exp\left(-\frac{\alpha_1 \Delta H_{ads}}{RT}\right) \tag{1.11}$$

$$\overleftarrow{k}_1 = \overleftarrow{k}_1^0 \exp\left(\frac{(1 - \alpha_1)\Delta H_{ads}}{RT}\right) \tag{1.12}$$

where $\overrightarrow{k}_1^0$ and $\overleftarrow{k}_1^0$ are the rate constants when $\Delta H_{ads} = 0$. Similarly the rate constants of reaction (1.5) may be related to the enthalpy of that reaction, ΔH_2. However, the sum of this enthalpy and that of reaction (1.3) is the enthalpy of the overall reaction, which is independent of the nature of the catalyst. Thus it is convenient to express the rates of reaction (1.5) in terms of the enthalpy of reaction (1.3) and include the latter in the constant terms. The analogues of Eqs. (1.8), (1.9), (1.11), and (1.12) for reaction (1.5) can then be written:

$$\overrightarrow{v}_2 = \overrightarrow{k}_2 p_B \theta \tag{1.13}$$

$$\overleftarrow{v}_2 = \overleftarrow{k}_2 p_{AB} (1 - \theta) \tag{1.14}$$

$$\overrightarrow{k}_2 = \overrightarrow{k}_2^0 \exp\left(\frac{\alpha_2 \Delta H_{ads}}{RT}\right) \tag{1.15}$$

$$\overleftarrow{k}_2 = \overleftarrow{k}_2^0 \exp\left(-\frac{(1 - \alpha_2)\Delta H_{ads}}{RT}\right) \tag{1.16}$$

Note that $\overrightarrow{k}_2^0$ and $\overleftarrow{k}_2^0$ in (1.15) and (1.16) are the rate constants for $\Delta H_{ads} = 0$.

The rate of the overall reaction can be obtained for three limiting conditions (the intermediate regions occur under small regions of conditions):

(a) Adsorption is fast and remains in equilibrium; reaction (1.5) controls the rate.
(b) Adsorption controls the rate and reaction (1.5) remains in equililibrium.
(c) Both reactions control the rate \and the back reactions may be neglected.

For case (a) equation (1.6) may be substituted in equation (1.13) and then substituting (1.7) and (1.15) in the result gives

$$v = \frac{\left(\overrightarrow{k}_2^0 / K_{ads}\right) p_A p_B \exp\left[-(1 - \alpha_2)\Delta H_{ads}/RT\right]}{1 + (p_A / K_{ads}) \exp(-\Delta H_{ads}/RT)} \tag{1.17}$$

From this it can be seen that $\ln v$ increases linearly with ΔH_{ads} at large negative values of ΔH_{ads}, goes through a maximum at

$$\Delta H_{ads} = RT\left(\frac{p_A \alpha_2}{K_{ads}(1 - \alpha_2)}\right) \tag{1.18}$$

and then decreases at large positive values of ΔH_{ads}. If ΔH_{ads} is eliminated between (1.7) and (1.18) it can be seen that the maximum rate in fact occurs when $\theta = (1 - \alpha_2)$

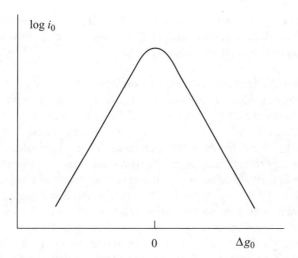

FIGURE 1.2   Form of relation between exchange current at hydrogen electrode and standard free energy of adsorption of hydrogen on electrode surface assuming that adsorbed atoms obey a Langmuir adsorption isotherm.

or at about half coverage. Similar results are obtained for conditions (b) and (c). The above can be seen in Figure 1.2. This shows that volcano curves are obtained whatever the detailed mechanism of the reaction.

It seems as though most workers on catalysis had little interest in volcano curves, perhaps because real mechanisms of catalytic reactions are more complex, involving more than one adsorbed species. Also, until the development of ultra-high-vacuum technology, it was difficult to obtain decisive data on clean surfaces. The above analysis was published in 1975 [11] (and even then in a rather inaccessible place) nearly 20 years after the comparable work on electrochemical reactions.

1.3   ATTEMPTS TO EXPLAIN EFFECTS OF NATURE OF ELECTRODE ON HYDROGEN EVOLUTION

It is generally accepted that hydrogen evolution can be explained in terms of three partial reactions:

1. The discharge reaction $H^+ + e^- \rightarrow H$, where H^+ is a solvated proton, e is an electron in the metal electrode, and H is a hydrogen atom adsorbed on the electrode surface. This was originally assumed to be the rate-determined reaction by Erdey Gruz and Volmer [7].
2. The recombination reaction $H + H \rightarrow H_2$, which was originally proposed as the rate-determining reaction by Tafel [12].

3. The alternative desorption reaction $H^+ + H \rightarrow H_2$ proposed by Heyrovsky [13] and later by Horiuti and Okamoto [14], who assumed that this would occur via the formation of an intermediate H_2^+.

While Bonhoeffer [15] noted the parallelism between the catalytic recombination of gaseous hydrogen and the rate of electrolytic hydrogen evolution, the first systematic attempt to study the effect of the nature of the metal electrode on the kinetics of hydrogen evolution was made by Bockris [16]. But, as with the catalysts at this time, it was difficult to ensure that the metal surface was not contaminated. Nevertheless, he made an attempt to correlate his results with other physical properties on metals [17]. This was apparently based on the theoretical approach to charge transfer due to Gurney [18] and modified by Butler [19] since it introduced the electronic work function as a main influence on the reaction rate. Although Bockris expected three types of behavior based on the above three partial reactions being rate determining, his plots of the logarithm of the exchange current (the rate of the reaction at equilibrium), versus the electronic work function showed two groups. Most of the metals lay on one line while the three least active metals (Tl, Pb, Hg) lay on a line parallel to it. However, it was shown later [20] that without strong interaction with the electrode the rate of a simple redox reaction is independent of the nature of the electrode material. This was confirmed experimentally [21, 22]. Consequently it is unlikely that the work function of the electrode has a direct influence on the kinetics of an electrode reaction. Of course it may relate directly to the formation of the bond between the electrode and an adsorbed species such as H.

A further attempt to correlate the effect of the metal on the rate of hydrogen evolution was made by Rüetschi and Delahay [23]. They calculated the strength of the metal–hydrogen bond from the strengths of the H–H bond and that of the metal atom in the metal M–M using the Pauling equation [24]:

$$D(M\text{–}H) = \tfrac{1}{2}[D(M\text{–}M) + D(H\text{–}H)] + 21.06\,(X_M\text{–}X_H)^2 \text{ kcal mol}^{-1} \qquad (1.19)$$

Where X_i is the electronegativity of atom i. On the basis of the observation by Eley [25] that the polarity of the M–H bond is small, they neglected the electronegativity term. Their plot of the experimental values of the hydrogen overpotential at a current density of $1\,\text{mA cm}^{-2}$ versus the energy of adsorption of H showed a linear relation for a number of metals having a negative slope with a few of the most active metals Pt, Pt, Au as well as Mo and Ta deviating. This slope would be expected if the discharge reaction is rate determining, as in the Horiuti and Polanyi model. Since they estimated the value of $D(M\text{–}M)$ in terms of the energy of sublimation of the metal, this amounts to a correlation with this quantity. Conway and Bockris [26] criticized their neglect of the electronegativity term. They repeated the calculation of $D(M\text{–}H)$ using values of electronegativity from Pritchard and Skinner [27] and Gordy and Thomas [28].

They found they there was still some evidence for a linear relation, but one with the opposite slope to that found by Rüetschi and Delahay (Fig. 1.3). They this interpreted in terms of a rate-determining ion+atom reaction which involved the desorption of the H atom so that the overpotential is greater for the more strongly bound H. The

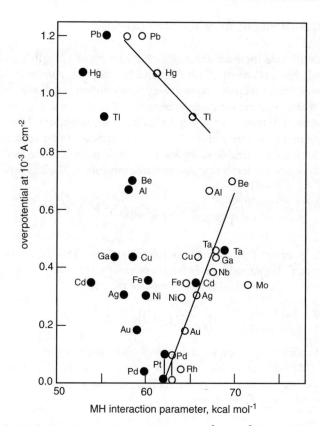

FIGURE 1.3 Plot of hydrogen overpotential at $10^{-3}\,A\,cm^{-2}$ as function of metal–H interaction energy parameter. (•) Rüetschi and Delahay's values [23]. (o) True values of D_{M-H} calculated from Pauling's equation using known electronegativity values. Values η at $10^{-3}\,A\,cm^{-2}$ are calculated from the experimental Tafel parameters recorded by Conway and Bockris [26].

exceptions to this were Hg, Pb, and Tl, which fell on a line with negative slope, in accord with the expectation for a rate-determining discharge reaction.

An approach which takes account of the kinetics of the three proposed component reactions in hydrogen evolution was made by Gerischer [29] following the work of Parsons and Bockris [30]. He attempted to predict the relative rates of these reactions as a function of the energy of adsorption of hydrogen. This led to the conclusion that under normal conditions (current density of 1 mA cm^{-2}), at metals weakly adsorbing hydrogen, the mechanism was discharge followed by the ion + atom reaction with the former controlling the rate. For very strongly adsorbing metals the mechanism would be the same but the ion + atom reaction would control the rate. There was a small region in between these two where the recombination reaction was rate controlling. He did not consider the back reactions and so could not include a discussion of the exchange current.

1.4   ELECTROCHEMICAL VOLCANO CURVE

Gerischer did include the back reaction in his next paper [31]. Independently and simultaneously he and Parsons [32] showed that the exchange current of the hydrogen reaction plotted versus the adsorption energy would follow a volcano-shaped curve. The special feature of electrochemical reactions is that their rate is dependent on the electrode potential. It is also frequently found that the equilibrium state is accessible where the rate of the forward and reverse reactions is equal. In the special case of the hydrogen reaction the probable existence of a single adsorbed intermediate atomic hydrogen can be assumed. It is first assumed that this follows the Langmuir isotherm:

$$\left(\frac{\theta^*}{1-\theta^*} \right)^2 = p_{H_2} \exp\left(-\frac{\Delta g^0}{kT} \right) \tag{1.20}$$

where θ is the fraction of the available surface occupied by H and the asterisk denotes that it is the equilibrium value, p is the pressure of molecular hydrogen in the gas phase, and Δg^0 is the standard Gibbs energy for the reaction:

$$H_2 + 2M \leftrightarrow 2H - M \tag{1.21}$$

the standard states being 1 atm for the gaseous hydrogen and $\theta = \frac{1}{2}$ for the adsorbed atoms. The Gibbs energy of adsorption of hydrogen molecules and their coverage of the surface are assumed negligible. For the discharge reaction the rates of the discharge reaction may be written as

$$\vec{v}_1 = \left(\frac{kT}{h} \right) a_{H^+} (1-\theta) \exp\left(\Delta \vec{g}_1^0 + \frac{\alpha_1 \phi e_0}{kT} \right) \tag{1.22}$$

and

$$\overleftarrow{v}_1 = \left(\frac{kT}{h} \right) \theta \exp\left(-\Delta \overleftarrow{g}_1^0 + \frac{(1-\alpha_1)\phi e_0}{kT} \right) \tag{1.23}$$

where a_{H^+} is the activity of protons in solution (strictly not measurable but will be assumed constant) and α_1 is defined as in eq. (1.1), φ is the inner potential difference between the metal and the solution, and $\Delta \vec{g}_1^0$ and $\Delta \overleftarrow{g}_1^0$ are the standard Gibbs energies of activation for the forward and reverse reactions, respectively, when $\varphi = 0$. These three quantities are not accessible to experiment or to theoretical calculation. Equations (1.22) and (1.23) can be rearranged in terms of accessible quantities following Tëmkin [33].

At equilibrium, $\vec{v}_1 = \overleftarrow{v}_1$, where

$$\exp\left(\frac{\phi^* e_0}{kT} \right) = a_{H^+} \left(\frac{1-\theta^*}{\theta^*} \right) \exp\left(\frac{\Delta \overleftarrow{g}_1^0 - \Delta \vec{g}_1^0}{kT} \right) \tag{1.24}$$

The potential $\varphi = \eta + \varphi^*$, where η is the overpotential. Then rate equations (1.22) and (1.23) may be written as

$$\vec{i}_1 = i_{0,1}\theta \exp(-\alpha_1 e_0 \eta) \tag{1.25}$$

$$\overleftarrow{i}_1 = i_{0,1}(1 - \theta) \exp[(1 - \alpha_1) e_0 \eta] \tag{1.26}$$

where i_0 is the exchange current of the discharge reaction and for $\theta = \theta^*$, where θ^* is the coverage at equilibrium

$$i_{0,1} = e_0 \left(\frac{kT}{h}\right) a_{H^+}^{1-\alpha} (\theta^*)^\alpha (1 - \theta^*)^{1-\alpha} \exp\left(-\frac{(1-\alpha)\Delta \vec{g}_1^0 + \alpha \Delta \overleftarrow{g}_1^0}{kT}\right) \tag{1.27}$$

The quantity in the exponential of (1.27) is experimentally accessible as the Gibbs energy of activation at the equilibrium potential and theoretically as the height of the Gibbs energy barrier when the Gibbs energy levels of the initial and final states are equal [33, 34]. The effect of the strength of adsorption of H on the exchange current is expressed in (1.27) entirely through the value of the equilibrium coverage θ^*. The dependence of the exchange current upon the standard Gibbs energy of adsorption is obtained by eliminating θ^* between (1.27) and (1.20):

$$i_{0,1} = a_{H^+}^{1-\alpha} \left[p_{H_2} \exp\left(-\frac{\Delta g^0}{2}kT\right)\right] \left[1 + p_{H_2} \exp\left(-\frac{\Delta g^0}{2}kT\right)\right]^{-1} G_1 \tag{1.28}$$

where

$$G_1 = e_0 \left(\frac{kT}{h}\right) \exp\left(-\frac{(1-\alpha)\Delta \vec{g}_1^0 + \alpha \Delta \overleftarrow{g}_1^0}{kT}\right) \tag{1.29}$$

From (1.28) it follows that when $\Delta g^0/2kT \gg 1$, that is adsorption is weak, the exchange current increases exponentially with a decrease of Δg^0. There is a maximum at $\Delta g^0 = 0$ and the volcano curve is as shown in Figure 1.2.

The other two partial reactions can be analyzed in the same way. The rate equations for the ion + atom reaction are

$$\vec{v}_2 = \left(\frac{kT}{h}\right) a_{H^+} \theta \exp\left[-\frac{\Delta \vec{g}_2^0 + \beta e_0 \phi}{kT}\right] \tag{1.30}$$

$$\overleftarrow{v}_2 = \left(\frac{kT}{h}\right) p_{H_2}(1 - \theta) \exp\left[-\frac{\Delta \overleftarrow{g}_2^0 - (1 - \beta)e_0 \phi}{kT}\right] \tag{1.31}$$

and the resulting volcano curve is given by

$$i_{0,2} = a_{H^+}^{1-\beta} \left[p_{H_2}^{\beta/2} \exp\left(-\frac{\beta \Delta g^0}{2} kT\right) \right] \left[1 + p_{H_2}^{\beta/2} \exp\left(-\frac{\Delta g^0}{2} kT\right) \right]^{-1} G_2 \quad (1.32)$$

where

$$G_2 = e_0 \left(\frac{kT}{h}\right) \exp\left(-\frac{(1-\beta)\,\Delta\vec{g}_2^0 + \beta\Delta\overleftarrow{g}_2^0}{kT}\right) \quad (1.33)$$

Clearly (1.32) has the same form as (1.28) and in fact they are identical if $G_1 = G_2$ and $\alpha = \beta$.

For the recombination reaction the rate equations are

$$\vec{v}_3 = \left(\frac{kT}{h}\right) \theta^2 \exp\left(-\frac{\Delta\vec{g}_3^0}{kT}\right) \quad (1.34)$$

$$\overleftarrow{v}_3 = \left(\frac{kT}{h}\right) p_{H_2}(1-\theta)^2 \exp\left(-\frac{\Delta\overleftarrow{g}_3^0}{kT}\right) \quad (1.35)$$

In this case the Gibbs energies of activation are independent of the electrode potential just as in an ordinary catalytic reaction, but of course they depend on the Gibbs energy of adsorption of H and this can be expressed using Eq. (1.2):

$$\Delta\vec{g}_3 = \Delta\vec{g}_3^0 - \gamma\,\Delta g^0 \quad (1.36)$$

$$\Delta\overleftarrow{g}_3 = \Delta\overleftarrow{g}_3^0 + (1-\gamma)\,\Delta g^0 \quad (1.37)$$

where γ depends on the relative slopes of the energy curves at their intersection in the same way as α and β. The term Δg_3^0 is the Gibbs energy of activation when the Gibbs energies of the initial and final states are equal. These lead to

$$i_{0,3} = p_{H_2} \exp\left(-\frac{(1-\gamma)\Delta g^0}{kT}\right) \left[1 + p_{H_2}^{1/2} \exp\left(-\frac{\Delta g^0}{kT}\right)\right]^{-2} G_3 \quad (1.38)$$

$$G_3 = 2e_0 \left(\frac{kT}{h}\right) \exp\left(-\frac{\Delta\vec{g}_3^0}{kT}\right) \quad (1.39)$$

Equation (1.38) has the same form as (1.28) and (1.32), but if $\alpha = \beta = \gamma$, the slopes of the branches of the plots away from the maximum are twice those for the other

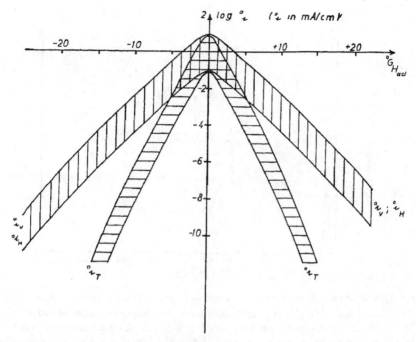

FIGURE 1.4   Standard exchange current dependence on $°G_{Had}$ at equilibrium conditions for three partial reactions. (Data adapted from Parsons and Bockris [30].)

two reactions. Gerischer [30] reached this conclusion by similar arguments, and his result is shown in Figure 1.4, in which he assumed that the maximum was the same for the three partial reactions. Parsons [32] used the well-established experimental observation that the rate of hydrogen evolution was the greatest on Pt, which must then be at the peak of the volcano. Also, the experimental results of Frumkin Dolin and Ershler [35] and those of Azzam and Bockris [36] suggest that the exchange current is the greatest for the discharge reaction and the least for the ion+atom reaction. Parsons also used the Tëmkin model [37] for a heterogeneous surface which leads to an adsorption isotherm and kinetic equations which differ from the Langmuir equations in the region of moderate adsorption. The resulting volcano curves are shown in Figure 1.5. They show a flat maximum in the region of moderate adsorption, but away from that they are identical to the Langmuir-based curves. Given the values of the exchange currents for the three partial reactions, it is straightforward to calculate the current–potential curves, and these are shown in Figure 1.6 for the labeled positions on the volcano curve. It is necessary to take account of the degree of coverage in doing this so that the correct version of the kinetics, that is, the Langmuir or Tëmkin, is used. The regions covered by the latter are indicated on each plot. The relation to experiment was discussed qualitative. Thus the very high overpotential metals Pb, Cd, and Tl and possibly Zn and Sn, which adsorb H weakly, would be expected to have the form shown

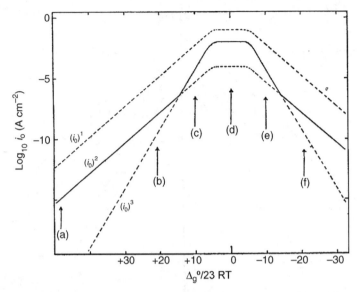

FIGURE 1.5 Proposed relation between exchange current of three partial reactions at hydrogen electrode and standard free energy of adsorption of hydrogen on electrode surface: $(i_0)^1$, exchange current for discharge reaction; $(i_0)^2$, exchange current for ion atom reaction; $(i_0)^3$, exchange current for combination reaction. Full line gives the observable exchange current. (Data adapted from Parsons [32].)

in Figure 1.7(a). In most cases the lower part of this curve might not be accessible because the currents are so small. Metals like NI, Fe, Mo, W, and Ta adsorb H strongly and so would be expected to follow Figure 1.6(f); thus their current–potential characteristic is much the same as that of the weakly adsorbing metals in the observable region. Copper and Ag seem to behave like Figure 1.6(b) and Pt was shown by Azzam and Bockris [36] to behave like Figure 1.6(d) when very high current densities were studied. Trasatti [38] made a detailed assessment of the available data used to plot the volcano curve (Fig. 1.7). Like Krishtalik [39], he came to the conclusion that there was no evidence for the flat top. This might mean that the metals near the peak of the volcano had uniform surfaces. They were all polycrystalline, and no systematic evidence has been gathered for well-defined surfaces. Conway and Tilak [40] in a comprehensive survey of the behavior of hydrogen at electrodes pointed out that on Pt hydrogen coverage was approximately complete at the equilibrium potential. They concluded that some other species of adsorbed hydrogen must be the intermediate on Pt and suggested that it was adsorbed on a surface already covered with the well-known adsorbed hydrogen. In a theoretical review of the hydrogen electrode reaction, Krishtalik [39], included the possibility of activationless and barrierless reactions in addition to the three reactions considered above.

Some further considerations of the simple model were published by Parsons, including the reverse reaction and the adsorption capacity in a generalized version

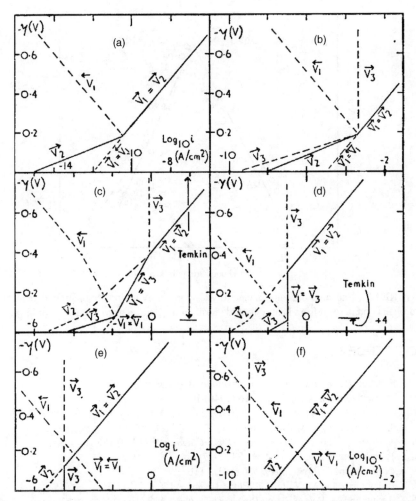

FIGURE 1.6 Tafel lines calculated using of exchange currents from Figure 1.5. Full line gives the observable Tafel line. (Data adapted from Parsons [32].)

[20] and the variation of the coefficients α, β, and γ with potential or with adsorption energy as a result of the parabolic shape of the potential energy curves [41]. In the same paper he set up a model of an electrode with two patches with different strengths of adsorption of hydrogen to estimate whether such electrodes might be better electrocatalysts, but this did not lead to any hope of improved performance. He also tried to use the kinetic approach to predict the yield of products in a branched electrochemical reaction [42]. However, this was a hypothetical scheme and it has not yet found application. Recently Nørskov et al. [43] have used density functional theory to calculate the energy of the chemisorption of H and rediscovered the volcano curve. Their kinetic analysis was criticized by Schmickler and Trasatti [44].

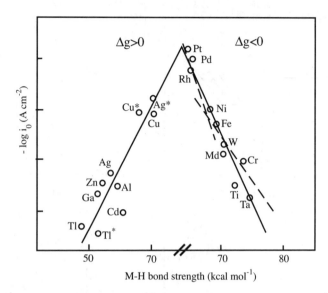

FIGURE 1.7   Exchange currents for electrolytic hydrogen evolution versus strength of metal–hydrogen bond derived from heat of hydride formation. (Data adapted from Trasatti [38].)

REFERENCES

1. F. Sabatier, *La catalyse en chimie organique*, Berauge, Paris 1920.
2. A. A. Balandin, *Zhur. Fiz. Khim.*, 16 (1946) 793.
3. A. A. Balandin, in *Advances in Catalysis*, Vol. 19, D. D. Eley, H. Fines, and P. B. Weisz (Eds.), Academic, New York, 1969, p. 1.
4. J. N. Brønsted and K. J. Pederson, *Z. Phys. Chem.*, 108 (1924) 185.
5. L. P. Hammett, *J. Am. Chem. Soc.*, 59 (1937) 66.
6. A. N. Frumkin, *Z. Phys. Chem.* 166 (1932) 116.
7. T. Erdey-Gruz and M. Volmer, *Z. Phys., Chem.*, A150 (1930) 203.
8. J. Horiuti and M. Polanyi, *Acta Physicochim. U.R.S.S*, 2 (1935) 505.
9. R. R. Dogonadze, Kuznetsov, and V. Levich, *Elektrokhimiya*, 3 (1967) 739.
10. D. D. Eley and E. K. Rideal, *Proc. Roy. Soc.*, A178 (1941) 429.
11. R. Parsons, in *Topics in Pure and Applied Electrochemistry*, S. K. Rangarajan (Ed.), SAEST Karaikudi, 1975, p. 91.
12. J. Tafel, *Zeit. Phys. Chem.*, (1905).
13. J. Heyrovsky, *Rec. Trav. Chim. Pays. Bas.*, 44 (1923) 500.
14. J. Horiuti and G. Okamoto, *Sci. Pap. Inst. Phys. Chem. Res. Tokyo*, 28 (1936) 231.
15. K. F. Bonhoeffer, *Zeit. Phys. Chem.*, B113 (1923) 199.
16. J. O'M. Bockris, *Faraday Discuss.*, 1 (1947) 95.
17. J. O'M. Bockris, *Zeit. Elektrochem.*, 55 (1951) 105.
18. R. W. Gurney, *Proc. Roy. Soc.*, A134 (1931) 133.
19. J. A. V. Butler, *Proc. Roy. Soc.*, A157 (1936) 423.
20. R. Parsons, *Surf. Sci.*, 2 (1964) 418.
21. A. Capon and R. Parsons, *J. Electroanal. Chem.*, 46 (1973) 215.
22. D. Galizzilo and S. Trasatti. *J. Electroanal Chem.*, 44 (1973) 367.

23. P. Rüetschi and P. Delahay, *J. Chem. Phys.*, 23 (1955) 195.
24. L. Pauling, *The Nature of the Chemical Bond*, Cornell University Press, Ithaca, NY, 1948, p. 60.
25. D. D. Eley, *J. Phys. Colloid Chem.*, 55 (1051) 1016.
26. B. E. Conway and J. O'M. Bockris, *J. Chem. Phys.*, 26 (1957) 532.
27. H. O. Pritchard and H. A. Skinner, *Chem. Rev.*, 55 (1955) 745.
28. W. Gordy and W. J. O. Thomas, *J. Chem. Phys.*, 24 (1956) 430.
29. H. Gerischer, *Z. Phys. Chem.*, N.F. 8 (1956) 137.
30. R. Parsons and J. O'M. Bockris, *Trans. Faraday Soc.*, 47 (1951) 914.
31. H. Gerischer, *Bull. Soc. Chim. Belg.*, 67 (1958) 506.
32. R. Parsons, *Trans. Faraday Soc.*, 54 (1958) 1053.
33. M. I. Tëmkin, *Zhur. Fiz. Khim.*, 22 (1948) 1081.
34. J. E. B. Randles, *Trans. Faraday Soc.*, 48 (1952) 828.
35. A. N. Frumkin, P. I. Dolin, and B. V. Ershler, *Acta Physicochim. U.R.S.S.*, 13 (1940) 779.
36. A. M. Azzam and J. O'M. Bockris, *Trans. Faraday Soc.*, 48 (1952) 145.
37. M. I. Tëmkin, *Zhur. Fiz. Khim.*, 15 (1941) 296.
38. S. Trasatti, *J. Electroanal. Chem.*, 39 (1972) 163.
39. L. I. Krishtalik, in *Advances in Electrochemisry and Electrochemical Engineering*, Vol. 7, P. Delahay (Ed.), Intersience, New York, 1970.
40. B. E. Conway and B. V. Tilak, *Electrochim. Acta*, 47 (2002) 3571.
41. R. Parsons, *Surface Sci.*, 18 (1969) 28.
42. R. Parsons, *Disc. Faraday Soc.*, 45 (1968) 40.
43. J. K. Nørskov, T. Bligard, A. Logardottir, J. R. Kitchin, J. G. Chen, S. Pandelov, and U. Stimming, *J. Electrochem. Soc.*, 152 (3) (2005) J. 23.
44. W. Schmickler and S. Trasatti, *J. Electrochem. Soc.*, 153 (2006) L31.

CHAPTER 2

Electrocatalysis: A Survey of Fundamental Concepts

ALEJANDRO JORGE ARVIA, AGUSTÍN EDUARDO BOLZÁN, and
MIGUEL ÁNGEL PASQUALE

Instituto de Investigaciones Fisicoquimicas Teoricas y Aplicadas (INIFTA),
Universidad Nacional de La Plata (UNLP), 1900 La Plata, Argentina

2.1 INTRODUCTORY ASPECTS TO FUEL CELL ELECTROCATALYSIS

A fuel cell is an electrochemical device in which electrochemical oxidation at the anode and a reduction reaction at the cathode take place and electrons that are released at the anode move to the cathode via the external circuit producing electrical work. This process is accompanied by the displacement of positive and negative ions through the conducting medium to the cathodic and anodic regions, respectively. The study of both the kinetics and mechanism of anodic and cathodic processes in either homogeneous or heterogeneous systems combines electrochemistry and catalysis knowledge [1–3]. Technical application of this knowledge to the development of fuel cells requires solving electrochemical engineering problems of design and optimization [4, 5]. The latter issue is beyond the scope of this chapter.

The term *electrocatalysis* was coined by N. Kobosev and W. Monblanowa [6] in 1934 and employed later, from 1963 onwards, when A. T. Grub established for the first time a correlation between hydrogen evolution reaction (HER) kinetic data on several metals and their respective parameters, such as metal–hydrogen bond energy and metal sublimation enthalpy [7]. The term is commonly employed to describe electrode processes where charge transfer reactions depend strongly on the nature of electrode material. A brief history of electrocatalysis and some cornerstone contributions from the last two centuries are given in Table 2.1.

The concept of electrocatalysis is applied to those electrochemical reactions that start from a dissociative chemisorption or a reaction step in which the electrode surface is involved. Accordingly, *electrochemical catalysis* and *heterogeneous catalysis* have

Catalysis in Electrochemistry: From Fundamentals to Strategies for Fuel Cell Development,
First Edition. Edited by Elizabeth Santos and Wolfgang Schmickler.
© 2011 John Wiley & Sons, Inc. Published 2011 by John Wiley & Sons, Inc.

TABLE 2.1   Brief Historical Survey of Fuel Cell Development

Date	Author(s)	Achievement	References
1802	H. Davy	Electricity from a chemical reaction in the cell carbon/aqueous nitric acid	8
1839	W. Groove	Development of the Groove cell, consisting of a hydrogen and oxygen electrode	9
1889	L. L. Mond and D. Langer	Improvement of Groove cell: oxygen from air and hydrogen from coal gas	10
1894	W. Ostwald	Fuel cell advantages are envisaged: heat-to-mechanical-energy conversion limitations; possibility of silent and clean devices	11
1896	W. W. Jacques	Development of Jacques cell: carbon/air cathode and carbon anode in molten NaOH	12
1921	E. Bauer, W. D. Treadwell, and G. Trumpler	Fe_3O_4 cathode and carbon anode in molten carbonate	13
1928	F. P. Bowden and E. Rideal	Catalytic effects of different metals on the HER	14
1931	N. Kobosev and W. Monblanowa	The term *electrocatalysis* was born	6
1935	J. Horiuti and M. Polanyi	H atom–metal interaction and proton discharge activation energy relationship	15
1946	O. K. Davtyan	Fuel cell progress in the former Soviet Union	16
1959	F. T. Bacon	Hydrogen/oxygen fuel cell at 40 atm and 200°C; dual-type pore electrode	17
1963	A. T. Grub	Correlation of HER kinetics of several metals with H atom/metal bond energy and metal sublimation energy data	7
1968	J. G. Broers	Fuel cell progress in the Netherlands	18
1970	E. Justi	Development of homoporous electrodes	19
2000		Fuel cell mass production in several countries	20–22

some common characteristics: (i) the substrate (electrode) activity depends on its electronic structure; (ii) substrate–adsorbate interactions due to either reactants or products are relevant; (iii) rate processes are sensitive to both the aspect ratio of catalyst particles and the mean coordination number of surface atoms; and (iv) the electrocatalyst lifetime depends on poisoning effects due to the accumulation of by-products as well as particle surface sintering and ripening phenomena at the electrode.

In contrast, there are remarkable differences between heterogeneous catalysis and electrocatalysis due to the presence of the electrochemical double layer (EDL) at the electrode–electrolyte interface, and the influence of the applied potential on the reaction rate and the change in the composition of the solution side of the EDL and the concentration of intermediates and products of reaction there. These are specific features of electrocatalysis that make the activation energy of electrocatalytic reactions

TABLE 2.2 Different Types of Nonbiological Fuel Cells

Type	Electrolyte	T / (K)	Fuel	Uses
AFC	Alkaline solutions	320–480	H_2	Space flights
PEM	Polymeric membranes	320–373	H_2	Space flights, cars, vehicles
PAFC	Phosphoric acid	~493	H_2	Residential supplies
MCFC	Molten carbonates	923	H_2, CH_4, natural gas	Residential supplies, stationary power systems
SOFC	Solid oxides	773–1273	H_2, CH_4, natural gas	Stationary power systems

depend considerably on the applied potential as well as, although usually to a lesser extent, on temperature. Then, the electrode potential is an important adjustable variable that can produce dramatic changes in the rate of the electrocatalytic reactions at low temperature. This enables a fine potential control at the electrochemical interface for handling the selectivity of electrocatalytic processes. Comparable effects in heterogeneous catalysis can be obtained by adjusting temperature (T), pressure (P), or reactant concentration (c) over a safe range in which no interference of side reactions occurs. The electrochemical modification of catalytic activity and selectivity control of conductive catalysts in contact with solid electrolytes or aqueous alkaline solutions is due to faradaic introduction of promoting ionic species at the catalytic surface. This is most often a reversible phenomenon of interest in electrocatalysis [23, 24].

From a technical standpoint, a fuel cell is fed with a fuel for the anodic reaction and oxygen/air from the atmosphere for the cathodic reaction, which are either continuously or discontinuously supplied from outside the housing, and wastes resulting from spent fuel are removed [2]. The electrochemical combustion involves either simple molecules such as those of hydrogen, alcohols, and alkanes or more complex ones, including biological materials. Hydrogen is frequently obtained from reforming processes. Typical fuel cells utilized for different applications are indicated in Table 2.2 [25, 26].

Hybrid devices that combine batteries, fuel cells, and redox flow cells are also employed. This is the case of metal–air cells that combine a metal electrode reaction at the anode and the electrocatalytic reduction of oxygen at the cathode [27]. Redox flow cells usually consist of an external storage with the electrolyte and a liquid fuel stored locally in a closed loop. In these cells two electrolytes are used: a redox metal–ion system at the anode ($M^{z+}/M^{(z+n)+} + ne^-$) and an oxygen electrode at the cathode. Products from the anodic reaction ($M^{(z+n)+}$) are utilized as intermediates for fuel oxidation at low T. After chemical regeneration, the electrolyte is returned to the anode.

Besides chemical-to-electrical direct energy conversion, electrocatalysis becomes important in other areas such as in the development of cleaner technologies for production of substances, devices for medical applications, optimization of technical processes in which biological systems and nanomaterials are combined, as in the protection of materials in different biological environments, energy conversion in mitochondria, enzymatic systems, and processes in the brain. Applications of electrocatalysis are surveyed in the literature [3, 28, 29].

2.2   ELECTROCHEMICAL ENERGY CONVERSION

The overall electrochemical reaction at the electrochemical cell involves reactants and products that either remain in the solution phase (homogeneous reaction) or interact with the electrode surface (heterogeneous reaction). In any case, electrons participate in the partial reaction at the respective electrode, either as reactants in the reduction reaction (at the cathode) or as products in the oxidation reaction (at the anode). For a fuel cell, heterogeneous electrocatalysis becomes more relevant. The reaction in the cell involves the conversion of Gibbs free energy (ΔG_r) into electrical work (w).

To illustrate the essential differences between thermal catalysis and electrocatalysis, let us consider the thermal and electrochemical combustion of hydrogen. In both cases, energy conversion proceeds via the conversion of hydrogen into water. In the gas phase the reaction is

$$H_2(g) + \tfrac{1}{2} O_2(g) \rightarrow H_2O(l) + w \tag{2.1}$$

This process occurs under the constraints imposed by the Carnot cycle. The useful work (w) is proportional to the enthalpy change of the reaction (ΔH_r),

$$w = \epsilon_c \, \Delta H_r \tag{2.2}$$

and the efficiency of the thermal machine (ϵ_c) is in the range 0.35–0.40.

On the other hand, the electrocatalysis at the $H_2(g)/O_2(g)$ acid fuel cell under standard conditions, that is, $P_{O_2} = P_{H_2} = 1$ atm, $T = 298$ K, and acid activity $a_{H^+} = 1$, implies the direct conversion of the chemical energy into electricity via the participation of aqueous (aq) ionic species, as follows:

1. At the anode

$$H_2(g) \xrightarrow{H_2O} H_2(aq) \Rightarrow \text{gas dissolution} \tag{2.3}$$
$$H_2(aq) \longrightarrow 2H^+(aq) + 2e^- \Rightarrow \text{HEOR} \tag{2.4}$$

2. At the cathode

$$\tfrac{1}{2}O_2(g) \xrightarrow{H_2O} \tfrac{1}{2}O_2(aq) \Rightarrow \text{gas dissolution} \tag{2.5}$$
$$\tfrac{1}{2}O_2(aq) + H_2O(l) + 2e^- \longrightarrow 2OH^-(aq) \Rightarrow \text{OERR} \tag{2.6}$$
$$2H^+(aq) + 2OH^-(aq) \rightleftharpoons 2H_2O(l) \Rightarrow \text{ionic equilibrium of water} \tag{2.7}$$

where HEOR and OERR stand for the hydrogen and oxygen electrochemical oxidation and reduction reactions, respectively.

Under virtual conditions, that is, at null current flow (infinite external resistance), the chemical energy conversion to electricity is given by the change in the standard Gibbs free energy (ΔG_r°) of the cell reaction,

$$w = \Delta G_r^\circ = -nFE_{\text{cell}}^\circ = \Delta H_r^\circ - T\,\Delta S_r^\circ \tag{2.8}$$

n being the charge transferred in the stoichiometric reaction, F is the Faraday constant (96,500 C), E_{cell}° is the standard potential of the cell, and $T\,\Delta S_r^\circ$ is the heat exchanged with the surroundings, which is usually small. Accordingly, the maximum efficiency (ϵ_M) of the electrochemical energy conversion is

$$\epsilon_M = \frac{\Delta G_r}{\Delta H_r} = -\frac{nFE_{\text{cell}}^\circ}{\Delta H_r} \tag{2.9}$$

Then, the energy spontaneously produced utilizing electrochemical cells results in the form of electricity and, consequently, processes at the fuel cell are not limited by the Carnot cycle [3, 30].

For the $H_2(g)/O_2(g)$ fuel cell in aqueous electrolyte under standard conditions, $E_{\text{cell}}^\circ = 1.23$ V [normal hydrogen electrode (NHE)] at $T = 298$ K, the expected value of ϵ_M is 0.83, although in real systems it lies in the range $0.50 \leq \epsilon_M \leq 0.70$. The difference between the potential of the electrochemical cell (E_{cell}°) and its value (E_{cell}^j) under the current density flow (j) is due to polarization effects at the anode and cathode, as described further on. The heat loss is given by the potential difference $E_{\text{cell}}^j - E_{\text{cell}}^\circ = T\,\Delta S_r/nF$. For $E_{\text{cell}}^j = 0.8$ V, the cell efficiency is about 50%.

2.2.1 Cell Overpotentials

To maintain a fuel cell under the stationary current density flow regime (j_{ss}), energy is used to overcome different resistive components in the cell. This energy loss represents a *polarization overpotential* (η_p). For each electrode, either anode (a) or cathode (c), the polarization overpotential (η) is defined as the difference between the electrode potential E^j and its equilibrium value E_{eq}, both measured against the same reference electrode.

Overpotential contributions at both the anode and cathode are (i) the activation overpotential (η_a) due to the slowness of the electron transfer process either by itself of by processes at the electrode surface level that precede or follow it; (ii) the concentration overpotential η_c due to the slowness of transport of reactants and products either toward or outward the electrode–solution interface (by migration, diffusion, and/or convection); and (iii) the ohmic overpotential due to ohmic resistance drops across the cell (η_Ω).

Under controlled stirring, a steady-state current produces a stationary local concentration gradient of reactants and products from the bulk of the solution to the electrode surface at both the anode and the cathode. Correspondingly, the concentration polarization overpotential (η_c) at each electrode surface is given

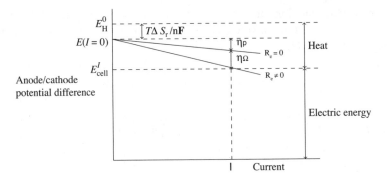

FIGURE 2.1   Linearized current–potential plots of a fuel cell with $\Delta S_r < 0$ under stationary current flow regime ($I = I_{ss}$) and different R_c; η_p represents the overpotentials of both anode and cathode; R_e and η_Ω are the Ohmic resistance and overpotential, respectively; E_H° the standard potential difference referred to the heat of combustion for the $H_2(g)/O_2(g)$ fuel cell; E_{cell}^I is the working potential difference for a current I_{ss}.

by the Nernst equation

$$\eta_c = \frac{RT}{nF} \ln \frac{a_{i,s}}{a_{i,0}} \tag{2.10}$$

where $a_{i,s}$ and $a_{i,0}$ are the activity of either reactants or products (i) at each electrode surface (s) and at the bulk of the solution (0), respectively, and R is the universal gas constant.

The ohmic overpotential (η_Ω) of the electrochemical cell under the I_{ss} regime comprises stationary electrical resistance that is the sum of resistances from the electrodes, the electrolyte, and the electrical contacts interposed between the anode and cathode,

$$\eta_\Omega = I_{ss} \sum_i R_i \tag{2.11}$$

where R_i stands for the resistance of component i.

The activation overpotential is considered more extensively in Section 2.3. The scheme of a current potential plot for the H_2/O_2 fuel cell is shown in Figure 2.1. In this case, it is assumed that $\eta_\Omega > \eta_p$ and the current–potential plot approaches a linear relationship.

2.2.2   Efficiency of Electrochemical Cell

The efficiency of a fuel cell under steady-state conditions is expressed as the faradaic yield and the energy efficiency. The faradaic yield is the ratio between the moles of product obtained in the cell for the passed charge Q_F and that expected from the stoichiometry of the cell reaction per Faraday (Q_{th}). Therefore, at constant potential, the percent faradaic yield (ϵ_F) is

$$\epsilon_F = \frac{100 \times Q_F}{Q_{\text{th}}} \tag{2.12}$$

and Q_F stands for the total electric charge transferred through the cell at constant I, that is,

$$Q_F = \int_0^t I \, dt \tag{2.13}$$

Under nonstationary conditions, $I = I(t)$, the value of ϵ_F offers the possibility of evaluating the instantaneous concentration of reactants or products at the electrode surface. For this purpose, different relaxation techniques are extensively described in the literature [31–33].

The energy efficiency of a cell is the ratio between the practical and theoretical power resulting at a preset current and voltage. Because of polarization effects, the energy efficiency decreases as I_{ss} increases. The percent energy efficiency (ϵ_E) at E_{cell} is

$$\epsilon_E = 100 \times \frac{E_{\text{cell}}^I}{E_{\text{cell}}^\circ} = \frac{100[E_{\text{cell}}^\circ - (\eta_a + \eta_c + \eta_\Omega)]}{E_{\text{cell}}^\circ} \tag{2.14}$$

Values of ϵ_F and ϵ_E can be determined at either constant current (galvanostatic control), charge (coulostatic control), or potential (potentiostatic control).

The power input–output of the cell (P_{cell}) is defined as

$$P_{\text{cell}} = I \times E \tag{2.15}$$

It permits evaluation of the power density P_d,

$$P_d = \frac{P_{\text{cell}}}{A} \tag{2.16}$$

where A refers to the electrode area.

Analogously, for three-dimensional (3D) electrodes the power density per unit volume (P_v) is

$$P_v = \frac{P_{\text{cell}}}{V_c} \tag{2.17}$$

where V_c is the electrode volume. Another useful expression is the specific power S_p,

$$S_p = \frac{P_{\text{cell}}}{\sum m_i} \tag{2.18}$$

$\sum m_i$ being the sum of the total mass of the fuel cell, including fuels, utilities, and so on, although often S_p is also referred to the electrocatalyst unit mass.

2.3 PHENOMENOLOGY OF ELECTROCHEMICAL REACTIONS

2.3.1 Stationary Single-Electron Transfer Kinetics

For a single-electron transfer cathodic reaction at a finite rate involving soluble reactants (R^+) and product (P) at a solid electrode under activation control such as

$$R^+(aq) + e^- \rightleftharpoons P(aq) \tag{2.19}$$

the reaction rate in terms of I can be expressed as the algebraic sum of the partial current in the forward ($\vec{I}$) and backward ($\overleftarrow{I}$) directions according to the Butler–Volmer equation [1],

$$I = \vec{I} - \overleftarrow{I} = j_0 A \left[\exp\left(-\frac{\beta F \eta_a}{RT} \right) - \exp\left(\frac{(1-\beta)F\eta_a}{RT} \right) \right] \tag{2.20}$$

j_0 and β being the exchange current density and the symmetry factor assisting the reaction in the forward direction, respectively, and A the electrode area. Following International Union of Pure and Applied Chemistry (IUPAC) recommendations, cathodic currents are taken as negative and vice versa.

For large and negative activation overpotentials the cathodic current density approaches a Tafel relationship [34],

$$j_c = -j_0 \exp\left(-\frac{\beta \eta_a F}{RT} \right) = -j_0 \exp\left(-\frac{\eta_a}{b_c} \right) \tag{2.21}$$

Similarly, for large and positive activation overpotentials, it results in

$$j_a = j_0 \exp\frac{(1-\beta)\eta_a F}{RT} = j_0 \exp\frac{\eta_a}{b_a} \tag{2.22}$$

j_c and j_a being the cathodic and anodic current densities and b_c and b_a the cathodic and anodic Tafel slopes for single-electrode transfer, respectively. It has been found that a large number of electrochemical processes relatively far from equilibrium, at constant T and P, fit Tafel relationships, that is, linear log $j_{c/a}$ versus η_a plots with slopes either b_c or b_a, respectively [1, 3].

On the other hand, when η_a is a few millivolts only, either negative or positive, after a series expansion of the exponential terms [Equation (2.20)], a linear $|\eta_a|$ versus $j_{c/a}$ relationship is obtained,

$$|\eta_a| = \frac{RT|j_{c/a}|}{F j_0} \tag{2.23}$$

The value of j_0 determines the rate of the electrochemical reaction at the equilibrium potential E_{eq}, and the ratio RT/Fj_0 is the polarization resistance of reaction (2.20).

2.3.2 Multielectron Transfer Kinetics

A multielectron transfer process, that is, a single reaction with $n > 1$, can be described with a rate equation of the same form as Eq. (2.20) utilizing a transfer coefficient α associated with the forward reaction ($\vec{I}$):

$$I = \vec{I} - \overleftarrow{I} = j_0 A \left[\exp\left(-\frac{\alpha n F \eta_a}{RT}\right) - \exp\left(\frac{(1-\alpha)n F \eta_a}{RT}\right) \right] \qquad (2.24)$$

n being the number of electrons involved in the electrochemical reaction. If the rate equation is referred to the potential of zero charge (E_{pzc}), the rate of a cathodic multielectron transfer reaction at a high cathodic overpotential results in

$$-j_c = n F c^s k^\circ \exp\left[\frac{n(1-\alpha)F E_{pzc}}{RT}\right] \exp\left[-\frac{n\alpha F(\eta_a - E_{pzc})}{RT}\right] \qquad (2.25)$$

c^s being the reactant concentration at the electrode surface and k° the rate constant at $\eta_a = 0$.

According to Eq. (2.25) the electrochemical reaction rate can be expressed as the product of two exponential terms, the first one that depends on the intrinsic properties of the material, as occurs in heterogeneous catalysis, and the second one that varies according to both the properties of the electrode material and the electric potential–current flow regime. Therefore, Eq. (2.25) becomes important for adequately handling the rate of electrocatalytic processes via either thermal or electrochemical activation. For instance, 10^7 V cm^{-1} potential gradient at the electrochemical interface is equivalent to 10^8 kcal mol^{-1} (1 eV mol$^{-1} = 23.060$ kcal mol^{-1}) producing a more efficient activation than a thermal one. This makes electrocatalysis useful for developing processes at temperatures lower than those required for thermal activation.

From the transition theory of rate processes [35], the expression of j_0 is

$$j_0 = \kappa \frac{n F k T}{\nu h} \left(\prod_i^k a_i\right) \exp\left(-\frac{\Delta H^{\circ *}}{RT}\right) \exp\left(\frac{\Delta S^{\circ *}}{R}\right) \exp\left(-\frac{n\alpha F E_{eq}}{RT}\right) \qquad (2.26)$$

where κ is the transmission coefficient, ν is the vibration frequency of the activated complex, h is the Planck constant, and $\Delta H^{\circ *}$ and $\Delta S^{\circ *}$ are the activation enthalpy and entropy change of the reacting species at E_{eq}, respectively. The latter is a thermodynamic quantity in which the galvanic potential that depends of the metal nature is included [3].

Equation (2.26) is useful for comparing the electrocatalytic activity of different materials for a particular reaction considering either values of $j_{c/a}$ at constant η_a or values of η_a at constant $j_{c/a}$, as described in Section 2.4.6.

2.3.3   Multistep Steady-State Kinetics

The rate (v) of a steady-state multistep reaction can be written as

$$v = w_1 c_1 - w_{-1} c_{x(1)} \tag{2.27}$$

c_1 being the initial concentration of species 1 and $x(1)$ representing the product from the first step that becomes a reactant for the second step and so on. Then, for a set of n consecutive reactions involving n values of v, and consequently $(n - 1)w$ rate constant terms, the solution under steady state is [36, 37]

$$\frac{1}{v_1} = \frac{1}{w_1} + \frac{w_{-1}}{w_1 w_2} + \frac{w_{-1} w_{-2}}{w_1 w_2 w_3} + \cdots + \frac{w_{-1} w_{-2} \cdots w_{-n-1}}{w_1 w_2 w_3 \cdots w_n} \tag{2.28}$$

and for the reverse reaction it is

$$\frac{1}{v_{-1}} = \frac{1}{w_{-n}} + \frac{w_n}{w_{-n} w_{-(n-1)}} + \cdots + \frac{w_n w_{n-1} \cdots w_2}{w_{-n} w_{-(n-1)} \cdots w_{-1}} \tag{2.29}$$

This set of equations can be reduced to a single one containing, for example, either the forward or backward reaction only assuming it as a single rate-determining step (RDS) of the overall reaction, those steps preceding and following the RDS being in pseudoequilibria. However, when the reactions following the RDS do not reach equilibrium, both the RDS and the following reactions, being then coupled, undergo at the same rate. In this case, assuming pseudoequilibrium [38], the overall rate is

$$j = j_0 \left[\exp \left\{ -\left(\frac{\vec{\gamma}}{v} + r\beta \right) \frac{F\eta_a}{RT} \right\} - \exp \left[\left(\frac{n - \vec{\gamma}}{v} - r\beta \right) \frac{F\eta_a}{RT} \right] \right\} \tag{2.30}$$

$\vec{\gamma}$ being the number of electrons involved in the forward reaction before the RDS, v the stoichiometric coefficient, that is, the number of times the RDS occurs for the overall reaction undergoing once, and r the number of electrons in the RDS ($0 \le r \le 1$).

Comparing Eq. (2.30) with the Butler–Volmer equation [Eq. (2.20)] it results in

$$\overleftarrow{\alpha} = \frac{n - \vec{\gamma}}{v} - r\beta \tag{2.31}$$

$$\vec{\alpha} = \frac{\vec{\gamma}}{v} + r\beta \tag{2.32}$$

$$\vec{\alpha} + \overleftarrow{\alpha} = \frac{n}{v} \tag{2.33}$$

$\bar{\alpha}$ and $\tilde{\alpha}$ being the transfer coefficients for the forward and backward reactions, respectively. These relationships are useful for analyzing the mechanism of complex electrode reactions. They have been particularly applied to determine the likely mechanism of the oxygen electrode reaction in aqueous solutions on different substrates [3, 39].

2.4 OTHER VARIABLES INFLUENCING ELECTROCHEMICAL RATE PROCESS

2.4.1 Effect of Temperature

Due to the limitations for determining the temperature dependence of the rate of electrochemical reactions at constant galvanic potential difference, an ideal heat of activation is usually evaluated at constant overpotential provided that the reversible electrode, against which the potential is measured involves a reaction similar to that under study [3]. For a single-step process at $\eta = 0$, the heat of activation ($\Delta H^{\circ *}$) can be obtained from expression (2.26), written as

$$\ln j_0 = \ln k_0 - \frac{\Delta H^{\circ *} - \beta \Delta E_{eq}}{RT} \tag{2.34}$$

where k_0 involves all E-independent terms. Then, the temperature effect on the rate process at the reversible potential is given by

$$\frac{\partial \ln j_0}{\partial (1/T)} = \frac{\Delta H^{\circ *} - \beta \Delta H^{\circ}}{R} = \frac{\Delta E^*}{R} \tag{2.35}$$

where $\Delta H^{\circ *}$ is the ideal heat of activation at E_{rev} [1].

Equation (2.35) involves some limitations resulting from the time dependence of β and the preexponential factor as well, as extensively discussed in the literature [3, 40–42].

2.4.2 Effect of Pressure

To analyze the effect of pressure on the electrode kinetics, let us consider the electrochemical oxidation of molecular hydrogen on a conducting substrate M that occurs according to the following consecutive two-stage process:

$$2M + H_2 \underset{k_{-1}}{\overset{k_1}{\rightleftharpoons}} 2MH \tag{2.36}$$

$$H_2O + MH \underset{k_{-2}}{\overset{k_2}{\rightleftharpoons}} H_3O^+ + M + e^- \tag{2.37}$$

When step (2.37) is RDS and $\theta_H \to 0$, the overall reaction under steady-state conditions is

$$\tfrac{1}{2} H_2 + H_2O \rightleftharpoons H_3O^+ + e^- \tag{2.38}$$

and the total anodic current density is given by

$$j_a = 2Fk_1 P_{H_2}(1 - \theta_H) \exp\left(\frac{(1 - \beta)\eta_a F}{RT}\right) \tag{2.39}$$

where

$$\eta_a = {}^m\Delta^s\phi - E_{eq} \tag{2.40}$$

Accordingly, from Eq. (2.39) the pressure dependence of j_a at constant η and T results in

$$\left(\frac{\partial \ln j_a}{\partial P_{H_2}}\right)_{\eta,T} = \left(\frac{\partial \ln k_1}{\partial P_{H_2}}\right)_T + \frac{(1 - \beta)F}{RT}\left(\frac{\partial E_{eq}}{\partial P_{H_2}}\right)_T \tag{2.41}$$

The second partial derivative on the right side of Eq. (2.41) is equal to ΔV_0, the volume change per electron for the overall reaction, whereas the term on the left side is

$$\Delta V_a^* = \Delta V_{t1}^* - (1 - \beta)\Delta V_0 \tag{2.42}$$

where ΔV_a^* is the apparent volume of activation for step 1 and ΔV_{t1}^* is the corresponding true value.

In aqueous solution, when $\theta_H \to 0$, step (2.36) becomes rate determining, and at $P_{H_2} = 10^4$ bars, there is a 20% pressure effect on the reaction rate, and problems in the evaluation of ΔV_{t1}^* become similar to those arising from the evaluation of the activation energy from the T dependence of current density at a given η_a [43–46].

On the other hand, when stage (2.37) is the RDS and $a_{H_2O} = 1$, the dependence of j_a on P_{H_2} results in

$$\left(\frac{\partial \ln j_a}{\partial P_{H_2}}\right)_{\eta,T} = \frac{\Delta V_a^*}{RT} = \frac{\Delta V_{t,2}^*}{RT} + \left(\frac{\partial \ln \theta_H}{\partial P_{H_2}}\right)_{\eta,T} + \frac{(1 - \beta)\Delta V_0}{RT} \tag{2.43}$$

The P_{H_2} dependence of θ_H can be obtained from the following HER mechanism:

$$H_3O^+ + e^- + M \underset{k_{-1}}{\overset{k_1}{\rightleftharpoons}} M(H) + H_2O \tag{2.44}$$

$$M(H) + H_3O^+ + e^- \underset{k_{-2}}{\overset{k_2}{\rightleftharpoons}} H_2 + H_2O + M \tag{2.45}$$

step (2.45) being rate determining. From this mechanism,

$$\theta_{H^+} = \frac{K_1 c_{H^+} \exp(-\beta\, {}^m\Delta^s\phi F/RT)}{1 + K_1 c_{H^+} \exp((1 - \beta)\, {}^m\Delta^s\phi F/RT)} \tag{2.46}$$

where $K_1 = k_1/k_{-1}$, and the partial derivative from Eq. (2.46) is

$$\left(\frac{\partial \ln \theta_{\mathrm{H}}}{\partial P}\right)_{\eta, P} = -\frac{\Delta V_1}{RT} + \frac{\Delta V_0}{RT} - \frac{\theta_{\mathrm{H}}}{RT}(\Delta V_0 - \Delta V_1) = (1 - \theta_{\mathrm{H}})\left(\frac{\Delta V_0 - \Delta V_1}{RT}\right)$$

(2.47)

where ΔV_1 is the volume change in step (2.44) and $\Delta V_0 - \Delta V_1$ is equal to the adsorption volume of hydrogen from $\frac{1}{2}\mathrm{H}_2$. For platinum, the latter is equal to $7.4 \pm 2 \ \mathrm{cm}^3 \ \mathrm{mol}^{-1}$ [47].

Values of the apparent and true activation volumes for various possible HER mechanisms are consistent with compressibility contributions of the primary hydration shell, although the greatest effect in the apparent activation volume should be related to the changes in the inner shell of the electrochemical double layer. Extensive discussions on the significant effect of pressure on the kinetics of electrode reactions are given elsewhere [46, 48].

2.4.3 Adsorption of Reacting Species

From the phenomenological standpoint, the interaction energy of the electrode surface with either species from the solution or intermediates produced in the course of multistep electrocatalytic reactions can be interpreted by means of adsorption isotherms. To deal with this problem, information provided by spectroscopic, nanoscopic, and surface techniques in general contributes to obtaining a more realistic description of the structure of the electrochemical interface as well as the reaction pathways of the proper rate processes.

Let us consider an anodic process under activation control without mass transport limitation consisting of a single charge transfer yielding an adsorbed intermediate (X) that subsequently participates in an adsorption equilibrium with species X_2 in the solution phase:

$$\mathrm{Substrate} + \mathrm{R}^- \underset{k_{-1}}{\overset{k_1}{\rightleftharpoons}} \mathrm{substrate(X)} + \mathrm{e}^-$$

(2.48)

$$2 \ \mathrm{Substrate(X)} \underset{k_{-2}}{\overset{k_2}{\rightleftharpoons}} \mathrm{X}_2 + \mathrm{substrate}$$

(2.49)

An electrochemical isotherm for X more realistic than a Langmuirian one is that in which the dependence of the adsorption free energy of X species on θ is considered, such as

$$\Delta G^\circ(\theta_{\mathrm{X}}) = \Delta G^\circ + r\theta_{\mathrm{X}}$$

(2.50)

where $r = \partial \Delta G^\circ(\theta_{\mathrm{X}})/\partial \theta_{\mathrm{X}}$ represents an interaction parameter where $\Delta G^\circ(\theta_{\mathrm{X}})$ becomes less negative as θ_{X} increases [49, 50]. An equation such as (2.50) has also been derived assuming a heterogeneous electrode surface [51]. Its linearity occurs up to $\theta_{\mathrm{X}} \approx 0.5$. For electrode interactions with surface ions or dipoles the equation

$$\Delta G^\circ(\theta_{\mathrm{X}}) = \Delta G^\circ + r\theta_{\mathrm{X}}^n$$

(2.51)

with $\frac{2}{3} \leq n \leq 3$ [3] and $r = \partial \Delta G^\circ(\theta_{\mathrm{X}})/\partial \theta_{\mathrm{X}}^n$ has been proposed [51].

From Eq. (2.50), the adsorption isotherm results in

$$\frac{\theta_X}{1 - \theta_X} = K c_{R^-} \exp\left(\frac{\eta_a F}{RT}\right) \exp\left(-\frac{r\theta_X}{RT}\right) \tag{2.52}$$

where $K = \left[\exp(\Delta G^\circ(\theta_X) + r\theta_X)/RT\right]$.

In the range $0.2 \le \theta_X \le 0.8$, $\theta_X/(1 - \theta_X) \approx 1$, Eq. (2.52) leads to the Temkin logarithmic isotherm [1, 2]

$$r\theta_X = RT \ln K + RT \ln c_{R^-} + F\eta_a \tag{2.53}$$

The simplest rate equation from the Volmer–Tafel proton discharge mechanism [52], assuming a constant hydrogen atom heat of adsorption, $\theta_H \ll 1$, and $K c_{H^+} \exp(-\eta_a F/RT) < 1$, is

$$j_c = 2F K^2 c_{H^+}^2 \exp\left(-\frac{2\eta_a F}{RT}\right) \tag{2.54}$$

The Tafel slope derived from Eq. (2.54) is $b_c = -RT/2F$.

Conversely, for a θ_H-dependent heat of adsorption, $\theta_H/(1 - \theta_H) \approx 1$, and the hydrogen combination reaction as RDS

$$j_c = 2F k_{H/H_2} \exp\left(\frac{2\alpha F\eta_a}{RT}\right) \tag{2.55}$$

and for $\alpha = 0.5$, the Tafel slope is $b_c = -RT/F$ [38].

Therefore, for a consecutive reaction pathway, the presence of intermediates interacting with the electrode under quasi-equilibrium has a definite influence on the j_c-versus-η_a plot. The contribution of intermediates in determining the kinetics of the molecular OERR depends on the RDS, as has been extensively analyzed elsewhere [53–60].

2.4.4 Electron Transfer and Adsorption Processes

To study electron transfer and adsorption processes, nonstationary electrochemical techniques at potential, current, or charge control are utilized. Potential sweep techniques, either as single or cyclic scanning or as combined routines, are usually employed. These techniques, initially developed to investigate diffusion processes [61] under rather complex conditions, were later applied to reactions in which adsorbed species are involved [62], either in the absence of or under mass transport control for distinguishing potential windows at which different processes undergo and for studying steady-state and transient kinetics and mechanisms of multistep reactions.

The adsorption of species, either neutral or charged, produced from a charge transfer process is equivalent to a charge storage at the electrochemical interface, associated with a pseudocapacitance (C_{ad}) that is distinct from double-layer capacitance (C_{dl}) [1, 32].

To evaluate C_{ad}, knowledge of the electrochemical adsorption isotherm of the ith species is required. In typical reversible adsorption process such as

$$A^- + \text{solid surface} \underset{k_c}{\overset{k_a}{\rightleftharpoons}} \text{solid surface(A)} + e^- \qquad (2.56)$$

species A is an adsorbate produced at a free surface site from a single-electron transfer oxidation of dissolved A^- under a linear potential scan $E = E_i + vt$ from the starting potential E_i at a constant scan rate $v = dE/dt$. At equilibrium, the degree of surface coverage by adsorbate A is θ_A, and the electrochemical adsorption isotherm can be expressed as

$$\frac{\theta_A}{1 - \theta_A} = K \exp(r\theta_A) \exp\left(\frac{EF}{RT}\right) \qquad (2.57)$$

where r is an interaction parameter and $K = k_a/k_c$, k_a and k_c being the rate constants of the oxidation and reduction reactions, respectively.

Considering $c_A \approx 1$ at the outer Helmholtz plane (OHP), irrespective of E, $n = 1$ and $\beta = 0.5$, j_a is given as [62]

$$j_a = k_a(1 - \theta_A) \exp\left(\frac{EF}{2RT}\right) \exp\left(-\frac{r\theta_A}{2}\right) - k_c\theta_A \exp\left(-\frac{EF}{2RT}\right) \exp\left(\frac{r\theta_A}{2}\right) \qquad (2.58)$$

For the reversible process, the current–potential scan is characterized by an anodic current peak that increases linearly with v,

$$j_p = \frac{Q_{ML}F}{4RT} v \qquad (2.59)$$

the value of Q_{ML} being the charge density required to form an adsorbate monolayer (ML) of A.

In this case, the value of j_p appears at a fixed potential (E_p),

$$E_p = -\frac{RT}{F} \ln K \qquad (2.60)$$

For reaction (2.56) under highly irreversible conditions,

$$j_a = k_a(1 - \theta_A) \exp\left(\frac{EF}{2RT}\right) \exp\left(-\frac{r\theta_A}{2}\right) \qquad (2.61)$$

and $dj_a/dt = 0$ leads to a peak value j_p that increases linearly with v,

$$j_p = \frac{Q_{ML}F}{e2RT} v \qquad (2.62)$$

and E_p increases linearly with $\ln v$,

$$E_p = \frac{2RT}{F} \ln \frac{Q_{ML}}{2k_aRT} + \frac{2RT}{F} \ln v \qquad (2.63)$$

with the slope $\partial E_p / \partial \ln v = 2RT/F$. Accordingly, the apparent adsorption pseudo-capacitance C_{ad} results in

$$C_{ad} = \frac{j_p}{v} = \frac{Q_{ML}F}{e2RT} \qquad (2.64)$$

The change from the reversible to irreversible behavior can be followed by plotting the potential at one-half peak current ($E_{p,1/2}$) versus $\log v/k$ (either k_a or k_c or assuming $k_a = k_c = k$). The transition is characterized by either a decrease of C_{ad} with $\log v/k$ or an increase of E_p with v, and an increase to a limiting value for $\Delta E_{1/2}$, the half-width peak potential [62].

For discovering fast surface electrochemical processes triangular modulated voltammetry [63] is a useful technique [63–66]. Briefly, it consists of a slow, either linear or triangular potential scan modulated with a relatively fast triangular potential signal of frequency f and small amplitude A_m. The shape of the current–potential envelope and its dependence on f and A_m offers the possibility of finding out the optimal coupling between the modulating signal and the rate of the electrochemical reaction to estimate kinetic parameters.

On the other hand, when soluble species participate as either intermediates or products, rotating disc [67] and disc-ring electrodes [32, 68] have successfully been applied for identification and yield determinations.

2.4.5 Work Function of Metal Electrode

The metal work function Φ^m, defined as the negative value of the real potential of the electron in the metal [69–72], depends upon the chemical potential (μ_l^m) and the electron overlap potential difference (ξ_l^m). In contrast to μ_l^m, the magnitude of ξ_l^m changes with the crystallography of the surface.

For aqueous electrochemical reactions, under nonequilibrium and null charge, the relevant work function ($\Phi^{m/s}$) is related to the jump of an electron through a water layer. For aqueous electrochemical reactions at null charge,

$$\Phi^{m/s} = -\mu_e^m + Fg^m - Fg^s \qquad (2.65)$$

where the g's represent dipole potential differences at the metal–solution (m/s) interface. For a charged metal surface

$$\Phi^{m/s} = -(\mu_e^m - Fg^m + Fg^s) + F^m \Delta^s \psi \qquad (2.66)$$

$^m\Delta^s\psi$ being the outer potential difference at the m/s interface. When the electron jump occurs from a metal to a solvated electron as the final state in the solution,

$$\Phi^{m/s} = \Phi^m - \Phi^s + F^m\Delta^s\psi \tag{2.67}$$

The value of $\Phi^{m/s}$ depends on the surface crystallography at the atomic level. It becomes an anisotropic property when the polycrystalline metal surfaces exhibit a distribution of patches (i) of different crystal faces. Correspondingly, the average value of $\Phi^{m/s}$ is considered,

$$\langle\Phi^m\rangle = \sum_i \theta_i \Phi_i^m \tag{2.68}$$

θ_i denoting the fraction of area with the true work function Φ_i^m [73, 74].

2.4.6 Compensation Effects

At heterogeneous chemical reactions, the variation in the preexponential factor $\mathcal{A}$ tends to compensate changes in the activation energy E_a^*. The rate of a multielectron transfer reference process at the equillibrium potential undergoing on different substrates obeys a linear $\log\mathcal{A}$-versus-E_a^* relationship [35],

$$-\log j_0 = -\log(nFc_s k_s) = -\log\mathcal{A} \approx E_a^* \tag{2.69}$$

The value of $\mathcal{A}$ depends on the electrical double-layer structure, the electron density at the solid electrode surface, the degree of irreversibility of the process, and the adsorption entropy of intermediates. For potential values near E_{pzc} and in the absence of specific adsorption, the potential drop at the diffuse double layer ($^{OHP}\Delta^s\phi$) determines the surface concentration of species i (c_i) at the reaction interface. Both $^{OHP}\Delta^s\phi$ and E_{pzc} depend linearly on Φ^m, the metal electrode work function. Then, as $^{OHP}\Delta^s\phi = K(E - E_{pzc})$,

$$\frac{\partial \prod_1^n c_i}{\partial\, ^{OHP}\Delta^s\phi} = -\frac{(n-\beta)F}{RT} \tag{2.70}$$

and

$$\frac{\partial \ln\mathcal{A}}{\partial\, ^{OHP}\Delta^s\phi} = -\frac{K(n-\beta)F}{RT} \tag{2.71}$$

where K is a proportionality constant. For concentrated electrolytes the term Kn is small and the ratio $\partial \ln\mathcal{A}/\partial\,^{OHP}\Delta^s\phi$ becomes negligible.

For semiconductor electrodes the electron density contribution at the electrocatalyst surface becomes important. In this case, the charge carrier surface concentration

(c_p) is given by the proportionality

$$c_p \approx \exp\left[-\frac{\beta(U' - U_F)}{RT}\right] \tag{2.72}$$

where U_F is the Fermi potential and $U' - U_F$ is the width of the forbidden band, β being a fraction of it. Equation (2.72) explains the influence of c_p on the electrocatalytic process in the presence of semiconductor films thicker than 2 nm, whereas for thinner films, an equivalent influence is related to the tunnelling effect. According to Eq. (2.72), the rate of electrocatalytic processes on semiconductor electrodes becomes smaller than on metals. This deficiency is to some extent compensated by the high density of states close to the edge of the conduction band that is responsible for the electron transfer. Transition metals have effective electron densities at the Fermi level higher than those at sp metals. Consequently, the rate of certain electrocatalytic processes at transition metals becomes higher than at sp metals.

Considering that generally the transmission coefficient in the rate equation (2.26) is close to 1, a shared bond model for proton transfer would lead to an increase in the adsorption enthalpy, that is, a decrease in E_a^*, and to an increase in $\mathcal{A}$ as the thickness of the energy barrier decreases. At high proton discharge overpotentials proton tunneling contribution becomes relatively small.

Generally, the value of j_0 is taken as a measure of the electrocatalytic activity of an electrode material for a particular reaction. Practically, this is done by comparing either values of j of a particular process on different electrocatalysts at a preset overpotential (η_a) or values of η_a at a fixed value of j. For the electrochemical discharge of a proton yielding a hydrogen adatom on a metal substrate as RDS, the value of ΔG_{ad}^* is calculated from the adsorption enthalpy (ΔH_{ad}^*). The entropy change, either ΔS_{ad}^* or ΔS_{ad}, is included in the value of $\mathcal{A}$. For a reversible adsorbate formation, assuming a Langmuirian adsorption equilibrium for hydrogen adatoms, the fraction of free substrate surface is

$$1 - \theta_i = \left[1 - K_e \exp\left(-\frac{\Delta G_{ad}^\circ}{RT}\right)\right]^{-1} \tag{2.73}$$

where θ_i is the hydrogen adatom surface coverage and ΔG_{ad}^* is its standard adsorption free energy. Then

$$j_0 = K' \frac{\exp\left(-\dfrac{\beta^m \Delta^s \phi F + \Delta G_{ad}^\circ}{RT}\right)}{1 + K_e \exp\left(-\dfrac{\Delta G_{ad}^\circ}{RT}\right)} \tag{2.74}$$

where K' contains all adsorption-independent terms and K_e is the adsorption equilibrium constant. At constant $^m\Delta^s\phi$, Eq. (2.74) predicts a j_0-versus-ΔG_{ad}° *volcano*

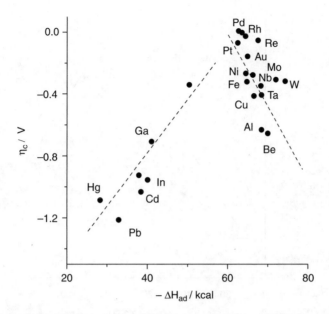

FIGURE 2.2   Plot of HER cathodic activation overpotential on different metals in acid solution versus metal–hydrogen standard adsorption enthalpy at 1×10^{-3} A cm^{-2}. (Data from Appleby et al. [28].)

plot (Fig. 2.2) in which both the ascending and descending portions of the curve depend on ΔG_{ad}°. Accordingly, for softer metals (Hg, Cd) j_0 increases exponentially with the M–H adsorption bond strength, whereas for transition metals (Pt, Pd, Rh) the reverse dependence is observed [15, 75–77]. A comparable relationship between j_0 and the d character of the metal has also been obtained [77]. For the HER, the volcano plot depends also on the mechanism of the intermediate formation (Fig. 2.3), although calculated and experimental volcano plots show some differences (Figs. 2.2 and 2.3). Accordingly, in terms of bulk metals the best electrocatalysts for the HER are platinum and palladium. There are, in principle, limitations to utilize less efficient, cheaper, bulk metals for the HER. The volcano plot (Fig. 2.2) approaches linear plots with either positive or negative slopes depending on whether ΔH_{ad}° is greater or lower than 0.60 kcal mol^{-1}, respectively.

A more complex reaction scheme is

$$O^{+} + e^{-} \underset{v_{-1}}{\overset{v_1}{\rightleftharpoons}} (A) \tag{2.75}$$

$$(A) + B \underset{v_{-2}}{\overset{v_2}{\rightleftharpoons}} P \tag{2.76}$$

involving species in solution (O^{+}, B, and P) and adsorbate (A), v_1, v_2, v_{-1}, and v_{-2} being the rate of each process in the forward and backward directions, respectively. At the equilibrium potential, under Langmuirian adsorption, in the absence of

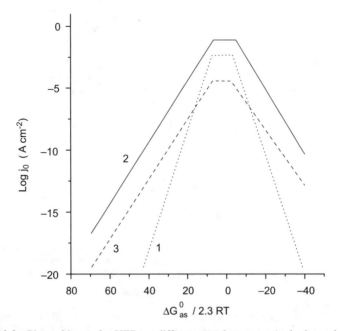

FIGURE 2.3 Plots of log j_0 for HER on different metals versus standard metal–hydrogen free energy of adsorption calculated for following RDS: (1) $2H_{ad} = H_2$; (2) $H^+ + e^- = H_{ad}$; (3) $H_{ad} + H^+ + e^- = H_2$ considering $\alpha = 0.5$. The maximum value of log j_0 appears within the same range of ΔG_{ad}°, irrespective of the RDS (From Parsons [77] by permission of The Royal Society of Chemistry.)

double-layer effects, assuming $v_1 = v_{-1}$ and $\theta_A = \theta_{A,eq}$, Eq. (2.26) results in [3]

$$\ln j_0 = (1 - \alpha_1) \ln a_{O^+} + \alpha_1 \ln \left[K \exp\left(\frac{-\Delta G_{ad}^\circ}{RT} \right) \right]$$
$$- \ln \left[1 + K \exp\left(-\frac{\Delta G_{ad}^\circ}{RT} \right) \right] + Q \qquad (2.77)$$

where Q includes all ΔG_{ad}°-independent terms. Equation (2.77) predicts a maximum value of j_0 for

$$K_1 \exp\left(-\frac{\Delta G_1^\circ}{RT} \right) = \frac{\alpha_1}{1 - \alpha_1} \qquad (2.78)$$

that is, for large ΔG_1°, either positive or negative, j_0 becomes small, this behavior being qualitatively the same irrespective of the isotherm considered.

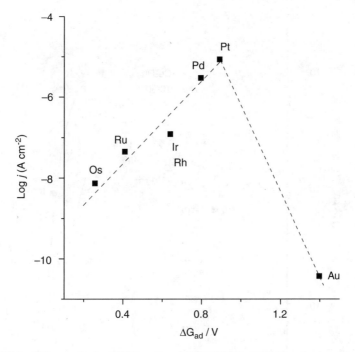

FIGURE 2.4   Volcano plot for OERR at $E = 0.8$ V (NHE) in 85% phosphoric acid at 298 K. (Data from Appleby [78].)

Similar volcano plots have also been made for the OERR at 0.8 V (NHE) in 85% phosphoric acid (Fig. 2.4) and for complex organic oxidation reactions such as ethylene on different metals and alloys (Fig. 2.5).

All these examples confirm the general validity of the compensation effect [28].

To identify the metal properties that govern the adsorption enthalpy, several correlations based on different electronic or structural properties of metals, including a *reactivity scale* to estimate the electrocatalytic activity of *unknown* materials for certain processes, have been proposed. An example of this is the use of the electron work function to predict the activity of metals for the HER [80].

Conversely, *synergetic effects* produced by reciprocal influence among a number of components in the electrocatalyst permit obtaining materials whose activity exceeds that of a single component only. In these cases, the electrocatalytic efficiency is the result of strong changes in the electronic structure of each component yielding optimal interactions between the electrocatalyst surface and reaction intermediates.

In conclusion, the phenomenology of the compensation effect may be caused by (i) changes in the mechanism of the process with T; (ii) changes in the effective area of the electrode; (iii) changes of the surface crystallography and phase distributions of active surface sites (or faces) that depend on both T and E; (iv) changes in the degree of surface coverage by reaction intermediates; (v) transport processes at rough and

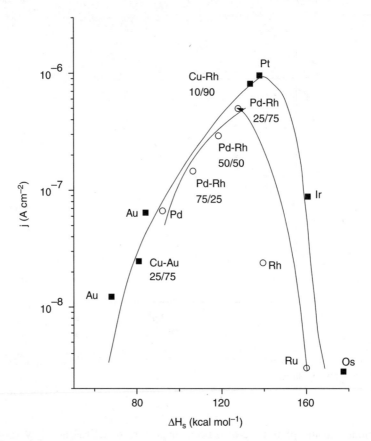

FIGURE 2.5   Log of current density versus standard enthalpy of sublimation (ΔH_s) volcano plot for ethylene oxidation at $\eta_a = 0.6$ V on different metals and alloys. (Curve drawn from data given in Khun et al. [79].)

porous surfaces; and (vi) limitations due to the poor knowledge of the true activation energy of the process.

2.5   ELECTROCATALYST SURFACE

Specific characteristics of solid electrocatalysts, such as surface heterogeneity, degree of dispersion, and surface dynamics, play an important role in determining the optimization of the electrocatalytic process. These characteristics are considered in electrocatalyst design aiming to produce a material with a high density number of active centers on a large electrode surface area that should be easily accessible for reactants. From a practical standpoint, cheap, handily machinable, and environmentally friendly materials are desirable [81]. All these requirements have to be evaluated for long-term efficient operation. Preliminary indications of electrocatalytic potentialities can be derived from nonstationary process data, although a full evaluation is required at technical scales to vouch for real applications.

Electrocatalysts can be used at low, intermediate, and high dispersion conditions. The distribution of crystallographic planes at the surface can be varied from that corresponding to a smooth, single-crystal surface to those of rough polycrystalline particles dispersed in a conducting substrate. The number, density, and energy distribution of active sites available at the electrode surface depends on the surface roughness structure of solid electrocatalysts. The concept of roughness is related to the geometry and distribution of reacting centers, the size and morphology of dispersed electrocatalyst particles, and their aspect ratios. Capillary effects appear through porosity, the facile surface diffusion of species interacting with surface atoms, and relaxation of the proper solid surface. Handling this set of properties becomes essential for producing efficient electrocatalysts.

2.5.1 Metal Single-Crystal Surfaces

Single-crystal metal electrodes can be prepared by different techniques and are utilized as reference solid surfaces in which the kinetics of electrochemical reactions undergoes on a well-defined crystallography. Details on the preparation and use of metal single-crystal electrodes in electrocatalysis can be obtained elsewhere [83–89]. At room temperature, the surface order of real single crystals is usually restricted to atomically smooth terraces of the order of some hundreds of square nanometers. Under different applied potentials, the interactions of metal single-crystal surfaces with OH species and water produce superstructures due to surface reconstruction [90, 91].

The use of well-defined single-crystal surfaces represents an important step for understanding the influence of surface crystallography on the kinetics of heterogeneous electrocatalysis, for determining the structure of adlayers at the atomic and molecular levels, and for evaluating adsorbate–substrate interactions and substrate dynamics [92].

A perfect monoatomic solid surface can be considered as a limiting case of a real *smooth surface* involving the periodic irregularities imposed by its own crystallographic lattice. As an example, the surface crystallography of gold (100), initially held at 0.442 V in aqueous perchloric acid at room temperature [Fig. 2.6(a)], changes when the applied potential is negatively shifted from previously flat 1×1 terraces into corrugated domains [Fig. 2.6(b)]. Similarly, for gold (111) surface reconstruction occurs by changing the applied potential E to either $E < E_{pzc}$ or $E > E_{pzc}$ (Fig. 2.7). In the former case, the gold $22 \times \sqrt{3}$ superstructure is formed, whereas for the latter the typical gold (111) topography remains, although covered at random by one gold atom high islands. This applied potential- and adsorbate-induced surface reconstruction is reversible [93]. This process has been investigated by Monte Carlo simulation assuming that reconstruction is an order–disorder phase transition [94, 95].

More complex situations arise when single-crystal surfaces of high Miller indices are considered. In this case, surface irregularity increases with step density, as can be seen, for instance, by plotting the first-layer relaxation percentage versus roughness

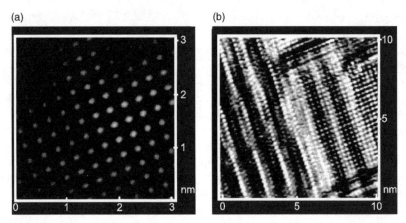

FIGURE 2.6 STM images of gold (100) in 0.1 M HClO$_4$: (a) 3 × 3-nm top view image at $E = 0.442$ V (NHE). Tunneling conditions: tip current 20 nA, bias voltage 6 mV. (b) After shifting E from 0.442 to -0.142 V and remaining there for 10 min. (From Gao et al. [82] by permission of The American Physical Society.)

referred to the ideal (111) crystallographic plane (Fig. 2.8) [92]. On the other hand, the presence of surface defects such as steps, vacancies, bumps, and holes as well as adsorbates introduces a certain degree of disorder at the surface.

2.5.2 Polycrystalline Electrodes

2.5.2.1 Facetted Metal Surfaces

Polycrystalline metal electrodes consist of ensembles of randomly oriented crystallites, that is, small single-crystal units with a random distribution of crystallographic faces (Fig. 2.9). Two or more crystallites form aggregates and grain boundaries. These crystallites are characterized by their dimension, shape, and grain size distribution. The redistribution of crystallographic faces produces facetted, surface domains. The electrochemical facetting of solid metal surfaces, such as that of polycrystalline platinum, can be brought about by applying a potentiodynamic cycling covering the O adatom adsorption–desorption monolayer range (Fig. 2.10) [97, 98]. Facetted surfaces resulting from this treatment behave as weakly disordered Euclidean surfaces. These surface structures were investigated by conventional surface techniques and nanoscopies [99, 100]. The change in the hydrogen adatom electrochemical adsorption–desorption voltammogram in going from polycrystalline to facetted platinum in acid solution is also shown in Figure 2.10. Further details about the pretreatment of polycrystalline metal electrocatalysts are available in the literature [101].

The quality of either the smoothness or roughness of a solid surface depends on the probing scale length [102]. Stepped metal surfaces are weakly disordered, as can be seen, for instance, by the crystallographic arrangement of crystals around the [111] pole of a gold polyfacetted single-crystal sphere (Fig. 2.11) that consists of

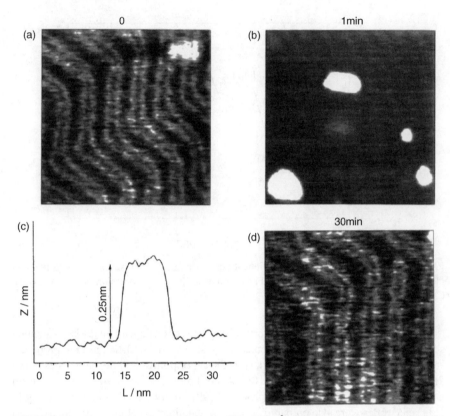

FIGURE 2.7 Sequential in situ STM images (38×38 nm^2) of a gold (111) surface domain in aqueous 0.5 M NaClO$_4$ + 0.01 M HClO$_4$ at 298 K to follow the (1 × 1) gold (111) $\rightleftharpoons$ gold (22 $\sqrt{3}$) surface reconstruction. The formation and disappearance of islands as E is changed from 0.20 to 0.35 V and backward (E_{pzc} = 0.25 V) are shown. A cross section of a one-gold-atom island is shown. (From Martín et al. [93] by permission of The American Chemical Society.)

small terraces and steps intersecting at 60° and a certain disorder caused by structural defects.

2.5.2.2 Dispersed Electrocatalysts

Often active metal electrocatalysts consist of crystallites dispersed in a chemically inert, supporting conductor. In this case, two or more crystallites separated by grain boundaries form aggregates of different size and shape. Then, the size distribution of crystallites and their average aspect ratio, for instance, the height-to-width ratio of crystallites, are interesting parameters for determining the efficiency of an electrocatalyst for a particular process [104].

On the other hand, for an electrocatalyst made of dispersed particles, the surface of each particle may consist of terrace domains surrounded by monoatomic or a few atoms in height steps. As an example, for small palladium crystals on C(0001)

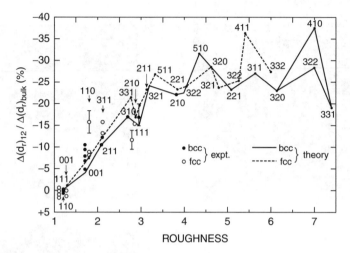

FIGURE 2.8 Experimental and theoretical percent first-layer relaxation versus roughness plot for several bcc and fcc surfaces. (From Somorjai [96] by permission of The American Chemical Society.)

(Fig. 2.12), the cross section of each crystallite is direction dependent [98]. For stepped surfaces the degree of disorder depends on the size of the yardstick used to probe the surface. With a rather large yardstick size, the set of points defining step edges in the 2D space behaves as a fractal [105, 106]. Accordingly, for a stepped surface, the term *marginal fractal surface* has been proposed [99]. Small palladium clusters on gold (111) electrodes were also prepared utilizing a scanning tunneling microscope (STM) [107]. Crystal engineering of metallic nanoparticles is reviewed by Creus et al. [108]. Recently, high-resolution transmission electron microscopy (TEM) micrographs of platinum nanocrystals were utilized for controlling the shape and size of these catalysts in the crystal range of 2–5 nm [109].

2.5.2.3 Rough Electrodes
A procedure for the electrochemical roughening of metals consists of the application of preset potential cycling by conveniently adjusting the potential routine parameters

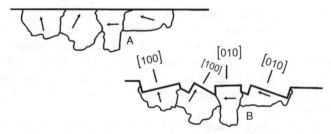

FIGURE 2.9 Bidimensional scheme of a polycrystalline surface profile before and after a mono-orientation process. The vectors indicate the preferred orientation of the crystallites.

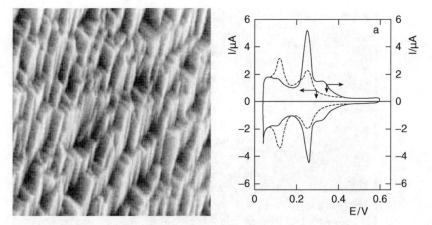

FIGURE 2.10 Voltammogram run at 0.1 V s^{-1} in 1 M aqueous sulfuric acid, at 298 K, and SEM micrograph of a polyfacetted platinum electrode after a 10-h repetitive square wave potential scanning from 0.05 to 1.50 V at $f = 6$ kHz. The full and dashed traces in the voltammogram correspond to the treated and the untreated platinum surface, respectively. SEM micrograph scale 1 μm. E values are referred to the NHE scale. (From Visintin et al. [103] by permission of Elsevier.)

so that the oxidation half cycle yields a preset amount of metal hydrous oxide on the surface and subsequently the oxide layer is electroreduced to produce a highly rough metal surface. The roughness of the surface $\mathcal{R}$ increases with the thickness of the hydrous metal oxide layer (h) and the potential cycling time (t).

Polycrystalline surfaces, particularly those prepared at fast potentiodynamic routines, exhibit a strong disorder at the micrometer scale and are described by the fractal geometry. Depending on whether these surfaces display either isotropic or anisotropic characteristics, they behave as either self-similar or self-affine fractals, respectively. Their irregular topography influences the kinetics and mechanism of surface processes undergoing on these materials.

The dynamic scaling analysis, as seen further on in Section 2.1.5.4, indicates that the term *rough surface* represents an irregular surface without overhangs [110]. The simplest irregular surface is that generated by a 3D structure exhibiting a preferred single crystallographic orientation in the distribution of material. In this case, overhangs do not dominate the scaling properties of the object. The absence of overhangs determines the self-affine character of rough surfaces, that is, their anisotropic surface disorder. The topography of rough surfaces is often stochastic but exhibits self-resemblance over a certain range of scale because they are formed by random processes (diffusion, surface reactions, fractures) that are typical of self-affine fractals [100].

The facetting and roughening of metal surfaces produced by applying fast potentiodynamic routines occurs via a place exchange processes at the electrode surface in which adsorbed anions are involved [112]. In the oxidation scan, metal atoms are removed from the lattice position to attain higher degree of coordination with OH- and O-containing species, and in the reduction scan, the nucleation of adatoms

FIGURE 2.11   STM image of a polyfacetted platinum single-crystal cut normally in the [111] plane direction and flame annealed for 5 min at $\sim$ 1470 K. (From Salvarezza and Arvia [111] by permission of Elsevier.)

initiates island, vacancy, and hole formation on the metal surface (Fig. 2.13) [113, 114].

The STM topography of irregular metal surfaces (platinum, gold, palladium) produced by the above electrochemical roughening [116] consists of a brushlike or a columnlike metal structure of $\sim$ 10 nm in average diameter dominated by most compact crystallographic planes [117–119]. Such metal electrodeposits exhibit roughness factors ($\mathcal{R}$) up to about 10^3 [120, 121]. The growth of these surfaces results in steady self-affine fractal patterns [100].

The value of $\mathcal{R}$ can be estimated from conventional voltammetry by comparing charges of either electrooxidation or electroreduction of a surface layer at the electrodes before and after electrochemical treatment. Thus, for rough platinum electrodes, $\mathcal{R}$ can be determined by comparing the hydrogen adatom voltammetric charge to the reference of a smooth platinum surface of the same geometric area, temperature, and environment [101]. For gold electrodes, these procedures are utilized by comparing charges from the oxide electroreduction charge at the rough surface (Q_f) and that from the smooth gold surface prior to anodization (Q_i), that is, $\mathcal{R} = Q_f / Q_i$. Similar rough surfaces were obtained for palladium [119] and rhodium [122] electrodes. A scheme of electrochemical facetting and roughening is given in Figure 2.14.

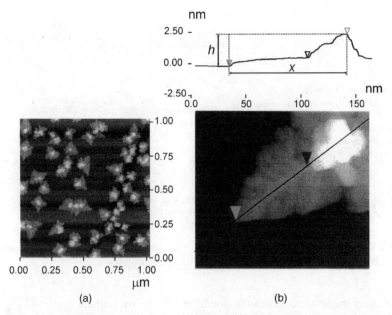

FIGURE 2.12 (a) STM images of branched palladium islands electrodeposited on C(0001) at $= -0.100$ V and 298 K. (b) Detail of single-island topography and cross section of branch is shown. The aspect ratio is defined as $\langle h/2x \rangle$. (From Gimeno et al. [115] by permission of The American Chemical Society.)

2.5.3 Determination of Real Surface Area

Most of the electrochemical kinetic and double-layer parameters are extensive quantities that have to be referred to the unit real (active) electrode area. Therefore, determination of the real active area of solid electrodes is of outmost importance for

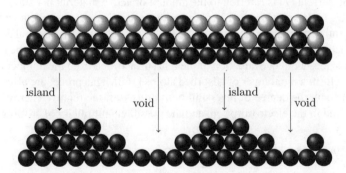

FIGURE 2.13 Schematic diagram of the place–exchange process accompanying electrochemical oxidation–reduction cycles. Metal atoms (dark grey) move out of lattice position attaining higher coordination with O-containing adsorbates (white) upon oxidation. After electroreduction, adatoms nucleate to form islands on and the vacancies nucleate to produce holes in the original surface upon reduction.

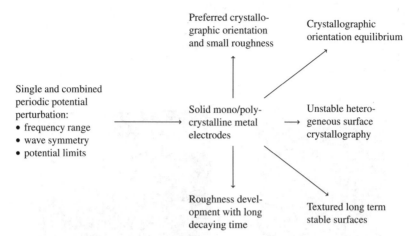

FIGURE 2.14 Scheme of electrochemical facetting and roughening possibilities applicable to metals of electrocatalytic interest utilizing aqueous electrolytes.

establishing the optimal operational conditions of electrochemical devices [123]. The real area refers to that part of a solid electrode surface that is accessible to reactants without being affected by product formation. For this determination, several ex situ and in situ methods have been proposed [124] (Table 2.3).

Real surface area measurements using either adsorption from the gas phase or electrosorption techniques are good tests to evaluate the degree of disorder at metal surfaces. Thin-metal films used in microelectronics, optical devices, and protective coatings, requiring defect-free surfaces and a small area of metal single-crystal surfaces, exhibit a smooth real surface area approaching their geometric area. Areas of this type are taken as reference for defining $\mathcal{R}$. Then, in this case $\mathcal{R}/\mathcal{R}_{smooth} = 1$, a figure generally related to a surface with a very low density of defects. For dispersed metals, the current (I) is referred to the amount of active material per surface unit.

For a few metals, electrochemistry offers the possibility of using cyclic voltammetry to determine the real surface area [139]. In this case, the underpontential deposition (UPD) electroadsorption–electrodesorption charge of several metals on the unknown surface provides a direct evaluation of the real surface area from the number of atoms required to form a monolayer of adsorbed atoms [160]. This procedure also gives information about the degree of reversibility of processes, the relative energy of active sites involved in the electrosorption, and the possible multiplicity of adsorbed states.

2.5.4 Dynamics of Solid Electrode Surfaces

Electrocatalysis is benefited not only from improvements in the preparation and the quality control of electrode materials but also from a better understanding of the dynamics of the surface structure. Changes in the electrocatalytic activity are due to a change in the surface roughness and also the coalescence of electrocatalyst particles. Both processes limit the lifetime of the electrocatalyst.

TABLE 2.3   Some Methods for Determination of Real Areas of Rough and Porous Electrodes

Method	References
In Situ Methods	
Measurements of double-layer capacitance	125–132
Drop weight (or volume)	
Capacitance ratio	
Measurements based on Gouy–Chapman–Stern theory to determine	
diffuse double-layer capacitance	133
Parsons–Zobel plot	
Measurements of extent of monolayer adsorption of indicator species	134–142
Hydrogen adsorption from solution	
Oxygen adsorption from solution	
Underpotential deposition of metals	
Adsorption of probe molecules from solution	
Voltammetry	143–145
Open-circuit potential relaxation	146, 147
Negative adsorption	148
Ionexchange capacity	149, 150
Mass transfer	151
Scanning tunnelling microscopy (STM)	152, 153
Atomic force microscopy (AFM)	
Ex Situ Methods	
Gravimmetric methods	154–156
Volumetric methods	154–156
Adsorption of probe molecules from gas phase	154
Weighing of saturated vapor adsorbed on solid	154
Hysteresis of adsorption isotherm	31, 157, 158
Thermodesorption	157, 158
Porosimetry	156
Liquid permeability and displacement	156
Gas permeability and displacement	156
Wetting heat (Harkins–Jura method)	157
Surface potential of thin pure metal films	31
Metal dissolution rate	31
SEM, STM, AFM, profilometer and stereoscan method	159
Diffuse light scattering	155
X-ray diffractometry	155
Nuclear magnetic resonance (NMR) spin–lattice relaxation	155
Radioisotopes	156, 157

The dynamics of roughness decay can be followed using high-sensitivty surface analysis techniques combined with different nanocopies [161]. Experimental data of rough surfaces in real space nanoscopy imaging (either by ex situ or in situ STM and AFM) offer the possibility of fractal surface characterization in the vertical dimension, yielding directly 3D patterns from the nanometer up to the hundreds of micrometer level [162]. In this scenario, dynamic scaling theory becomes useful for accounting for surface relaxation processes.

2.5.4.1 Dynamic Scaling Theory

Let us consider the development of a simple rough contour without overhangs that grows from a flat surface of length L with N_S sites ($L \ll N_S$) at time $t = 0$ [100] [Fig. 2.15(a)]. For $t > 0$ the contour of the growing surface in the direction normal to L displaces in height (h) with respect to the roughening-free contour at $t = 0$, and the instantaneous value of h is described by a single-valued function of x and t.

The instantaneous width $w(L, t)$ taken as a measure of surface roughness is defined by the root-mean-square of height fluctuations [163],

$$w(L, t) = \left[\frac{1}{N_s} \Sigma [h(x_i) - \langle h \rangle]^2 \right]^{1/2} \tag{2.79}$$

$\langle h \rangle$ being the average height normal to the direction of L at time t. Initially, $w(L, t)$ increases with t as random fluctuations do, according to

$$w(L, t \rightarrow 0) \simeq t^{\beta^*} \tag{2.80}$$

where the exponent β^* describes the kinetics of the contour formation.

After a long time t, a steady-state contour is reached [Fig. 2.15(b)], that is,

$$w(L, t \rightarrow \infty) \simeq L^{\alpha^*} \tag{2.81}$$

α^* being the roughness exponent, and the surface becomes a scale-invariant self-affine fractal [100]. However, as a self-affine fractal is not invariant at all scale lengths, when the surface is rescaled by a factor b in the horizontal direction, it should be rescaled by a factor b^{α^*} in the direction perpendicular to the surface to get the similarity between the original and the rescaled surface. Globally considered, when the self-affine surface is probed with a yardstick larger than the interface width, it appears flat, whereas for the reverse situation it behaves as a rough surface characterized by the exponent α^* [97, 100]. Small and large values of α^* are associated with jagged and well-correlated smooth-textured surfaces, respectively.

Parameters derived from the dynamic scaling analysis provide information about the degree of surface disorder (α^*), the mechanism of surface dynamics (β^*), and the critical time ($t_c = L^z$, where $z = \alpha^*/\beta^*$) for the steady-state roughness regime transition. More recently, the dynamic scaling analysis was extended to systems whose size evolves in time [164, 165]

The evaluation of the fractal dimension (D_f) depends on the geometric method used [97, 100]. For columnar structured electrodes, D_f can be evaluated from nanoscopy imaging data [166] utilizing the *lakes-and-islands* simulation method. For platinum and palladium electrodes with columnar surface structures, $D_f = 2.7 \pm 0.05$ [162].

The fractal analysis of rough electrode surfaces can also be attempted using a diffusion-controlled reaction in solution, such as

$$[Fe(CN)_6]^{4-}(aq) = [Fe(CN)_6]^{3-}(aq) + e^-$$ (2.82)

as a test process. The deviation of the potentiostatic current transient from that expected for the ideal flat electrode surface permits the evaluation of D_f. For a fractal surface, the general expression of a potentiostatic current transient is [167]

$$I \simeq t^{-p}$$ (2.83)

the exponent p being related to D_f by

$$p = \frac{(D_f - 1)}{2}$$ (2.84)

In this case, the diffusion layer thickness scales the irregularities of the surface as a time-dependent yardstick. For $t \to 0$, the average diffusion layer thickness is much smaller than that of the smallest irregularities, whereas when $t \to \infty$, the diffusion layer thickness largely exceeds the size of the largest surface irregularities. The fractal-to-nonfractal transition crossover critical time t_c is [167]

$$t_c = \frac{\lambda^2}{D_i}$$ (2.85)

where λ is the size of the largest surface irregularity and D_i is the diffusion coefficient of the reactant i in the solution. In the fractal regime, the diffusion kinetics is characterized by values of $p > 0.5$, in contrast to $p = 0.5$ for Euclidean smooth flat electrode surfaces. For instance, for a columnar structured gold electrode it results in $p = 0.76 \pm 0.05$ for $t < 1$ s and $D_f = 2.5$, whereas for stepped gold single-crystal $p = 0.50 \pm 0.05$ and $D_f = 2.0$ [168].

2.5.5 Surface Atom Diffusion and Roughness Decay

Roughness development [Fig. 2.15(a)] starting from a linear profile evolves either to attain a steady-state regime [Fig. 2.15(b)] or continuously increasing roughness without reaching saturation [Fig. 2.15(c)]. Conversely, roughness decay plays an important role for determining how long the specific characteristics of the solid surface could be maintained in the course of the process [Fig. 2.15(d)]. Very often surface diffusion participates in the the mechanisms of surface roughness evolution [169, 170].

Both highly dispersed porous and disordered solid fractal surfaces involve sites of different energy and, correspondingly, surface energy gradients driving surface atoms to attain a minimum energy. This process decreases both the roughness and the value of D_f and changes the particle size distribution at the surface, as can be concluded from in situ STM imaging data [104, 118]. These surface energy gradients

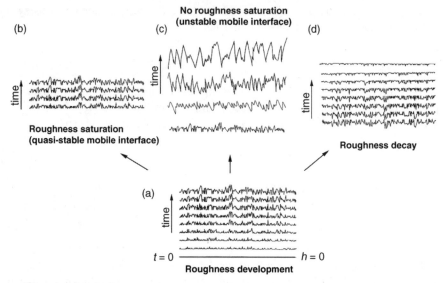

FIGURE 2.15 Scheme of evolution of surface profiles: (a) roughness development of smooth profiles (exponent β^* can be calculated); (b) evolution of rough profile to attain quasi-stable roughness regime (exponent α^* can be evaluated); (c) continuous roughness increase (no steady state is reached; only exponent β^* can be estimated); (d) roughness decay to approach a quasi-smooth profile (decay kinetics can be determined).

at electrochemical interfaces mean fields of about 10^7–10^8 V cm^{-1} that are of the order of magnitude of intraatomic fields. These electric fields can also assist in the displacement of surface atoms (surface diffusion). This effect becomes even more complex in inhomogeneous electric fields [171].

For a substrate considered as a static participant in which a frozen potential landscape is felt by diffusing species, neglecting interactions among surface species, individual adatom diffusion to a surface site can be interpreted by Fick's second law. Accordingly, the mean adatom surface diffusion velocity ($\langle v_d \rangle$) is proportional to the reciprocal of the square of the atom jump length l [169],

$$\langle v_d \rangle \simeq \frac{D_s}{l^2} \tag{2.86}$$

D_s being the surface adatom diffusion coefficient. This static approach is improved after both field ion microscopy (FIM) and STM results, and first-principle calculations demonstrate that adatom diffusion on some face-centered-cubic (fcc) (100) metal surfaces can proceed via a place-exchange mechanism [169] as earlier observed on fcc (110) surfaces, in which even for the diffusion of single adatoms many body cooperative interactions operate, as is the case of collective slip diffusion mechanisms for single nanocrystals [172]. This situation, of interest in electrocatalysis, appears when the substrate plays an active part in the surface process and diffusion jumps normally result from concerted displacements of diffusing species and substrate neighbors.

Then, an exchange mechanism occasionally operates in which a diffusing species pushes out substrate atoms and takes their position, while the substrate atom starts to diffuse [169]. On the other hand, when many particles are diffusing at a time, they may interact with each other by manifold nonadditive, generally anisotropic, lateral forces that can be attractive, repulsive, or oscillatory short or long range.

2.5.6 Clusters and Islands

Recent studies have shown that lateral interactions, even at low coverages, yield atom clusters that at later stages may lead to overlayer condensation and Ostwald ripening phenomena. Then, besides atoms, surface diffusion may also promote the formation of either clusters or islands. Accordingly, the number and complexity of diffusion pathways will depend on the degrees of freedom of the diverse diffusion species [112, 173]. For diffusion of clusters and atomic islands different mechanisms are distinguished, namely a sequential displacement of individual atoms (tetramers) on fcc (100); a dimer-shearing mechanisms for diffusion and dissociation of clusters of less than nine atoms on fcc (100); clusters gliding at the surface as a whole from one position to the other; a motion of a loose adatom along island edges; an evaporation/condensation (interchange) of adatoms between different islands and steps producing island coarsening; a dimer diffusion in atomic channels; an adatom reaccommodation moving out from a dislocation to a new location; or a rolling mechanism [174–177].

Surface atom diffusion in postdeposition surface roughness of column-structured metal films assists in the coalescence of nanometer-sized metal particles in aqueous electrolytes yielding large particles. This aging effect at the solid surface usually produces a decrease in the electrocatalytic efficiency of dispersed and nanostructured metals.

2.5.6.1 Ostwald Ripening

Islanding processes are of importance for understanding the driving forces operating at solid surfaces in which changes of strained-layer lattices are produced [178]. These mismatched lattices involve a rather uniform layer a few atoms in height under nonequilibrium conditions favoring cluster formation. Then, since thermodynamics drives the system to a cluster configuration, islanding phenomena become relevant in dealing with solid electrocatalysts and their performance.

The Ostwald ripening model considers that clustering occurs as a result of the Gibbs–Thomson effect, which describes the equilibrium concentration of adatoms (c_i) at the island surface [173]. Assuming a spheroidal or flat cylindrical island of radius r remains constant, a general expression for the radial growth of clusters is

$$r = K(n)t^{1/n} \tag{2.87}$$

the exponent n depending on the RDS. For surface diffusion processes with long-range surface concentration gradients, $n = 4$. For surface diffusion with either a constant surface concentration between islands or an energy barrier for atomic separation at the island surface, $n = 3$.

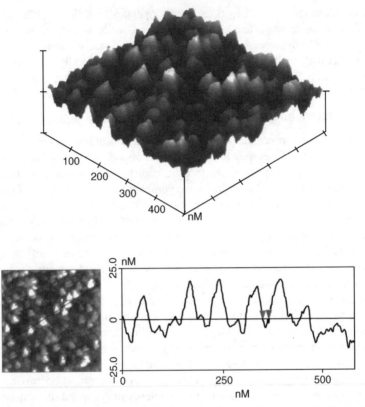

FIGURE 2.16 Ex situ STM image (500×500 nm^2) of an electrochemically roughened platinum electrode ($\mathcal{R} = 15$) and cross-sectional analysis. Electrode treatment: symmetric square-wave scanning ($\tau = 560$ μs) from 0.40 to 2.40 V (NHE) for $t = 3$ min and subsequently 10 h aging at open circuit. (From Martins et al. [179] by permission of Elsevier.)

The topography of column-structured metals at the nanometer–micrometer range can roughly be described as a matrix arrangement of columns and voids at the micrometer scale and the column tips, at the nanometer scale, consisting of cluster aggregates (Fig. 2.16). Accordingly, the rate processes at smooth domains of size smaller than the column diameter can be handled as an Ostwald ripening kinetics. Then, the apparent surface diffusion coefficients of metal adatoms and clusters under different conditions can be estimated, including the presence of adsorbates [104, 118]. The coalescence of gold particles of average radius $\langle r_x \rangle$ in the x direction obeys the rate equation $\langle r_x \rangle \simeq t^{1/4}$ [118]. Accordingly, when gold surface atom diffusion becomes the RDS, $\langle r_x \rangle$ increases with the relaxation time t [180],

$$(\langle r_x \rangle - \langle r_0 \rangle)^4 = 2\langle \gamma \rangle a^4 D_s \frac{t}{kT} \qquad (2.88)$$

$\langle r_0 \rangle$ being the initial average radius of particles in the rough topography, $\langle \gamma \rangle$ is the average surface tension of the metal in the environment, a is the metal lattice parameter, and D_s is the average adatom surface diffusion coefficient.

For a simpler columnar model of average columnar height $\langle h \rangle$ and average column radius r_c, being $\langle h \rangle \gg r_c$, the following instantaneous roughness proportionality occurs [181]:

$$\mathcal{R} \simeq \frac{\langle h \rangle}{3 \langle r_c \rangle} \tag{2.89}$$

Equation (2.89) can be used to estimate $\langle r_c \rangle$ from $\mathcal{R}$, the roughness factor obtained from electrochemical experiments. Values of $\langle r_c \rangle$ derived from Eq. (2.89) are in good agreement with those obtained from STM imaging data. The value of $\langle h \rangle$ can also be derived from the cross-sectional analysis of the metal deposit obtained from scanning electron micrographs.

The spontaneous decrease in $\mathcal{R}$ can also be followed by voltammetry [104, 181] and by capacity measurements utilizing concentrated solutions at low frequency [160]. The roughness relaxation of the electrochemically produced column-structured gold and platinum layers in contact with aqueous acids obeys a reasonable linear $\mathcal{R}^4$-versus-t plot under either potential control or open-circuit conditions [104, 181]. Using the simple columnar model and further assuming that $d\langle h \rangle / dt \ll dr_c/dt$, Eqs. (2.88) and (2.89) lead to [104]

$$\mathcal{R} \simeq \frac{\langle h \rangle}{2\gamma(a^4 D_s t/kT + r_0)^{1/4}} \tag{2.90}$$

Then, by plotting $\mathcal{R}^4$ versus t, values of D_s for surface diffusion of atoms and clusters on several rough metals in aqueous electrolytes can be estimated provided that the value of γ is known. The latter depends on both the solution composition and the presence of adsorbates. The change in the value of D_s caused by strongly adsorbed anions [104] and molecules [181] becomes negligible. Values of D_s under an open circuit in 1 M aqueous sulfuric acid and 298 K are $D_s = 5 \times 10^{-14}$ cm^2 s^{-1} for gold and 10^{-18} cm^2 s^{-1} for platinum [181]. The temperature dependence of D_s in the range 273–325 K fulfils an Arrhenius plot with apparent activation energies of 14 ± 2 kcal mol^{-1} for gold and 19 ± 2 Kcal mol^{-1} for platinum [181]. The preexponential factors derived from Arrhenius plots are 0.68×10^{-3} cm^2 s^{-1} for gold and 0.8×10^{-4} cm^2 s^{-1} for platinum. The difference in these values could be related to the entropic contributions in the respective surface diffusion mechanisms.

The metal atom diffusion jump length is related to the surface diffusion coefficient by

$$l = \sqrt{2 D_s t} \tag{2.91}$$

and D_s is related to both the surface activation entropy and entalphy, that is, ΔS^* and ΔH^*,

$$D_s = K' \exp\left(\frac{\Delta S^*}{RT}\right) \exp\left(-\frac{\Delta H^*}{RT}\right) \tag{2.92}$$

K' denoting the preexponential factor resulting from the theory of rate processes [35]. As $\Delta H^* \simeq 0.2 \, \Delta H_s$, ΔH_s being the sublimation enthalpy, and $T = T_m$, the melting temperature,

$$D_s \simeq D_0 \, \exp\left(-\frac{0.2 \, \Delta H_s}{R T_m}\right) \qquad (2.93)$$

where $D_0 = K' \exp(\Delta S^*/R T_m)$. Taking into account the relationship between $\mathcal{R}$ and l and Eqs. (2.91) and (2.93),

$$(\mathcal{R})^2 \simeq D_0 \, \exp\left(\frac{0.2 \, \Delta H_s}{R T_m}\right) \qquad (2.94)$$

This proportionality predicts a linear dependence of $\ln \mathcal{R}$ on either ΔH_s or T_m^{-1} [182]. This dependence of the electrode roughness on the physical properties of the metal explains the decrease of the real surface of the metal electrode in the long-time range.

Chakraverty's model for particle growth can also be utilized to interpret roughness decay data [123]. It considers a population of 3D hemispherical nuclei placed on a flat 2D substrate with a certain nuclei size distribution. To minimize the surface free energy, roughness decay occurs via adatom incorporation into the nuclei–substrate interface from the preferred dissolution of smaller nuclei yielding larger ones. This kinetic description is, in principle, applicable to roughness relaxation at column-structured solid surfaces in which small flat rounded nuclei constitute the upper part of large columns where the transport of adatoms from small dissolving tip sites to coalescing column domains takes place. Surface mass transport undergoes exclusively among contacting small column domains so that larger column tips tend to grow further at the expense of smaller ones leaving deep pores and crevices in between large columns. This explains why no complete roughness decay to flat surface structures can be attained [181].

A scheme of the dynamic properties of rough surfaces in which clustering phenomena are emphasized is illustrated in Figure 2.17.

2.5.7   Adsorption and Dynamics of Adsorbates on Solid Rough Electrodes

As a first approximation, the surface of a solid electrocatalyst can be considered as a frozen potential energy landscape felt by diffusing atoms and molecules. The interaction of irregular solid surfaces with adsorbable species involves the uncertainty in the definition and relative contribution of specific surface properties [105, 106, 183]. In fact, already in the simplest case of Langmuirian adsorbate–substrate interactions, the adsorbate–substrate site cross-sectional ratio turns out to be a parameter that dominates adsorption features at irregular surfaces. In this case, the adsorbate (A) becomes the yardstick probing the substrate (S) and the scaling of the surface is affected by an exclusion volume in the evaluation of the thermodynamic adsorption parameters

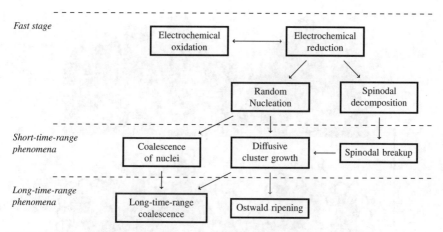

FIGURE 2.17 Scheme related to formation of metal solid phases by fast electroreduction of oxygen-containing adlayer previously produced by cyclic potential routine as described for electrochemical facetting and roughening of Pt, Au, Pd, and Rh.

[105, 184, 185], particularly when the size of the probing molecule approaches the size of irregularities (Fig. 2.18). Correspondingly, thermodynamic data derived from adsorption isotherms depend on the irregularity of the substrate, and it can be related to the adsorbate–average irregularity cross-sectional ratio.

At constant concentration of A (c_A) and T, the adsorption equilibrium on a stabilized fractal solid surface is determined by substrate–adsorbate and adsorbate–adsorbate specific interactions, which in turn depend on the topographic features (size, geometry, and distribution of irregularities) and, at the electrochemical interface, on the applied potential.

The influence of topographic features on molecular adsorption can be derived from an adequately selected model with a well-characterized solid substrate topography interacting with adsorbates of different size. In all cases, the adsorbate–substrate irregularity average size ratio (A/S) and the degree of surface coverage of the substrate by A determine the nonaccessible area of S or, more generally, the excluded volume for molecular adsorption. This approach is equivalent to scaling S with yardsticks of different sizes such as an atom, a molecule, or a radical species. Accordingly, three limiting situations for a rough surface can be described: (i) when $A/S \to 0$, the entire substrate surface becomes accessible to the adsorbate; (ii) when $0 < A/S < \infty$, the excluded volume effect due to the surface fractality appears and, for a 2D space, the excluded volume results in the difference between the real contour of the surface and that defined by the centers of adsorbate A; and (iii) when $A/S \to \infty$, only the tips of irregularities become accessible to adsorption, that is, a smooth surface is approached.

Situation (i) is observed for H atom and O atom electrosorption on rough platinum [186], O electrosorption on rough gold [187], carbon monoxide adsorption and electroadsorption on rough platinum [188, 189], underpotential deposition of cadmium

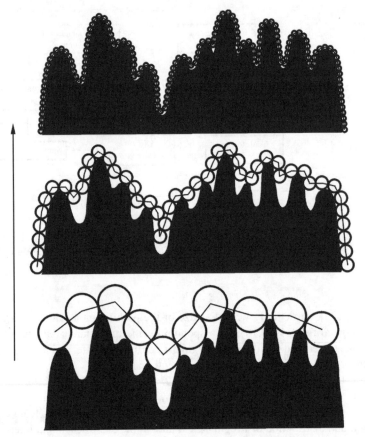

FIGURE 2.18 Principle of real surface area measurement by monolayer adsorption technique. The result depends on the yardstick size of the adsorbate. Larger molecules ignore the fine details of surface morphology. The arrow indicates the increase in the resulting value of roughness. (From Gómez et al. [184] by permission of The American Chemical Society.)

and lead on silver dendrites [166], glucose electroadsorption on rough gold [190], and so on. The value of $\mathcal{R}$ determined from the electroadsorption charge becomes almost the same irrespective of the yardstick size.

Situation (ii) is approached by the electrochemical adsorption of 1,10-phenanthroline in aqueous 0.1 M perchloric acid on rough gold electrodes [184]. In this case, the adsorption isotherms, plotted as θ_A versus c_A, where θ_A is the degree of surface coverage by 1,10-phenanthroline species and c_A is its concentration in solution, are estimated from the gradual reduction in the O atom desorption voltammetric charge. For columnar gold electrodes in the range $1 \leq \mathcal{R} \leq 55$, the adsorption data plots approach a Frumkin-type isotherm [2, 38],

$$\beta' c_A = \theta_A \, \exp\!\left(-\frac{2a\theta_A}{1 - \theta_A}\right) \tag{2.95}$$

which are shifted toward lower values of θ as $\mathcal{R}$ is increased.

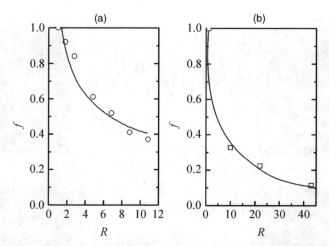

FIGURE 2.19 (a) Monte Carlo simulations of pyridine adsorption on rough electrodes with columns of height h and width $d = 4.8 \times 10^{-7}$ cm spaced 1.6×10^{-7} cm. Plot of the phenomenological function f versus $\mathcal{R}$. $f = K\mathcal{R}^n$ ($K \approx 1$ and $n = 0.44$) and adsorption probability 0.98. The value of $\mathcal{R}$ was obtained from the ratio between the length of a rough substrate for different values of h and that of the flat substrate. (b) Plot of f versus $\mathcal{R}$ for pyridine adsorption on rough gold electrodes from aqueous pyridine-containing 1 M HClO4 at $E = -0.4$ V (vs. NHE) and 298 K. (From Gómez et al. [192] by permission of Elsevier.)

In Eq. (2.95), a is a complex interaction parameter and β' is the adsorption coefficient given by

$$\ln \beta' = -\frac{\Delta G^*_{ads}}{RT} \tag{2.96}$$

where ΔG^*_{ads} is the adsorption free energy. Accordingly, straight-line plots are obtained by plotting $\ln \theta_A/c_A$ versus $\theta_A/(1 - \theta_A)$, their slopes increasing with the value of $\mathcal{R}$, but β for $\theta_A = 0$ remains the same irrespective of $\mathcal{R}$. The value of ΔG^*_{ads} results in -26 ± 1.5 kJ mol^{-1} at 298 K, a fact that indicates weak 1,10-phenanthroline–columnar gold surface interactions. The value of a increases with $\mathcal{R}$ from $a = -10$ to $a = -14$, that is, there are repulsive interactions among neighboring adsorbates.

Situation (iii) is approached by the adsorption of a molecule whose average size is larger than the average void size of the rough electrode surface. This is the case of the adsorption of thymol blue from aqueous solution on rough gold electrodes. The average cross section of this molecule is greater than 10^{-12} cm^2, exceeding the average roughness features of S. Then, in this case the degree of surface coverage approaches that of a smooth surface, as expected for $A/S \to \infty$ [162].

The above adsorption data were compared with those obtained from Monte Carlo simulations on Langmuirian adsorption on columnar gold surfaces considering steric and barrier effects (Fig. 2.19) [191–193]. As shown in Figure 2.19, by introducing in Eq. (2.95) the correction function $f = K\mathcal{R}^n$ with $K \approx 1$ and $n = 0.44$, a reasonable correlation between experimental and Monte Carlo simulation data is obtained.

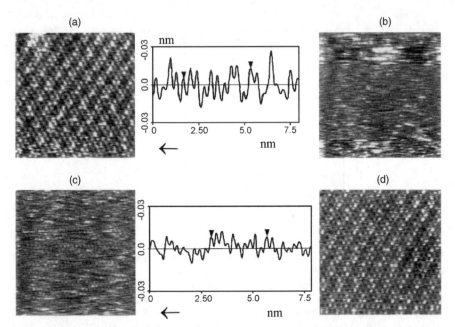

FIGURE 2.20 In situ STM images (8×8 nm^2) of pyridine adsorbed on gold (111). (a) The 4×4 Py adlayer observed at $E = 0.70$ V (cross section included). (b) Blurred image with domains exhibiting the 1×1 gold (111) lattice after stepping from $E = 0.70$ V to $E = 0.96$ V. (c) Image (and cross section) recorded immediately after stepping from $E = 0.96$ to 0.70 V. (d) STM image taken 3 min after (c): The 4×4 Py adlayer reappears. Corrugation and nearest-neighbor distances are 0.01 and 0.29 nm for (d) and 0.02 and 0.38 nm for (a), respectively. (From Andreasen et al. [195] by permission of Elsevier.)

STM images of pyridine (Py) adsorbates on Au(111) terraces in aqueous environment at 298 K and $E < E_{pzc}$ show isolated Py molecules adsorbed flat on the Au(111) terraces and an interesting surface dynamics [194, 195]. In this case, the adsorption of Py increases the corrugation of the Au(111) surface, that is, these adsorbates behave as chemical defects that increase surface disorder at the submonolayer level. On the other hand, the same process at $E > E_{pzc}$ results in ordered arrays of Py molecules adsorbed perpendicular to the Au(111) surface forming (4×4) hexagonal patterns with the intermolecular distance $d = 0.38$ nm [194, 195]. However, the adlattice order is far from complete as vacancies in the adsorbate layer are clearly resolved. As the adsorption time increases, these vacancies tend to be filled, so that the surface corrugation and disorder decrease (Fig. 2.20).

Similar surface disorders are also present in self-assembled monolayers (SAMs) of alkyl thiols on Au(111) at room temperature [196–201]. STM images of these SAM patterns often reveal a large number of randomly distributed single-atom deep holes and compact $\sqrt{3} \times \sqrt{3} \times R$ 30° domains coexisting with 4×2 domains. The simultaneous presence of different SAM structures and holes results in heterogeneous topography with a significant degree of disorder. Real-time imaging reveals fluctuations in the average size of holes accompanied by $R\ 30°c(4 \times 2)$ surface structure transitions that occur within the range of a few seconds [199].

The dynamics of molecular adsorbates depends on the applied potential, solution composition, P, and T. Then, the overall process is complicated by the participation of further interactions at the electrochemical interface with the solution components, which may contribute to the reconstruction of the electrode surface. A general discussion of the dynamics of the electrode surface is given elsewhere [202] as well as fundamental aspects of the interactions of water with solid surfaces [91, 203].

ACKNOWLEDGMENT

Financial support from the Consejo Nacional de Investigaciones Científicas y Técnicas (CONICET), Agencia Nacional de Promoción Científica y Tecnológica (ANPCYT, PICT 34530), and Comisión de Investigaciones Científicas de la Provincia de Buenos Aires (CICPBA) of Argentina is acknowledged.

REFERENCES

1. J. O. Bockris and K. N. Reddy, *Modern Electrochemistry*, MacDonald, London, 1970.
2. E. Gileadi, *Electrode Kinetics for Chemists*, Chemical Engineering and Materials Scientists, VCH, New York, 1993.
3. J. O. Bockris and S. H. Khan, *Surface Electrochemistry*, Plenum, New York, 1993.
4. D. Pletcher and F. C. Walsh, *Industrial Electrochemistry*, 2nd ed., Chapman and Hall, London, 1990.
5. E. Heinze and G. Kreysa, *Principles of Electrochemical Engineering*, VCH, Weinheim, 1986.
6. N. Kobosev and W. Monblanowa, *Acta Physicochim. URSS*, 1 (1934) 611.
7. A. T. Grub, *Nature*, 198 (1963) 883.
8. H. Davy, *Nicholson* J., (1802) 144.
9. W. Groove, *Philos. Mag.*, 14 (1839) 297.
10. L. L. Mond and D. Langer, *Proc. Roc. Soc.*, A46 (1889) 296.
11. W. Ostwald, *J. Electrochem. 1*, (1894) 122.
12. W. W. Jacques, *Harpers Mag.*, 94 (1896) 144.
13. E. Bauer, W. D. Treadwell, and G. Trumpler, *Z. Elektrochem.*, 27 (1921) 199.
14. F. P. Bowden and E. Rideal, *Proc. R. Soc. A.*, 120 (1928) 59.
15. J. Horiuti and M. Polanyi, *Acta Physicochim. USSR*, 2 (1935) 505.
16. O. K. Davtyan, *Bull. Acad. Sci. USSR, Classic Sci. Tech.*, 197 (1946) 125.
17. F. T. Bacon, in *Fuel Cells*, Vol. 1, G. J. Young (Ed.), Reinhold, 1960.
18. J. G. Broers, *Abh. Saech. Akad. Wiss. Leipzig*, 49 (1968) 395.
19. E. Justi, *A Solar-Hydrogen Energy System*, Plenum, New York, 1987.
20. N. M. Sammes and R. Boersma, *J. Power Sources*, 86 (2000) 98.
21. G. Cacciola, V. Antonucci, and S. Freni, *J. Power Sources*, 100 (2001) 67.
22. M. Farooque and H. C. Maru, *J. Power Sources*, 160 (2006) 827.
23. C. G. Vayenas, S. Bebelis, and S. Ladas, *Nature*, 625 (1990) 343.

24. E. P. M. Leiva, C. Vázquez, M. I. Rojas, and M. M. Mariscal, *J. Appl. Electrochem.* 38 (2008) 1065.
25. J. Larminie and A. Dick, *Fuel Cells Systems Explained*, Wiley, Chichester, 2000.
26. W. Vielstich, A. Lamm, and H. A. Gasteiger, *Handbook of Fuel Cells: Fundamentals, Technology, Applications*, Vol. 1, Wiley, Chichester, 2003.
27. O. Haas, F. Holtzer, K. Müller, and S. Müller, in *Handbook of Fuel Cells: Fundamentals, Technology, Applications*, W. Vielstich (Ed.), Wiley, Chichester, 2003, pp. 282–408.
28. A. J. Appleby, in *Comprehensive Treatise of Electrochemistry*, Vol. 7, B. E. Conway, J. O'M. Bockris, E. Yeager, S. U. Khan, and R. E. White (Eds.), Plenum, New York, 1983, p. 230.
29. J. Lipkowski and P. Ross, *Electrocatalysis*, Wiley-VCH, 1998.
30. W. Vielstich, *Fuel Cells*, Wiley, New York, 1970.
31. A. J. Salkind, in *Techniques of Electrochemistry*, Vol. 1, A. J. Salkind and E. Yeager (Eds.), Wiley, New York, 1972.
32. A. J. Bard and L. R. Faulkner, *Electrochemical Methods*, Wiley, New York, 1980.
33. A. T. Kühn, *Techniques in Electrochemistry, Corrosion and Metal Finishing*, Wiley, New York, 1987.
34. J. Tafel, *Z. Phys. Chem.*, 50 (1905) 641.
35. S. Glasstone, K. J. Laidler, and H. Eyring, *The Theory of Rate Processes*, McGraw-Hill, New York, 1941.
36. C. A. Christiansen, *Z. Physik. Chem.*, B28 (1935) 303.
37. N. M. Rodiguin and E. N. Rodiguina, *Consecutive Reactions*, Van Nostrand, Princeton, NJ, 1964.
38. B. E. Conway, *Theory and Principles of Electrode Processes*, Ronald, New York, 1965.
39. A. Damjanovic, in *Modern Aspects of Electrochemistry*, Vol. 5, J. O'M. Bockris and B. E. Conway (Eds.), Plenum, New York, 1965.
40. R. W. Gourney, *Proc. Roy. Soc.*, A134 (1932) 137.
41. S. Srinivasan, H. Wroblowa, and J. O'M. Bockris, *Adv. Catal.*, 17 (1967) 351.
42. E. J. Calvo, in *Comprehensive Chemical Kinetics*, Vol. 26, C. H. Bamford and R. G. Compton (Eds.), Elsevier, Amsterdam, 1986, p. 1.
43. R. Dogonatze, *Dokl. Akad. Nauk. SSSR*, 142 (1961) 1108.
44. R. Dogonatze and Y. A. Chizmadzhev, *Dokl. Akad. Nauk. SSSR*, 144 (1962) 1077.
45. R. Dogonatze and Y. A. Chizmadzhev, *Dokl. Akad. Nauk. SSSR*, 145 (1962) 848.
46. B. E. Conway and J. C. Currie, *J. Electrochem. Soc.*, 125 (1978) 252.
47. J. O'M. Bockris, M. J. Mannan, and A. Damjanovic, *J. Chem. Phys.*, 48 (1968) 1898.
48. B. E. Conway and J. C. Currie, *J. Electrochem. Soc.*, 125 (1978) 257.
49. N. Frumkin, *Z. Physik. Chem.*, 116 (1925) 466.
50. M. I. Temkin, *Zh. Fiz. Khim.*, 15 (1941) 296.
51. B. Damaskin, O. Petrii, and V. Batrakov, *Adsorption of Organic Compounds on Electrodes*, Plenum, New York, 1971.
52. A. K. Vetter, *Electrochemical Kinetics*, Academic, New York, 1967.
53. A. Damjanovic and V. Brusic, *Electrochim. Acta*, 12 (1967) 615.
54. D. Jovancicevic and J. O'M. Bockris, *J. Electrochem. Soc.*, 133 (1986) 1799.
55. L. M. Vracar, D. B. Sepa, and A. Damjanovic, *J. Electrochem. Soc.*, 133 (1986) 1835.
56. D. B. Sepa, M. V. Vojnovic, M. Stojenovic, and A. Damjanovic, *J. Electrochem. Soc.*, 134 (1987) 845.
57. L. M. Vracar, D. B. Sepa, and A. Damjanovic, *J. Electrochem. Soc.*, 134 (1987) 1695.
58. K. C. Pillai and J. O'M. Bockris, *J. Electrochem. Soc.*, 131 (1984) 568.
59. M. Ghoneim, S. Closer, and E. Yeager, *J. Electrochem. Soc.*, 132 (1985) 1160.

60. E. Yeager, *J. Mol. Catal.*, 38 (1986) 5.
61. R. S. Nicholson and I. Shain, *Anal. Chem.*, 36 (1964) 706.
62. S. Srinivasan and E. Gileadi, *Electrochim. Acta*, 11 (1966) 321.
63. B. E. Conway, H. Angerstein-Kozlowska, F. C. Ho, J. Klinger, B. MacDougall, and S. Gottesfeld, *Disc. Faraday Soc.*, 56 (1973) 210.
64. A. Bolzán, A. C. Chialvo, and A. J. Arvia, *J. Electroanal. Chem.*, 179 (1984) 71.
65. J. O. Zerbino, N. R. de Tacconi, and A. J. Arvia, *J. Electrochem. Soc.*, 125 (1978) 1266.
66. N. R. de Tacconi, J. O. Zerbino, and A. Arvia, *J. Electroanal. Chem.*, 79 (1977) 287.
67. V. G. Levich, *Physicochemical Hydrodynamics*, Prentice-Hall, Englewood Cliffs, NJ, 1962.
68. W. J. Albery and M. L. Hitchman, *Ring-Disc Electrodes*, Clarendon, Oxford, 1971.
69. R. Parsons and J. O'M. Bockris, *Trans. Faraday Soc.*, 47 (1951) 914.
70. R. Parsons, in *Modern Aspects of Electrochemistry*, Vol. 1, J. O'M. Bockris and B. E. Conway (Eds.), Butterworths, London, 1959, Chapter 1.
71. S. Trasatti, in *Comprehensive Treatise of Electrochemistry*, Vol. 1, J. O'M. Bockris, B. E. Conway, and E. Yeager (Eds.), Plenum, New York, 1979, p. 45.
72. L. I. Krishtalik, in *Comprehensive Treatise of Electrochemistry*, Vol. 7, B. E. Conway, J. O'M. Bockris, E. Yeager, S. U. Khan, and R. E. White (Eds.), Plenum, New York, 1983, p. 87.
73. J. P. Badiali, M. L. Rosinberg, and J. Goodisman, *J. Electroanal. Chem.*, 130 (1981) 31.
74. S. Trasatti, in *Trends in Interfacial Electrochemistry*, Vol. 179, A. F. Silva (Ed.), NATO ASI Series, Riedel, Dordrecht, 1986.
75. R. W. Gurney, *Proc. Roy. Soc.*, A134 (1932) 137.
76. J. O'M. Bockris and S. Srinivasan, *J. Electrochem. Soc.*, 111 (1964) 844.
77. R. Parsons, *Trans. Faraday Soc.*, 54 (1958) 1053.
78. A. J. Appleby, *Surf. Sci.*, 27 (1971) 225.
79. A. T. Khun, H. Wroblowa, and J. O'M. Bockris, *Trans. Faraday Soc.*, 63 (1967) 1458.
80. S. Trasatti, *J. Electroanal. Chem.*, 39 (1972) 163.
81. B. Scrosati, *J. Appl. Electrochem.*, 38 (2008) 1.
82. X. Gao, A. Hamelin, and M. Weaver, *Phys. Rev. Lett.*, 67 (1991) 618.
83. J. Clavilier, R. Fauré, G. Guinet, and R. Durand, *J. Electroanal. Chem.*, 107 (1980) 205.
84. J. Clavilier, R. D. G. Guinet, and R. Fauré, *J. Electroanal. Chem.*, 127 (1981) 281.
85. M. P. Soriaga (Ed.), *Electrochemical Surface Science: Molecular Phenomena at Electrode Surfaces*, American Chemical Society, Washington, DC, 1988.
86. K. Kinoshita and P. Stonehart, in *Modern Aspects of Electrochemistry*, Vol. 12, J. O'M. Bockris and B. E. Conway (Eds.), Plenum, New York, 1977.
87. K. Kinoshita, in *Modern Aspects of Electrochemistry*, Vol. 14, J. O'M. Bockris and B. E. Conway (Eds.), Plenum, New York, 1982.
88. J. Solla-Gullón, P. Rodríguez, H. E. Herrero, A. Aldaz, and J. M. Feliú, *Phys. Chem. Chem. Phys.*, 10 (2008) 1359.
89. N. M. Markovic and J. P. N. Ross, *Surf. Sci. Rep.*, 45 (2002) 117–230.
90. K. Bedürftig, S. Völkening, Y. Wang, J. Wintterlin, K. Jacobi, and G. Ertl, *J. Chem. Phys.*, 111 (1999) 11147.
91. M. A. Henderson, *Surf. Sci. Rep.*, 46 (2002) 1.
92. G. Somorjai, *Chemistry in Two Dimensions: Surfaces*, Cornell University Press, Ithaca, NY, 1981.
93. H. Martín, P. Carro, A. H. Creus, S. González, G. Andreasen, R. C. Salvarezza, and A. J. Arvia, *Langmuir*, 16 (2000) 2915.
94. F. Nieto, C. Uebing, and V. Pereyra, *Surf. Sci.*, 416 (1998) 152.

95. A. V. Myshlyavtsev and V. P. Zhdanov, *J. Chem. Phys.*, 92 (1990) 3909.
96. G. Somorjai, *J. Phys. Chem.*, 94 (1990) 1013.
97. B. B. Mandelbrot, *The Fractal Geometry of Nature*, W. H. Freeman, New York, 1982.
98. L. Vázquez, A. H. Creus, P. Carro, P. Ocón, P. Herrasti, C. Palacio, J. M. Vara, R. C. Salvarezza, and A. J. Arvia, *J. Phys. Chem.*, 96 (1992) 10454.
99. M. V. Berry and Z. F. Lewis, *Proc. Roy. Soc. Lond. A*, 370 (1980) 459.
100. F. Family, *Physica A*, 168 (1990) 561.
101. A. J. Arvia, J. C. Canullo, E. Custidiano, C. Perdriel, and W. E. Triaca, *Electrochim. Acta*, 31 (1986) 1359.
102. L. Vázquez, R. C. Salvarezza, P. Ocón, P. Herrasti, J. M. Vara, and A. J. Arvia, *Appl. Surf. Sci.*, 70–71 (1993) 413.
103. A. Visintin, J. C. Canullo, W. E. Triaca, and A. J. Arvia, *J. Electroanal. Chem.*, 239 (1988) 67.
104. P. García, M. M. Gómez, R. C. Salvarezza, and A. J. Arvia, *J. Electroanal. Chem.*, 347 (1993) 237.
105. P. Pfeifer and M. Obert, in *The Fractal Approach to the Heterogeneous Chemistry*, D. Avnir (Ed.), Wiley, New York, 1989.
106. D. Farin and D. Avnir, in *The Fractal Approach to the Heterogeneous Chemistry*, D. Avnir (Ed.), Wiley, New York, 1989, p. 271.
107. G. E. Engelmann, J. C. Ziegler, and D. M. Kolb, *J. Electrochem. Soc.*, 145 (1998) L33.
108. A. H. Creus, Y. Gimeno, R. C. Salvarezza, and A. J. Arvia, *Encyclopedia of Nanoscience and Nanotechnology*, Vol. 2, American Scientific, New York, 2004.
109. M. Schrinner, M. Ballauff, Y. Talmon, Y. Kauffmann, J. Thun, M. Möller, and J. Breu, *Science*, 323 (2009) 617.
110. J. F. Gouyet, M. Rosso, and B. Sapoval, in *Fractals and Disordered Systems*, A. Bo and S. Havlin (Eds.), Springer-Verlag, Berlin, 1991, p. 231.
111. R. C. Salvarezza and A. J. Arvia, *Surf. Sci.*, 335 (1995) 378–388.
112. M. A. Pasquale, F. J. R. Nieto, and A. J. Arvia, *Surf. Rev. Lett.*, 15 (2008) 1.
113. A. Patrykiejew, S. Sokolowski, and K. Binder, *Surf. Sci. Rep.*, 37 (2000) 207.
114. F. J. R. Nieto, M. A. Pasquale, C. R. Cabrera, and A. J. Arvia, *Langmuir*, 22 (2006) 10472.
115. Y. Gimeno, A. H. Creus, P. Carro, S. González, R. C. Salvarezza, and A. J. Arvia, *J. Phys. Chem.*, B 106 (2004) 4232.
116. A. J. Arvia, R. C. Salvarezza, and W. E. Triaca, *Electrochim. Acta*, 34 (1989) 1071.
117. J. M. G. Rodríguez, L. Vázquez, A. Baró, R. C. Salvarezza, J. M. Vara, and A. J. Arvia, *J. Phys. Chem.*, 96 (1992) 347.
118. C. Alonso, R. C. Salvarezza, J. M. Vara, A. J. Arvia, L. Vázquez, A. Bartolomé, and A. M. Baró, *J. Electrochem. Soc.*, 137 (1990) 2161.
119. T. Kessler, A. Visintín, A. E. Bolzán, G. Andreasen, R. C. Salvarezza, W. E. Triaca, and A. J. Arvia, *Langmuir*, 12 (1996) 6587.
120. A. C. Chialvo, W. E. Triaca, and A. J. Arvia, *J. Electroanal. Chem.*, 146 (1983) 93.
121. M. L. Marcos, J. M. Vara, J. G. Velasco, and A. J. Arvia, *J. Electroanal. Chem.*, 224 (1987) 189.
122. A. C. Chialvo, W. E. Triaca, and A. J. Arvia, *J. Electroanal. Chem.*, 237 (1987) 237.
123. B. K. Chakraverty, *J. Phys. Chem. Solids*, 28 (1967) 2401.
124. S. Trasatti and O. A. Petrii, *Pure Appl. Chem.*, 63 (1991) 711.
125. D. C. Grahame, *J. Am. Chem. Soc.*, 79 (1957) 701.
126. V. I. M. Gaikasian, V. V. Voronchikhina, and E. A. Zakharova, *Elektrokhimiya*, 12 (1968) 1420.

127. J. Broadhead, G. J. Hills, and D. R. Kinnibrugh, *J. Appl. Electrochem.*, 1 (1971) 147.

128. E. M. L. Valeriote and R. G. Barradas, *Chem. Instrum.*, 1 (1968) 153.

129. M. Breiter, *J. Electroanal. Chem.*, 81 (1977) 275.

130. I. M. Novoselskii, N. I. Konevskikh, L. Y. Egorov, and E. P. Sidorov, *Elektrokhimiya*, 7 (1971) 893.

131. E. Robert, *Oberfläche Surf.*, 22 (1981) 261.

132. T. Ohmori, *J. Electroanal. Chem.*, 157 (1983) 159.

133. R. Parsons and F. R. G. Zobel, *J. Electroanal. Chem.*, 9 (1983) 333.

134. F. G. Will and C. A. Knorr, *Z. Elektrochem.*, 64 (1960) 258.

135. S. Gilman, *J. Phys. Chem.*, 67 (1963) 78.

136. S. Gilman, *J. Electroanal. Chem.*, 7 (1964) 382.

137. S. Gilman, *J. Phys. Chem.*, 71 (1967) 4339.

138. D. V. Sokolskii, B. Y. Nogerbekov, N. N. Gudeleva, and R. G. Mustafina, *Elektrokhimiya*, 22 (1986) 1185.

139. R. Woods, *J. Electroanal. Chem.*, 49 (1974) 217.

140. M. Breiter, K. Hoffmann, and C. Knorr, *Z. Elektrochem.*, 61 (1957) 1168.

141. A. Vashkyalis and A. Demontaite, *Elektrokhimiya*, 14 (1978) 1213.

142. H. Siegenthaler and K. Jüttner, *J. Electroanal. Chem.*, 163 (1984) 327.

143. G. Singh, M. H. Miles, and S. Srinivasan, in *Electrocatalysis in Nonmetallic Surfaces*, A. D. Franklin (Ed.), Special Publication No. 455, National Bureau of Standards, Washington, DC, 1976.

144. B. V. Tilak, C. G. Rader, and S. K. Rangarajan, *J. Electrochem. Soc.*, 124 (1977) 1879.

145. K. Micka and I. Rousar, *Electrochim. Acta*, 32 (1987) 1387.

146. L. Bai, L. Gao, and B. E. Conway, *J. Chem. Soc. Faraday Trans.*, 89 (1993) 235.

147. L. Bai, L. Gao, and B. E. Conway, *J. Chem. Soc. Faraday Trans.*, 89 (1993) 243.

148. R. K. Schofield, *Nature*, 160 (1974) 480.

149. A. Kozawa, *J. Electrochem. Soc.*, 106 (1959) 552.

150. C. H. Giles, A. P. D. Silva, and A. S. Tuvedi, *Proceedings of the International Symposium "Surface Area Determination"*, Butterworths, London, 1969.

151. J. C. Puippe, *Oberfläche Surf.*, 26 (1985) 6.

152. R. Sonnenfeld, J. Schneir, and P. K. Hansma, in *Modern Aspects of Electrochemistry*, Vol. 13, R. E. White, J. O'M. Bockris, and B. E. Conway (Eds.), Plenum, New York, 1990, p. 159.

153. P. Lustenberger, H. Röhrer, R. Cristoph, and H. Siegenthaler, *J. Electroanal. Chem.*, 243 (1988) 225.

154. S. Brunauer, P. H. Emett, and E. Teller, *J. Am. Chem. Soc.*, 60 (1938) 309.

155. J. L. Lemaitre, P. G. Menon, and F. Delannay, in *Characterisation of Heterogeneous Catalysts*, F. Delannay (Ed.), Marcel Dekker, New York, 1984.

156. S. J. Gregs and K. S. W. Sing, *Adsorption, Surface Area and Porosity*, Academic, London, 1982.

157. A. W. Adamson, *Physical Chemistry of Surfaces*, Wiley, New York, 1982.

158. J. M. Thomas and W. J. Thomas, *Introduction to the Principles of Heterogeneous Catalysis*, Academic, London, 1967.

159. A. J. Arvia and R. C. Salvarezza, *Electrochim. Acta*, 39 (1994) 1481.

160. R. C. Salvarezza, L. Vázquez, P. Ocón, P. Herrasti, J. M. Vara, and A. J. Arvia, *Europhys. Lett.*, 20 (1992) 727.

161. W. J. Lorenz and W. Plieth (Eds.), *Electrochemical Nanotechnology*, Wiley VCH, Weinheim, 1998.

162. R. C. Salvarezza and A. J. Arvia, in *Modern Aspects of Electrochemistry*, Vol. 28, B. E. Conway, J. O'M. Bockris, and R. E. White (Eds.), Plenum, New York, 1995.

163. A. Bunde and S. Havlin (Eds.), *Fractals and Disordered Systems*, Springer Verlag, Berlin, 1991.

164. J. Galeano, J. Buceta, K. Juárez, B. Pumarinio, J. de la Torre, and J. M. Iriondo, *Europhys. Lett.*, 63 (2003) 83.

165. J. M. Pastor and J. Galeano, *Central Eur. J. Phys.*, 5 (2007) 539.

166. J. M. Gomez-Rodríguez, L. Vázquez, A. Baró, R. C. Salvarezza, J. M. Vara, and A. J. Arvia, *J. Phys. Chem.*, 96 (1992) 347.

167. T. Pajkossy and L. Nyikos, *Electrochim. Acta*, 34 (1989) 171.

168. P. Ocón, P. Herrasti, L. Vázquez, R. C. Salvarezza, J. M. Vara, and A. J. Arvia, *J. Electroanal. Chem.*, 319 (1991) 101.

169. M. Tringuides (Ed.), *Surface Diffusion: Atomistic and Collective Processes*, NATO ASI Series B, Physics Vol. 360, Plenum, New York, 1997.

170. C. B. Duke and E. W. Plummer (Eds.), *Frontiers in Surface and Interface Science, Surface Dynamics Growth and Etching*, Vol. 500, Elsevier, 2002.

171. G. Ehrlich, *Surf. Sci.*, 299/300 (1994) 628.

172. G. Ge and L. E. Brus, *Nano Lett.*, 1 (2001) 219.

173. M. Zinke-Allmang, L. C. Feldman, and M. H. Grabow, *Surf. Sci. Rep.*, 16 (1992) 377.

174. Z. Zhang, Z.-P. Shi, and K. Haug, in *Surface Diffusion: Atomistic and Collective Processes*, M. C. Tringuides (Ed.), NATO ASI Series B: Physics Vol. 360, Plenum, New York, 1997, p. 103.

175. H. Brune, K. Bromann, and K. Kern, in *Surface Diffusion: Atomistic and Collective Processes*, M. C. Tringuides (Ed.), NATO ASI Series B: Physics Vol. 360, Plenum, New York, 1997, p. 135.

176. C. Duport, J. Villain, and C. Priester, in *Surface Diffusion: Atomistic and Collective Processes*, M. C. Tringuides (Ed.), NATO ASI Series B: Physics Vol. 360, Plenum, New York, 1997, p. 191.

177. P. Jensen, P. Deltour, L. Bardotti, and J.-L. Barrat, in *Surface Diffusion: Atomistic and Collective Processes*, M. C. Tringuides (Ed.), NATO ASI Series B: Physics Vol. 360, Plenum, New York, 1997, p. 403.

178. M. Zinke-Allmang, L. C. Feldman, and S. Nakahara, *Appl. Phys. Lett.*, 51 (1987) 975.

179. M. E. Martins, F. J. R. Nieto, R. C. Salvarezza, G. Andreasen, and A. J. Arvia, *J. Electroanal. Chem.*, 477 (1999) 14.

180. H. P. Bonzel, in *Surface Physics of Materials*, J. M. Blackely (Ed.), Academic, New York, 1975.

181. C. Alonso, R. C. Salvarezza, J. M. Vara, and A. J. Arvia, *Electrochim. Acta*, 35 (1990) 1331.

182. R. C. Salvarezza, C. Alonso, J. M. Vara, E. Albano, H. Martin, and A. J. Arvia, *Phys. Rev. B*, 41 (1990) 12502.

183. W. Rudzinski and D. H. Everett, *Adsorption of Gases on Heterogeneous Surfaces*, Academic, London, 1992.

184. M. M. Gómez, M. P. García, J. S. Fabián, L. Vázquez, R. C. Salvarezza, and A. J. Arvia, *Langmuir*, 12 (1996) 818.

185. M. M. Gómez, M. P. García, J. S. Fabián, L. Vázquez, R. C. Salvarezza, and A. J. Arvia, *Langmuir*, 13 (1997) 1317.

186. A. C. Chialvo, W. E. Triaca, and A. J. Arvia, *J. Electroanal. Chem.*, 146 (1983) 43.

187. A. C. Chialvo, W. E. Triaca, and A. J. Arvia, *J. Electroanal. Chem.*, 71 (1984) 303.

188. A. M. C. Luna, M. C. Giordano, and A. J. Arvia, *J. Electroanal. Chem.*, 259 (1989) 173.

189. M. L. Marcos, J. M. Vara, J. G. Velasco, and A. J. Arvia, *J. Electroanal. Chem.*, 224 (1987) 189.
190. A. M. C. Luna, A. E. Bolzán, M. F. L. de Mele, and A. J. Arvia, *Z. Phys. Chemie* (*NF*), 160 (1988) 25.
191. M. M. Gómez, J. M. Vara, J. C. Hernández, R. C. Salvarezza, and A. J. Arvia, *Electrochim. Acta*, 44 (1998) 1255.
192. M. M. Gómez, J. M. Vara, J. C. Hernández, R. C. Salvarezza, and A. J. Arvia, *J. Electroanal. Chem.*, 474 (1998) 74.
193. F. J. R. Nieto, M. Pasquale, M. Martins, F. Bareilles, and A. Arvia., *Monte Carlo Methods Appl.*, 12 (2006) 271.
194. F. T. Arce, M. E. Vela, R. C. Salvarezza, and A. J. Arvia, *Surf. Rev. Lett.*, 4 (1997) 637.
195. G. Andreasen, M. E. Vela, R. C. Salvarezza, and A. J. Arvia, *J. Electroanal. Chem.*, 467 (1999) 230.
196. M. D. Porter, T. B. Bright, D. L. Allara, and E. D. Chidsey, *J. Am. Chem. Soc.*, 109 (1987) 3559.
197. G. E. Poirier and E. D. Pylant, *Science*, 272 (1996) 1145.
198. G. E. Poirier, *Langmuir*, 13 (1997) 2019.
199. F. T. Arce, M. E. Vela, R. C. Salvarezza, and A. J. Arvia, *Electrochim. Acta*, 44 (1998) 1053.
200. F. T. Arce, M. E. Vela, R. C. Salvarezza, and A. J. Arvia, *Langmuir*, 14 (1998) 7202.
201. F. T. Arce, M. E. Vela, R. C. Salvarezza, and A. J. Arvia, *J. Chem. Phys.*, 109 (1998) 5703.
202. F. Discussion, *The Dynamics Electrode Surface*, Vol. 121, The Faraday Divison of the Royal Society of Chemistry, London, 2002.
203. P. A. Thiel and T. E. Madey, *Surf. Sci. Rep.*, 7 (1987) Nos. 7–8.

CHAPTER 3

Dynamics and Stability of Surface Structures

MARGRET GIESEN and GUILLERMO BELTRAMO

Jülich Forschungzentrum, Institute of Complex Systems, D-52425 Jülich, Germany

3.1 INTRODUCTION

Many of the characteristic chemical and physical properties of electrodes are related to the boundary of the material bulk to the surrounding liquid phase—the electrode surface. For instance, electronic surface states may considerably influence the electron charge density of the electrode at its surface and hence, among others, may also influence adsorption phenomena, catalysis, electrical conductivity, growth, and corrosion [1, 2]. Despite their importance, early data on surface chemistry and electrocatalysis neglected the contribution of defect sites on surface chemical reactions, corrosion, and catalysis. Electrode surfaces were considered merely as flat. Later on, the identification of defect structures on solid surfaces led to a precise characterization of surface structures with high defect density and their influence on chemisorptions [3] and surface reactions [4]. However, the structured surface was still considered as an entirely static object [4]. Throughout the 1990s, however, it became obvious that surface defects are mobile and the mobility may considerably influence physical and chemical properties of solid electrodes and their interaction with the environment [5, 6]. In nonequilibrium situations, the mobility of surface defects may lead to severe instabilities of surface structures. In particular in an electrochemical environment, the sudden contact of a solid electrode with the liquid is an abrupt change in environmental conditions, and hence the electrode surface may experience a sudden jump from an equilibrium to a nonequilibrium situation. Here, severe mass transport on the electrode surface is the inevitable consequence.

For a long time, it has furthermore been neglected that defects on surfaces do not move entirely randomly. Examples are the correlated diffusion of single-surface adatoms due to a short-range pair-interaction via electronic surface states [7] and

Catalysis in Electrochemistry: From Fundamentals to Strategies for Fuel Cell Development,
First Edition. Edited by Elizabeth Santos and Wolfgang Schmickler.

long-range interactions between infinitively long steps that lead to directed atom mobility [8]. The latter may have consequences ranging from mere small local variations in the interstep distance [9–11] over step bunching where the local step density is dramatically increased, leaving the rest of the surface free of steps [12–14] up to a complete faceting of the surface [15]. In the latter, the original surface crystal orientation vanishes in favor of the formation of local, energetically more favorable crystal facets.

Among others, it is the achievement of the scanning tunneling microscope [16], the electrochemical version of this microscope [17, 18], and today's ultrafast video-scanning tunneling microscope [19, 20] that we are able to visualize the surface structure with atomic resolution and simultaneously track the dynamic properties of surface structure and defects.

In order to understand the structure and dynamics of solid surfaces, theoretical methods from statistical thermodynamics set out to conquer surface science studies in vacuum as well as in electrolyte and the analysis of atom and defect migration [6]. Simultaneously, modern enhanced computer capacity enabled to calculate the relevant formation and migration energies using ab initio calculations such as density functional theory [21]. Simulations of surface diffusion processes [5] also had their input into these research studies. At least in the case of vacuum studies, the theoretical input led to a tremendous new understanding of surface structure, stability, and dynamics.

Whereas surface structure and dynamics on solid surfaces in contact with vacuum or a gas phase are quite well understood today, technically more relevant systems such as solid surfaces in an electrochemical environment still pose a challenge to experiment as well as to theory. The reason lies in the presence of the liquid, the ions contained in the electrolyte, and their interaction with the solid via direct chemical interactions, adsorption phenomena, and charge transfer between the solid and the liquid. To date only few people have dedicated their scientific work to atom migration at the solid–liquid interface and to quantitative analysis and understanding of diffusion processes on solid surfaces in contact with a liquid [e.g., 6, 22–33].

This chapter gives an overview of previous and recent achievements in the understanding of electrode surfaces—their structure, their stability, and the underlying mechanisms. Our focus will lie on the structuring of surface electrodes by surface defects as steps and islands, their mobility, and as one consequence the stability of electrode surfaces. We will then discuss the consequences of defects on the electrochemistry of structured electrodes and how classical electrochemical analyses such as voltammetry and capacity measurements may be used to learn about the electrochemical properties of those defects.

3.2   STRUCTURE AND STABILITY OF ELECTRODE SURFACES

3.2.1   Equilibrium Crystal Shape as a Surface Free-Energy Contour

We would like to start this chapter with a *Gedankenexperiment*: Consider a three-dimensional crystal in the macroscopic limit. Now we apply energy to the crystal and separate the crystal volume into two, thereby creating two two-dimensional surfaces of equal size. The energy used to create the two surfaces is the total surface energy of the system. Ideally, the system's condition represents a minimal energy situation and the created surfaces are stable. When a lower surface energy is possible for a different surface orientation or structure, the surface will reorganize and the initially created surface is unstable. The thermodynamically relevant quantity to define a crystal's equilibrium state is rather the surface free energy, which is the surface energy minus the entropy term. Entropy is increased by introducing defects to the surface. Depending on T and the formation energy E_{form} of those defects, the defect concentration is given by $\sim \exp(-E_{form}/k_B T)$. As long as T remains below the roughening temperature T_R [34, 35], the crystal surface orientation is still well defined. Below T_R, the height-to-height correlation function $\langle (h_{ij} - h_{kl})^2 \rangle$ (h_{ij} and h_{kl} are the surface heights with respect to the flat, defect-free crystal plane at two points on the surface with coordinates ij, respectively kl) is a constant when the distance r_{ijkl} between those points is infinitively large. Above T_R, $\langle (h_{ij} - h_{kl})^2 \rangle$ shows a logarithmic divergence with $r_{ijkl} \to \infty$ [36] and the particular surface is called *thermodynamically rough*. A crystal plane below T_R is called a *facet* and is identified in equilibrium shapes of three-dimensional crystals as flat regions [Fig. 3.1(a)]. Above T_R, the facet vanishes and the equilibrium shape is rounded. The latter observation of flat and rounded crystal regions is rather phenomenological. And due to the logarithmic divergence it may be hard to decide whether a crystal plane is still a real facet or whether it is above its roughening transition. The decision is easily made when looking at the anisotropy

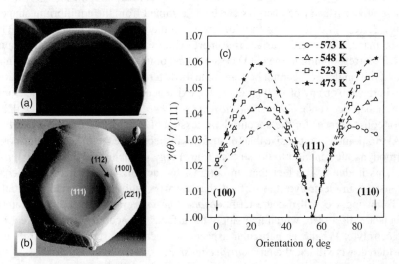

FIGURE 3.1 (a) Scanning electron microscope (SEM) image of equilibrated lead particles on graphite after annealing at 473 K and imaging at room temperature [38]. (b) Scanning tunneling microscope (STM) image of Pb crystallite on Ru(001) at 323 K, showing (111) facet, average facet radius ~140 nm, and smaller side facets [39]. (c) Corresponding relative surface free energy vs. azimuthal angle for Pb: [109] zone. (Data originally published in ref. 38.)

of the surface free energy, that is, the variation of the surface free energy with the polar angle θ.

In equilibrium, the surface free energy of a small crystal with volume V and a number of atoms N (of one chemical component) must be minimized, that is, $dF = 0$ (T, V constant). If A is the surface area of the crystal and γ the specific surface free energy, we find

$$dF = \gamma \, dA = 0 \qquad (3.1)$$

or

$$\int \gamma \, dA = \min \qquad (3.2)$$

From Eq. (3.2) one obtains a minimum total surface free energy. Equilibrium therefore corresponds to a crystal with $\{hkl\}$-oriented facets at the polar angle θ_{hkl} exposed such that $\int \gamma_{hkl} \, dA_{hkl}$ is minimized. Hence, from the anisotropy of $\gamma(\theta)$, one may deduce the equilibrium shape of a crystal. This is performed via the famous Wulff construction, which offers an elegant way to directly relate a crystal's equilibrium shape to its surface free-energy contour [37]. As examples, Figures 3.1(a),(b) show lead particles on graphite after annealing to 473 K and subsequent cooling to room temperature [38] respectively on Ru(001) at 323 K [39]. One obviously sees the $\{111\}$ facets at the crystal top, respectively in the 60° direction. Figure 3.1(c) shows the corresponding anisotropy of $\gamma(\theta)$ [38] along the $\langle 110 \rangle$ and $\langle 100 \rangle$ crystallographic zones. Note, that only relative surface free energies can be determined from the equilibrium shape. In Figure 3.1(c), all data are given with respect to the surface free energy of the (111) crystal plane. The specific surface free energy shows a pronounced, sharp minimum at $\sim54.7°$ corresponding to the (111) crystal plane of a face-centered-cubic (fcc) material. At this angle, $\gamma(\theta)$ cannot be continuously differentiated and is called the *cusp*. A cusp in the anisotropy of $\gamma(\theta)$ indicates the existence of a crystal facet in the equilibrium shape. Vice versa, *if there is no cusp in $\gamma(\theta)$, there is no facet!* This result will have major consequences for the upcoming discussion of the stability of stepped surfaces.

At this point we emphasize that for surfaces in electrolyte the corresponding correct thermodynamic function to the surface free energy in vacuum is the surface tension [2]. This is due to the fact that, in contrast to vacuum, where the surface charge is constant, in electrolyte the surface charge varies with the electrode potential. In the following we will use the term *surface free energy* where general theoretical aspects are discussed, which may apply to surfaces in vacuum as well to electrodes in electrolyte. Whenever particular aspects of electrode surfaces in electrolyte are considered, we will use the term *surface tension*.

As we have discussed before, defects play a major role in the structure and stability of surfaces at finite temperatures. Therefore, in the following sections we will discuss some important defect structures as observed on solid electrodes and which have a major influence on the electrochemistry of surfaces. Due to their role as nanostructured surfaces with well-defined properties, we will focus in particular on regularly stepped surfaces as well as on islands on surfaces which are easily created on metal electrodes by either lifting of surface reconstructions or by corrosion/deposition processes.

3.2.2 Defects on Solid Surfaces

3.2.2.1 Infinite Steps

What is a *defect* on a surface? As a definition every deviation from a perfect, atomically flat surface is called a defect. Practically this means that an individual *adatom* on top of the flat substrate is called a defect as well as an individual *vacancy*. Mesoscopic defects are *steps* of monatomic or multiple-layer height. Steps originate from bulk dislocations [40] when the dislocation penetrates the surface. However, steps can also be created artificially by cutting a crystal in a small angle (*vicinal*) with respect to a low-Miller-index crystal plane such as, for instance, the (001) or (111) plane of a fcc crystal (Fig. 3.2).

Whereas the artificially created steps on vicinal surfaces can be considered to be infinite, pinned steps or steps originating from a screw dislocation in the crystal bulk have a starting or/and an end point. Close inspection of STM images such as the one shown in Figure 3.2(b) reveals that defects such as steps are not static objects but rather undergo equilibrium fluctuations due to atomic motion at their contour. Due to the finite scan speed of the scanning tunneling microscope tip, defect contours on metal surfaces appear frizzy [41], a phenomenon frequently observed in vacuum [6] as well as in electrolyte [6, 22, 42]. The frizziness is indicative of rapid kink motion at the defect contour.

Steps may also form closed loops and are then denoted *islands* and *vacancy islands* if they are formed from an ensemble of individual adatoms, respectively vacancies (Fig. 3.3).

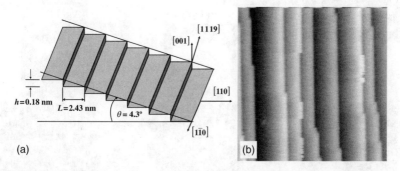

(a)   (b)

FIGURE 3.2 (a) Model of stepped (vicinal) fcc $(1,1,19)$ surface cut at small angle θ with respect to low-Miller-index (001) crystal plane, that is, $(1,1,19)$ surface constituted of narrow (001)-oriented, flat terraces seperated by monatomic high steps with average orientation along $\langle 1\bar{1}0 \rangle$ crystal direction. (b) STM image of Cu$(1,1,19)$ surface at 290 K (displayed area $19 \times 19\,\text{nm}^2$) with step-downward direction from left to right [43]. Protrusions within the step edges belong to a further defect category called *kinks* which indicate the starting point, respectively end point, of an additional atomic row at the step edge.

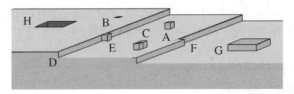

FIGURE 3.3 Model of surface with various defects such as single adatoms (A), vacancies (B), dimers (C), infinite steps (D), step adatoms (E), kinks (F), adatom islands (G), and vacancy islands (H).

3.2.2.2 Closed-Loop Steps: Islands

Adatom islands on surfaces are formed by material deposition whereas vacancy islands are generated by ion erosion. Here, we do not discuss the physics of island nucleation and growth [8]. Rather, we focus on the equilibrium and nonequilibrium properties of islands after material deposition has stopped and island coarsening leads to a low-energy equilibrium state.

Figure 3.4 shows atomic ball models and STM images of islands on fcc (001) and (111) surfaces. On fcc (001), the island perimeter constitutes one and the same step type [forming a (111) microfacet with the underlying terrace] due to the fourfold symmetry of the substrate. On fcc (111), the threefold symmetry leads to alternating steps along the island perimeter forming (001), respectively (111) microfacets. These

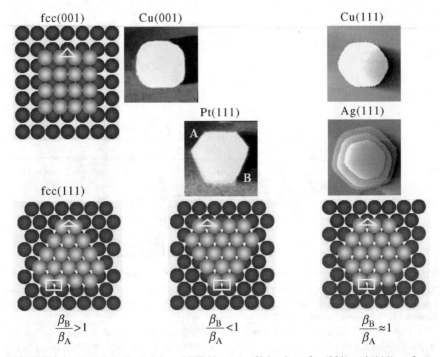

FIGURE 3.4 Atomic ball models and STM images of islands on fcc (001) and (111) surfaces.

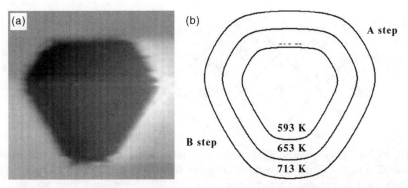

FIGURE 3.5 (a) STM image of vacancy island on Pt(111) at 633 K in vacuum. The displayed area is $98 \times 98 \, \text{nm}^2$. (b) Equilibrium shape of vacancy islands on Pt(111) for various temperatures after averaging over several hundred individual island shapes as extracted from STM data. (Data from ref. 45.)

steps are called A and B steps, respectively. Depending on the *step free energy* β associated with formation of those steps, the islands on (111) surfaces may assume various shapes from almost perfect hexagons to truncated triangles. Islands on fcc (001) are almost square shaped. The island edges are rounded and reveal no one-dimensional facets (see Section 3.1.3). Furthermore, atomic motion at the island perimeter gives rise to contour fluctuations and a frizzy appearance of island edges [Fig. 3.5(a)]. In order to measure the equilibrium shape of islands, one therefore must average over many (in general several hundred!) individual island shapes as deduced from STM data. Figure 3.5(b) shows equilibrium shapes of vacancy islands on Pt(111) in vacuum for various temperatures. Obviously, the island perimeter straightens with decreasing temperature. For all temperatures, the A segments are shorter than the B segments, indicating that A steps have the larger step free energy compared to B steps. The ratio β_A/β_B can be directly derived by measuring the ratio between the island radii r_A/r_B. For experimental data from Cu(100), Cu(111), Ag(111), and Pt(111) surfaces in vacuum see [44, 45].

Detailed information and statistical analysis of the equilibrium shape of defects and their equilibrium dynamics as well as their coarsening behavior enable to determine defect formation energies, such as the step free energy, kink energies, as well as major mass transport processes. The employed procedures will be discussed in Chapter 3.

3.2.3 Smoluchowski Effect

As an inherent feature, any defect on a metal surface is characterized by an electron charge modulation due to the Smoluchowski effect, thereby smoothing out strong surface corrugations as introduced by defects [46]. As visualization, Figure 3.6 shows a model surface with corrugation, including a defect and the resulting smoothing caused by the Smoluchowski effect (grey shaded area). The charge redistribution leads to electron accumulation at the bottom of the defects and to an electron reduction at the

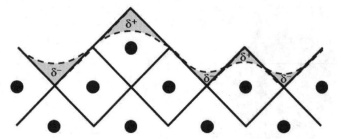

FIGURE 3.6 Model of adatom or step defect on otherwise flat metal surface. The corrugation is smoothed by an electron charge modulation near the defect, reducing negative charge at the defect's top and increasing it at its bottom.

top of the defects. As a consequence, defects on the surface are associated with a dipole moment with the positive charge pointing outward and its size depending on the defect type and the metal substrate involved. This is significant because the dipole moments of the defects are oriented in opposite direction than the dipole of the flat surface.

The dipole moments at defects reduce the local work function, in particular of vicinal surfaces [47]. The work function of a metal electrode (Φ) and the potential of zero charge (ϕ_{pzc}) are related via [48]

$$\phi_{pzc} = \frac{1}{e}(\Phi - \Phi_{Ref}) + \delta\chi_M - \delta\chi_{Sol} \qquad (3.3)$$

where $\delta\chi_M$ and $\delta\chi_{Sol}$ denote the changes in the surface potentials of the metal and the solution when the two phases are brought into contact and Φ_{Ref} is the work function of the reference electrode. For not (or little) specifically adsorbing electrolyte, the changes in the surface potentials are small, and ϕ_{pzc} shows a linear dependence on the work function. From that one finds that the equivalent of the reduction in the work function caused by the step dipole moment is the shift in pzc [2, 48–50]. For regularly stepped Au and Ag surfaces, shifts in the pzc have been studied by Hamelin et al. and Lecoeur et al. [51–53] and by our group [50, 54, 55].

Expressed in terms of the dipole moment of the defect (μ_{def}), the shift in pzc, $\Delta\phi_{pzc}$, is

$$\Delta\phi_{pzc} \equiv \phi_{pzc,0} - \phi_{pzc,def} = \frac{\rho\mu_{def}}{\varepsilon_0} \qquad (3.4)$$

where, $\phi_{pzc,0}$ and $\phi_{pzc,def}$ denote the pzc of the surface without and with defects, ρ is the defect concentration, and ε_0 the vacuum permittivity. Since we will discuss the special case of a step defect on a regularly stepped (vicinal) surface in more detail throughout this chapter, the equation as it reads for a step is also given:

$$\Delta\phi_{pzc} = -\frac{p_z}{\varepsilon_0 a_{||} L} \qquad (3.5)$$

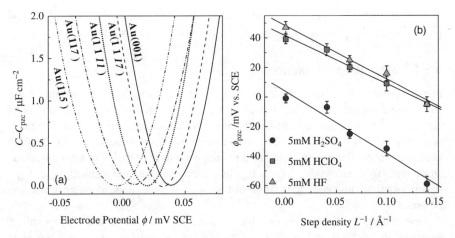

FIGURE 3.7 (a) Capacitance as obtained near potential of zero charge on Au(001) and various vicinal Au($11n$) surfaces in 5 mM $HClO_4$. The capacity curves were obtained using a frequency of 20 Hz and an amplitude of 5 mV. (b) Potential of zero charge ϕ_{pzc} (as determined from minimum in capacity curve in (a)) vs. step density on vicinal Au(001) surfaces in various electrolytes. (Data from ref.50.)

with p_z the step dipole moment per step atom, $a_\parallel$ the atomic length parallel to the step orientation, and L the mean distance between steps on a vicinal surface as defined in Figure [3.2].

The shift in pzc is toward negative potentials since the dipole moments of steps, adatoms, and vacancies point with their positive ends away from the metal surface [56] and is experimentally well documented [50, 55]. Note that the flat surface has a dipole with the positive end toward the metal. In these studies, it is also shown that the dipole moments roughly correspond to the dipole moments measured for surfaces in vacuum. As an example, Figure 3.7(a) shows the shift in pzc for stepped Au(001) surfaces [50].

According to Eq. (3.5), ϕ_{pzc} decreases linearly with the step density and from the slope one directly obtains the step dipole moment p_z. This is shown in Figure 3.7(b) for stepped Au(001) surfaces in various electrolytes. The step dipole moments as determined for Au and Ag surfaces are summarized in Table 3.1.

As a very important consequence of the Smoluchowski effect, the modified electronic structure of defects also affects the reactivity of surfaces [57] and the modified charge density distribution causes the ion cores to relax to new configurations with strained bonds [58]. The local strain field around steps leads to a specific repulsive

TABLE 3.1 Dipole Moment on Stepped Au and Ag(001) Surfaces in Various Electrolytes

	Au(001) [50]			Ag(001) [54]
Electrolyte	5 mM H_2SO_4	5 mM $HClO_4$	5 mM HF	10 mM $KClO_4$
$p_z/10^{-3}\ e\text{Å}$	6.8 ± 0.8	5.2 ± 0.4	5.8 ± 0.5	3.5 ± 0.5

interaction between steps on vicinal surfaces that stabilize regular arrays of steps [5, 6]. This will be discussed in more detail in the following section.

3.2.4 Step–Step Interactions

Step–step interactions have attracted considerable interest in the past, since they have an influence on the step mobility and the stability of surfaces. As was already mentioned in Section 3.2.1, they furthermore play a crucial role in the equilibrium shape of crystals [59, 60] as well as in the roughening transition of surfaces [35, 61–63]. First, we discuss the general principles of step–step interactions. Then we introduce an example from electrode surfaces in a liquid environment and show that the step–step interaction and hence the stability of stepped surfaces depend on the electrode potential.

One distinguishes between two basic forms of step–step interactions: explicit *energetic* interactions and *entropic* interactions. Examples for *energetic* step–step interactions are first dipole–dipole interactions arising from the electron distribution near step edges. Dipole–dipole interactions may be of *attractive* or *repulsive* nature depending on the orientation of the dipole moment. A second origin of energetic interactions is elastic interactions, which arise from the lattice distortions at step edges. Elastic interactions are always repulsive [64]. *Entropic* interactions arise from the *noncrossing condition* of steps, which is analogous to interpreting steps as fermion particles. The crossing of steps is forbidden, because it would lead to overhangs, which have a high energy. In all three cases, the interaction potential $V(x)$ decays with $1/x^2$ [65]. Other possible interactions have been proposed which follow a different distance dependence [66, 67].

3.2.4.1 Mean-Field Approximation

The simplest way to describe step–step interactions is by a mean-field model where a step is considered to wander in the mean potential created by all other steps. For explicit repulsive interactions, beyond the noncrossing condition, the potential is quadratic in the distance s and its harmonic approximation around $s = 0$ is given by

$$V(s) = \frac{A}{\langle L \rangle^2} \left(\sum_{n=1}^{\infty} \frac{2}{n^2} \right) + \frac{\pi^4}{15} \frac{A}{\langle L \rangle^4} s^2 + \dots \tag{3.6}$$

with A the *interaction constant*. The index n counts the steps forming the mean-field potential where $n \equiv 1$ corresponds to the nearest-neighbor step. Then, the step–step distance distribution is a Gaussian [68],

$$P(s) = \frac{1}{\sigma\sqrt{2\pi}} \exp\left(-\frac{(s-1)^2}{2\sigma^2} \right) \tag{3.7}$$

with the variance σ given as

$$\sigma^2 = \sqrt{\frac{b^2(T)k_BT}{48Aa_{\parallel}}} \qquad (3.8)$$

For strong repulsive interactions, the mean-field approximation is an appropriate model to describe step–step interactions. For weaker repulsive or even attractive interactions, the mean-field model fails and other models become more appropriate.

3.2.4.2 Free-Fermion Model

An elegant way to theoretically describe noncrossing, infinite steps is the one-dimensional *free-fermion* model. Here, step positions are interpreted as noninteracting free fermions and the step contour is considered as a world line of the fermion. This model can be extended so that energetic repulsive and attractive interactions are included [69]. In all cases, the terrace width distribution is asymmetric. Figure 3.8 shows the scaled terrace width distribution for the free-fermion case (solid line), repulsive interactions (dashed line), and attractive interactions (dot-dashed line) [69]. For entropic repulsion and energetic attraction, the terrace width distribution has a high skewness and the Gaussian approximation [Eq. (3.7)] obviously fails. In the case of strong repulsive interactions, however, the Gaussian description is a good approximation.

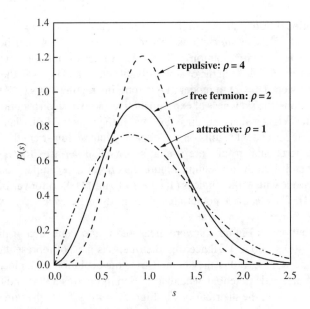

FIGURE 3.8 Scaled terrace width distribution for repulsive interactions (dashed line), no interactions (solid line), and attractive interactions (dot-dashed line) between steps [69].

3.2.4.3   Generalized Wigner–Ibach Surmise

More recently, the application of *random matrix theory* [70, 71] provided a new insight into the analysis of terrace width distributions. Inspired by an early heuristic formula by Harald Ibach (see Giesen [72] and Figure 3 in Einstein [73]), Einstein and Pierre-Louis [74] showed that the terrace width distribution of fluctuating steps can be approximated by

$$P_\rho(s) = a_\rho s^\rho \exp\left(-b_\rho s^2\right) \tag{3.9}$$

The parameter ρ is related to the interaction strength A via

$$A = \frac{b^2(T)}{a_{||}} \frac{\rho(\rho - 2)}{4} \tag{3.10}$$

where $b^2(T)$ is the *diffusivity* and is closely related to the kink concentration (see Section 3.3.1). The constants a_ρ and b_ρ in Eq. (3.9) are determined by the normalization and the unit mean of the terrace width distribution and are given in terms of gamma functions [74, 75]:

$$a_\rho = \frac{2b_\rho^{(\rho+1)/2}}{\Gamma\left[(\rho + 1)/2\right]} \quad b_\rho = \left(\frac{\Gamma\left[(\rho + 2)/2\right]}{\Gamma\left[(\rho + 1)/2\right]}\right)^2 \tag{3.11}$$

Analytic solutions of Eq. (3.9) are available for $\rho = 1, 2, 4$ [76–78]: $\rho = 1, 4$ correspond to attractive and repulsive interactions, respectively. For $\rho = 2$, A vanishes and $P_2(s)$ equals the free-fermion distribution. For all other values of ρ, Eq. (3.9) serves as an extrapolation scheme between the exactly solvable cases and holds in the range from noninteracting steps ($\rho = 2$) to moderately strong interacting steps ($\rho \sim 4$) [74]. For strong interactions, the skewness of experimental terrace width distributions is almost zero. Then, the Wigner–Ibach surmise fails and the symmetric Gaussian distribution [Eq. (3.7)] yields excellent results on the step–step interaction strength.

Various experimental results are available for metal surfaces [6] as well as for silicon surfaces [5], both in vacuum. Figure 3.9 shows an example from a stepped Au(001) electrode with Miller indices (1,1,17) in 10 mM H_2SO_4 for various electrode potentials [14]. The electrode potential is given as the difference $\Delta\phi_{pzc}$ with respect to pzc.

The distributions in Figure 3.9 were measured in the same area of the electrode surface (size $400 \times 400\,\text{Å}^2$). Hence, the distributions are not representative for the entire electrode surface; rather they demonstrate the evolution of the local step–step distance with electrode potential. The distribution in Figures 3.9(a) and (b) are for a reconstructed surface, the distributions in Figures 3.9(c)–(f) for the unreconstructed surface. The distributions are normalized with respect to the mean terrace width where the mean $\langle L(\phi)\rangle$ of the unnormalized distribution at a given potential ϕ was used rather than the nominal mean $L = 24.5\,\text{Å}$. The actual mean of all distributions was slightly

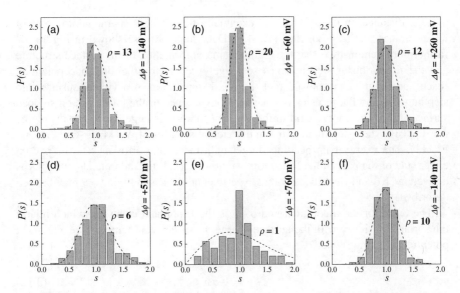

FIGURE 3.9 Terrace width distribution on Au(1, 1, *17*) in 10 mM H$_2$SO$_4$ for various electrode potentials ϕ given with respect to pzc [−10 mV saturated calomel electrode (SCE)]. The dashed line is a fit according to Eq. (3.9) [14].

smaller than 24.5 Å, indicating that the distributions do not represent the statistical average over the surface. For $\Delta\phi = -140, +60, +260$ mV, the distributions are narrow, with the width of the distribution at +60 mV even smaller than for $\Delta\phi = -140$ mV. This is corroborated by the values of ρ as determined by fitting the distributions to Eq. (3.9), where ρ is very large ($\rho = 13, 20, 12$ for $\Delta\phi = -140, +60, +260$ mV, respectively). Hence, the shape of the distribution is that of a Gaussian and the step–step interaction potential is strongly repulsive. For higher potentials than +260 mV, the distributions systematically broaden and the skewness increases, indicated also by a decreasing ρ ($\rho = 6, 1$ for $\Delta\phi = +510, +760$ mV, respectively). At $\Delta\phi = +760$ mV, the value $\rho = 1$ indicates that the step–step interaction potential has become attractive. The broadening of the distribution, however, is reversible. When the potential is stepped back to $\Delta\phi = -140$ mV [Fig. 3.9(f)], the width of the distribution decreases again, ρ is now 10, comparable to Figure 3.9(a). This is surprising since the data in Figure 3.9(f) is for the unreconstructed rather than for the reconstructed surface as in Figure 3.9(a). Thus it seems as if the distribution width is not influenced by the presence of reconstruction lines but rather depends basically on the electrode potential itself, more correctly, exclusively on the absolute difference with respect to pzc—regardless of the sign!

3.2.4.4 Potential Dependence of Step–Step Interactions

Theoretical considerations show that regular step arrays on electrode surfaces in contact with a liquid are unstable [13]: In the course of thermal equilibrium fluctuations steps merge to form multiple steps with a mean distance closer than the average distance on the surface and eventually the surface separates into areas of large terraces and step bunches [12, 14]. Data such as those shown in Figure 3.9 furthermore demonstrate that the equilibrium conformation of the surface structure and the formation of step bunches sensitively depend on the electrode potential. Ibach and Schmickler showed that the phase separation of a vicinal surface into step bunches and flat terraces is a natural consequence of the lower pzc on stepped surfaces [13]. The energy gain in the phase separation is large enough to overcome the repulsive step–step interaction on the vicinal surfaces of Ag(111), Au(111), and Pt(111). The energy gain becomes larger with increasing difference to the pzc on either side of the pzc. In the following, we briefly recall the theoretical equations as derived in [13] and we will show that experimental data as in Figure 3.9 may be well described.

The basis for the theoretical approach is the Lippmann equation, which relates the surface tension γ with the surface charge σ (for constant temperature, chemical potential, and strain) via

$$d\gamma = -\sigma\phi \tag{3.12}$$

Because of Eq. (3.12), the surface tension γ has a maximum at the potential of zero charge ϕ_{pzc} with respect to a variation of the potential if the interface remains in equilibrium with the electrolyte. The potential dependence of the surface tension is obtained from experiment by integration over the surface charge,

$$\gamma(\phi) = \gamma_{pzc} - \int_{\phi_{pzc}}^{\phi} \sigma(\phi')\, d\phi' \tag{3.13}$$

In the cases of interest here, ϕ_{pzc} can be determined experimentally from the minimum in the capacitance $C(\phi_{pzc})$ versus potential in sufficiently dilute electrolytes. The surface charge is the integral over the capacitance from pzc to the potential in question. To lowest order, a parabolic shape of the surface tension versus potential results:

$$\gamma(\phi) = \gamma_{pzc} - \tfrac{1}{2} C\left(\phi_{pzc}\right)\left(\phi - \phi_{pzc}\right)^2 + \cdots \tag{3.14}$$

A fourth-order term adds to Eq. (3.14) if the potential dependence of the Gouy–Chapman capacitance is taken into account. The opening of the parabola is determined by the interfacial capacitance which, for dilute electrolytes, is entirely determined by the electrolyte, not by the material properties of the electrode.

The surface tension of the stepped surface in contact with the electrolyte can be calculated in terms of the surface tension of the flat surface at pzc and the step line tension at pzc, denoted as β_{pzc}. For small θ one obtains [13]

$$\gamma(\theta, \phi) = \gamma_{pzc} + \frac{\beta_{pzc}}{h}\tan\theta - \tfrac{1}{2}C(\phi - \phi_{pzc} - \Delta\phi_{pzc}^{(\theta)})^2 \tag{3.15}$$

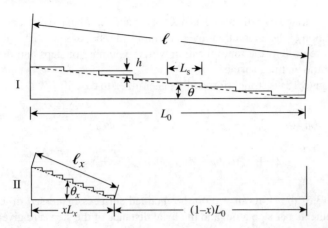

FIGURE 3.10 Terrace geometry and definitions for two states of stepped surface in electrolyte [13].

where h is the step height. Ibach and Schmickler calculated the surface tension of stepped surfaces with elastic step–step repulsion included [13]. They considered two limiting cases (Fig. 3.10): Case I is the standard vicinal surface with a regular step array and in case II the steps are bunched into stripes covering a fraction x of the surface area while a fraction $1 - x$ remains step free. It is assumed that the step bunches contain many steps and finite-domain size effects are neglected. The macroscopic lengths of the stepped surfaces ℓ and ℓ_x are given by (Fig. 3.10)

$$\ell = \frac{L_0}{\cos \theta} = L_0(1 + \kappa) \quad \ell_x = \frac{xL_0}{\cos \theta_x} = xL_0(1 + \kappa_x) \tag{3.16}$$

with

$$\kappa = \sqrt{1 + \tan^2\theta} - 1 \quad \kappa_x = \sqrt{1 + \tan^2\theta/x^2} - 1 \tag{3.17}$$

The total work to create a surface in state I per length perpendicular to the plane of view in Figure 3.10, denoted as Π_I, is given as

$$\Pi_I = \ell\gamma(\theta) = L_0\gamma_{pzc} + \beta_{pzc}\frac{L_0}{L_s}$$

$$-\frac{1}{2}C(1 + \kappa)L_0\left(\phi - \phi_{pzc} - \Delta\phi_{pzc}^{(\theta)}\right)^2 + \frac{\pi^2 AL_0}{3L_s^3(1 + \kappa)^2} \tag{3.18}$$

It has been shown that steps as such do not introduce a notable change in the capacity [26, 79]. The capacity of the inclined surface is therefore simply scaled to the increased length by multiplication with $1 + \kappa$. The last term in Eq. (3.18) describes the elastic repulsion between steps [64]. According to Eq. (3.6), the interaction energy between two steps separated by a distance L is A/L^2. The factor $\pi^2/3$ in the last term arises from the summation of the interactions over an infinite number of steps. The multiplication with L_0/L_s represents the multiplication with the number of steps and the scaling to the area of the inclined surface.

Equation (3.18) is straightforwardly generalized to case II:

$$\Pi_{II} = L_0 \gamma_{pzc} + \beta_{pzc} \frac{L_0}{L_s} - \frac{1}{2}(1-x)CL_0(\phi - \phi_{pzc})^2$$
$$- \frac{1}{2}xC(1+\kappa_x)L_0(\phi - \phi_{pzc} - \Delta\phi_{pzc}^{(\theta_x)})^2 + x\frac{\pi^2 A L_0}{3(xL_s)^3(1+\kappa_x)^2}$$

(3.19)

With Eq. (3.5) the shifts in pzc on the inclined surfaces expressed in terms of the dipole moments per step atom and the local distance of the steps are given as

$$\Delta\phi_{pzc}^{(\theta)} = -\frac{p_z}{\varepsilon_0 a_{||} L_s(1+\kappa)}$$

(3.20)

$$\Delta\phi_{pzc}^{(\theta_x)} = -\frac{p_z}{\varepsilon_0 a_{||} x L_s(1+\kappa_x)}$$

(3.21)

The difference in the surface tensions for cases I and II, $\Delta\gamma = (\Pi_{II} - \Pi_I)/L_0(1+\kappa)$, to the lowest relevant order in $\tan\theta$ is then

$$\Delta\gamma = -C\tan^2\theta \underbrace{\left(\frac{p_z^2}{2\varepsilon_0^2 a_{||}^2 h^2} + \frac{1}{4}(\phi - \phi_{pzc})^2 \right)}_{C_1} \left(\frac{1}{x} - 1 \right)$$
$$+ \underbrace{\frac{\pi^2 A \tan^3\theta}{3h^3}}_{C_2} \left(\frac{1}{x^2} - 1 \right) \cdots$$

(3.22)

Higher order contributions are small and can be neglected [13].

The first term C_1 in Eq. (3.22) is an effective attractive interaction since it is always negative. The potential dependent term arises from the fact that the macroscopic surface and thence the total capacity is larger in case II. The attractive interactions are balanced by the repulsive elastic interactions described in the second term C_2. Here, $\Delta\gamma$ has a minimum for $x < 1$ at $x_{min} = 2C_2/C_1$ if $C_1 < 2C_2$.

The data for Au(1,1,17) electrodes as shown in Figure 3.9 may be explained by the influence of the electrode potential on the step–step interaction potential and hence on the stability of the step array according to Eq. (3.22). Obviously, for $|\phi - \phi_{pzc}| \sim 510\,mV$ there is a driving force to increase the terrace width distribution. Hence, one may safely conclude that $\Delta\gamma$ in Eq. (3.22) is negative. On the other hand, at $|\phi - \phi_{pzc}| \sim 260\,mV$ the distribution is narrow and $\Delta\gamma$ in Eq. (3.22) must be positive. From that one may find a lower and an upper limit for the interaction constant A. Figure 3.11 shows the calculation of Eq. (3.22) for Au(1,1,17)

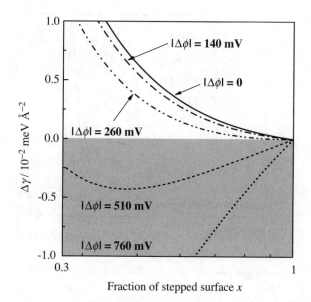

FIGURE 3.11 Calculation of difference between surface tensions of regularly stepped Au(1,1,17) surface in 10 mM H_2SO_4 and a surface with step bunches according to Eq. (3.22) and ref. 13 for various potentials and using step–step interaction strength $A = 14$ meV Å. The grey shaded area indicates the regime, where the surface undergoes step bunching [14].

10 mM H_2SO_4 for various values of $\Delta\phi$ using $p_z = 0.006$ eÅ for Au($11n$) vicinal surfaces ($n = 5, 7, 9, 11, 17$) [50, 54], $C = 40$ μF cm^{-2} [50], $h = 2.04$ Å, $\tan\theta = 0.0832$, and $a_\perp = 2.88$ Å. The interaction constant A has been chosen such that the interaction potential is repulsive at $|\phi| = 260$ mV and below, and attractive for $|\phi| = 510$ mV and above, in accordance with the data in Figure 3.9. From those considerations one finds a minimal value $A_{min} = 14$ meV Å (used in Fig. 3.11) and a maximum value $A_{max} = 32$ meV Å. Hence, the interaction constant on Au(1,1,17) in 10 mM H_2SO_4 is slightly larger, yet in comparable order of magnitude to stepped metal surfaces in vacuum where values between 6 and 10 meV Å are found for Cu and Ag surfaces [6]. The reason for the slightly higher interaction constant may lie in the presence of specifically adsorbed sulfate anions on the gold surface.

Another contribution to the rather complex stability behavior of stepped Au($11n$) electrodes is the potential-dependent presence of the surface reconstruction layer [80, 81]. Indeed, stepped Au($11n$) surfaces are reconstructed for $n \geq 7$ after flame annealing. Figure 3.12 shows cyclic voltammograms of Au($11n$), $n = 5, 7, 9, 11, 17$, and for comparison the voltammogram for the nominally flat Au(001) surface in 5 mM H_2SO_4 [55]. Safe for the curve for the (115) surface, all electrodes display a clear reconstruction peak, albeit the peak may be shifted considerably to lower electrode potentials compared to nominally flat Au(001). In addition, Au(1,1,17) shows a double peak at +280 and +360 mV SCE, which could be explained by the formation of step bunches on the surface. Similar step bunches are also observed for (117), (119) and

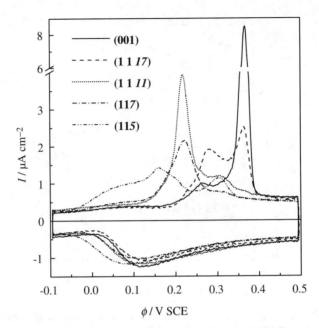

FIGURE 3.12 Cyclic voltammograms of stepped Au($11n$) electrodes [vicinal to (001)] and for comparison that of nominally flat Au(001) surface. The Au($11n$) with $n \geq 7$ display clear reconstruction peaks. The Au(115) with a mean terrace width of 2.5 atomic distances reveals no reconstruction [55].

(1,1,11). Here, however, the main peak lies around +220 mV SCE and a small shoulder is observed around +300 mV SCE. No pronounced peak at +300 mV SCE is found for (115), which in fact is unreconstructed.

Figure 3.13 shows an STM image of Au(119) in 10 mM H_2SO_4 after flame annealing and merging into the electrolyte at −150 mV SCE. The surface shows a phase separation of the nominal (119) plane into (115) and (1,1,33) domains. This observation is in accordance with previous studies which reported the (115) orientation to be a particular stable one ("magical surface") [83, 84, 85]. Currently, detailed investigations of the time evolution of this phase transition and the preferred bunch orientations in the reconstructed and the unreconstructed regime of stepped Au(001) are performed [14, 86].

While the effect of the reconstruction on the stability of vicinal surfaces is well documented, the reverse effect of the steps on the reconstruction unit cell is rarely mentioned. First evidence for a possible influence of steps on the buckling periodicity was reported by Yamazaki et al. [87]. Watson et al. found enlarged reconstruction periods in the stepped sections of a 1.7° miscut Au(100) surface that showed phase separation into (100) facets and "magic" ∼2° vicinals [85]. In a theoretical work, Bartolini et al. postulated the existence of a reconstruction-free region (*dead zone*) about 2.5 atoms in length (=0.72 nm) at the foot of the steps [83]. A similar observation was made by Binnig et al. in early STM data on reconstructed Au(001). These authors

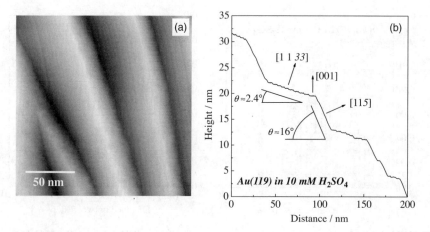

FIGURE 3.13 (a) STM image of Au(119) in 10 mM H₂SO₄ at −150 mV SCE. (b) Height profile analysis of similar image of same surface which shows clear phase separation of (119) plane into (115) and (1,1,33) domains [82].

mentioned that a region about 1.5 atoms from the step foot is not reconstructed [88]. More recently, Bombis and Ibach showed that the reconstruction stripes are always strictly parallel to step edges and that furthermore the stripes of one atom-layer-high islands are 180° out of phase with the stripes on the terraces below [89], in perfect agreement with the prediction of Bartolini et al.

While the existence of a reconstruction-free zone is understood insofar as it is reproduced by theory, the enlarged reconstruction period on stepped surfaces as mentioned by Watson et al. [85] is not. The expansion is particularly puzzling in view of the fact that the reconstruction period next to steps on a nominally flat Au(100) surface remains practically unchanged. In a recent study on the surface morphology on nominal Au(119), Au(1,1,*11*), and Au(1,1,*17*) surfaces in ultrahigh vacuum (UHV) and in electrolyte [82], it was found that the repeat distance of the reconstruction-induced buckling changes from 1.44 nm [80, 90] on flat Au(001) to 1.8–1.9 nm on stepped surfaces. Furthermore, a "magical rule" for the observed terrace widths was reported. It was found [82] that the distance between adjacent steps are multiples k of the buckling periodicity. As a consequence, the terrace width L in units of atom diameters is described in terms of a multiple of the corrugation period. Figure 3.14 shows the terrace width distribution as obtained for (a) Au(1,1,*17*) in electrolyte and (b) Au(1,1,*11*) in vacuum.

Obviously, the terrace width distribution is not a simple Gaussian [Eq. (3.7)] or a Wigner [Eq. (3.9)] function as typically found for (unreconstructed) fcc metal surfaces [6, 11]. Rather, the probability of finding particular distances is significantly increased, leading to a number of apparently equidistant maxima in the distributions.

The distribution maxima are directly related to the presence of a surface reconstruction on the Au(11*n*) surfaces. Figure 3.15 shows the distance Δ_{max} of maxima in the terrace width distribution $P(s)$ as a function of the number k of the reconstruction

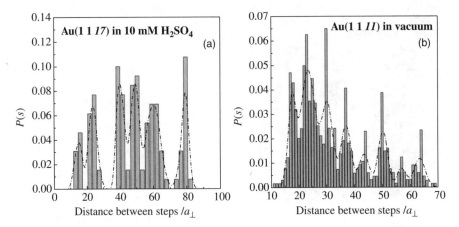

FIGURE 3.14   Terrace width distribution as measured on (a) Au(1,1,*17*) in 10 mM H_2SO_4 at −150 mV SCE and (b) Au(1,1,*11*) in vacuum at room temperature [82].

lines for the data presented in Figure 3.14. The data points for the Au(1,1,*11*) surface in UHV follow a linear fit to $\Delta_{max} = (l_d + kl_p)a_\perp$ with $l_d = 3.92 \pm 0.21$ and $l_p = 6.61 \pm 0.03$ (dashed line) and $a_\perp$ the distance between densely packed atomic rows. The nonzero offset of about $4a_\perp$ is indicative of a reconstruction-free zone at the step edges. For the nominal Au(1,1,*17*) surface in electrolyte, the data follow the same relation with a smaller reconstruction periodicity and a smaller offset ($l_d = 2.16 \pm 1.77$ and $l_p = 5.88 \pm 0.22$, dash-dotted line). Both data sets (for UHV as well as for electrolyte) lie significantly above the theoretical curve to be expected if the terraces had the same (5×20) surface reconstruction known from Au(001) (dotted line). In other words, the reconstruction unit cells observed on stepped Au(11*n*) are considerably larger than for Au(001). From the slopes in Figure 3.15 one directly obtains the repeat distance of the reconstruction lines on Au(11*n*) to 1.91 ± 0.01 nm (UHV) and 1.7 ± 0.1 nm (electrolyte).

Moiseeva et al. [82] showed that the enlarged periodicity is related to the presence of periodic step arrays on the surface. Corroborated by earlier studies by Watson et al. [85] the reported evidence points to a nonlocal effect of steps on the reconstruction period, namely stress field originating in the displacement of atoms at steps from their lattice positions on vicinal surfaces with a periodic arrangement of flat regions and surface areas with step bunches.

3.2.4.5   Elastic Step–Step Interactions and Surface Stress Effects

Periodic domains as observed for stepped Au(11*n*) electrodes have a considerable effect on the surface stress of stepped surfaces [91–93]. As was mentioned above, elastic contributions may have a considerable effect on the step–step interaction potential. Surface stress effects on the step–step interaction are presented briefly in the following.

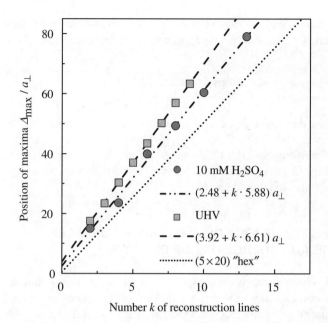

FIGURE 3.15 Relation between locations $\Delta_{\max}$ of maxima of distribution $P(s)$ in Figure 3.14 and number k of reconstruction lines as found for Au(1,1,*17*) in 10 mM H_2SO_4 (dark grey circles) and for Au(1,1,*11*) in UHV (light grey squares). The dotted line represents the theoretical curve as expected for a (5×20)-quasi-hex reconstruction [82].

The elastic energy of a stressed body with a periodic profile is given by [91]

$$W = \frac{2\left(1 - \nu^2\right)(s_1 - s_2)^2 + 4s_1 s_2 \sin^2\left(\theta_x/2\right)}{\pi Y \ell} \ln\left[\frac{\ell}{\pi a} \sin\frac{\pi \ell_x \cos\left(\theta_x - \theta\right)}{\ell}\right] \quad (3.23)$$

where, ν is Poisson's ratio, Y is Young's modulus; s_1, s_2 are the force components acting along the surfaces of the new facets, a is an atomic unit factor, and the other quantities are defined in Figure 3.16. The logarithmic dependence of the elastic interaction energy has been described elsewhere [64].

For the special case of a periodic one-dimensional surface undulation with a periodic height profile $h(x) = h_0 \cos\left(2\pi x/\lambda\right)$ (h_0 the height of the flat surface, that is,

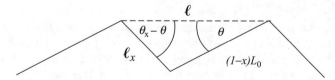

FIGURE 3.16 Model of initially flat surface which has separated into two domains with angles θ and $\theta_x - \theta$ with respect to original plane orientation. The height undulation has a periodicity ℓ; the notation here is the same as in Figure 3.10.

the vicinal surface without step bunching in our case) and a uniaxial stress, one finds the elastic energy contribution to the surface energy due the periodic domains as [91]

$$\delta W_{el} = -\frac{\sigma^2 \pi h_0^2}{\lambda} \frac{1 - \nu^2}{Y} \qquad (3.24)$$

where σ is the diagonal element of the uniaxial stress tensor and δW_{el}, the difference in elastic energy between the nominally flat surface and the surface with periodic domains of step bunches. Eq. (3.24) would be an additional elastic term to be considered in Eq. (3.22) in the case of potential-dependent step bunching on stepped electrode surfaces in a liquid environment.

3.3   MASS TRANSPORT ON ELECTRODE SURFACES: DYNAMICS AND ENERGETICS OF DEFECTS

In this section we focus on some aspects of the theoretical description of the thermodynamics and kinetics of steps on solid surfaces as defined in Figure 3.3: infinite steps (D) and steps forming a closed loop as in adatom and vacancy islands (G, H). Furthermore, the basic energy parameters that govern the equilibrium and nonequilibrium properties of steps, such as the *step free energy β*, the *step edge stiffness β̃*, or the *kink formation energy, ε*, are introduced. We start with the discussion of the equilibrium fluctuations of infinite steps followed by the equilibrium fluctuations of the step perimeter in islands, their relation to the step free energy, and how the latter can be derived from the equilibrium island shape. Then, relaxation of protrusions in infinite steps and the relaxation of nonequilibrium shapes of islands in Smoluchowski ripening (island coalescence) are discussed. This is followed by a discussion of another nonequilibrium coarsening phenomena, the Ostwald ripening of islands.

3.3.1   Step Dynamics in Thermodynamic Equilibrium

3.3.1.1   Equilibrium Dynamics of Infinite Steps
As described in detail elsewhere [5, 6] and references therein, the equilibrium fluctuations of a step can be described by a step correlation function

$$G(y, t) = \left\langle (x(y, t) - x(y_0, t_0))^2 \right\rangle \qquad (3.25)$$

which defines the probability of finding a step edge at a position $x(y)$ at time t when it has been located initially at a position $x(y_0)$ at a starting time t_0, x denoting the axis along the mean orientation of the step and y the perpendicular direction [Fig. 3.1(a)]. First, we discuss the two cases of pure spatial and pure time fluctuations of infinite steps; that is, the step correlation function is given by $G(y)$ and $G(t)$, respectively. The first case is fulfilled when kink motion is negligible on the time scale of an STM image, the second when steps appear frizzy in STM images and the kink mobility is high even on the time scale of scanning a single pixel. A theoretical consideration of the intermediate time scale when the step correlation function depends on both y and t is given in the literature [94, 95].

Spatial Correlation Function   The spatial correlation function $G(y)$ is given by

$$G(y) = \left\langle (x(y) - x(y_0))^2 \right\rangle = \frac{k_B T}{\tilde{\beta}} |y - y_0| \qquad (3.26)$$

where $\tilde{\beta}$ is the temperature-dependent *step edge stiffness*. The step edge stiffness is related to the *diffusivity* b^2 [68] via

$$\tilde{\beta} = \frac{k_B T}{b^2} a_{\|} \qquad (3.27)$$

with $a_{\|}$ the nearest-neighbor atomic distance. For a square lattice, the diffusivity can be expressed in terms of the kink formation energy ε,

$$b^2 = 2a_{\perp}^2 \exp\left(-\frac{\varepsilon}{k_B T}\right) \qquad (3.28)$$

if $k_B T \ll \varepsilon$, that is, for low temperatures. Here, $a_{\perp}$ is the distance between densely packed atomic rows. According to Eqs. (3.26)–(3.28), one can deduce the kink formation energy from measuring the spatial step correlation function $G(y)$. An example is shown in Figure 3.1(b) for Au(111) in 0.1 M H_2SO_4 + 0.5 mM HCl [96].

Step Correlation Function in Time   Following scaling arguments [97] the time correlation function $G(t)$ obeys a scaling law:

$$G(t) = c(T, \phi) L^{\delta} t^{\alpha} \qquad (3.29)$$

where L is the mean distance between steps on the surface as defined in Figure 3.2(a). $c(T, \phi)$ is a prefactor containing all activation energies and attempt frequencies. For surfaces in electrochemical environment it depends on temperature and electrode potential ϕ. The scaling exponents α and δ may assume values $\alpha = \frac{1}{2}, \frac{1}{3}, \frac{1}{4}$ and $\delta = \pm \frac{1}{2}$ and $\frac{1}{4}$ depending on the dominant atomic mass transport on the surface. For details on the different mass transport situations, see Jeong and Williams [5] and Giesen [6]. As an example atomic motion restricted to step edges, generally called *periphery diffusion*, is discussed in the following. For periphery diffusion, $G(t)$ is given by [98]

$$G(t) \approx 0.464 \left(\frac{(k_B T)^3 a_{\|}^3 a_{\perp}^2 \Gamma_h t}{\tilde{\beta}^3} \right)^{1/4} \qquad (3.30)$$

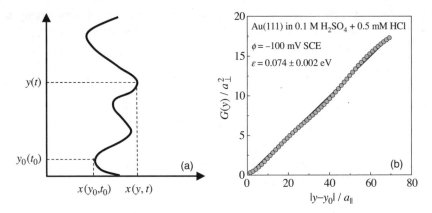

FIGURE 3.17 (a) Sketch of step contour to define step correlation function $G(y, t)$. (b) Spatial correlation function $G(y)$ (grey circles) as measured from STM images of Au(111) in 0.1 M $H_2SO_4 + 0.5$ mM HCl at electrode potential -100 mV SCE and at room temperature [96]. From the slope of the linear fit (black line) one obtains the kink formation energy ε.

where Γ_h is the hopping factor, $\Gamma_h = \Gamma_0 \exp\left(-E_{\Gamma_h}/k_B T\right)$ and Γ_0 is the preexponential factor and E_{Γ_h} the total activation energy for atomic diffusion along a step. The microscopic interpretation of Γ_h and in particular of the activation energy E_{Γ_h} may be difficult and depends on the particular step considered.

So far only infinite steps in thermodynamic equilibrium have been considered and a theoretical representation of spatial and time fluctuations in real space has been described. The fluctuation amplitudes of the step contour as shown in Figure 3.17(a) may also be considered as a protrusion which in principle increases locally the chemical potential of the step and therefore must eventually decay such that the step contour straightens. The time evolution of the shape of the protrusion may be described by the time evolution of the wavenumber components as found from a Fourier transformation of the step correlation function $G(t)$. Although the Fourier representation is also valid in the case of a step in thermodynamic equilibrium, it is applied in Section 3.3.3 exclusively to the nonequilibrium case.

3.3.1.2 Equilibrium Fluctuations of Closed-Loop Steps (Islands)

Island shape fluctuations are analyzed by comparing the island perimeter in a distinct STM image with the equilibrium shape at the same temperature. The procedure is illustrated in Figure 3.18. Here, $R(\theta_p, j)$ and $r(\theta_p)$ are vectors that define the perimeter of an island in an STM frame j and the equilibrium island shape, respectively, with θ_p the polar angle. The relative difference of $R(\theta_p, j)$ and $r(\theta_p)$ is denoted as [99–101]

$$g(\theta_p, j) = \frac{R(\theta_p, j) - r(\theta_p)}{r(\theta_p)} \tag{3.31}$$

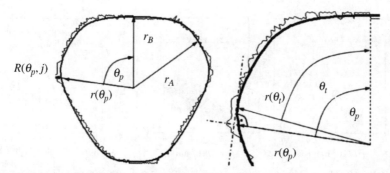

FIGURE 3.18 Equilibrium shape (bold solid line) and island perimeter as obtained from individual island in particular STM image (thin solid line) [45].

A measure of the island perimeter fluctuations in equilibrium is

$$G(\bar{r}, j) = \frac{\bar{r}^2}{2\pi} \int_0^{2\pi} g^2(\theta_p, j)\, d\theta_p \qquad (3.32)$$

where $\bar{r}$ is the mean radius of the equilibrium shape:

$$\bar{r} = \frac{1}{2\pi} \int_0^{2\pi} r(\theta_p)\, d\theta_p \qquad (3.33)$$

For the isotropic case where the island equilibrium shape is a circle, the ensemble-averaged fluctuation function can be written as [101]

$$\langle G(\bar{r}) \rangle_j = \frac{3k_B \bar{r}\, T}{4\pi \bar{\beta}} \qquad (3.34)$$

where $\bar{\beta}$ is the mean step free energy per length (in general the line tension [2]). Hence, $\bar{\beta}$ may be obtained by measuring the averaged perimeter fluctuations for different mean island sizes. This has been performed for various metal surfaces in vacuum: Ag(111) [99], Cu(111) [99, 101], Pt(111) [45] and also for Au(001) electrodes in electrolyte [102–104]. Figure 3.19 shows experimental data for the perimeter fluctuation function $\langle G(\bar{r}) \rangle$ versus the product of the island mean radius $\bar{r}$ and temperature T from Au(001) in (a) 50 mM H_2SO_4 [103] and (b) 100 mM $HClO_4$ + 1 mM HCl [104]. Figure 3.19(c) is a plot of the corresponding step line tension versus the electrode potential [104].

Equation (3.34) holds for the isotropic case; however, it proves to be a good approximation for island fluctuations on Cu(111), Ag(111) where the equilibrium shape is close to a hexagon and the anisotropy of the step free energy is small [44, 99, 101]. The deviation of an island shape from a circle is considered when using Eq. (3.35)

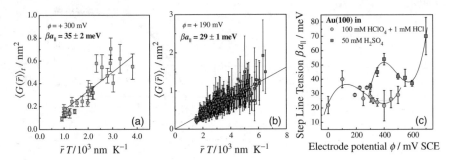

FIGURE 3.19 Analysis of step line tension using Eq. (3.34) on Au(001) electrodes in 50 mM H_2SO_4 [103] and 100 mM $HClO_4$ + 1 mM HCl [104].

rather than Eq. (3.34):

$$\langle G(\bar{r}) \rangle_j = \frac{k_B \bar{r} T}{2\pi \bar{\beta}} \sum_{|n|>1} \frac{1}{n^2 - \eta} \tag{3.35}$$

where, η is an *anisotropy factor* which is defined as [99]

$$\eta = \frac{1}{2\pi} \int_0^{2\pi} \frac{\beta(\theta_p)\sqrt{r^2(\theta_p) + \left(\partial r(\theta_p)/\partial \theta_p\right)^2}}{\bar{\beta}\,\bar{r}} d\theta_p \tag{3.36}$$

and η equals 1 for a circle, 1.028 for Cu(111) in vacuum at 313 K [99], and 1.060 for Pt(111) in vacuum at 653 K [45]. The sum in Eq. (3.35) yields $\frac{3}{2}$, 1.507, and 1.516, respectively. Hence, for Cu(111) as well as for the highly anisotropic islands on Pt(111), the isotropic model [Eq. (3.34)] is a good approximation. Figure 3.20(a) displays the anisotropy factor for Au(001) in 50 mM H_2SO_4 versus the electrode potential, which has been determined from the fourfold symmetric equilibrium shape of Au islands on the (001) electrode surface (e.g., [102]). The sum in Eq. (3.35) then yields 1.514 ± 0.003 as an average value over all potentials and is comparable to the anisotropy factor for trigonal Pt(111) islands; still its value is close to $\frac{3}{2}$ for a circular island. From the values of η and from a 2D Wulff analysis [37] of the island equilibrium shape, one obtains the angular anisotropy of the relative step line tension $\beta(\theta_p)/\beta(\theta_p = 0)$ as plotted in Figure 3.20(b) [105] [compare also Fig. 3.1(b) for a three-dimensional Wulff plot]. Note that in Figure 3.20(b) the relative step line tension $\beta(\theta_p)/\beta(\theta_p = 0)$ is plotted versus the polar angle θ_p rather than versus the tangential angle θ_t as obtained from a Wulff plot [Fig. 3.18 and Eq. (3.39)]. For symmetry reasons, $\beta(\theta_p)/\beta(\theta_p = 0)$ is plotted merely in the θ_p range between $0°$ and $45°$. In contrast to vacuum studies [e.g., for Pt(111)] [45], where the island anisotropy decreases with increasing temperature, for Au(001) in 50 mM H_2SO_4 the anisotropy is highest at large potentials. This is in accordance with the finding that the mean step line tension as obtained from the perimeter fluctuation function is highest for large potentials (Fig. 3.19).

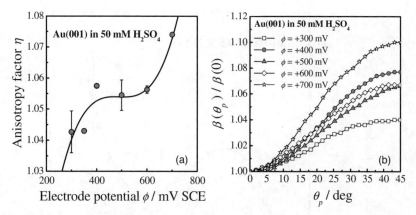

FIGURE 3.20 (a) Anisotropy factor for Au(001) in 50 mM H_2SO_4 vs. electrode potential [Eq. (3.36)] and (b) angular anisotropy of step line tension. (Data from ref. 105.)

3.3.1.3 Anisotropy of Absolute Step Free Energy

According to [99], there is a direct relation between the mean step free energy $\bar{\beta}$ and the angle-dependent, absolute step free energy $\beta(\theta)$:

$$\bar{\beta} = \frac{1}{2\pi} \int_0^{2\pi} \left[\frac{\beta(\theta_p)}{\bar{r}} \frac{r^4(\theta_p)}{\left[r^2(\theta_p) + \left(\partial r(\theta_p)/\partial\theta_p \right)^2 \right]^{3/2}} \right] d\theta_p \qquad (3.37)$$

Equation 3.37 can be used to determine absolute values for the free energy of steps along high-symmetry directions: The perimeter of an island on an fcc (100) surface, for instance, consists mainly of B steps. For fcc (111), the island edges are constituted from A and B steps (Fig. 3.4). Since islands on fcc (100) surfaces have a fourfold symmetry, the angular anisotropy of $\beta(\theta)$ is periodic in $\pi/2$. In the following, we define $\beta(\theta_p = 0) = \beta_B$ as the step free energy of an B step. Dividing both sides of Eq. (3.37) by $\beta(\theta_p = 0) = \beta_B$ and multiplying the integrand by r_B^4/r_B^4 (r_B being the island radius where the island edge is a perfect B step, that is, oriented along $\langle 110 \rangle$) yield

$$\frac{\bar{\beta}}{\beta_B} = \frac{1}{2\pi} \int_0^{2\pi} \left[\frac{\beta(\theta_p)/\beta_B}{\bar{r}/r_B^4} \frac{r^4(\theta_p)/r_B^4}{[r^2(\theta_p) + (\partial r(\theta_p)/\partial\theta_p)^2]^{3/2}} \right] d\theta \qquad (3.38)$$

Equation 3.38 enables to determine absolute values of step free energies: $\beta(\theta_p)/\beta(\theta_p = 0)$ is determined using the angular anisotropy of the step free energy $\beta(\theta_t)$ as obtained from the island equilibrium shape via the inverse Wulff construction [37], where θ_t is the tangential angle rather than the polar angle! (see Fig. 3.18. for

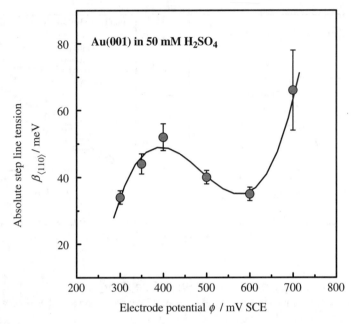

FIGURE 3.21 Absolute step line tension for $\langle 110 \rangle$-oriented steps on Au(001) in electrolyte [105]. The data as shown here are corrected for angular anisotropies in the step line tension according to Eqs. (3.35) and (3.36).

the definition of θ_t.) There is a simple relation between θ_p and θ_t [106]:

$$\theta_p = \theta_t + \arctan\left(\frac{\partial\beta(\theta_t)/\partial\theta_t}{\beta(\theta_t)}\right) \qquad (3.39)$$

An experimental result is shown in Figure 3.21 where the absolute step line tension for the B step ($\beta_{\langle 110 \rangle}$) for Au(001) in 50 mM H_2SO_4 is plotted versus electrode potential [105]. The data as shown here are corrected for angular anisotropies in the step line tension according to Eqs. (3.35) and (3.36). Comparison with Figure 3.19(c) shows that the data for the step line tension are hardly affected by the angular anisotropy of $\beta(\theta)$ or by the noncircular shape of the islands. Therefore, neglecting the anisotropy of $\beta(\theta)$ in Eq. (3.34), that is, using a mean line tension $\bar\beta$, and approximating the sum in Eq. (3.35) by a factor of $\frac{3}{2}$ are reasonable assumptions even for highly anisotropic islands.

Finally it is emphasized that the complete angular dependence of the step free energy can only be obtained if the two-dimensional equilibrium shape of islands has no facets or sign changes in the curvature. So far, facets have never been observed on islands. Thus, steps seem to behave as one-dimensional objects, which are thermodynamically rough at any finite temperature. Furthermore, islands of three- and fourfold symmetry studied so far all retain the same sign of the curvature along the perimeter.

3.3.2   Step Line Tension and Capacity Measurements on Stepped Electrodes

The step line tension is a very important parameter which governs the formation and stability of electrode surfaces. As we have seen before, it determines the equilibrium fluctuations of infinite steps as well as of closed-loop steps as found in adatom and vacancy islands. Furthermore, as we will discuss in Section 3.3.4, it also controls island-coarsening processes at elevated temperatures, with respect to both their shape and their size distribution.

Despite its importance little effort has been spent so far to study and understand the structure and properties of vicinal single-crystal electrode surfaces in an electrolyte environment. Experimental studies dealt with the potential, respectively charge dependence, of the step line tension: Using impedance spectroscopy the step line tension was calculated as the difference in the surface tensions of stepped and flat surfaces while the surface tensions are obtained from the integrated charge densities, respectively double integration of the capacity [50, 54]. In the following we briefly review the theoretical basis for this analysis.

By definition, the step line tension for $(11n)$ surfaces is

$$\beta_{(11n)}(\phi) = \lim_{L \to \infty} L \left\{ \gamma_{(11n)}(\phi) - \gamma_{(001)}(\phi) \right\} \tag{3.40}$$

where L is the distance between steps on the $(11n)$ surfaces as defined in Figure 3.2(a) and $\gamma_{(11n)}(\phi)$ and $\gamma_{(001)}(\phi)$ are the surface tensions of the $(11n)$ and (001) surfaces. The relative surface tensions can be calculated from the surface charge density $\sigma(\phi)$ by integrating the Lippmann equation [Eq. (3.12), Section 3.2.4]. The determination of the step line tension $\beta(\phi)$ thus requires the measurement of potential of zero charge pzc and the capacitance $C(\phi)$ on vicinal surfaces and the (nominally) step free (001) surface. Since experimentally merely the relative surface tension is accessible, the step line tension is always given with respect to a reference potential, in our case pzc.

For a qualitative understanding of the potential dependence of the step line tension, it is useful to consider the case where the capacitance of a stepped surface is shifted on the potential scale according to the shift in the pzc, $\Delta\phi_{pzc}$, and corrected by a factor that accounts heuristically for the different screening and polarizability at step sites [50, 107]:

$$C_{(11n)}(\phi) = C_{(001)}(\phi - \Delta\phi_{pzc}) \left(1 + \frac{l_c}{L} \right) \tag{3.41}$$

where l_c is a length that characterizes the line capacitance of a step. The contribution of steps to the surface capacitance (line capacitance) is positive (negative) if $l_c > 0$ ($l_c < 0$). Taking l_c as independent of the potential requires that the isotherm for specific adsorption of ions is identical for step and terrace sites except for the shift in the pzc. As is demonstrated in detail in [107], steps exclusively influence the Helmholtz capacitance and leave the Gouy–Chapman capacitance unchanged. There-fore, l_c as defined in Eq. (3.41) becomes concentration dependent and has not a direct physical interpretation. Nevertheless, for a given electrolyte concentration, it nicely parameterizes the effect of steps on the capacitance in the entire potential range [50].

The shift in the pzc, $\Delta\phi_{pzc}$, can be expressed in terms of the dipole moment per step atom p_z [as in Eq. (3.5), Section 2.3]. The shift in pzc is toward negative potentials, as experimentally well documented [2, 50, 52, 108]. Hence, the dipole moments of steps (and also adatoms and vacancies) point with their positive ends away from the metal surface [56].

Then the charge density can be expanded as

$$
\sigma_{(11n)}(\phi) \cong \left[\sigma_{(001)}(\phi - \Delta\phi_{pzc})\right] \left(1 + \frac{l_c}{L}\right)
$$
$$
\cong \sigma_{(001)}(\phi) - \frac{\partial\sigma_{(001)}}{\partial\phi}\Delta\phi_{pzc} + \sigma_{(001)}(\phi)\frac{l_c}{L}
$$

(3.42)

From that one finds the approximation [50]

$$
\beta(\phi) \cong \beta(\phi_{pzc}) - \frac{p_z}{\varepsilon_0 a_\parallel}\sigma_{(001)}(\phi) + \left[\gamma_{(001)}(\phi) - \gamma_{pzc,(001)}\right] l_c
$$

(3.43)

The experimental determination of the potential dependence of the step line tension therefore requires measurements of the capacitance versus the potential ϕ on vicinal surfaces and the flat surface. The ϕ_{pzc} is determined from the minimum in the capacitance $C(\phi)$ versus the potential in sufficiently dilute electrolytes.

Figure 3.22(a) shows the capacitance curves of Ag(001) and Ag(117) in 10 mM KClO$_4$ as solid and dashed lines, respectively [54]. The capacitance curves of the flat surface have a shape similar to that of the vicinal surface. The minimum in the capacitance curve, and hence the pzc, for Ag(117) is shifted to more negative potentials than that for Ag(001). Compared to Au(117) [Fig. 3.22(b)] the capacitance curve of Ag(117) is slightly broadened around the pzc, which indicates that the potential dependence of the capacitance near steps may differ from the flat surface.

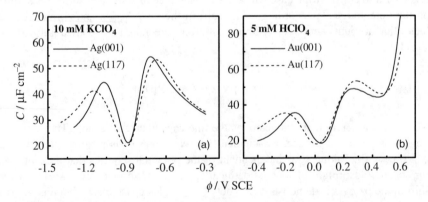

FIGURE 3.22 Capacitance curves for (001) (solid lines) and (117) (dot-dashed lines) surfaces of (a) Ag electrodes in 10 mM KClO$_4$ and (b) Au electrodes in 5 mM HClO$_4$ [50, 54]. The capacitance curves were obtained using frequency 20 Hz and 5 mV amplitude.

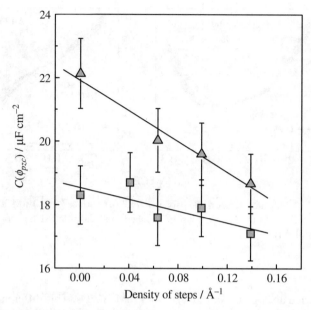

FIGURE 3.23 Capacitance $C(E_{pzc})$ at potential of zero charge vs. step density determined from capacitance curves for Ag in 10 mM KClO$_4$ (triangles) and Au in 5 mM HClO$_4$ (squares) [50, 54].

Figure 3.23 displays the capacitance at pzc, $C(E_{pzc})$, versus the step density for Ag(11n) in 10 mM KClO$_4$ (triangles) together with the results for Au(11n) in 5 mM HClO$_4$ (squares) [50]. In both cases, the capacitance decreases with increasing step density and the decrease appears to be approximately linear in L^{-1}, in agreement with Eq. (3.41). A step density–independent characteristic length l_c can therefore be assigned to both the gold and the silver surfaces. For both materials, the characteristic length l_c is negative, that is, the capacitance is reduced by the presence of steps.

Table 3.2 summarizes the step capacitance $C(E_{pzc})$ and the characteristic length l_c at the pzc for Au(11n) and Ag(11n) in various electrolytes [50, 54, 107]. (The corresponding dipole moments per step atom are shown in Table 3.1, Section 3.2.3). The errors are taken as the single variance obtained from fitting a straight line through the data points. According to Table 3.2, the capacitance at the pzc is somewhat larger for silver than for gold. The difference is mostly to be attributed to the higher electrolyte concentration [49]. The characteristic length l_c for Ag(11n) electrodes in 10 mM

TABLE 3.2 Capacity at pzc $C(\phi_{pzc})$ and Characteristic Length l_c [Eq. (3.41)] on Stepped Au and Ag(001) Surfaces in Various Electrolytes

	Au(001) [50]			Ag(001) [54]
Electrolyte	5 mM H$_2$SO$_4$	5 mM HClO$_4$	5 mM HF	10 mM KClO$_4$
$C(\phi_{pzc})/\mu Fcm^{-2}$	30.6 ± 1.0	18.6 ± 0.3	13.7 ± 0.5	22.0 ± 0.9
$l_c/\text{Å}$	-0.9 ± 0.2	-0.5 ± 0.2	-0.7 ± 0.4	-1.1 ± 0.1

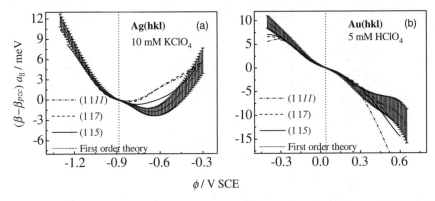

FIGURE 3.24 Step line tension, $\beta a_{\parallel}$ for (a) Ag(11n) surfaces in 10 mM KClO$_4$ and (b) Au(11n) surfaces in 5 mM HClO$_4$ as solid, dashed, and dash-dotted line for $n = 5, 7, 11$, respectively [50, 54].

KClO$_4$ is twice as large as for Au(11n) in 5 mM HClO$_4$. This difference cannot be attributed to the difference in the concentrations.

Figure 3.24(a) displays the step line tension $\beta a_{\parallel}$ calculated using the definition [Eq. (3.40)] for Ag(115), Ag(117), and Ag(1,1,11) as solid, dashed, and dash-dotted lines, respectively. The approximate dependence calculated from Eq. (3.43) with the dipole moments p_z and the characteristic length l_c taken from Tables [3.1] and [3.2] is shown as a dotted line with estimated error bars. The data are referenced to the (unknown) line tension at the pzc. The respective results for Au(11n), $n = 5,7,11$ [50], are shown for comparison in Figure 3.24(b).

For both the Ag(001) and the Au(001) steps the potential dependence of the step line tension is small compared to the values at the pzc. At the pzc, the step line tension should be approximately equal to the step formation energy at the solid–vacuum interface for which values 130 meV [109] and 170 meV [89] were determined. Considering the step line tension as the driving force for the decay of nanostructures on surfaces, the decay rate should merely weakly depend on the potential. However, the actual decay rate depends also on the speed of the diffusion processes, and hence on the activation energies for diffusion. Just as the potential dependence of the step line tension depends on the dipole moment of the step atoms, the formation energy of adatoms carrying the diffusion current depends on the dipole moment of the adatoms [110, 111], which amounts to $0.075e$Å and $0.095\ e$Å on Ag(001) and Au(001), respectively [56]. These dipole moments are about a factor of 20 larger than the step dipole moments and will dominate the potential dependence of decay processes on the (001) surfaces. The situation is different for the (111) surfaces of silver and gold for which the step dipole moments are a factor of 5–10 higher than for the (001) surfaces. The resulting dramatic consequences for the potential dependence of the defect formation energies and the activation energies for atom migration and thus for the process of *electrochemical annealing* are discussed in more detail in Section 3.3.5.

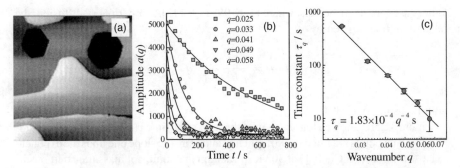

FIGURE 3.25 (a) STM image of step on Cu(111) at 303 K in vacuum immediately after monatomic high Cu island merged with step to form step bump. Scanwidth is 60 nm. (b) Decay curves of various wavenumber components. (c) Log–log plot of time constant vs. wavenumber. (Figures originally published in ref. 114.)

3.3.3 Step Fluctuations Out of Thermodynamic Equilibrium

3.3.3.1 *Step Relaxation of Infinite Steps: Decay of Step Bumps*

Let us consider a bump in the step that was formed at a point $t_0 = 0$ in time. Figure 3.25(a) shows an example of such a bump in a monatomic high step on Cu(111) in vacuum at 303 K which was formed when a monatomic high island in close proximity to a step eventually merged with the step edge. The step correlation function $G(y,t)$ can be expressed in a Fourier representation by using the Fourier components x_q of the spatial coordinate x:

$$x(y, t) = \int_q x_q(t)\, e^{iqy}\, dy \quad q = \frac{2\pi n}{L_y} \tag{3.44}$$

where q is the wavenumer and L_y the length of the step. In a step bump such as displayed in Figure 3.25(a), L_y is of the order of the base of the bump. This approximation will be used again in the discussion of the relaxation of coalesced islands in the next paragraph. Using Eq. (3.44) in Eq. (3.25) one finds for the step correlation function

$$G_q(t) = a(q) \left[1 - e^{-|t-t_0|/\tau(q)} \right] \tag{3.45}$$

where $a(q)$ is the amplitude of the fluctuations and $\tau(q)$ the relaxation time. Figure 3.25(b) shows $a(q)$ for various wavenumbers versus time. When the step bump decays, small wavenumbers decay slower than large values of q. The amplitude $a(q)$ is related to the step edge stiffness via

$$a(q) = \frac{2k_B T}{\tilde{\beta} L_y q^2} \tag{3.46}$$

As before, the relaxation time $\tau(q)$ depends on the dominant mass transport mechanism on the surface. Again, we will discuss merely the case of periphery diffusion.

Other cases are described elsewhere [5, 6, 98, 112, 113]. For periphery diffusion, the relaxation time is given by

$$\tau(q) = \frac{k_B T}{\tilde{\beta} a_\perp^3 a_\parallel^2 \Gamma_h} q^{-4} \tag{3.47}$$

with Γ_h as in Eq. (3.30).

Figure 3.25(c) shows a log–log plot of the relaxation time $\tau(q)$ versus the wavenumber. The data points are fitted to a straight line and from the slope one finds the exponent to be -4. Hence, on Cu(111) at 303 K in vacuum the dominant mass transport mechanism during bump decay is periphery diffusion. If one measures the decay also as a function of temperature, one can also deduce the hopping factor Γ_h which was demonstrated for a special case of bump decay during an island coalescence event (see below) on Cu(100) in vacuum [115].

3.3.3.2 Step Relaxation of Closed-Loop Steps: Smoluchowski Ripening

Similar to the case of an island merging with a step edge as discussed before, the island could also merge with another island. Figure 3.26(a) shows a series of STM images of two vacancy islands on Cu(001) in vacuum before and after coalescence [115]. A similar coalescence event observed on Au(001) in 100 mM $HClO_4$ + 1 mM HCl at +190 mV SCE [104] and room temperature is displayed in Figure 3.26(b). Due to the crystal symmetry of Cu and Au(001) surfaces the islands assume a fourfold symmetry in equilibrium.

Immediately after coalescence, new larger islands are formed with a shape far from equilibrium. The narrow neck relaxes in time and eventually the new islands assume again the equilibrium shape. This coarsening phenomena is called *Smoluchowski ripening* [116, 117] or island *coalescence*. In analogy to the relaxation of a step bump as discussed above, the relaxation of the neck can be described in terms of the Fourier representation of the step correlation function.

Here, the step length L_y is defined as the region of positive step curvature [Fig. 3.26(a)]. Using Eqs. (3.44) and (3.47) one finds

$$\tau(q) \propto \frac{L_y^4}{a_\perp^3 a_\parallel^2 \Gamma_h} \frac{k_B T}{\tilde{\beta}} \tag{3.48}$$

The step length L_y scales with the island area A via the relation

$$A \propto L_y^2 \tag{3.49}$$

Hence, for the relaxation time one obtains

$$\tau(q) \propto \frac{A^2}{a_\perp^3 a_\parallel^2 \Gamma_h} \frac{k_B T}{\tilde{\beta}} \tag{3.50}$$

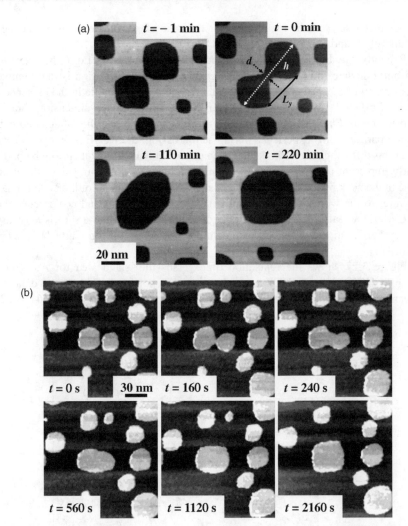

FIGURE 3.26 (a) STM images of two vacancy islands on Cu(001) at 323 K in vacuum before and after coalescence [45]. (b) STM images of coalescence event on Au(001) in 100 mM HClO$_4$ + 1 mM HCl at +190 mV SCE and 293 K [104].

Equation (3.50) holds for the case that the island shape relaxation is predominated by periphery diffusion. Including all mass transport situations the generalization of Eq. (3.50) is

$$\tau(q) \propto \frac{A^{\delta/2}}{a_{\perp}^{\delta-2} a_{\parallel}^2 \Gamma_h} \frac{k_B T}{\tilde{\beta}} \tag{3.51}$$

where $\delta = 4, 3, 2$ refers to the mass transport cases periphery diffusion and slow and fast terrace diffusion, respectively.

For metal surfaces in vacuum, so far the case of periphery diffusion has been exclusively found experimentally [115, 118–120].

Recently Liu and Evans [121] showed that the exponent in Eq. (3.51) may significantly deviate from the value $\delta = 4$ even if periphery diffusion is the dominant mass transport mechanism during relaxation. The reason may lie in the presence of a *kink–Ehrlich–Schwoebel energy* (KESE) [122, 123], the one-dimensional analog of the famous *Ehrlich–Schwoebel barrier* [124, 125]. While the latter is an extra energy barrier atoms experience during step crossing, the KESE is an extra energy for atoms diffusing around kinks in a step edge. The KESE may therefore be particularly important for mass transport during Smoluchowski ripening since the island edge in the neck region of the merged island has a large number of kinks. For details we refer to the work of Liu and Evans [121]. Analysis of island coalescence data on Cu(001) reveals a KESE of the order of 0.2 eV [115]. This is a surprisingly large energy barrier in the view that the adatom diffusion barrier on Cu(001) is 0.4 eV [126, 127].

Figure 3.27 shows the relaxation time versus the island area for Au(001) in 100 mM $HClO_4 + 1$ mM HCl at various electrode potentials. Fitting the data to a linear curve one finds a mean exponent for all electrode potentials of $\delta = 1.64 \pm 0.28$.

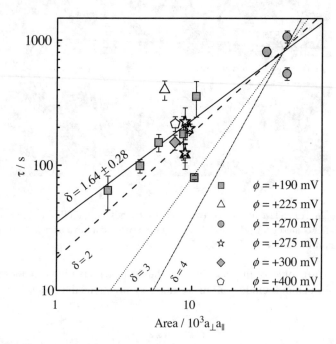

FIGURE 3.27   Log–log plot of coalescence relaxation time τ [Eq. (3.51)] vs. total island area A. From a linear fit to the data (solid line) one finds a scaling exponent $\delta = 1.64 \pm 0.28$. For comparison the theoretical curves for exponents $\delta = 4$ (perimeter diffusion), $\delta = 3$ (diffusion-limited terrace diffusion), and $\delta = 2$ (detachment-limited terrace diffusion) are plotted. (Data from ref. 104.)

For comparison Figure 3.27 contains additionally the theoretical exponents for the different mass transport situations of perimeter diffusion ($\delta = 4$), diffusion-limited terrace diffusion ($\delta = 3$), and detachment-limited terrace diffusion ($\delta = 2$). The experimental value is close to the theoretical value of 2 and the island coalescence process is apparently dominated by terrace diffusion where the time-limiting step of the transport process is the detachment, respectively reattachment, of atoms at the island edge. As a caveat we should emphasize that the relevant transport path after detachment may also include diffusion through the adjacent liquid before reattachment at the island edge, as has been observed for Ag(111) in electrolyte [128, 129]. In this case one would also expect δ to be of the order of 2 [97].

The data for the relaxation time τ versus the total island area A were obtained at different electrode potentials. In principle, one cannot a priori assume that the coalescence kinetics is independent of the electrode potential. On the contrary, one should expect the mass transport on the surface to depend on the electrode potential, as has been shown in studies on equilibrium step fluctuations and island coalescence on metal electrodes [6, 96, 110]. Then, if τ is a function of the area A as well as of the potential ϕ, the coalescence relaxation exponent δ [Eq. (3.51)] might also be a function of ϕ. A strong indication, however, that the dominant mass transport path does not change within the relevant potential range +190 and 400 mV SCE for the data as shown in Figure 3.27 is the fact that all data points fall on the same linear curve. The case of Au(001) in electrolyte is indeed the first case where island coalescence mediated by terrace diffusion has been observed.

A further point worth mentioning is that the continuum theory for island coalescence presented here is strictly valid only for large islands and for perfect corner-to-corner coalescence. Liu and Evans [121] showed that the coalescence exponent δ may substantially decrease if islands coalesce side to side. For instance, if the dominant mass transport during equilibration would be diffusion along the island perimeter, the measured relaxation exponent would be $\delta = 3$ rather than $\delta = 4$. In most merging events as studied in [104], the Au islands were slightly shifted along $\langle 110 \rangle$ with respect to each other. Hence, the initial island configuration before coalescence was an intermediate compared to the extreme cases of corner-to-corner and side-to-side coalescence. This may account for the slightly smaller value of $\delta = 1.64$ compared to the theoretical value 2 for detachment-limited terrace diffusion-mediated island coalescence.

3.3.4 Theory of Island Decay (Ostwald Ripening)

Original literature on Ostwald ripening deals with analytical models to describe the coarsening in colloidal liquids [130, 131]. More recently, similar models have been reported for the coarsening of islands during island decay [6, 132–134]. These analytical models are formulated for particular boundary conditions of an isotropic problem where a circular island with radius r_i is surrounded by a mean adatom concentration field. Experimentally, this mean field approach is realized if an adatom island is located in the center of a larger almost circular vacancy island with radius r_o (Fig. 3.28). The chemical potential of an island is determined by its radius r via the

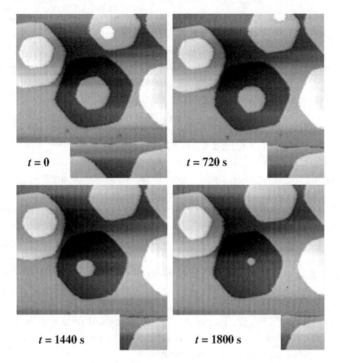

$t = 0$ $t = 720$ s

$t = 1440$ s $t = 1800$ s

FIGURE 3.28 STM images of Cu adatom island within larger vacancy island on Cu(111) at 303 K in vacuum [6]. Scan width 60 nm.

generalized *Gibbs–Thomson relation*

$$\mu(r) = \kappa \, \Omega \frac{\beta}{r} \tag{3.52}$$

where, β is the step line tension as used before, Ω is the area of an atom, and κ is a shape factor with $\kappa = \sqrt{2\sqrt{3}/\pi}$ for a hexagonal lattice and $\kappa = 2/\sqrt{\pi}$ for the square lattice of an fcc substrate. The shape factor κ relates an island to an equivalent circular island with the same area [135]. The gradient in the chemical potential between the edge of the inner and the outer island radius gives rise to a mass diffusion current between the islands, balancing the attachment and detachment of atoms at the island edges, and leading to a variation of the island area A with time (for a detailed derivation see, e.g., Giesen [6]):

$$\frac{dA}{dt} = -2\pi\nu_0\Omega \exp[-(E_{\mathrm{ad}} + E_{\mathrm{diff}})/(k_BT)] \frac{\exp\{[\kappa\beta\Omega/(k_BT)](1/r_i - 1/r_o)\}}{\ln|r_o/r_i| + a/sr_i} \tag{3.53}$$

where ν_0 is the preexponential attempt frequency; E_{ad}, E_{diff} the adatom formation and terrace diffusion barrier, respectively; and r_i, r_o the radii of the inner and outer island perimeters. The sticking factor s is defined as the ratio between the sticking frequency of atoms at island edges and the diffusion rate on the terrace. For $s = 1$, no additional sticking energy barrier is present; for $s = 0$, the barrier is infinitively large. Approximations of Eq. (3.53) for two limiting cases are described in the following.

3.3.4.1 Diffusion Limit: $s = 1$

For $s = 1$, the sticking of atoms at the island edge is fast compared to the diffusion between the islands. With the additional, not necessarily fulfilled approximations $a/r_i \ll \ln|r_o/r_i|$ and $\exp(\kappa\beta\Omega/r_{i,o}) \approx \kappa\beta\Omega/r_{i,o}$ one gets

$$\frac{dA}{dt} \approx -\frac{2\pi\nu_0\Omega}{\ln|r_o/r_i|} \exp[-(E_{ad} + E_{diff}/k_BT)] \frac{\kappa\beta\Omega}{k_BT}\left(\frac{1}{r_i} - \frac{1}{r_o}\right) \qquad (3.54)$$

By further neglecting the time dependence of $\ln|r_o/r_i|$,[1] one has $dA/dt \propto 1/r_i$. With $A = \pi r_i^2$ one has

$$A(t) \propto (t - t_0)^{2/3} \qquad (3.55)$$

Hence, if the sticking probability is 1 and diffusion on the terrace is slow compared to the detachment/attachment of atoms at the inner island edge (*diffusion limit*), the time dependence of the island area scales approximately as $t^{2/3}$. Note that the effective exponent in Eq. (3.55) decreases if the step line tension β increases. This is a consequence of the approximation $\exp(\kappa\beta\Omega/r_{i,o}) \approx \kappa\beta\Omega/r_{i,o}$ used in Eq. (3.54). Therefore, smaller exponents may be measured in experiment, in particular when small islands are analyzed where β is large.

3.3.4.2 Detachment Limit: $s \ll 1$

For $s \ll 1$, a finite energy barrier reduces the sticking probability of atoms at the island edge such that atom attachment/detachment is slow compared to terrace diffusion. Then, with the assumption $\exp(\kappa\beta\Omega/r_{i,o}) \approx \kappa\beta\Omega/r_{i,o}$ as before one obtains

$$\frac{dA}{dt} \approx -2\pi\nu_0 s\Omega \exp[-(E_{ad} + E_{diff}/k_BT)] \frac{\kappa\beta\Omega}{ak_BT}\left(1 - \frac{r_i}{r_o}\right) \qquad (3.56)$$

that is, $dA/dt \propto$ const. The time dependence of A and r_i is therefore

$$A(t) \propto (t - t_0) \qquad (3.57)$$

Hence, if diffusion on the terrace is fast compared to the detachment/attachment of atoms at the inner island edge (*detachment limit*), the island area decays linearly in time. It is noted that the classical Ostwald theory and hence the limiting cases

---

[1] Note that this approximation may be not justified since the radii of the inner and the outer island perimeter change in time and in particular the ratio of the radii may depend significantly on the time.

[Eqs. (3.55) and (3.57)] are valid only under the following assumptions: First to fulfill the mean-field approximation of the classical Ostwald theory, the island/vacancy island should be located within another vacancy island or an equivalent configuration. Furthermore, the island must be located at the center of the outer island. Second, the ratio of the outer and inner island radii must be approximately time independent. And finally, the islands must be large such that the assumption of a small β is justified. These are very strict conditions which are rarely fulfilled in real experiments. Therefore, experimental scaling exponents may differ from those in classical Ostwald theory.

Whereas the Ostwald theory describes the situation of an individual island within defined boundary conditions (Fig. 3.28) in real experiments—in particular in electrolyte—a large number of adatom islands in the vicinity of other adatom islands and step edges are created (Fig. 3.29).

Accordingly, island area decay curves as found for such islands exhibit a time exponent and features not described by the classical Ostwald theory: Due to the loss of mass of smaller atoms, large islands grow initially (see, e.g., decay curves of islands 3 and 4 in Fig. 3.30). Furthermore, when small islands (island 1 in Fig. 3.30) vanish in the vicinity of larger islands (island 2 in Fig. 3.30), the area of the latter increases suddenly and the decay curve of the large island displays a sudden "hiccup" (arrow in Fig. 3.30) [135]. In order to correctly analyze such decay events, one would have to solve the full Laplace equation of diffusion, including all islands in the image (for details see, e.g., refs. [135 and 136]). If the decay of an individual island is, however,

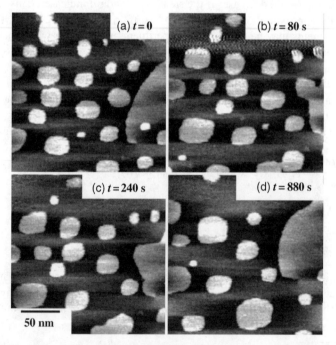

FIGURE 3.29 Time series of STM images of monatomic high-Au islands on Au(001) in 100 mM $HClO_4$ + 1 mM HCl at +400 mV SCE [111].

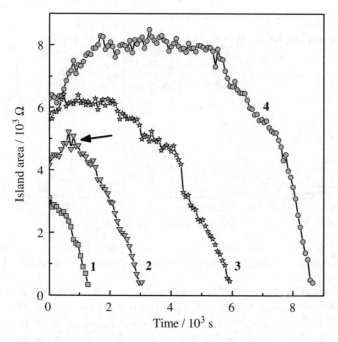

FIGURE 3.30 Decay curves of Au islands on Au(001) in 100 mM $HClO_4$ + 1 mM HCl at +190 mV SCE. The numbers indicate specific islands as discussed in the text. The island area is given in atoms (the area of an atom is $\Omega = 0.0832\ nm^2$) [111].

influenced by its vicinity, the dominant mass transport on the surface during island decay must be diffusion limited. In the case of detachment-limited decay, the local environment of an island plays no role and the decay curve is always linear [137].

3.3.5 Electrochemical Annealing

Classical electrochemical reactions involve the transfer of charged particles through the interface between an electrode and an electrolyte solution. The driving force for such reactions is the product of the charge on the particle and the difference in the electrostatic potential across the interface, which is proportional to the electrode potential. The reaction rates depend exponentially on the electrode potential, since the energies of activation vary linearly with the driving force—a principle known as a linear free-energy relation in chemical kinetics. In electrochemical STM studies, processes are investigated that occur on the electrode surface itself and do not involve the transfer of charge across the interface. Examples are the diffusion of single adatoms or vacancies, their generation from kink sites, and the interlayer transport of atoms. Since the initial and final states of such processes are both on the surface, their driving force is not directly affected by the electrode potential. Nevertheless, the rates of many coarsening phenomena, which involve a combination of the processes listed above, have been found to depend exponentially on the potential [6, 110, 111]. In most

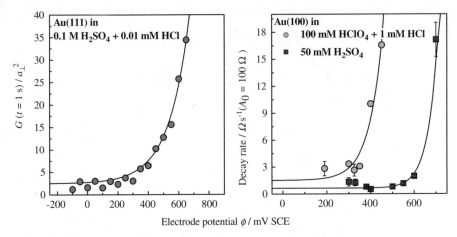

FIGURE 3.31 Exponential increase of mass transport rates with electrode potential: (a) step fluctuation amplitude on Au(111) [96] and (b) island decay rate on Au(001) [111].

cases, the rate grows as the potential is raised toward more positive values. The thereby induced smoothening of the surface has been termed *electrochemical annealing* [138] in order to stress the apparent similarity between the effect of the potential and a rise in temperature. The first quantitative investigations on electrochemical annealing concern the smoothening of rough gold and platinum electrodes [139–141]. Later, the filling of STM-induced indentations and the decay of deposited islands were studied on Ag(100) and Au(100) electrodes [31, 33, 142]. Quantitative investigations on structurally well-defined systems include equilibrium fluctuations of monatomic steps on Cu, Au, and Ag surfaces (for a review see ref. 6) and the decay of islands [110, 111]. All studies report an approximately exponential increase of the rate of transport processes with the electrode potential. Examples for step fluctuations on Au(111) [96] and island decay on Au(001) [111] are shown in Figure 3.31. Because of the very different nature of the investigated processes, which involve different combinations of atomic processes with different activation energies, there is a common, rather general cause for the exponential dependence irrespective of the specific process, surface, and electrolyte which is discussed in the following.

The thermodynamics of electrode surfaces differs from the thermodynamics of surfaces in vacuum because of the constraint of constant electrode potential. The correct thermodynamic potential for the surface energy, the surface tension γ, defined as the work per area required for generating a surface, is

$$\gamma = f_s + \sigma \frac{\partial f_s}{\partial \sigma} \equiv f_s - \sigma \phi \tag{3.58}$$

in which f_s is the area specific free energy, σ is the area specific surface charge [26], and ϕ is the electrode potential on the vacuum scale. Free surface energy and surface tension coincide only for uncharged surfaces [2]. Under the boundary condition of constant ion concentrations in the electrolyte and constant temperature, the surface tension is expressed in terms of the potential dependence of the surface charge density $\sigma(\phi)$ according to the Lippmann equation [Eq. (3.12)]. These considerations can be expanded to the thermodynamics of defect generation and migration [110].

As has been discussed before (Section 3.2.3) defects, point defects, and steps carry a dipole moment which in most cases points with its positive end toward the electrolyte. The work required to generate the defect with a concentration ρ on an electrode surface held at constant potential ϕ_{def} is the difference in the surface tension per defect for a surface with defects γ_{def} and without the defects γ_0; hence

$$E_{\text{def}}(\phi) = \lim_{\rho \to 0} \frac{1}{\rho} \{\gamma_{\text{def}}(\phi) - \gamma_0(\phi)\} \tag{3.59}$$

The surface tension is then expressed in terms of the charge densities of the surface with and without the defects:

$$E_{\text{def}}(\phi) = \Delta F_{\text{def}} - \lim_{\rho \to 0} \frac{1}{\rho}$$
$$\times \left\{ \int_{\phi_{\text{pzc}, 0}}^{\phi} [\sigma_{\text{def}}(\phi') - \sigma_0(\phi')] \, d\phi' + \int_{\phi_{\text{pzc, def}}}^{\phi_{\text{pzc, def}} + \Delta\phi} \sigma_{\text{def}}(\phi') d\phi' \right\} \tag{3.60}$$

where ΔF_{def} is the Helmholtz free energy (or the work) required to create a defect on an uncharged surface in the same electrochemical environment. This quantity is calculated and measured for the uncharged surfaces in vacuum surface physics. The second integral, from $\phi_{\text{pzc,def}}$ to $\phi_{\text{pzc,def}} + \Delta\phi$, is proportional to $\Delta\phi^2$ and therefore proportional to ρ^2. It vanishes in the limit $\rho \to 0$. The thermodynamically correct expression for the work required to generate a defect at the solid–electrolyte interface held at constant potential is therefore

$$E_{def}(\phi) = \Delta F_{def} - \lim_{\rho \to 0} \frac{1}{\rho} \left\{ \int_{\phi_{\text{pzc}, 0}}^{\phi} [\sigma_{\text{def}}(\phi') - \sigma_0(\phi')] \, d\phi' \right\} \tag{3.61}$$

Hence, the potential dependence of the defect formation energy is given by the integral over the difference in the charge densities on a surface with and without the defects, $\sigma_{\text{def}}(\phi) - \sigma_0(\phi)$.

In order to make use of Eq. (3.61), one needs the charge densities as a function of the potential. The charge density of the defect free surface $\sigma_0(\phi)$ is easily measured for Ag and Au for instance according to standard procedures in electrochemistry. The experimental determination of $\sigma_{\text{def}}(\phi)$ for the case of vicinal surfaces with a regular array of step defects has been discussed in Section 3.3.2. For other defects, for example, for adatoms or vacancies on terraces, $\sigma_{\text{def}}(\phi)$ must be taken from theory. Knowing that defects carry a dipole moment that shifts the pzc, the simplest possible approach to calculate $\sigma_{\text{def}}(\phi)$ is to assume that $\sigma_{\text{def}}(\phi)$ differs from $\sigma_0(\phi)$ only

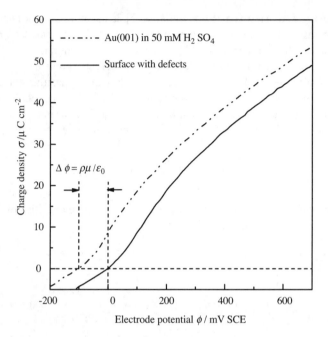

FIGURE 3.32 Measured surface charge density as function of electrode potential for unreconstructed Au(100) surface in 50 mM H_2SO_4 (solid line) and model charge density shifted by $\Delta\phi = \rho\mu/\varepsilon_0$ (dashed line) [110].

because of that shift in the pzc. To illustrate the point, the surface charge density on an unreconstructed Au(100) surface in 50 mM H_2SO_4 is plotted in Figure 3.32.

The dashed line is the charge density for a particular concentration ρ of defects carrying a particular dipole moment μ. The assumption that the charge density is merely shifted on the potential axis is a good approximation in the absence of specific adsorption of ions. This was explicitly shown for steps by Ibach and Schmickler [26] by a self-consistent solution of the Schrödinger–Poisson equation for a metal in the jellium approximation and the Poisson–Boltzmann equation for the electrolyte. The result should be pertinent also to other defects.

As previously discussed, the defect merely shifts the charge density $\sigma_{def}(\phi)$ on the potential scale by $\Delta\phi$ [110]. Hence, the charge density can be expanded around this shift as

$$\sigma_{def}(\phi) \approx \sigma_0(\phi + \Delta\phi) \approx \sigma_0(\phi) + \frac{\partial\sigma_0}{\partial\phi}\Delta\phi + \frac{1}{2}\frac{\partial^2\sigma_0}{\partial\phi^2}(\Delta\phi)^2 \cdots \qquad (3.62)$$

The third term is again proportional to ρ^2 and vanishes in the limit $\rho \to 0$. One then obtains the remarkable simple equation for the potential dependence of the defect formation energy [110]

$$E_{\text{def}}(\phi) = \Delta F_{\text{def}} - \frac{\mu_{\text{def}}}{\varepsilon_0}\sigma_0(\phi) \tag{3.63}$$

The same type of equation applies to the energy of a defect in an activated state, for example, in the transition state for diffusion. The activation energy for the diffusion is therefore [110]

$$E_{\text{act}}(\phi) = \Delta F_{\text{act}}(\phi_{\text{pzc}}) - \frac{\mu^+ - \mu_{\text{def}}}{\varepsilon_0}\sigma_0(\phi) \tag{3.64}$$

with ΔF_{act} the difference in the free energy of the defect in the transition and the ground state and μ^+ the dipole moment of the defect in the transition state. Since σ_0/ε_0 is the electric field at the surface without defects, Eqs. (3.63) and (3.64) have a very simple interpretation: The extra energy term is the electrostatic energy of the relevant dipole moment in the electric field at the surface. By virtue of this electrostatic energy, all defect formation energies and all activation energies involved in surface migration processes become a linear function of the surface charge density. In other words, the model predicts that the logarithms of atom transport rates at solid–electrolyte interfaces should depend linearly on the surface charge density, regardless of the nature of the rate-determining step. This is indeed observed for mass transport phenomena as demonstrated in Figure 3.31. There the data are well fitted by exponential functions (solid lines), in accordance with a linear potential dependence of the data on a logarithmic scale.

3.4 INFLUENCE OF ADSORBATES

Changes in the chemical properties of surfaces, in particular due to chemical adsorption, may have profound effects on the dynamics and stability of surfaces as discussed in previous sections. The reason lies in the considerable influence of chemical adsorption on energetic parameters such as defect formation energies, activation barriers, or the diffusing species. Despite the tremendous importance of adsorbates on the dynamics and stability of surfaces, only few studies can be found in the literature dealing with the quantitative analysis of adsorbate-induced dynamics and stability phenomena. The reason may lie in the fact that these studies still pose a number of challenges to experimentalist as well as to theorists. In the following we will briefly review and discuss studies on the influence of adsorbates or molecules on surface dynamics and stability.

3.4.1 Adsorbates in Electrolyte

Adsorbates on surface terraces may directly affect the structure and distribution of surface defects as well as their kinetics and thermodynamics. An example is the formation of an ordered adlayer of specifically adsorbed anions [143]. Particularly well-studied examples are the ordered Cl⁻, Br⁻, or sulfate adlayers on Cu(100) or (111) electrode surfaces [23, 143, 144]. Figure 3.33(a) shows an STM image of Cu(001) in Cl-containing electrolyte in the potential regime where the anion adlayer is disordered and only the (1×1) lattice structure of the clean Cu(001) surface can be resolved [145]. Clearly seen is the frizziness of the contour of the monatomic high step indicative of rapid kink motion at the step edge [42, 146]. The reason for the high kink mobility is the orientation of the step edge, which is along the $\langle 001 \rangle$ direction. For fcc Cu(001) surfaces, this direction corresponds to a step edge with a kink concentration of 100%; that is, the step contour exhibits one kink per atomic site. As has been quantitatively analyzed in the literature [106, 147], the step fluctuation amplitude increases with increasing number of geometrically enforced kinks. Under these potential conditions, the step dynamics on Cu(001) in Cl⁻ solution reveals a behavior similar to the bare metal surface in vacuum. This is evidence that the influence of the disordered anion adlayer is small for dilute electrolyte concentration or weakly adsorbing anions [145]. As a caveat we would like to emphasize that this is not necessarily the case: On Cu(111) in vacuum, step edge perimeter diffusion (time scaling $t^{1/4}$) and diffusion-limited terrace diffusion (time scaling $t^{1/2}$) is observed in various temperature ranges [114]. Around room temperature, perimeter diffusion is predominant. For Cu(111) in 1 mM HCl, however, between $\phi = -500\,\mathrm{mV}$ SCE and $\phi = -100\,\mathrm{mV}$ SCE and between 285 and 305 K a time scaling $\sim t^{1/3}$ is found which corresponds to detachment-limited terrace diffusion [148]. Interestingly this is the only case reported so far where a time scaling $\sim t^{1/3}$ is observed.

When in the case of Cu(001) in Cl-containing electrolyte [145], the potential is increased to the range where the ordered chloride $c(2 \times 2)\,R45°$ adlayer is formed, a new surface morphology is observed [Fig. 3.33(b)]. Since the ordered adlayer is

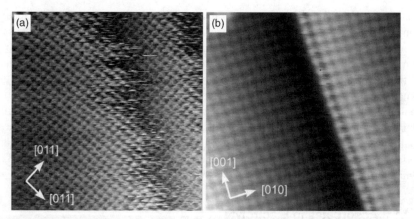

FIGURE 3.33 In situ STM images of Cu(100) showing (a) frizzy steps on surface covered by disordered adlayer phase (observed lattice is (1×1) lattice of Cu substrate) and (b) $\langle 001 \rangle$-oriented, straight steps in presence of a $c(2 \times 2)$ Cl adlayer. Scan size $7 \times 7\,\mathrm{nm}$ [145].

rotated by 45° with respect to the unit cell of the clean Cu(001) surface, the densely packed direction of the Cl adstructure is the ⟨001⟩ direction of the substrate. The new step morphology reveals perfectly straight, almost unkinked steps along ⟨001⟩ and the steps are predominantly of double- or even multiple-layer height. The position of the step edge is always localized between the maxima of the $c(2 \times 2)$ lattice, indicating that the structure of the steps is well defined on the atomic scale.

The reorientation of step edges as those found for the ordered Cl adlayer on Cu(001) has also been observed for other halide adstructures on Cu and Au surfaces [22, 149–153] and for sulfur structures on Cu(111) [154, 155].

It is not exclusively the surface and step morphology which is affected by the adsorption of ordered adlayers on electrode surfaces; also the surface mass transport itself and its underlying mechanisms are influenced by the presence of surface adsorption. As has been discussed in detail in Section [3.3], step fluctuations as well as island ripening scale characteristically with time, depending on the underlying dominant mass transport. Furthermore, it has been shown in various studies that the dominant surface mass transport may depend on the electrode potential [23, 96, 148, 156–158]. Giesen and Kolb [96] demonstrated that step fluctuations are considerably influenced by the type of specifically adsorbing anion. To discuss these observations in more detail, here we compare data on Cu(001) as presented in Figure 3.33 with other data as published for the bare Cu surface. According Giesen et al. [157], the step fluctuations on Cu(001) in 0.05 M H_2SO_4 are dominated by two different mass transport mechanisms depending on the electrode potential: Up to potentials around −490 mV SCE, the time-scaling exponent α in Eq. (3.29) is $\frac{1}{4}$ and there is no dependence of the time correlation function on the step–step distance, that is, $\delta = 0$ [see Eq. (3.29)]. Hence, in this potential range the equilibrium step fluctuations are due to perimeter diffusion [97]. This is also observed for stepped Cu(001) surfaces in vacuum around room temperature [159]. Furthermore, in this potential range, the fluctuation amplitude is comparable to that found in vacuum. Between −490 and −390 mV, on the other hand, the exponent α changes to $\frac{1}{2}$ and δ is now $\frac{1}{2}$ [157]. Hence, the dominant mass transport mechanism in this potential range includes exchange of atoms with adjacent terraces as well as with neighbor steps with no additional step-crossing barrier present [97]. For potentials above −390 mV Cu dissolution sets in. If one compares these results with the case of Cu(001) in the presence of an ordered Cl adlayer [145], one finds rather that the dominant transport mechanism is perimeter diffusion [143]. The reduction in the degree of freedom of the surface mass transport compared to Cu(001) in sulfuric acid solution is a direct consequence of the adsorbate structure on the terrace which likely hinders the diffusion of Cu adatoms on the terraces.

A similar observation is made for step fluctuations on Au(111) for various electrolytes [96]. In the potential range up to +300 mV SCE, where the Au(111) is reconstructed [81], the dominant mass transport is atom exchange between steps and adjacent terraces. This result is in accordance with vacuum studies which show that the reconstructed Au(111) surface reveals a high atom mobility on the terraces due to the presence of the herring-bone reconstruction [160]. For potentials above +300 mV, the surface is unreconstructed and anions are specifically adsorbed. In the case of chloride-containing electrolytes as used elsewhere [96], the Cl coverage amounts up to 0.5 monolayers [161]. Then, the dominant mass transport mechanism changes to perimeter diffusion [96]. Although there is not yet an ordered Cl adlayer formed on the Au(111) surface in this potential range (which is formed at coverages higher than 0.5 and potentials >0.6 V SCE [161]), it is reasonable to assume that local patches of (mobile and highly dynamic) ordered Cl layers [162] are generated which influence the dominant mass transport mechanism. However, a change in the diffusing species from Au atoms to an Au–Cl complex as proposed in [96] as the reason for the observation of a change in the mass transport mechanism cannot be excluded.

For Au(111), step fluctuations depend sensitively on the specifically adsorbed anion. As was shown by Giesen and Kolb [96], the fluctuation amplitude increases exponentially with electrode potential for electrolytes with an Cl^- concentration of 0.5 mM and higher (Fig. 3.34, dark grey circles). The exponential increase is less steep for lower Cl^- concentration (Fig. 3.34, grey squares). Hardly any increase of the fluctuation amplitude with potential is observed when no Cl^- is present (Fig. 3.34, light grey triangles).

As has been described in Section 3.3.5, the exponential increase of step fluctuations is explained by a linear dependence of all defect formation and migration energies on electrode potential [see Eq. (3.63)] [110]. A change of the exponential increase for different Cl^- concentrations as found for Au(111) in Figure 3.34 is therefore indicative of a change of relevant migration and defect energies with the anion concentration.

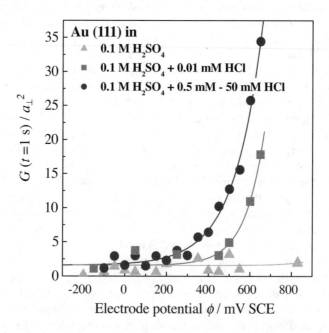

FIGURE 3.34 Plot of time correlation function $G(t)$ [Eq. (3.29)] at reference value $t = 1\,s$ for Au(111) in various electrolytes vs. electrode potential. The fluctuation amplitude depends sensitively on the amount of Cl^- in the electrolyte. (Data originally published in ref. 96.)

Using Eq. (3.63), the relevant mass transport energies could be determined for the case of step fluctuations on Au(111), as shown in Figure 3.34. There, activation energies for the fluctuations of 0.8 ± 0.1 eV and 0.7 ± 0.1 eV for Cl^- concentrations of 0.5 mM and higher (circles in Fig. 3.34) and for 0.01 mM (squares in Fig. 3.34), respectively, are found [96]. This is a very small difference which does not allow any further conclusions. Energy values for the data of Cl^- -free electrolyte (triangles in Fig. 3.34) could not be obtained since an increase of $G(t = 1$ s) with electrode potential is hardly visible.

For island decay on Au(001) electrodes, however, a detailed analysis of activation and formation energies according to Eq. (3.63) for Cl^- -containing and Cl^- - free electrolytes could be performed [111]: Island decay on Au(001) in 50 mM H_2SO_4 [163] as well as in 100 mM $HClO_4$ + 1 mM HCl [111] is diffusion limited (see Section 3.3.4.1). According to Eq. (3.54) the island decay is hence governed by the sum of the adatom formation energy E_{ad} on the terrace and of the surface diffusion barrier E_{diff}. Using Eq. (3.63), which describes the linear relationship between defect and migration energies and electrode potential, $E_{ad} + E_{diff}$ was determined for both electrolytes [111]. The result is shown in Figure 3.35. Obviously $E_{ad} + E_{diff}$ is much larger for 50 mM H_2SO_4 than for 100 mM $HClO_4$ + 1 mM HCl. This explains the slower Ostwald ripening of Au islands on Au(001) in sulfuric acid compared to chloride-containing electrolyte [111].

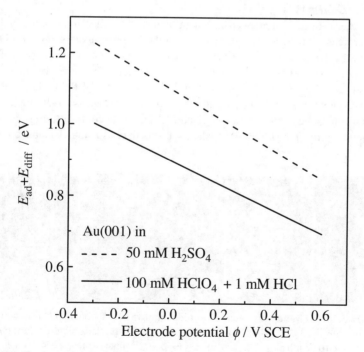

FIGURE 3.35   Potential dependence of $E_{ad} + E_{diff}$ for Au(001) in 50 mM H_2SO_4 (dashed line) and in 100 mM $HClO_4$ + 1 mM HCl (solid line) according to Pichardo – Pedrero et al. [111].

It is reasonable to assume that the situation is similar in the case of step fluctuations on Au(111): The relevant formation and migration energies might be larger in sulfuric acid solution compared to Cl$^-$-containing electrolytes, which then would explain the larger fluctuation amplitude for the latter.

The experimental data as presented here are in accordance with the general understanding that chloride serves as a brightener in metal plating, that is, enhances surface mass transport and decreases relevant transport energy barriers [110].

It is worth mentioning that a similar speed-up of ripening processes as observed for Cl$^-$-ions in electrolyte is also observed for sulfur adsorbates on Cu surfaces in vacuum [164].

A rather interesting influence of Cl$^-$ on the island shape has been observed for Cu islands on Cu(111) in 10 mM HCl [23]. In contrast to Cu(111) in vacuum, where the equilibrium island shape has a hexagonal symmetry (see Fig. 3.4), in HCl, the islands are trigonal. Furthermore, they exhibit pronounced sharp corner edges which have never been observed in vacuum studies and escapes any theoretical description so far. From an energetic point of view, sharp corners are extremely unfavorable and it is not clear what the driving force for the formation of those corners in the case of Cu(111) in HCl might be.

3.4.2   Adsorbates in Vacuum

3.4.2.1   CO/Pt(111)

Another interesting example of adsorbate influence on the thermodynamics and kinetics of surfaces stems from vacuum studies on the homoepitaxial growth on Pt(111). Kalff et al. [165] found that the growth shape and orientation of Pt island formed during deposition of 0.15 ML Pt at 400 K were dramatically altered by traces of CO on the surface (Fig. 3.36). Due to the triangular symmetry of the Pt(111) substrate, the island shapes in equilibrium reveal also a triangular shape with alternating A and B steps forming the contour line (see Section 3.2.2.2). In contrast to Figures 3.4 and 3.5, where *equilibrium shapes* of Pt islands, respectively vacancy islands, on Pt(111)

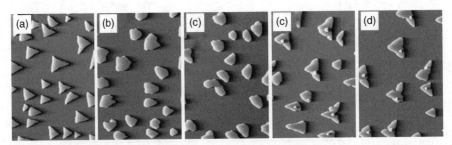

FIGURE 3.36   STM images after deposition of 0.15 ML Pt on Pt(111) at 400 K with deposition rate $5 \times 3 \ 10^{-3}$ ML s^{-1} and with different CO partial pressures during growth: (a) $p_{CO} < 5 \times 10^{-12}$ mbar ("clean"); (b) $p_{CO} = 1 \times 10^{-10}$ mbar; (c) $p_{CO} = 4.7 \times 10^{-10}$ mbar; (d) $p_{CO} = 9.5 \times 10^{-10}$ mbar; (e) $p_{CO} = 1.9 \times 10^{-9}$ mbar. The surface was exposed to CO 100 s prior to Pt deposition. Scan size 170×250 nm. (Data from ref. 165.)

are shown, Figure 3.36 shows *growth shapes* whose contour line is predominantly determined by kinetics rather than by thermodynamics. Figure 3.36(a) shows the growth shape of Pt islands on Pt(111) surface for an ambient CO pressure below 5×10^{-12} mbar, which presumably is sufficiently small to obtain a CO-free "clean" Pt(111) surface. Under these conditions, A steps on Pt(111) grow faster than B steps. As a consequence, the contour of Pt islands during growth is formed exclusively by B steps. When the ambient CO pressure is increased, small amounts of CO are adsorbed at the Pt steps [165, 166] (and references therein), which cause a drastic change in the island growth shape. Figure 3.36(b)–(e) show the growth shapes of Pt islands for various ambient CO pressures during Pt deposition. In (b) and (c) the island are no longer triangular but have compact irregular shapes and the island contour contains A-step segments. The islands in Figure 3.36(d)–(e) exhibit a reversed orientation, indicating that the predominant step type forming the island contour is now the A step rather than the B step. In addition, with increase of the CO concentration the step edge barrier increases from 0.08 to about 0.2 eV [165, 167], which leads to the nucleation of second-layer islands in (e) and (f).

The reason for the dramatic change in growth shapes is due to the fact that the binding energy of CO at A steps is higher than at B steps [166]. Hence, possible binding sites for Pt adatoms at A steps are passivated by CO molecules and Pt adatoms are preferentially attached to B steps during growth. As a consequence, growth shapes of Pt islands in the presence of CO reveal A steps rather than B steps.

3.4.2.2 O/Cu(111)

To our knowledge, the first quantitative study of the influence of adsorbates on step fluctuations was one performed by our group on oxygen adsorption on stepped Cu(111) surfaces, a study unfortunately never published.

Figure 3.37(a) shows STM images of the clean stepped Cu(111) surface at 323 K (upper left) and of the oxygen-covered surface for three different coverages. Obviously, the step fluctuation amplitude is dramatically reduced with increasing O coverage. At 20 L, the step edges are almost straight. A measure for the decrease of the step fluctuations is the coverage dependence of the time correlation function $G(t)$ [Eq. (3.29)] as plotted in Figure 3.37(b) for a data series obtained at 303 K.

The reason this study never has been published is the fact that, in addition to a reduction in the step fluctuation amplitude, a formation of double steps is observed—a phenomenon which has been mentioned for stepped surfaces in electrolyte already in Section 3.2.4.4 and is also observed for stepped Pt and Ni surfaces under the influence of oxygen and will be addressed once more in Section 3.4.3. Due to the overlap of two different effects, it has been impossible so far to separate the influence of oxygen adsorption on $G(t)$ from the contribution of double-step formation.

3.4.2.3 C_{60}/Ag(111)

Figure 3.38 shows STM images of Ag islands on Ag(111) surface (a) for clean steps [99] and where all step edges, in particular the island edges, are covered by a compact chain of C_{60} molecules [169]. If one compares the island shape in the presence of C_{60} with that of the clean Ag(111) surface, one immediately sees that the presence of C_{60}

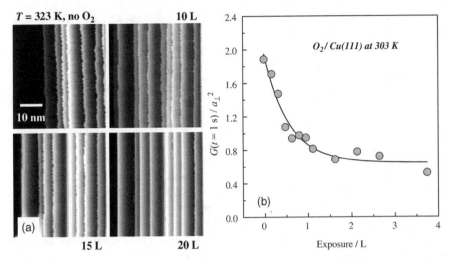

FIGURE 3.37 (a) STM images of stepped Cu(111) surfaces at 323 K for clean surface with frizzy fluctuating steps (upper left) and for different oxygen coverages. (b) Plot of time correlation function value at $t = 1\,\mathrm{s}$ [Eq. (3.29)] for data set obtained at 303 K vs. coverage [168].

dramatically alters the equilibrium island shape and hence the step free energy (see Section 3.3.1.2). Since in experiments as reported by Tao et al. [169] C_{60} molecules are exclusively adsorbed at the step edges and no further adsorption on terraces is observed, $C_{60}/Ag(111)$ provides a model system in which the observable collective fluctuations of the metal–molecule interface can be readily interpreted in terms of fundamental time constants and free energies and the results can be directly compared with the adsorbate-free corresponding system.

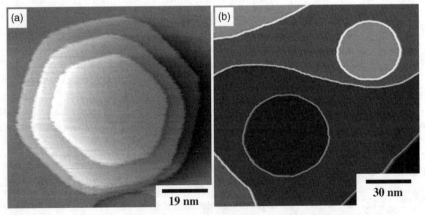

FIGURE 3.38 STM images of Ag islands on Ag(111) around 300 K with (a) clean, undecorated steps [99] and (b) steps decorated by C_{60} molecules [169].

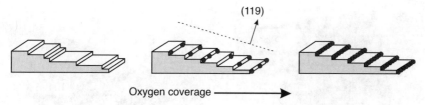

FIGURE 3.39 Schematic illustration of changes in Ni(119) surface upon oxygen adsorption [175].

Following the procedures to analyze island perimeter fluctuations as described in Section 3.3.1.2 a mean step free energy per atom $a_{\parallel}\tilde{\beta} = 0.19\,\text{eV}$ is found.[2] This value is lower than the value $a_{\parallel}\tilde{\beta} = 0.23\,\text{eV}$ for the clean Ag(111) surface; however, it is of comparable order of magnitude [99]. The lower value, however, is indicative of a smaller angular anisotropy of the step free energy [45, 99], in accordance with the observation of almost circular islands in the case of adsorbed C_{60} compared to the islands with hexagonal symmetry for the clean Ag(111) surface.

The influence of adsorbed C_{60} molecules on the thermodynamics of Ag(111) surfaces is also corroborated by studies of the step fluctuations for this system: The step fluctuation amplitude as found at room temperature and $t = 1\,\text{s}$ is about $G(t) \sim 0.075\,\text{nm}^2$ [170]. Comparing with data from the clean Ag(111) surface [Fig. 16(b) in ref. 6] one finds a fluctuation amplitude about an order of magnitude larger [$G(t) \sim 0.6\,\text{nm}^2$], which is clear evidence that C_{60} adsorbed to Ag(111) has a strong influence on surface mobility, in contrast to what is claimed by Tao et al. [170].

3.4.3 The Stability of Stepped Surfaces in Presence of Adsorbates

Adsorbates on surfaces may also have a dramatic effect on the structure and stability of surfaces. A well-studied example is the oxygen-induced step doubling on Pt(997) [3, 171–173], a vicinal surface to Pt(111) with a regular array of B steps (see Fig. 3.4). Similar observations are made for low-coverage oxygen and sulfur adsorption on stepped Cu and Ni surfaces [174, 175]. Figure 3.39 shows schematically the oxygen-induced transition on Ni(119) with increasing oxygen coverage in the range $0.03 \le \theta \le 0.15\,\text{ML}$, where the low-energy electron diffraction (LEED) pattern reveals a (9×1) superstructure [175]. Upon additional exposure to oxygen, the (9×1) pattern gradually shifts to a $p(2 \times 2)$ pattern like that observed for oxygen on flat Ni(001) [176].

Finally, we would like to mention a rather interesting example for surface restructuring under the influence of adsorbates, the faceting of stepped Cu(001) surfaces upon

---

[2] Tao et al. [169] in fact performed a Fourier analysis similar to the one described in Sections 3.3.3.1 and 3.3.3.2 which can be easily rewritten in the form of the fluctuation analysis in Section 3.3.1.2. From the mode analysis one determines the mean step edge stiffness [compare, e.g. Eq. [3.46]] rather than the mean step free energy. For circular islands, however, the step edge stiffness equals the step free energy (see, e.g., ref. [45].)

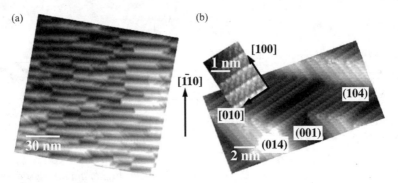

FIGURE 3.40 (a) STM images of Cu(119) after oxidation (1000 L) around 500 K. (b) Formation of {104} facets [177].

oxidation around 500 K [177] (Fig. 3.40). Whereas on the clean Cu(119) surface the steps are oriented along ⟨110⟩, the steps on the oxidized surface run preferentially along ⟨100⟩—very similar to what is found for Cu(001) in Cl⁻-containing electrolyte (Fig. 3.33) [145]. Furthermore, the surface reveals new facets identified as {104} facets. This faceting is driven by the formation of Cu–O–Cu chains [178] on the stepped surface.

3.5 CONCLUDING REMARKS

In order to successfully use nanostructured electrodes in technical applications, it is essential to understand the dynamics and stability of electrode surfaces. Throughout the last two decades, statistical methods and theoretical considerations of surface and step free energies have proven to provide an enormous understanding of the dynamics and stability of surfaces in vacuum as well as in electrochemical studies. Naturally, the studies in electrolyte in most cases include the influence of specifically adsorbed molecules on the thermodynamics and kinetics on nanostructured electrode surfaces. Detailed understanding of the experimentally determined energy parameters, though often comparable to those found in vacuum studies, is still lacking. This is mainly due to the difficulties in performing appropriate theoretical model calculations fully describing the electrolyte–electrode interface. Similarly, experimental studies on the influence of adsorbates in vacuum studies are still scarce, although theoretical work on energy parameters in the presence of adsorbates is available for vacuum situations. Hence, a direct comparison of experimental data in vacuum and electrolyte is still impossible. It is our hope for the future that experimentalists as well as theorists might overcome this gap.

REFERENCES

1. H. Lüth, *Solid Surfaces, Interfaces and Thin Films*, 4th ed., Springer, Berlin, 2001.
2. H. Ibach, *Physics of Surfaces and Interfaces*, Springer-Verlag, Berlin, 2006.

3. B. Lang, R. W. Joyner, and G. A. Somorjai, *Surf. Sci.*, 30 (1972) 454–474.

4. G. A. Somorjai, R. W. Joyner, and B. Lang, *Proc. Roy. Soc. Lond., Seri. A (Math. Phys. Sci.)*, 331 (1972) 335–346.

5. H.-C. Jeong and E. D. Williams, *Surf. Sci. Rep.*, 34 (1999) 171.

6. M. Giesen, *Prog. Surf. Sci.*, 68 (2001) 1.

7. J. Repp, F. Moresco, G. Meyer, K.-H. Rieder, P. Hyldgaard, and M. Persson, *Phys. Rev. Lett.*, 85 (2000) 2981.

8. T. Michely and J. Krug, *Islands, Mounds and Atoms: Patterns and Processes in Crystal Growth Far from Equilibrium*, Springer Series in Surface Science, Vol. 42, Springer, Heidelberg, 2004.

9. J. Frohn, M. Giesen, M. Poensgen, J. F. Wolf, and H. Ibach, *Phys. Rev. Lett.*, 67 (1991) 3543.

10. M. Giesen, *Surf. Sci.*, 370 (1997) 55–63.

11. M. Giesen and T. L. Einstein, *Surf. Sci.*, 449 (2000) 191.

12. S. Baier, H. Ibach, and M. Giesen, *Surf. Sci.*, 573 (2004) 17–23.

13. H. Ibach and W. Schmickler, *Surf. Sci.*, 573 (2004) 24.

14. M. Alshakran, G. Beltramo, M. Giesen, and H. Ibach, *Surf. Sci.*, 605 (2011) 232.

15. M. Giesen, U. Linke, and H. Ibach, *Surf. Sci.*, 389 (1997) 264–271.

16. G. Binnig, H. Rohrer, and C. Gerber, *Appl. Phys. Lett.*, 40 (1982) 178.

17. R. Sonnenfeld and P. K. Hansma, *Science*, 232 (1986) 211.

18. H. Liu, F. F. Fan, C. W. Lin, and A. J. Bard, *J. Am. Chem. Soc.*, 108 (1986) 3838.

19. L. Kuipers, R. W. M. Loos, H. Neerings, J. t. Horst, G. J. Ruwiel, A. P. d. Jongh, and J. W. M. Frenken, *Rev. Sci. Instrum.*, 66 (1995) 4557.

20. L. Zitzler, B. Gleich, O. M. Magnussen, and R. J. Behm, *Proc. Electrochem. Soc.*, 99–28 (2000) 29.

21. W. Kohn and L. J. Sham, *Phys. Rev.*, 140 (1965) A1133.

22. O.M. Magnussen and M. R. Vogt, *Phys. Rev. Lett.*, 84 (2000) 357.

23. P. Broekmann, M. Wilms, M. Kruft, C. Stuhlmann, and K. Wandelt, *J. Electroanal. Chem.*, 467 (1999) 307.

24. Y. He and E. Borguet, *J. Phys. Chem. B*, 105 (2001) 3981.

25. W. Schmickler and U. Stimming, *J. Electroanal. Chem.*, 366 (1994) 203.

26. H. Ibach, M. Giesen, and W. Schmickler, *J. Electroanal. Chem.*, 544 (2003) 13–23.

27. M. Del Popolo, *Surf. Sci.*, 597 (2005) 133–155.

28. M. G. D. Pópolo, E. P. M. Leiva, and W. Schmickler, *Ang. Chem. Int. Ed.*, 40 (2001) 4674–4676.

29. O. Magnussen, *Faraday Discuss.*, 121 (2002) 43–52.

30. T. Tansel and O. M. Magnussen, *Phys. Rev. Lett.*, 96 (2006) 026101.

31. N. Hirai, K. Watanabe, and S. Hara, *Surf. Sci.*, 493 (2001) 568.

32. N. Hirai, K. Watenabe, A. Shiraki, and S. Hara, *J. Vac. Sci. Technol. B*, 18 (2000) 7.

33. N. Hirai, H. Tanaka, and S. Hara, *Appl. Surf. Sci.*, 130–132 (1998) 506–511.

34. J. Villain, D. R. Grempel, and J. Lapujoulade, *J. Phys. F*, 15 (1985) 809–834.

35. H. v. Beijeren and I. Nolden, in *Structure and Dynamics of Surfaces*, Vol. 2, W. Schommers and P. v. Blanckenhagen (Eds.), Springer, Berlin, 1987.

36. T. Engel, in *Chemistry and Physics of Solid Surfaces* Vol. 7, R. Vanselow and R. F. Howe (Eds.), Springer Verlag, Berlin, 1988, pp. 407–428.

37. G. Wulff, *Z. Kristallgr. Mineral.*, 34 (1901) 449.

38. J.C. Heyraud and J. J. Metois, *Surf. Sci.*, 128 (1983) 334.

39. C. Bombis, A. Emundts, M. Nowicki, and H. P. Bonzel, *Surf. Sci.*, 511 (2002) 83–96.

40. W. Schilling, in *Bergmann Schaefer: Lehrbuch für Experimentalphysik Band 6 Festkörper*, R. Kassing (Ed.), de Gruyter, Berlin, 2005, pp. 219–289.
41. J. F. Wolf, B. Vicenzi, and H. Ibach, *Surf. Sci.*, 249 (1991) 233.
42. M. Dietterle, T. Will, and D. M. Kolb, *Surf. Sci.*, 327 (1995) L495.
43. M. Giesen-Seibert and H. Ibach, *Surf. Sci.*, 316 (1994) 205–222.
44. M. Giesen, C. Steimer, and H. Ibach, *Surf. Sci.*, 471 (2001) 80–100.
45. J. Ikonomov, K. Starbova, H. Ibach and M. Giesen, *Phys. Rev. B (Condens. Matter and Mater. Phys.)*, 75 (2007) 245411–245418.
46. R. Smoluchowski, *Phys. Rev.*, 60 (1941) 661.
47. K. Besocke, B. Krahl-Urban, and H. Wagner, *Surf. Sci.*, 68 (1977) 39.
48. S. Trasatti, in *Advances in Electrochemistry and Electrochemical Engeneering*, H. Gerischer and C. W. Tobias (Eds.), Wiley Interscience, New York, 1977.
49. W. Schmickler, *Interfacial Electrochemistry*, Oxford University Press, New York, 1996.
50. G. L. Beltramo, H. Ibach, and M. Giesen, *Surf. Sci.*, 601 (2007) 1876.
51. A. Hamelin and J. Lecoeur, *Surf. Sci.*, 57 (1976) 771.
52. A. Hamelin, L. Stoicoveciu, L. Doubova, and S. Trasatti, *Surf. Sci.*, 201 (1988) L498.
53. A. Hamelin, *J. Electroanal. Chem.*, 142 (1982) 299–316.
54. G. L. Beltramo, H. Ibach, U. Linke, and M. Giesen, *Electrochim. Acta*, 53 (2008) 6818–6823.
55. G. Beltramo, M. Giesen, and H. Ibach, *Electrochim. Acta*, 54 (2009) 4305.
56. J. E. Müller and H. Ibach, *Phys. Rev. B*, 74 (2006) 085408.
57. K. Wandelt, *Surf. Sci.*, 251–252 (1991) 387–395.
58. M.W. Finnis and V. Heine, *J Phys. F*, 4 (1974) L37.
59. C. Herring, in *Structure and Properties of Solid Surfaces*, R. Gomer (Ed.), The University of Chicago Press, Chicago, IL 1953, p. 5.
60. M. Wortis, in *Chemistry and Physics of Solid Surfaces*, R. Vanselow and R. Howe (Eds.), Springer, New York, 1988, p. 367.
61. J. Lapujoulade, *Surf. Sci. Rep.*, 20 (1994) 191–250.
62. S. T. Chui and J. D. Weeks, *Phys. Rev. B*, 14 (1976) 4976.
63. E. H. Conrad, *Prog. Surf. Sci.*, 39 (1992) 65.
64. V. I. Marchenko and A. Y. Parshin, *Sov. Phys. JETP*, 52 (1981) 129.
65. P. Nozières, in *Solids Far from Equilibrium*, C. Godrèche (Ed.), Cambridge University Press, Cambridge, 1991, p. 1.
66. A. C. Redfield and A. Zangwill, *Phys. Rev. B*, 46 (1992) 4289.
67. W. W. Pai, J. S. Ozcomert, N. C. Bartelt, and T. L. Einstein, *Surf. Sci.*, 307–309 (1994) 747.
68. N. C. Bartelt, T. L. Einstein, and E. D. Williams, *Surf. Sci.*, 240 (1990) L591.
69. B. Joós, T. L. Einstein, and N. C. Bartelt, *Phys. Rev. B*, 43 (1991) 8153.
70. M. L. Mehta, *Random Matrices*, 2nd ed. Academic, New York, 1991.
71. T. Guhr, A. Müller-Groeling, and H. A. Weidenmüller, *Phys. Rep.*, 299 (1998) 189.
72. M. Giesen, *Surf. Sci.*, 370 (1997) 55.
73. T.L. Einstein, *App. Phys. A: Mater. Sci. Proc.*, 87 (2007) 375–384.
74. T. L. Einstein and O. Pierre-Louis, *Surf. Sci.*, 424 (1999) L299.
75. M. Giesen and T. L. Einstein, *Surf. Sci.*, 449 (2000) 191–206.
76. B. Sutherland, *J. Math. Phys.*, 12 (1971) 246.
77. B. Sutherland, *J. Math. Phys.*, 12 (1971) 251.
78. B. Sutherland, *Phys. Rev. A*, 4 (1971) 2019.
79. H. Ibach and W. Schmickler, *Phys. Rev. Lett.*, 91 (2003) 016106.

80. O. M. Magnussen, J. Hotlos, R. J. Behm, N. Batina, and D. M. Kolb, *Surf. Sci.*, 296 (1993) 310.

81. D. M. Kolb, *Prog. Surf. Sci.*, 51 (1996) 109–173.

82. M. Moiseeva, E. Pichardo-Pedrero, G. Beltramo, H. Ibach, and M. Giesen, *Surf. Sci.*, 603 (2009) 670–675.

83. A. Bartolini, F. Ercolessi, and E. Tosatti, *Phys. Rev. Lett.*, 63 (1989) 872.

84. M. Sotto and J. C. Boulliard, *Surf. Sci.*, 214 (1989) 97–110.

85. G. M. Watson, D. Gibbs, D. M. Zehner, M. Yoon, and S. G. J. Mochrie, *Surf. Sci.*, 407 (1998) 59–72.

86. H. Ibach, G. Beltramo, and M. Giesen, *Surf. Sci.*, 605 (2011) 240.

87. K. Yamazaki, K. Takayanagi, Y. Tanishiro, and K. Yagi, *Surf. Sci.*, 199 (1988) 595–608.

88. O. K. Binnig, H. Rohrer, C. Gerber, and E. Stoll, *Surf. Sci.*, 144 (1984) 321–335.

89. C. Bombis and H. Ibach, *Surf. Sci.*, 564 (2004) 201–210.

90. K. H. Rieder, T. Engel, R. H. Swendsen, and M. Manninen, *Surf. Sci.*, 127 (1983) 223–242.

91. P. Müller and A. Saúl, *Surf. Sci. Rep.*, 54 (2004) 157–258.

92. H. Ibach, *Surf. Sci. Rep.*, 29 (1997) 195–263.

93. H. Ibach, *Surf. Sci. Rep.*, 35 (1999) 71–73.

94. T. Ihle, C. Misbah, and O. Pierre-Louis, *Phys. Rev. B*, 58 (1998) 2289.

95. E. LeGoff, L. Barbier, L. Masson, and B. Salanon, *Surf. Sci.*, 432 (1999) 139.

96. M. Giesen and D. M. Kolb, *Surf. Sci.*, 468 (2000) 149–164.

97. A. Pimpinelli, J. Villain, D. E. Wolf, J. J. Métois, J. C. Heyraud, I. Elkinani, and G. Uimin, *Surf. Sci.*, 295 (1993) 143.

98. N. C. Bartelt, J. L. Goldberg, T. L. Einstein, and E. D. Williams, *Surf. Sci.*, 273 (1992) 252.

99. C. Steimer, M. Giesen, L. Verheij, and H. Ibach, *Phys. Rev. B*, 64 (2001) 085416.

100. S. V. Khare and T. L. Einstein, *Phys. Rev. B*, 54 (1996) 11752.

101. D. C. Schlösser, L. K. Verheij, G. Rosenfeld, and G. Comsa, *Phys. Rev. Lett.*, 82 (1999) 3843.

102. S. Dieluweit and M. Giesen, *J. Electroanal. Chem.*, 524–525 (2002) 194.

103. S. Dieluweit, H. Ibach, and M. Giesen, *Faraday Discuss.*, 121 (2002) 27.

104. E. Pichardo-Pedrero and M. Giesen, *Electrochim. Acta*, 52 (2007) 5659–5668.

105. S. Dieluweit, E. Pichardo-Pedrero, and M. Giesen, in preparation.

106. S. Dieluweit, H. Ibach, M. Giesen, and T. L. Einstein, *Phys. Rev. B*, 67 (2003) R121410.

107. G. Beltramo, M. Giesen, and H. Ibach, *Electrochim. Acta*, in press.

108. J. Lecoeur, J. Andro, and R. Parsons, *Surf. Sci.*, 114 (1982) 320–330.

109. B. D. Yu and M. Scheffler, *Phys. Rev. B*, 55 (1997) 13916.

110. M. Giesen, G. Beltramo, S. Dieluweit, J. Muller, H. Ibach, and W. Schmickler, *Surf. Sci.*, 595 (2005) 127–137.

111. E. Pichardo-Pedrero, G. Beltramo, and M. Giesen, *Appl. Phys. A*, 87 (2007) 461–467.

112. N. C. Bartelt, T. L. Einstein, and E. D. Williams, *Surf. Sci.*, 312 (1994) 411.

113. T. L. Einstein and S. V. Khare, in *Dynamics of Crystal Surfaces and Interfaces*, P. M. Duxbury and T. J. Pence (Eds.), Plenum Press, New York, 1997, p. 83.

114. M. Giesen and G. Schulze Icking-Konert, *Surf. Sci.*, 412–413 (1998) 645–656.

115. J. Ikonomov, K. Starbova, and M. Giesen, *Surf. Sci.*, 601 (2007) 1403–1408.

116. M. V. Smoluchowski, *Phys. Z. Sowjetunion*, 17 (1916) 585.

117. M. V. Smoluchowski, *Z. Phys. Chem.*, 92 (1917) 192.

118. W. W. Pai, A. K. Swan, Z. Zhang, and J. F. Wendelken, *Phys. Rev. Lett.*, 79 (1997) 3210.

119. J.- M. Wen, J. W. Evans, M. C. Bartelt, J. W. Burnett, and P. A. Thiel, *Phys. Rev. Lett.*, 76 (1996) 652–655.

120. M. S. Hoogeman, M. A. J. Klik, R. v. Gastel, and J. W. M. Frenken, *J. Phys. Condens. Matter*, 11 (1999) 4349.

121. D.-J. Liu and J. W. Evans, *Phys. Rev. B*, 66 (2002) 165407.

122. O. Pierre-Louis, M. R. D'Orsogna, and T. L. Einstein, *Phys. Rev. Lett.*, 82 (1999) 3661.

123. J. Kallunki and J. Krug, *Surf. Sci.*, 523 (2003) L53–L58.

124. G. Ehrlich and F. G. Hudda, *J. Chem. Phys.*, 44 (1966) 1039.

125. R. L. Schwoebel and E. J. Shipsey, *J. Appl. Phys.*, 37 (1966) 3682.

126. H. Dürr, J. F. Wendelken, and J.-K. Zuo, *Surf. Sci.*, 328 (1995) L527.

127. J. J. d. Miguel, A. Sánchez, A. Cebollada, J. M. Gallego, J. Ferrón, and S. Ferrer, *Surf. Sci.*, 189/190 (1987) 1062.

128. S. Baier and M. Giesen, *Phys. Chem. Chem. Phys.*, 2 (2000) 3675.

129. S. Baier, S. Dieluweit, and M. Giesen, *Surf. Sci.* 502–503 (2002) 463–473.

130. L. M. Lifshitz, and V. V. Slyozov, *J. Phys. Chem. Solids*, 19 (1961) 35.

131. B. K. Chakraverty, *J. Phys. Chem. Solids*, 28 (1967) 2401.

132. P. Wynblatt and N. A. Gjostein, in *Progress in Solid State Chemistry*, J. O. McCaldin and G. Somorjai (Eds.), Pergamon Press, Oxford, 1975, p. 21.

133. J. G. McLean, B. Krishnamachari, D. R. Peale, E. Chason, J. P. Sethna, and B. H. Cooper, *Phys. Rev. B*, 55 (1997) 1811.

134. G. Rosenfeld, K. Morgenstern, M. Esser, and G. Comsa, *Appl. Phys. A*, 69 (1999) 489.

135. G. S. Icking-Konert, M. Giesen, and H. Ibach, *Surf. Sci.*, 398 (1998) 37.

136. W. Theis, N. C. Bartelt, and R. M. Tromp, *Phys. Rev. Lett.*, 75 (1995) 3328.

137. C. Klünker, J. B. Hannon, M. Giesen, H. Ibach, G. Boisvert, and L. J. Lewis, *Phys. Rev. B*, 58 (1998) R7556.

138. M. S. Zei and G. Ertl, *Surf. Sci.*, 442 (1999) 19.

139. J. M. Doña and J. González-Velasco, *Surf. Sci.* 274 (1992) 205.

140. M. Hidalgo, M. L. Marcos, and J. González-Velasco, *Appl. Phys. Lett.*, 67 (1995) 1486.

141. J. J. Martinez Jubrias, M. Hidalgo, M. L. Marcos, and J. González-Velasco, *Surf. Sci.*, 366 (1996) 239.

142. K. Kubo, N. Hirai, T. Tanaka, and S. Hara, *Surf. Sci.*, 565 (2004) L271–L276.

143. O. Magnussen, *Chem. Rev.*, 102 (2002) 679–725.

144. M. Wilms, P. Broekmann, C. Stuhlmann, and K. Wandelt, *Surf. Sci.*, 416 (1998) 121.

145. O. M. Magnussen, M. R. Vogt, J. Scherer, A. Lachenwitzer, and R. J. Behm, *Mater. Corros. Werkstoffe Korros.*, 49 (1998) 169–174.

146. M. Poensgen, J. F. Wolf, J. Frohn, M. Giesen, and H. Ibach, *Surf. Sci.*, 274 (1992) 430–440.

147. M. Giesen and S. Dieluweit, *J. Mol. Catal. A: Chem.*, 216 (2004) 263.

148. S. Baier, S. Dieluweit, and M. Giesen, *Surf. Sci.*, 502–503 (2002) 463.

149. A. Hommes, A. Spaenig, P. Broekmann, and K. Wandelt, *Surf. Sci.*, 547 (2003) 239–247.

150. P. Broekmann, M. Anastasescu, A. Spaenig, W. Lisowski, and K. Wandelt, *J. Electroanal. Chem.*, 500 (2001) 241–254.

151. M. Wilms, P. Broekmann, M. Kruft, C. Stuhlmann, and K. Wandelt, *Appl. Phys. A*, 66 (1998) 473.

152. R. McHardy, W. H. Haiss, and R. J. Nichols, *Phys. Chem. Chem. Phys.*, 2 (2000) 1439.

153. O. M. Magnussen, L. Zitzler, B. Gleich, M. R. Vogt, and R. J. Behm, *Electrochim. Acta*, 46 (2001) 3725–3733.

154. A. Spanig, P. Broekmann, and K. Wandelt, *Electrochim. Acta*, 468 (2005) 149–164.

155. M. R. Vogt, F. A. Möller, C. M. Schilz, O. M. Magnussen, and R. J. Behm, *Surf. Sci.*, 367 (1996) L33.

156. M. Giesen, M. Dietterle, D. Stapel, H. Ibach, and D. M. Kolb, *Surf. Sci.*, 384 (1997) 168–178.

157. M. Giesen, R. Randler, S. Baier, H. Ibach, and D. M. Kolb, *Electrochim. Acta*, 45 (1999) 527.

158. D. M. Kolb, R. Ullmann, and J. C. Ziegler, *Electrochim. Acta*, 43 (1998) 2751.

159. M. Giesen-Seibert, F. Schmitz, R. Jentjens, and H. Ibach, *Surf. Sci.*, 329 (1995) 47.

160. J. de la Figuera, K. Pohl, O. R. de la Fuente, A. K. Schmid, N. C. Bartelt, C. B. Carter, and R. Q. Hwang, *Phys. Rev. Lett.*, 86 (2001) 3819–3822.

161. O. M. Magnussen, B. M. Ocko, R. R. Adzic, and J. X. Wang, *Phys. Rev. B*, 51 (1995) 5510.

162. M. L. Foresti, M. Innocenti, F. Forni, and R. Guidelli, *Langmuir*, 14 (1998) 7008–7016.

163. S. Dieluweit and M. Giesen, *J. Phys. Condens. Matter*, 14 (2002) 4211.

164. P. J. Feibelman, *Phys. Rev. Lett.*, 85 (2000) 606.

165. M. Kalff, G. Comsa, and T. Michely, *Phys. Rev. Lett.*, 81 (1998) 1255.

166. P. J. Feibelman, B. Hammer, J. K. Norskov, F. Wagner, M. Scheffler, R. Stumpf, R. Watwe, and J. Dumesic, *J. Phys. Chem. B*, 105 (2001) 4018–4025.

167. K. Kyuno and G. Ehrlich, *Phys. Rev. Lett.*, 81 (1998) 5592.

168. K. Starbova, J. Ikonomov, and M. Giesen, unpublished.

169. C. Tao, T. J. Stasevich, W. G. Cullen, T. L. Einstein, and E. D. Williams, *Nano Letters*, 7 (2007) 1495–1499.

170. C. Tao, T. J. Stasevich, T. L. Einstein, and E. D. Williams, *Phys. Rev. B*, 73 (2006) 125436.

171. G. Comsa, G. Mechtersheimer, and B. Poelsema, *Surf. Sci.*, 119 (1982) 159–171.

172. E. Hahn, H. Schief, V. Marsico, A. Fricke, and K. Kern, *Phys. Rev. Lett.*, 72 (1994) 3378.

173. G. Lindauer, P. Légaré, and G. Maire, *Surf. Sci.*, 126 (1983) 301–306.

174. R. C. Cinti, T. T. A. Nguyen, Y. Capiomont, and S. Kennou, *Surf. Sci.*, 134 (1983) 755–768.

175. H. E. Dorsett, E. P. Go, J. E. Reutt-Robey, and N. C. Bartelt, *Surf. Sci.*, 342 (1995) 261–271.

176. M. Bäumer, D. Cappus, H. Kuhlenbeck, H.-J. Freund, G. Wilhelmi, A. Brodde, and H. Neddermeye, *Surf. Sci.*, 253 (1991) 116–128.

177. N. Reinecke, S. Reiter, S. Vetter, and E. Taglauer, *Appl. Phys. A* 75 (2001) 1.

178. D. J. Coulman, J. Wintterlin, R. J. Behm, and G. Ertl, *Phys. Rev. Lett.*, 64 (1990) 1761–1764.

CHAPTER 4

Electrocatalytic Properties of Stepped Surfaces

JUAN M. FELIÚ, ENRIQUE HERRERO, and VÍCTOR CLIMENT

Instituto de Electroquímica, Universidad de Alicante, E-03080 Alicante, Spain

4.1 INTRODUCTION

In the Preface of his book *Interfacial Electrochemistry*, W. Schmickler attributed the progress of electrochemistry at the end of the twentieth century to four developments, the first one being "the use of single crystal electrodes" [1]. In many cases, the electrochemical response of polycrystalline electrodes is at least a weighted average of the individual contributions from the different orientations that have different adsorption energies and surface bonding. More often, the situation is even more complex since different neighbor parts of the surface interact among them, with the electronic properties of particular sites affecting adjacent areas and intermediate species originated in some parts diffusing and reacting in other parts. In these cases, the overall surface cannot be considered simply a combination of different patches. The use of single-crystal electrodes has simplified the study of the electrochemical interface because the metal side is as uniform as possible, fitting in uniformity the solution side.

The use of single-crystal electrodes started with the study of coinage metals (Au, Ag, Cu), but it was in 1980, after the publication of the flame-annealing technique for platinum [2], when the techniques for their preparation became available to laboratories worldwide. Most of the studies were focused on the so-called basal plane electrodes, which are located at the vertices of the stereographic triangle and thus should contain wide flat terraces essentially containing a single type of site (the unit cell contains only one atom) and lead to representative electrochemical responses corresponding to limiting situations. The treatment conditions to obtain bidimensionally ordered surfaces were warmly discussed as well as the nature of the species responsible for the charge transfer [3–5]. A lot of work involving imaging techniques was

Catalysis in Electrochemistry: From Fundamentals to Strategies for Fuel Cell Development,
First Edition. Edited by Elizabeth Santos and Wolfgang Schmickler.

necessary to understand the conditions in which the experimental electrode surfaces in contact with the electrolyte agreed with that expected from the Miller indices [6]. Nowadays, it is known how to achieve this objective in most cases, although questions still remain around the Pt(110) pole.

The importance of studies of well-defined basal plane platinum electrodes, especially on Pt(111), deals with the fact that they describe the interaction between the solution species with a single type of site. This leads to the knowledge of data of fundamental importance that can be used later as a reference in modeling. In spite of the enormous progress in the power of computers, there is no way to predict univocally the response of electrochemical systems and comparison with the experiment is still required [7]. In addition, satisfactory modeling of the electrode activity is expected to pose new questions on reactivity that would be elucidated by further refined, specific experiments that should be performed with single-crystal electrode surfaces in a cross feedback between theory and experiment.

Applied research has also benefited from single-crystal work in defining which is the most active surface arrangement for a particular reaction. However, basal plane electrodes have the conceptual problem that there are no practical electrodes with bidimensional domains larger than a modest number of atoms. To approach the behavior of practical, nanoparticle electrodes, the use of well-defined stepped surfaces has been, and still is, the experimental strategy. These studies belong to what we could classify within the name of fundamentals of electrocatalysis and have at least two main points.

First, it is believed that defect sites are generally the most active for a particular reaction. The way to measure the difference in reactivity of different sites under quantitative basis is to prepare series of surfaces having terraces of a particular symmetry separated by steps that also contain a single type of site. As this arrangement can be built at the atomic level, it is possible to study reactivity as a function of the density of defects, which becomes a parameter that can be controlled.

Second, it is observed in most cases that for a wide number of stepped surfaces the resulting reactivity of the electrode shows simple trends that can be understood as a linear combination of the terrace and step contributions. This line can be extrapolated either toward the most defective surfaces that have a high concentration of steps close to each other or in the other direction to zero defects. In the first case, the comparison with the experimental response of a heavily stepped surface would evidence the role of an additional term that arises from the interaction of the step dipoles and becomes nonnegligible at short distances. In the second case, the extrapolation to zero defects will give a measure of the reactivity of ideal bidimensional surfaces that can be compared to that measured with the basal plane electrode and highlight long-range effects. All experimental surfaces have defects and the studied reaction may take place on these defects or, conversely, competitive adsorption may inhibit the process. This happens, for example, for the oxidation of adsorbed CO [8] or for poison formation from formic acid on Pt(111) electrodes [9], two important and widely studied reactions. The extrapolated activity at zero defects will supply the relevant parameters that could finally be modeled in theoretical studies of reactivity, which do not consider the presence of defects in the calculation.

This chapter is devoted to the role of steps in electrocatalytic activity. In the following pages the reader will find a description of aspects related to the electrochemistry of stepped platinum surfaces. First, after the revision of some nomenclature issues, it will be shown how the voltammetric behavior in blank electrolyte solutions can be used to characterize the electrode surface with exquisite sensitivity. Next, a description will be given of the particular electronic properties induced by the presence of steps. The chapter ends with a revision of the reactivity of stepped platinum surfaces toward some selected reactions of particular interest in electrocatalysis, with emphasis on the mechanistic knowledge that can be gained from the use of these particular electrodes.

4.2 PHENOMENOLOGICAL DESCRIPTION OF STEPPED SURFACES

4.2.1 Nomenclature: Crystallographic Zones

Surfaces on the edges of the stereographic triangle of face-centered-cubic (fcc) metals have ideal structures composed of flat terraces separated by monoatomic steps. For these surfaces, the centers of the atoms in the terrace are located in a common plane, while the steps define a straight line. These ideal surfaces are very conveniently described by the Lang, Joyner, and Somorjai (LJS) [10] nomenclature:

$$M(hkl) = M(S)[n(h'k'l') \times (h''k''l'')] \tag{4.1}$$

where M is the symbol for the metal; S stands for stepped surface; (hkl), $(h'k'l')$ and $(h''k''l'')$ are the Miller indices of the stepped surface, the terrace, and the step orientations respectively; and n indicates the number of atomic rows in the terrace. Figure 4.1 summarizes the relation between Miller indices and LJS nomenclature. Other significant properties of the surface that can be calculated by this model are the step density, or number of steps per unit length, and the area of the unit cell. These properties can be easily calculated from the dimensions indicated in Table 4.1:

$$S = L \times W \tag{4.2}$$

$$N = L^{-1} \tag{4.3}$$

where L and W represent the dimensions of the unit cell (L is projected along the plane of the terrace), S is its area, and N is the step density. Defined in this way, the area of the unit cell has been projected on the plane of the terrace. For comparison with experimental values, this area should be divided by the cosine of the angle between the surface and the terrace to refer it to the plane of the stepped surface. Alternatively, the experimental charges can be projected on the plane of the terrace by dividing them by the cosine of the angle between both planes. This can be easily calculated from

$$\cos \beta = \frac{L}{\sqrt{L^2 + H^2}} \tag{4.4}$$

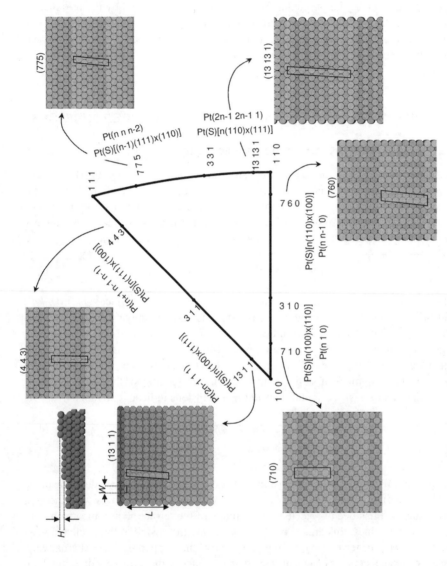

FIGURE 4.1 Representation of stereographic triangle for Pt with notation of stepped surfaces and their hard-sphere model.

TABLE 4.1 Relation between Miller Indices and LJS Notation of Stepped Surfaces and Dimensions of Unit Cell

Zone	LJS Notation	Miller Indices	L	W	H	S	$\cos\beta$
$[1\bar{1}0]$	Pt(S)[$(n-1)(111)\times(110)$]	Pt$(n,n,n-2)$	$\dfrac{\sqrt{3}d}{2}\left(n-\dfrac{2}{3}\right)$	d	$\sqrt{\dfrac{2}{3}}d$	$\dfrac{\sqrt{3}}{2}d^2\left(n-\dfrac{2}{3}\right)$	$\dfrac{3n-2}{\sqrt{9n^2-12n+12}}$
	Pt(S)[$n(110)\times(111)$]	Pt$(2n-1,2n-1,1)$	$\sqrt{2}d\left(n-\dfrac{1}{2}\right)$	d	$\dfrac{d}{2}$	$\sqrt{2}d^2\left(n-\dfrac{1}{2}\right)$	$\dfrac{4n-2}{\sqrt{16n^2-16n+6}}$
$[01\bar{1}]$	Pt(S)[$n(111)\times(100)$]	Pt$(n+1,n-1,n-1)$	$\dfrac{\sqrt{3}d}{2}\left(n-\dfrac{1}{3}\right)$	d	$\sqrt{\dfrac{2}{3}}d$	$\dfrac{\sqrt{3}}{2}d^2\left(n-\dfrac{1}{3}\right)$	$\dfrac{3n-1}{\sqrt{9n^2-6n+9}}$
	Pt(S)[$n(100)\times(111)$]	Pt$(2n-1,1,1)$	$d\left(n-\dfrac{1}{2}\right)$	d	$\dfrac{\sqrt{2}d}{2}$	$d^2\left(n-\dfrac{1}{2}\right)$	$\dfrac{2n-1}{\sqrt{4n^2-4n+3}}$
$[001]$	Pt(S)[$n(100)\times(110)$]	Pt$(n,1,0)$	$\dfrac{\sqrt{2}d}{2}n$	$\sqrt{2}d$	$\dfrac{\sqrt{2}d}{2}$	d^2n	$\dfrac{n}{\sqrt{n^2+1}}$
	Pt(S)[$n(110)\times(100)$]	Pt$(n,n-1,0)$	$d\left(n-\dfrac{1}{2}\right)$	$\sqrt{2}d$	$\dfrac{d}{2}$	$\sqrt{2}d^2\left(n-\dfrac{1}{2}\right)$	$\dfrac{2n-1}{\sqrt{4n^2-4n+2}}$

Note: The meanings of L, W, and H are indicated in Figure 4.1: H is the height of the step or the distance between two terrace atomic layers; d stands for the atomic diameter; S is the area of the unit cell projected on the plane of the terrace, $S = LW$; β is the angle between the plane of the surface and the terrace.

where H is the height of the step or the distance between two terrace layers. Using the Miller notations of the surface (hkl) and the terrace $(h'k'l')$, the cosine of the angle can also be calculated from the dot product of the vectors:

$$\cos \beta = \frac{h \cdot h' + k \cdot k' + l \cdot l'}{\sqrt{h^2 + k^2 + l^2}\sqrt{h'^2 + k'^2 + l'^2}} \tag{4.5}$$

Finally, the area of the unit cell can also be calculated as [11]

$$S = 2d^2\sqrt{h^2 + k^2 + l^2}\,p \quad \text{where } p = \begin{cases} \frac{1}{2} & \text{for } h, k, l \text{ not all odd} \\ \frac{1}{4} & \text{for } h, k, l \text{ all odd} \end{cases} \tag{4.6}$$

The real structure can be more complex and some faceting can occur, leading to the formation of higher steps that are compensated by wider terraces, in order to keep constant the overall crystallographic angle or, alternatively, terraces of different sizes are separated by monoatomic steps so that the average terrace width matches the nominal value. Also, a straight step can be transformed into a zigzag line if the new structure decreases the overall surface energy. The structure of real stepped surfaces have been monitored by low-energy electron diffraction (LEED) for surfaces prepared in ultrahigh vacuum (UHV) [12], and it was found that surfaces vicinal to (111) and (100) in the [1$\bar{1}$0] and [01$\bar{1}$] zones are very stable although they may suffer faceting after oxygen adsorption. Surfaces in the [001] zone are less stable and form multiple height steps. Also, Pt(S)$[n(110) \times (111)]$ surfaces show some faceting. The effect of flame annealing and different cooling conditions was studied by ex situ scanning tunneling microscopy (STM) [13, 14]. For (111) vicinal surfaces it was found that cooling after the flame annealing in a hydrogen–argon mixture provides surface structures close to that predicted by the ideal hard-sphere model. If the electrode is allowed to cool down in an oxygen-containing atmosphere, some faceting takes place, resulting in nonuniform terrace width and the presence of kinks. The stability of surfaces vicinal to the (100) basal plane was also studied by STM. For $[n(100) \times (111)]$ electrodes, good correspondence was found between the ideal and experimental structure when cooling after flame annealing is done in an H_2–Ar atmosphere. However, $[n(100) \times (110)]$ surfaces exhibit a corrugated structure with variable terrace width and irregular edges. The Pt(311) and Pt(310) electrodes have also been studied by in situ diffraction techniques [15, 16]. It has been found that the Pt(311) electrode has a missing row structure whereas the results for the Pt(310) are compatible with a pseudo-(1×1) structure. In general, faceting and disordering are more important when the electrode is cooled down in air or in a pure Ar atmosphere [17]. Other cooling conditions have also been tested for the preparation of single-crystal electrodes and it was found that cooling in a CO–N_2 atmosphere could also be used to obtain unreconstructed (1×1) surface structures [17].

4.2.2   Voltammetric Characterization

After introduction of the flame-annealing technique for preparation and decontamination of platinum single-crystal electrodes, it became clear that cyclic voltammetry is very sensitive to the crystallographic structure of the surface. Cyclic voltammograms have been considered for this reason as fingerprints of the electrode surface, allowing a fast and inexpensive characterization of the quality of single-crystal electrodes. The characterization can be made quantitative if the charge corresponding to adsorption processes on different surface sites can be integrated separately. Figure 4.2 shows voltammograms recorded in sulfuric acid solutions for different stepped surfaces in the $[1\bar{1}0]$ zone with (111) terraces and (110) steps. For Pt(111), two regions can be differentiated in the voltammogram. The low-potential region, between 0.06 and 0.30 V, is due to hydrogen adsorption, while the features above 0.30 V correspond to anion adsorption. In this particular electrolyte, sulfate is the anion that adsorbs, but equivalent behaviors are observed in electrolytes containing different anions, such as phosphate, acetate, oxalate, and chloride. The introduction of steps is reflected in the appearance of a peak at 0.12 V and a decrease of the current between 0.12 and 0.30 V and also of that due to anion adsorption above 0.35 V. The charge associated with the different regions of the voltammogram can be integrated separately by considering reasonable baselines. The peak associated with the steps can be separated from the current due to hydrogen adsorption on (111) terrace sites by considering a straight baseline, as indicated in Figure 4.2 for Pt(775). Charges integrated in this way can be compared with theoretical values predicted by the hard-sphere model.

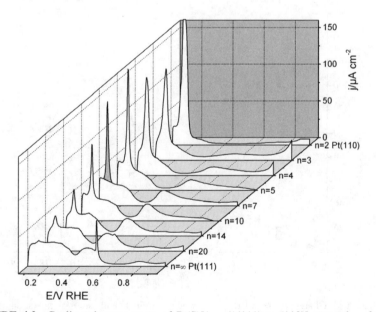

FIGURE 4.2   Cyclic voltammograms of Pt(S)$[(n-1)(111) \times (110)]$ stepped surfaces in 0.1 M H_2SO_4. Sweep rate: 50 mV s^{-1}.

The charges expected from the hard-sphere model of the surface can be obtained by dividing the number of sites of each geometry by the area of the unit cell. From the model of the (775) surface in Figure 4.1 it can be seen that, for this orientation, the unit cell contains $n - 2$ sites of (111) geometry and one site of (110) geometry. Then, the charge corresponding to one electron per (111) terrace site is [18]

$$q_{terr} = \frac{(n - 2)e}{S} = q_{111} - \frac{4 \, eN}{3 \, d} \tag{4.7}$$

where e is the elementary charge, d is the atomic diameter, and q_{111} is the charge corresponding to one electron exchange per surface atom in the perfect (111) surface:

$$q_{111} = \frac{2e}{\sqrt{3}d^2} \tag{4.8}$$

The step charge is given by:

$$q_{st} = \frac{e}{S} = \frac{Ne}{d} \tag{4.9}$$

According to these expressions, the validity of the hard-sphere model can be checked by plotting the charges integrated under different regions (projected on the plane of the terrace) as a function of the step density. This is shown in Figure 4.3. An excellent correspondence between theoretical and experimental charges is obtained for step sites, indicating that one electron per platinum step site is transferred during

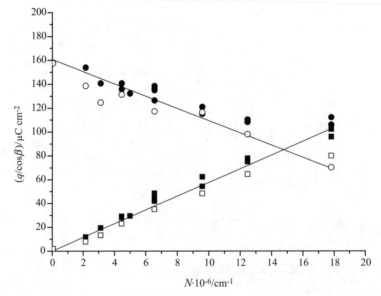

FIGURE 4.3 Charges integrated for terrace (circles) and step sites (squares) for Pt(S)[$(n - 1)(111) \times (110)$] stepped surfaces in 0.1 M HClO$_4$ (close symbols) and 0.5 M H$_2$SO$_4$ (open symbols). The lines represent the behavior predicted according to the hard-sphere model.

this adsorption process. Moreover, comparison of the voltammograms recorded in sulfuric and perchloric acid solutions revealed that sulfate does not adsorb in this potential range [19]. This allows assigning the peak at 0.12 V, almost exclusively, to hydrogen adsorption on step sites. For hydrogen adsorption on terrace sites, the slope is smaller than that predicted by Eq. (4.7). This fact reflects the smaller coverage of hydrogen on the flat (111) surface, which stands for two-thirds in the potential window in which the coverage can be easily measured, that is, at potentials more positive than the hydrogen evolution [20]. One important observation is the significant decrease of the sharp spike at 0.51 V corresponding to the phase transition from a disordered sulfate adlayer at lower potential to an ordered ($\sqrt{3} \times \sqrt{7}$) adlayer at higher potentials [21], due to the presence of steps. This indicates that the formation of the ordered adlayer is very sensitive to the long-range order of the surface. The peak completely disappears for terraces shorter than 10 atoms, giving an idea of the minimum distance necessary for the formation of the ordered domains. Also, associated with the presence of steps, a displacement of the anion adsorption process toward higher potentials is observed. This can be interpreted as a consequence of the diminution of collaborative interactions within the adlayer due to the introduction of steps.

As a second example, Figure 4.4 shows the voltammograms of different platinum stepped surfaces in the [01$\bar{1}$] zone. The lower part of the figure corresponds to (111)

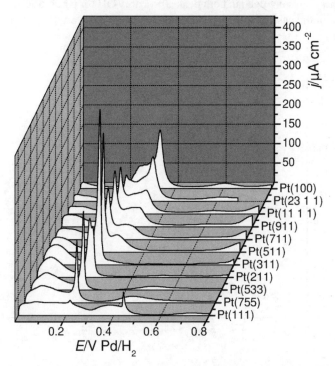

FIGURE 4.4 Cyclic voltammograms of platinum stepped surfaces in [01$\bar{1}$] zone in 0.1 M H$_2$SO$_4$. Sweep rate: 50 mV s. [Adapted from G.A. Attard, O. Hazzazi, P.B. Wells, V. Climent, E. Herrero, and J.M. Feliu *J. Electroanal. Chem.*, 568 (2004) 329.]

vicinal surfaces and the higher part to (100) vicinal surfaces. For the $[n(111) \times (100)]$ electrodes, changes in the voltammogram similar to those described before for the surfaces with (110) steps are observed after the periodic introduction of steps. The additional peak grows now at 0.22 V, associated with the presence of steps, while the hydrogen and anion adsorption on terraces decreases, as in the previous case. For $n < 4$ the position of the peak changes slightly, shifting in the positive direction. According to the hard-sphere model for these surfaces, the charge associated to one electron per step site is still given by Eq. (4.9), while the charge for the terrace is given now by [22]

$$q_{terr} = \frac{(n-1)e}{S} = q_{111} - \frac{2}{3}\frac{eN}{d} \tag{4.10}$$

The charges are plotted as a function of step density in Figure 4.5. Good correspondence is observed between theoretical and experimental values for the step. However, a significant difference is observed between the voltammograms recorded in sulfuric and perchloric acid solutions, suggesting that anion adsorption on step sites takes place during the process associated with the peak. Since the charge measured in perchloric and sulfuric acid solutions is the same (within the experimental error), anion adsorption is involved in both peaks (probably OH for perchloric acid solutions and sulfate in the case of sulfuric acid) and the charge transferred upon adsorption is equivalent.

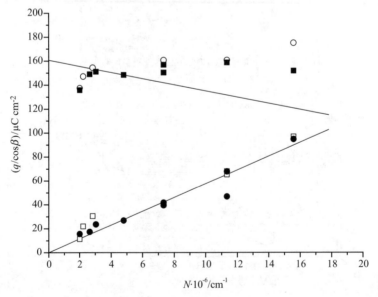

FIGURE 4.5 Charges integrated for step (circles) and terrace (squares) contributions for Pt(S)[$n(111) \times (100)$] stepped surfaces in 0.1 M HClO$_4$ (closed symbols) and 0.5 M H$_2$SO$_4$ (open symbols). The lines represent the behavior predicted according to the hard-sphere model.

The surface Pt(311) is the turning point of the zone and can be considered as either Pt(S)[2(111) × (100)] or Pt(S)[2(100) × (111)], that is, the unit cell contains the same amount of (111) and (100) sites. If we keep increasing the angle toward the (100) basal plane, the (111) terrace becomes the step and the (100) step becomes the terrace. From the evolution of the voltammetric profile as the angle is moved through the zone, the current below 0.20 V for Pt(S)[n(100) × (111)] surfaces can be assigned to the response from the steps and the contribution above 0.25 V is clearly assigned to adsorption on the (100) terraces. For the assignment of the new peak contribution, which in sulfuric acid appears at 0.27 V, a detailed charge analysis should be made. For these surfaces, the hard-sphere model indicates that the terrace charge should be

$$q_{terr} = \frac{(n-1)e}{S} = q_{100} - \frac{1}{2}\frac{eN}{d} \tag{4.11}$$

and the charge of the step should be given by [Eq. (4.9)]. The charge measured between 0.3 and 0.45 V, that is, the adsorption states clearly associated with the terrace, is in agreement not with Eq. (4.10), but with the expression [23]

$$q_{terr} = \frac{(n-2)e}{S} = q_{100} - \frac{3}{2}\frac{eN}{d} \tag{4.12}$$

which has been derived assuming that one of the rows of the terrace (the one close to the step) is not available to hydrogen adsorption. However, the charge measured upon CO adsorption on these surfaces at 0.2 V (vide infra), which gives information on the amount of hydrogen adsorbed at that potential, is in agreement with a hydrogen coverage on the terrace with a value of 1 [24]. Therefore, the adsorption of hydrogen on the row close to the step is involved in the peak at 0.27 V, which explains why its charge is proportional to the step density [23]. Finally, the peak also depends on the anion present on the surfaces, and therefore, the adsorption of anions should be involved. From these data, it can be proposed that this peak is also associated to the adsorption of the anion on the step, blocking not only the step but one additional row on the terrace. The desorption of the anion at 0.27 V in sulfuric acid leads to the adsorption of hydrogen only on the row of atoms of the terrace blocked by the anion, since the adsorption of hydrogen on the steps takes place below 0.2 V. Therefore, the charge of this peak contains the contribution of the anion adsorption on the step and hydrogen adsorption on the terrace. This situation is similar to that observed when adatoms are deposited on these surfaces, where the adatom adsorbed on the step blocks additional sites on the terrace [25].

When the contribution of the step is analyzed, it has been found that it has a smaller slope than that predicted by Eq. (4.9) with an intercept higher than zero. To explain this contribution, it has been proposed that even the best (100) electrode contains a nonnegligible density of steps originating during the lifting of the reconstruction that takes place during flame annealing. Since the final surface structure of the electrode is dependent on the thermal treatment and cooling conditions [17], the measured charge should also be dependent on that.

For the other stepped surfaces, the available information is much smaller, since good preparation conditions are not well established. It is worth noting that these surfaces are open and reconstruction and faceting are more facile. This is especially the case for the surfaces with (110) terraces. For these surfaces, cooling in CO–N$_2$ atmosphere could lead to unreconstructed surface structures, as has been found for the Pt(110) electrode [17]. The situation can be similar for the Pt(S)[n(110) × (111)] surface, where the atomic density of the step is very low and reconstruction of the step to give a zigzag line is thermodynamically favorable [14].

4.3   ELECTRONIC PROPERTIES OF STEP SITES

4.3.1   Potential of Zero Charge, Work Function, and Smoluchowski Effect

Step sites are expected to have significantly different electronic properties from terrace sites. Those properties arise from the lower coordination number of step atoms that make them especially reactive. The effect of the steps on the electronic density distribution can be understood from a simplified jellium model [26], in which the discrete ion cores are replaced by a uniform positive background charge and the electronic density is calculated by the density functional method. One important result obtained with this approach is the justification of the appearance of a surface dipole potential oriented with the negative end toward the vacuum as a result of the spillover of the electronic density beyond the limit of the positive background. When a step is introduced into the surface, this picture is affected so that the smooth electronic density is not able to follow the abrupt change in the positive potential defined by the ion cores in the step site and some negative charge is smeared on the bottom part of the step while some excess positive charge remains on the top part of the step (Fig. 4.6). This situation creates a surface dipole opposed to the negative dipole created by the spillover of electrons.

The experimental verification of the electronic effect described above comes from work function measurements. The work function of a crystal surface is the minimum energy required to remove an electron from the bulk of the metal to a point at a macroscopic distance outside the surface. The negative surface dipole created by the spillover of electrons increases the work function while the dipole due to

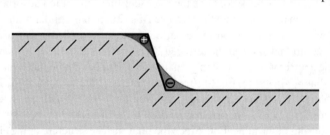

FIGURE 4.6   Schematic representation of step and electronic density with appearance of a surface dipole.

the introduction of steps decreases it. Experimentally, a linear decrease of the work function with the step density has been observed [27]. The slope of this plot allows calculating the magnitude of the dipole associated with the steps according to

$$\frac{\Delta \Phi}{e} = 3 \times 10^{-16} 4\pi N \mu \qquad (4.13)$$

where N is the step density (in cm^{-1}) and μ is the dipole moment associated with the steps (in D/cm).

In an electrochemical environment, the property equivalent to the work function is the potential of zero charge (pzc) [28, 29]. A linear decrease has also been found for the pzc of gold and silver stepped electrodes as a function of the step density and the equivalent dipole moments has been calculated [30–32]. For these metals, the pzc is easily measured by locating the minimum centred at the pzc in the double-layer capacity that, according to the Gouy–Chapman-Stern theory, should be observed in diluted solutions.

For platinum electrodes the situation is more complex due to the presence of adsorption processes, namely, hydrogen and anion adsorption. In general, it is not possible to separate by thermodynamic analysis the double-layer charge from the charge associated with adsorption processes. Then, the thermodynamically measurable quantity is the total charge, including both contributions, while the true charge in the double layer, called the free charge in the following, can only be accessed through consideration of a particular model for the location of charges at the interphase [33]. Accordingly, two values of the pzc should be considered, the potential of zero total charge (pztc) and the potential of zero free charge (pzfc). Experimentally, the problem is that the minimum in the diffuse capacity, which is deeper as the electrolyte concentration diminishes, is not normally observed and alternative methods should be used to determine both values of the pzc.

4.3.2 CO Displacement

The first method for determination of the pztc of platinum single-crystal electrodes was based on the CO charge displacement experiment. In this experiment, CO is potentiostatically adsorbed on the electrode surface while the current flow induced by the CO adsorption (displaced charges) is monitored [34–40]. Two reasonable premises are used for the interpretation of the value of CO-displaced charges. First, it is considered that CO is a neutral probe, that is, CO adsorption does not involve charge transfer by itself. Secondly, the charge remaining on the CO-covered interphase is assumed to be negligibly small. Under these premises the displaced charge can be equated to the opposite of the charge present at the interphase at the potential of the experiment:

$$q_{dis} = q_f - q_i \simeq -q_i \qquad (4.14)$$

The first equality comes from the first hypothesis stated above, where q_i and q_f are the total charges before and after the CO adsorption, respectively. The second equality comes from the second hypothesis, that is, from assuming that the remaining charge after adsorption, q_f, is negligible.

Then, the pztc can be equated with the potential where the displaced charge is zero. From a practical point of view, it is easier to obtain the charges from the integration of the voltammogram using a single CO-displaced charge value as the integration constant. In this way, a complete q–E curve can be calculated from which the pztc can be easily extracted. The validity of the previous premises has been tested experimentally. First, the results obtained with adlayers of known stoichiometry proved the validity of the method [34]. The second point that needs to be discussed is the magnitude of the remaining charge on the CO-covered surface. This is assumed to be very small on the basis of the small value of the capacity of the CO-saturated surface. The value of the remaining charge could be calculated from the integration of the voltammogram if the value of the pzc of the CO-covered surface was known. This latter property has only been estimated for the CO-covered Pt(111) surface from the work function of the CO + H_2O-covered surface [41, 42]. In this case, a value of the pzc around 1 V has been estimated. This value, combined with the small capacitance, gives charge density values around 13 μC cm^{-2} at the pztc that can be taken into account to correct the value of the pztc obtained with the CO charge displacement method. This charge is relatively small when compared with the large charges resulting from the high values of pseudocapacitance of platinum electrodes and the correction of the pztc is smaller than 50 mV.

This method has been used for the calculation of the pztc of platinum stepped surfaces vicinal to the (111) and (100) basal planes [19, 24, 39]. Figure 4.7 shows the variation of the pztc as a function of the step density for (111) vicinal surfaces with (111) and (100) steps. In both cases, there is a nearly linear decrease of the pztc with the step density at small step densities followed by a plateau and finally an increase at high step densities. This behavior is qualitatively very similar to the one exhibited by the work function of the same surfaces [43], although the numerical values of the slope and the plateau are very different. This is not surprising since the magnitude that should be compared with the work function is the pzfc and not the pztc. As stated before, values of the pzfc are not easily accessible from experimental results, although they can be estimated under some assumptions. This is done by considering that there is a potential region where the total charge is free from faradaic contributions, that is, it coincides with the free charge. This is assumed to happen in the double-layer region between hydrogen and anion adsorption regions, which are separated in perchloric acid solutions. The second consideration is to assume a constant value of double-layer capacity extrapolated into the hydrogen region. For this calculation, the CO remaining charge was considered negligible. Although these approximations render the resulting values as just a rough estimation, the result is interesting since the slope for the decrease of the pzfc with the step density is comparable with that of the work function decrease. It can be concluded from this result that a similar dipole value is associated with the steps in UHV and electrochemical environments. This would mean that the effect of solvent interaction with the step sites parallels that with the terraces in such a way that the net displacement is not affected by the presence of the water.

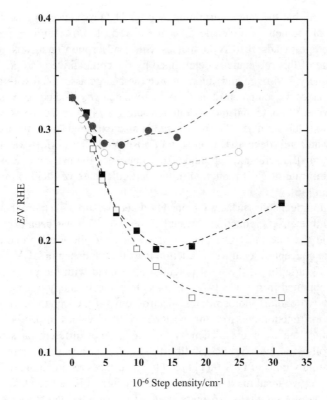

FIGURE 4.7 Plot of pztc of platinum stepped surfaces vicinal to (111) basal plane as function of step density. Circles: (100) step; squares: (111) step; open symbols: 0.5 M H_2SO_4; closed symbols: 0.1 M $HClO_4$.

When comparing the values of the changes in the pztc of the surfaces vicinal to the (111) and the (100) planes, the diminution for Pt(n, n, $n-2$) is much higher than that measured for Pt($n+1$, $n-1$, $n-1$) and Pt($2n+1$, 1, 1) (140, 50, and 40 mV, respectively, for a step density for 10^5 cm^{-1}) [24, 40]. This means that the effects of a similar step density on the properties of the surface are different, and the expected changes in the overall behavior with respect to the basal plane should be higher for the Pt(n, n, $n-2$) surfaces. As aforementioned, the changes in the work function for these surfaces follow a similar trend, but the differences for (n, n, $n-2$) and ($n+1$, $n-1$, $n-1$) surfaces are similar and higher than those measured for the ($2n+1$,1,1) surfaces [27, 44]. This means that the measured differences for the pztc not only have a contribution from the electronic properties of the step but also contain contributions from the adsorption properties of the step.

4.3.3 N_2O Reduction

Another method to estimate the pztc of platinum single-crystal electrodes is based on the use of N_2O as a probe sensitive to the charge present on the interphase. It has been proposed that the rate-determining step for N_2O reduction is preceded by the adsorption of the neutral molecule. The rate of reduction is therefore proportional to the coverage of adsorbed N_2O species. Since adsorption energy is very small, the coverage of this molecule is determined by the competitive adsorption of other species, namely, hydrogen and anions. Since the charge associated with adsorption processes makes the main contribution to the total charge, good correspondence was found between the pztc and the potential where the N_2O reduction is faster [45]. Besides, some sensitivity to the free charge is also expected, since the existence of polarized water will decrease the energy of adsorption. This effect was invoked to explain the small discrepancy (around 90 mV) between the pztc and the potential of the maximum rate of N_2O reduction in the particular case of Pt(111) in perchloric acid solution [46].

For nonhomogeneous surfaces (stepped or defective surfaces) more than one peak is observed in the N_2O reduction current [23, 45, 46]. This has been explained considering that the rate of N_2O reduction is sensitive to the local properties of the surface. The concept of local work function is well established in UHV. Experimental proofs of variation of the local potential associated with the presence of steps have been obtained from STM [47], photoelectron spectroscopy of adsorbed xenon (PAX), and two-photon photoemission spectroscopy [48]. Given the correspondence between work function and pzc, the existence of local values of pzc is expected in electrochemical environments. Then, N_2O is reduced at different rates on terraces and steps following the existence of different local charge values. Figure 4.8 shows cyclic voltammograms of $Pt(S)[n(111) \times (111)]$ surfaces recorded in N_2O-saturated solutions. The two local maxima in the reduction current have been taken as the indication of two local values of pztc, the pztc of the step located at lower potentials than the pztc of the terrace. To calculate an overall pztc of the surface, the local pztc values should be used together with the corresponding pseudocapacities associated with terrace and step sites. In this way, the overall charge can be calculated and the pztc corresponds to a situation where the negative charge on the terraces is compensated by positive charge on the steps, giving an overall net charge equal to zero. In this way, the overall surface can be considered a linear combination of terrace and step sites. With this picture, the decrease of the pztc when the number of steps is increased is easily understood, the pztc shifting from the value of the terrace toward the local value of the step. This also explains the smaller slope of the decrease for $Pt(S)[n(111) \times (100)]$ surfaces, since the pztc of the (100) steps is higher than the pztc for (110) steps. For high step densities, the model fails since steps are very close to each other and start to interact. In this case, the surface can no longer be considered a combination of terrace and step but should be considered as a new surface with its particular electronic properties.

4.3.4   Potential of Maximum Entropy

Another technique that has been revealed to be very useful to obtain information about the electronic properties of platinum single-crystal surfaces is the laser-induced

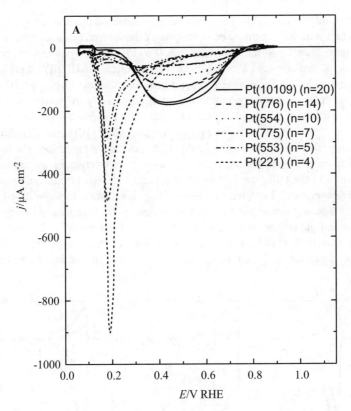

FIGURE 4.8 Cyclic voltammograms corresponding to Pt(S)[$(n-1)(111) \times (110)$] stepped surfaces in N_2O-saturated solutions. Sweep rate: $10\,mV\,s^{-1}$. [From V. Climent, G.A. Attard, and J.M. Feliu *J. Electroanal. Chem.*, 532 (2002) 67.]

temperature jump method [49–53]. With this technique, a sudden temperature jump is caused with a pulsed laser irradiation while the potential transient associated with the increase in temperature under coulostatic (open-circuit) conditions is monitored. This potential transient gives a measure of the thermal coefficient of the double-layer potential drop $\left(\partial\Delta_M^S\Phi/\partial T\right)_q$. This thermal coefficient is mainly determined by the effect of the temperature on the dipolar contribution due to the orientation of the solvent molecules in the interphase. Then, the sign of the potential transient when the interphase is heated is an indication of the orientation of water. For water oriented with the oxygen toward the metal, a positive transient is expected when the temperature is raised, since the thermal disorder implies the decrease of a negative dipolar contribution. Accordingly, and independently of the magnitude of the transients, positive transients are measured at high enough potentials. At low enough potentials, water will be oriented with the hydrogen toward the metal and negative transients are measured. The potential where the transient is zero can be identified as the potential where there is no net orientation of the water at the double layer.

From thermodynamic arguments it can be demonstrated that this corresponds with the potential where the entropy of formation of the interphase is maximum or there is maximum disorder in the water network. For Pt(111) the potential of maximum entropy is located at ~0.45 V reversible hydrogen electrode (RHE) (at pH3), that is, within the double-layer region. For stepped surfaces the situation has been revealed to be more complex, since two potentials of maximum entropy are measured, coinciding with the existence of two local values of the pztc [50, 53]. Figure 4.9 shows the laser-induced potential transients for stepped surfaces in the [110] zone. The experiment was performed in a 0.1 M KClO$_4$ + 1 mM HClO$_4$ electrolyte solution. The use of perchlorate anion in the supporting electrolyte avoids the interference of anion-specific adsorption. On the other hand. the use of a higher pH has two advantages for this particular experiment. First, it decreases the magnitude of the thermodiffusion potential that arises as a consequence of the temperature gradient that exits between the reference and the heated working electrode. More importantly, with a lower proton concentration, the rate of hydrogen adsorption is lowered, allowing the decoupling of the double-layer response and the adsorption processes. In this way, in the time scale

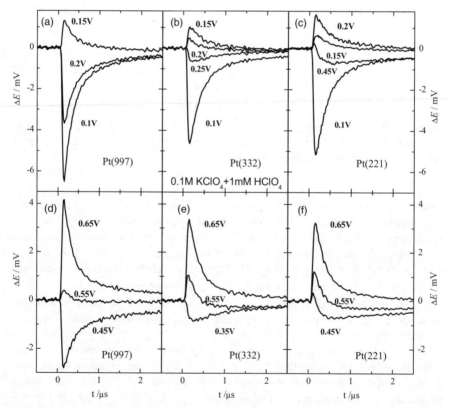

FIGURE 4.9 Laser-induced potential transients for three selected stepped surfaces in 0.1 M HClO$_4$ + 1 mM HClO$_4$. [From N. Garcia Araez, V. Climent, and J. Feliu *Electrochim. Acta*, 54(2009) 966]

of the experiment, only the double-layer response gives a significant contribution to the laser-induced potential transient. In all cases, at low enough potentials, transients have negative sign, indicating that water orientates with the hydrogen toward the metal. Around the potential corresponding to the local pztc of the step ($\sim$0.15 V at pH3), transients become positive, suggesting a change in the orientation of water. At higher potentials, transients become negative again, while if the potential is further increased, a new change in the sign takes place. The existence of two potentials where transients change sign from negative to positive indicates the existence of two potentials of maximum entropy (PMEs). The first one takes place at the potential region of the voltammetric peak associated with the steps. The second one takes place within the so-called double-layer region. The latter is similar to that of Pt(111) and can be attributed to the terraces, while the first one can be associated with the presence of steps. Remarkably, the PME related with the terrace coincides with the observation of a shoulder on the voltammograms recorded in N_2O-saturated solutions, reinforcing the idea that the rate of reduction of this weakly adsorbing molecule is sensitive to the existence of polarized water at the interphase.

The existence of two PMEs is unexpected since this property results from an overall contribution of the whole surface and no sensitivity to local contributions was anticipated. The positive transient when the potential is close to the step peak indicates that the overall dominant orientation of water, not only that at the steps, is with the oxygen toward the metal. This means that the process at the steps induces a change in the orientation of a significant portion of the water adlayer, due to strong cooperative effects, enough to change the sign of the overall response. However, this change in water orientation only survives in a narrow potential window. If the potential is increased above the peak potential, a new change of sign suggests that water returns to the hydrogen toward the metal orientation, either because the anion (hydroxyl) adsorption process on the steps forces a new change in the state of adsorbed water or because the increasingly positive charge on the steps results in a new configuration of the water with clusters of opposing orientation on the steps and on the terraces.

A closer look at the transients in Figure 4.9 reveals the existence of slow contributions to the response to the laser heating, resulting in bipolar potential transients, more clearly seen for Pt(221). Slow responses are normally due to slow kinetics of adsorption processes. However, the bipolar transients in Figure 4.9 are observed in the double-layer region, where no adsorption processes are expected. Then, it is suggested that the rupture of the bidimensional order due to the introduction of steps increases the time constant of the water reorganization process. The existence of water molecules strongly bonded to step sites interrupting the water network adsorbed on terraces can explain this observation.

4.4   ADSORPTION ON STEP SITES AND STEP DECORATION: ADATOMS ADSORPTION

Modification of platinum surfaces by the deposition of foreign metal adatoms has an important implication in electrocatalysis since bimetallic surfaces often exhibit improved performance for reaction of technological interest. There are two main approaches to achieve a controlled and selective modification of the surface. First, the underpotential deposition (UPD) is a reversible process that takes place when the metal cation in the solution is in equilibrium with the discharged atoms being deposited. Apart from its interest in electrocatalysis, UPD has a fundamental importance because it can serve to understand the initials steps in metal deposition.

The second approach is the irreversible adsorption of adatoms that remain adsorbed even after the adsorbing species is removed from the solution. The irreversible deposition can be easily achieved for some elements of the *p* block (Ge, Sn, Pb, As, Sb, Bi, S, Se, and Te), since they adsorb strongly on platinum when the electrode is dipped in a solution containing the appropriate ion of the element. In some cases, a reductive environment should be created with the introduction of hydrogen. Then, the electrode can be washed and transferred to an electrochemical cell for characterization. The amount of deposited adatoms can normally be calculated from the magnitude of a new redox process associated with the foreign element. One significant advantage of this approach is that coverage can be easily changed by selecting different deposition times and/or concentration of the ions in solution. Besides, the coverage for a given adlayer remains constant in a broad potential range, in opposition to the UPD process where the coverage is a reversible function of the electrode potential.

Interestingly, with stepped surfaces, some adatoms can be selectively deposited on step sites, reinforcing the idea that these sites have particularly enhanced reactivity. Based on this property, it is possible to create bimetallic surfaces with a regular pattern of monodimensional rows uniformly distributed, allowing a detailed study of the effect of the spatial distribution of the promoter on the catalytic performance of the electrode. Figure 4.10 shows the selective modification of a Pt(775) surface. This has been achieved by introducing a small quantity of Bi(III) in the solution ($\sim 5 \times 10^{-6}$ M). Under these conditions, the rate of deposition is controlled by the slow diffusion of the reactant to the electrode surface and not by a particular value of the applied potential. The deposition process can be easily monitored by following the evolution of the cyclic voltammograms [54, 55]. The selective deposition on the steps is inferred from the initial decrease of the voltammetric peak at 0.12 V, while the broad peak corresponding to anion adsorption on the terraces remains essentially unaltered. During this stage, a new peak at 0.25 V grows, linked to the decrease of the peak of the step. This new peak has been assigned to induced anion adsorption on platinum sites adjacent to the step site covered by the adatom. This conclusion is supported by the fact that this peak is not observed in perchloric acid solution. In a second stage, which starts when the step peak at 0.12 V is completely blocked, adatom deposition starts to take place on terrace sites, as evidenced by the progressive blockage of the hydrogen and anion adsorption processes on the terraces, the appearance of the redox peak characteristic of Bi oxidation on Pt(111) surfaces at 0.63 V, and the progressive disappearance of the adatom-induced anion peak at 0.25 V. The same qualitative behavior is observed with tellurium deposition and also with As and Sb, although, in these cases, deposition on the terrace starts slightly before the step is completely blocked, indicating that the selectivity for step sites is smaller with these adatoms.

Completely different behavior is observed with S and Se, as shown in Figure 4.11. With these adatoms, deposition on the terrace sites starts from the very beginning

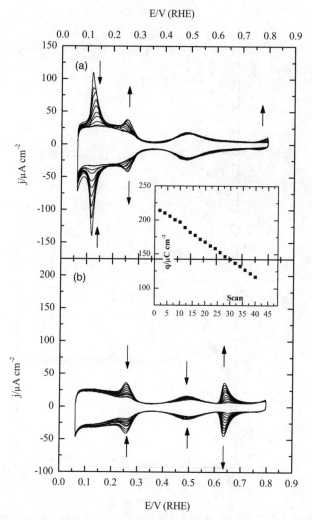

FIGURE 4.10 Bismuth deposition on Pt(775) electrode from 5×10^{-6} M Bi^{+3} + 0.5 M H_2SO_4 solution: (a) Scans 1–19; (b) scans 21–39. Only even scans are shown. The arrows indicate the evolution of the profile upon cycling. Inset: Charge for hydrogen and sulfate/bisulfate adsorption vs. scan number. (Reproduced from E. Herrero, V. Climent, and J.M. Feliu, *Electrochem. Comm.*, 2 (2000)636)

and no selectivity toward the step is observed. The different behavior exhibited by the studied adatoms has been rationalized in terms of their different work function or electronegativity values, relative to the platinum substrate. Those elements with lower electronegativity tend to create a surface dipole with excess of positive charge on the adatom. On the contrary, more electronegative adatoms will retain negative charge. These trends agree with the observations measured with the laser-induced temperature jump method [56], supporting the idea that adatoms of different electronegativity tend

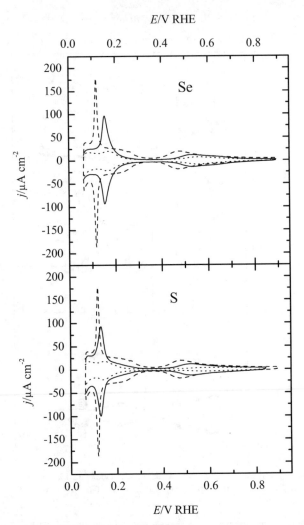

FIGURE 4.11   Different stages in Se and S deposition on Pt(775) electrode in 0.5 M H₂SO₄ solution. Dashed line: first scan; full line: 3rd scan; dotted line: 32nd scan. Sweep rate: 50 mV s⁻¹.

to orientate water differently, suggesting that a partial charge is retained on the adatom with a sign that depends on its electronegativity. The fact that electropositive adatoms interact more strongly with the steps reinforces the idea that these sites have a lower local pzfc and therefore retain negative charge in a broader potential range. Given the distribution of charge on step sites predicted by the Smoluchowsky model discussed above, it is easily understood that electropositive adatoms will tend to interact more strongly with the lower part of the step, which is partially negatively charged, while electronegative adatoms will deposit on the terrace, more likely near the upper part of the step, since this part is positively charged.

The effect of adatom decoration on the pztc of stepped surfaces has also been studied by the CO displacement method and the analysis of N_2O reduction currents [23, 55]. As discussed above, the introduction of steps tends to decrease the pztc of the electrode surface, which linearly shifts from the value of the terrace toward the value of the step. When the step sites are blocked by the adatom, the local charge associated with step sites is quenched, resulting in a positive shift of the overall pztc toward that of the terrace. The analysis of N_2O reduction currents supports this idea. While on the unmodified stepped surface N_2O shows two reduction maxima, once the steps sites are selectively blocked with bismuth, only the maxima associated with the terrace sites remain [23].

Another example of selective decoration of stepped surfaces is found with copper deposition. If copper concentration in solution is high enough to avoid diffusion limitations, the UPD process shows several adsorption/desorption peaks resulting from the different adsorption energy on different sites on which copper and anions react. For surfaces vicinal to the (111) basal plane, the deposition on steps normally takes place at a higher potential than on the terrace, indicating that attachment of copper to step sites is stronger [57–61]. Selective deposition on steps can be achieved as discussed previously by using a small concentration of copper in such a way that the deposition rate is controlled by diffusion. In this case, the potential should be cycled also in the overpotential deposition range and high potentials should be avoided since copper deposition is a reversible process and copper would be desorbed at potentials higher than the UPD peak. Similar to other adatoms, copper, being more electropositive than platinum, is expected to adsorb on the lower part of the step sites. Consistently with this, copper decoration of steps induces specific adsorption of anions on platinum sites next to it [60, 61].

A different situation happens with surfaces vicinal to the (100) surface. In this case adatoms interact more strongly with the terrace [62]. There are two reasons for this. First, the dipole created on the steps of this symmetry is smaller and, second, coordination with the square sites of the terrace is probably stronger. Consequently, for copper deposition on Pt(100) vicinal surfaces the peak at highest potentials has been assigned to adsorption on terrace sites while the experiments performed at low concentrations shows a preferential deposition on terrace sites. Another example shown in Figure 4.12 is the deposition of arsenic on Pt(19,1,1). As in the case of copper, the fast decrease of the voltammetric features due to adsorption on terraces is not paralleled by a similar decrease of step and step edge sites.

4.5 REACTIVITY OF STEPPED SURFACES

4.5.1 CO Oxidation

On platinum electrodes, it is now well accepted that CO oxidation takes place according to the following mechanism:

$$Pt + CO \rightarrow Pt - CO \tag{4.15}$$

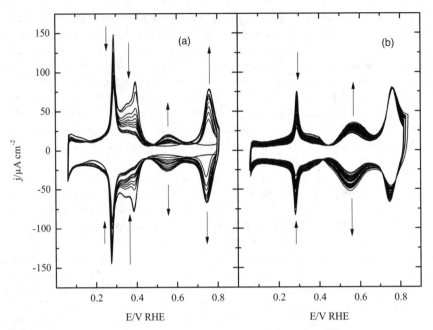

FIGURE 4.12 Arsenic deposition on Pt(19,1,1) electrode from 10^{-6} M As^{+3} + 0.1 M H$_2$SO$_4$ solution: (a) Scans 1–12, (b) scans 13–20. The arrows indicate the evolution of the profile upon cycling.

$$Pt + H_2O \rightarrow Pt - OH + H^+ + e \qquad (4.16)$$

$$Pt - CO + Pt - OH \rightarrow 2Pt + CO_2 + H^+ + e \qquad (4.17)$$

These three steps correspond to the irreversible adsorption of CO [reaction (4.15)], the reversible formation of adsorbed OH [reaction (4.16)], and finally the electrochemical reaction between the adsorbed OH and CO species to yield CO$_2$, which is the rate-determining step [reaction (4.17)]. This final reaction is a paramount example of a Langmuir–Hinselwood (LH) step, in which two adsorbed species react to yield the final product. The detailed kinetics of such surface reaction can take place according to two different models. The first one is the nucleation and growth mechanism in which, after some initial nuclei are created which correspond to the incipient adsorption of OH on the CO-covered surface, the reaction progresses outward from the initial nuclei, that is, after the adsorption of the initial OH nuclei, the CO-free patches created grow from the border. Eventually, these patches collapse with other patches and finally the whole CO layer is oxidized (in the absence of CO in the solution). The general rate equation for a process following such a model is [63, 64]:

$$v \propto at^{n-1} \exp\left(bt^n\right) \qquad (4.18)$$

where a and b are constants and n is a parameter in the model which depends on the nucleation step ($n = 2$ if the nucleation is instantaneous and $n = 3$ for a progressive nucleation).

The second model for the LH kinetics is the mean-field model. In this model, the adsorbed CO and OH species are uniformly distributed throughout the surface. In the case of flat surfaces, this model implies that the mobility of the CO and/or OH molecules is high, much higher than the reaction rate, so that a uniform distribution is always observed. The rate for this model is

$$v \propto \theta_{OH} \theta_{CO} \qquad (4.19)$$

Since the OH coverage is proportional to the fraction of the surface not covered by CO, this contribution is proportional to $1 - \theta_{CO}$ and therefore the rate can be written as

$$v = k\theta_{CO} (1 - \theta_{CO}) \qquad (4.20)$$

Experimentally, it has been found for the low-index platinum single crystals that chronoamperometric and voltammetric transients are better described using the mean-field approximation [65–67]. As can be easily inferred from Eq. (4.20), CO oxidation for an ideally perfect CO adlayer is impossible since there is no initial free site where the adsorption of OH can occur. Therefore, the presence of defects and steps on the platinum surface may alter significantly the reaction rate.

Studies with Pt(111) and Pt(100) electrodes had demonstrated that the oxidation on these surfaces takes place at higher potentials than on the polycrystalline surface, which suggests that the steps and defects certainly play a role on the oxidation kinetics. Additionally, it was found that the oxidation of the preadsorbed CO layer was very dependent on the electrode pretreatment for Pt(111) electrodes [68]. Different pretreatments lead to different amounts of defects on the surface and the results clearly indicate that the kinetics is the lowest for the surface with the lowest amount of defects.

Being so, it is clear that the steps should play a role in the oxidation kinetics. In order to determine the role of the steps in the oxidation mechanism, studies with stepped surfaces with (111) terraces and (110) steps were carried out [8, 67, 69, 70]. The studies with stepped surfaces can also provide information on the CO mobility on the surface. If the initial sites for the OH adsorption are the steps, the CO molecules should diffuse over the surface to reach these reaction sites. If the mobility of the CO is low, a "Cottrell-type" tail for longer times in the chronoamperometric transients should be observed for the wider terraces [71]. Figure 4.13 shows the chronoamperometric transients for CO oxidation recorded on several stepped surfaces. As can be seen in Figure 4.13, the tail is absent from all the transients, irrespective of the terrace width, implying that CO diffusion is always faster than the reaction rate and a homogeneous CO distribution is obtained. Two additional conclusions can be drawn from the experiments. First, the mean-field model reproduces the experimental results much better than the nucleation and growth model. Second, the rate for the CO

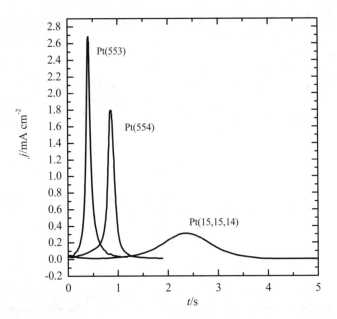

FIGURE 4.13 Stripping chronoamperometry current transients for oxidation of saturated CO monolayer on Pt(553), Pt(554), Pt(15,15,14). Potential is 0.855 V vs. NHE. 0.5 M H_2SO_4. [Data from N.P. Lebedeva, M.T.M. Koper, J.M. Feliu, and R.A.V. Santen *J. Phys. Chem. B*, 106 (2002) 12938.]

oxidation is strongly dependent on the step density, being the fastest for the surface with the highest step density.

From the chronoamperometric transients at different potentials, it was possible to determine the rate constant as a function of the applied potential [k(E)]. From the Tafel plots, a slope of 70–80 mV was obtained in all cases [8]. This slope is in good agreement with previous results for polycrystalline and single-crystal electrodes [65, 72, 73]. From these plots, it was possible to determine the rate constant at 0 V. This rate constant was plotted versus the step density (Fig. 4.14). As can be seen, the representation is linear and the extrapolation to zero step density is, within the experimental error, zero, which indicates that the reaction rate on the terrace sites is negligible with respect to that measured on the step. Therefore, it can be concluded that the CO oxidation reaction takes place exclusively on the step sites. Since the oxidation requires the adsorption of OH, this step should take place preferentially on the step site.

In the presence of CO in solution, the observed behavior for the stepped surfaces with (111) terraces and (110) steps follows the same trends as that observed for the oxidation of CO layers (Fig. 4.15) [74]. To control the diffusion condition of CO from the bulk to the surface, these experiments should be performed under a rotating-disk configuration. The voltammogram shows a clear hysteresis between the positive- and negative-going scans for all electrodes. In general, no current is detected below E_2. This situation corresponds to an electrode fully covered by CO and where the oxidation process is hindered since the number of free sites available for OH

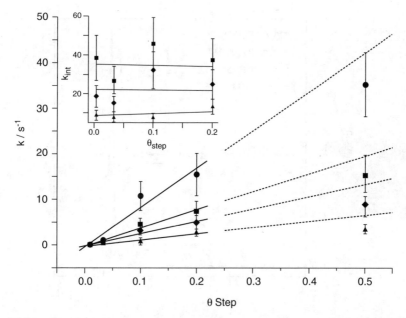

FIGURE 4.14 Dependence of apparent rate constants determined by fitting experimental transients with Eq. (4.5), on step fraction. Final potentials: 0.73 V (triangles), 0.755 V (diamonds), 0.78 V (squares), and 0.805 V (circles). Inset shows the independence of the apparent "intrinsic" rate constant per step. [Reproduced from N.P. Lebedeva, M.T.M. Koper, J.M. Feliu, and R.A.V. Santen *J. Phys. Chem. B*, 106 (2002) 12938.]

adsorption is very small. At this potential, the ignition potential, which is indeed more positive than the onset potential for oxidation of an adsorbed CO layer, a sharp peak is observed corresponding to the oxidation of the adsorbed CO layer. From that point, the measured current is, basically, that corresponding to the CO oxidation diffusion-controlled limiting current. In the negative scan, the limiting current is maintained until a fast decay in the current is observed at E_1. The decay occurs when the oxidation rate for CO is lower than its diffusion rate, so that CO is accumulated again on the surface. It has been shown that E_2 basically depends on the number of available sites for OH adsorption on the electrode surface, whereas E_1 is associated with the oxidation rate [74]. For the stepped surfaces, the E_1 and E_2 potentials depend linearly on the step density for terraces wider than five atoms (Fig. 4.16). This linear dependence of E_1 on the step density is a consequence of the dependence of the oxidation rate on the step density, as was observed in the chronoamperometric CO-stripping experiments. In the case of E_2, the number of initial sites for OH adsorption also depends on the step density, since OH adsorption is more favorable at the step. Additionally, it should be mentioned that the behavior of the Pt(111) electrode in sulfuric acid solutions markedly differs from the expected values extrapolated from the behavior of the stepped surfaces due to the formation of an ordered sulfate layer on the Pt(111) electrode.

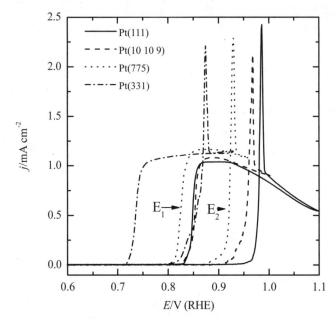

FIGURE 4.15 Saturated CO bulk oxidation voltammograms corresponding to Pt(111) and some stepped surface electrodes in 0.1 M H_2SO_4 at 20 mV s^{-1} and 600 rpm. The arrows mark the potentials for E_1 and E_2 for the Pt(775) electrode.

However, not all stepped surfaces behave in a similar way. Unlike electrodes with $n(111) \times (110)$ structure, the presence of (110) steps on the (100) terrace does not catalyze the oxidation of adsorbed CO (Fig. 4.17). In fact, two different types of behavior can be observed in the chronoamperometric transients, depending on the terrace width for these surfaces. For terraces wider than five atoms, the CO oxidation rate diminishes with step density, whereas for narrower terraces, the rate increases again. This indicates that there are two different types of behavior, one associated with long terraces and the second one with short terraces. The analysis of the chronoamperometric transients show two waves for the Pt(510) corresponding to both types of behavior, probably as the result of a distribution of terrace sizes around the mean value. In all cases, transients can be modeled according to the mean-field approximation.

The different behaviour of the $n(100) \times (110)$. surfaces for the CO oxidation reaction can be a consequence of the differences in the adsorption of OH and in general for anion adsorption. Probably, the OH adsorption process also affects the Tafel slope, since a higher Tafel slope ($\sim$120 mV as compared to 70–90 mV measured for the other electrodes) is obtained for the (100) vicinal surfaces with narrower terraces.

CO oxidation has also been studied on stepped surfaces with (111) terraces and (110) and (100) steps in alkaline media [75]. In high-pH media, the mobility of the CO is much lower since separate contributions to the voltammetric profile can be identified for the terrace, step, and kink sites. The charge density measured for each contribution depends on the scan rate, which clearly indicates that the mobility rate for

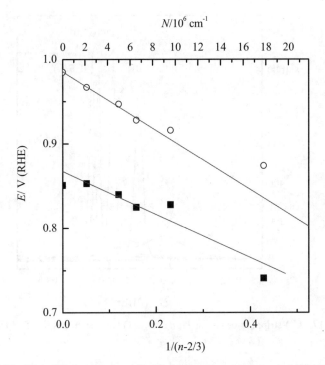

FIGURE 4.16 Plot of E_1 (squares) and E_2 (circles) measured in saturated CO bulk oxidation voltammograms corresponding to $Pt[n(111) \times (110)]$ electrodes in 0.1 M H_2SO_4 at 20 mV s^{-1} and 600 rpm.

CO is similar to the reaction rate. The behavior then depends on the number of terrace and step sites in a given surface and also on the time needed for the CO adsorbed on the terrace to diffuse to a step site.

4.5.2 Formic Acid Oxidation on Stepped Surfaces

The oxidation mechanism of formic acid to CO_2 consists of two parallel pathways, the first one going through an active intermediate and the second one having CO as poisoning intermediate [76]:

$$(4.21)$$

This scheme is typical for the oxidation of C_1 molecules. Among them, formic acid is considered a model molecule in electrocatalytic studies, since its oxidation involves only two electrons. In general, the shape of the voltammetric profile for the oxidation of C_1 molecules is very similar, showing a characteristic hysteresis

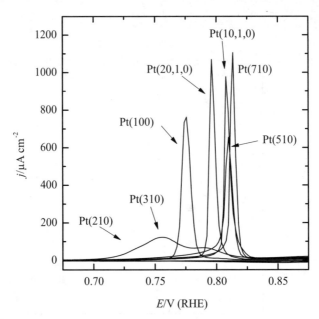

FIGURE 4.17   CO-stripping peaks for Pt[n(100) × (110)] electrodes in 0.5 M H$_2$SO$_4$. Scan rate 20 mV s^{-1}.

between the positive- and negative-going scans. As in the case of CO bulk oxidation, the hysteresis is due to the accumulation of adsorbed CO at low potentials formed from the dissociative adsorption (and dehydrogenation, except in the case of formic acid) of the C$_1$ molecule. In this particular case, formic acid dissociates at low potentials to yield adsorbed CO and a water molecule. The CO remains adsorbed on the surface up to potentials above 0.6 V, blocking the surface and hindering the oxidation through the active intermediate. In the negative scan, the surface concentration of adsorbed CO is lower, resulting in higher oxidation currents.

Formic acid oxidation reaction is very sensitive to the surface structure, as indicated by the huge differences of the voltammetric profiles recorded for the three basal planes [77, 78] and the stepped surfaces [79, 80]. The paramount case is that of the Pt(100) electrode, which has a very high poisoning rate. For that reason, the currents in the positive scan are almost negligible. However, currents in the negative scan are very high, because the activity of the surface for the direct oxidation is also very high. The lowest currents are obtained for a Pt(111) electrode, which in turn has a very low poisoning rate, as suggested by the small hysteresis. In fact, the reaction on this electrode is completely dominated by the kinetics of the oxidation process. Experiments carried out with a Pt(111) hanging-meniscus rotating-disk configuration showed that the voltammogram is independent of the rotation rate (formic acid concentration 0.2 M), which indicated that the reaction is controlled by the electron transfer process [81]. Poisoning rates and direct oxidation rates are obtained from chronoamperometric experiments, since the analysis of the results is simpler. Assuming that formic acid oxidation can be described as a process controlled by both charge transfer and mass transport, the rate constant was obtained for the basal planes and surfaces vicinal to the (100) plane [82, 83]. It was found that the Pt(110) had the highest activity and the surfaces vicinal to the (100) plane had a similar activity to the Pt(100) electrode.

The poison formation pathway can be studied separately by recording the amount of CO formed after immersion of the electrode in a formic acid solution and then transferring it to an electrochemical cell with just the supporting electrolyte. Moreover, the use of adatom-modified surfaces has been shown useful for the separation of the direct and poison formation pathways [84]. The poison formation reaction was extensively studied in the presence of different adatoms on Pt(111) and Pt(100) electrodes [85–90]. For Pt(100) electrodes, the irreversibly adsorbed adatoms always acted as a third body for the poison formation reaction, that is, poison formation requires an ensemble of sites to occur [86, 88]. On the other hand, two different types of behavior where found for the adatoms on the Pt(111) electrode. Electronegative adatoms with respect to platinum (such as Se and S) were simply acting as a third body [87], whereas very low coverage values of electropositive adatoms (such as Bi, Pb, Te, and As) were able to completely inhibit the poison formation reaction [85, 89, 90]. In the case of bismuth, the complete inhibition of poison formation was achieved at adatom coverage as low as 0.04. A simple estimation indicated that bismuth should extend its influence to inhibit poison formation at a distance of six to eight platinum atoms [85]. Such an effect cannot be produced by a simple modification of the electronic properties of the surface induced by the presence of an adatom since calculations indicate that these modifications only reach the second or third shell of the platinum atoms close to the adatom [91]. In order to get insight into this problem, stepped surfaces modified with adatoms were used.

The poison formation reaction was studied on both series of stepped electrodes vicinal to the Pt(111) surfaces modified with adatoms [9, 92–94]. As aforementioned, the initial deposition site is the step site for the electropositive adatoms on both the Pt(n, n, $n-2$) and Pt($n+1$, $n-1$, $n-1$) stepped surfaces [54] and the deposition process on the terraces only starts when all the step sites have been covered. In this way, it can be said that bismuth is able to decorate stepped surfaces. Such decorated surfaces are an excellent tool to prove the effect of the adatoms in electrocatalysis since its distribution on the surface is well known.

Depending on the terrace width, adatom-modified stepped surfaces show two distinct behaviors. For surfaces with wide (111) terraces, the total accumulation of poison on the surface remained constant and equal to that obtained for the unmodified surface when the adatom does not completely cover the step sites (Fig. 4.18) [92, 93]. Once the step has been completely decorated, poison formation decays sharply to a negligible amount. These results indicate that the poison formation only takes place on the step site, since the amount formed on surfaces with decorated steps is negligible. Moreover, the total poison accumulation for step surfaces with long (111) terraces depends on the symmetry of the step, the (110) steps being more active for poison formation than the (100) steps [92]. These results can explain the unusual long-range

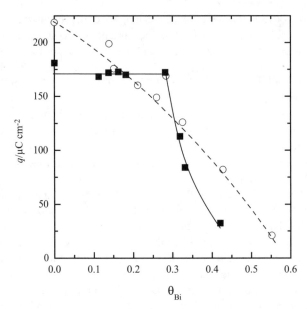

FIGURE 4.18   Total poison accumulation for two different stepped electrodes modified with bismuth vs. fraction of platinum sites blocked by adatom Bi–Pt(775) (squares) and Bi–Pt(553) (circles).

effect of the electropositive adatoms in the inhibition of CO formation on Pt(111) electrodes. It is well known that even the best quality Pt(111) electrodes have a certain amount of defects. These defects are responsible for the formation of poison on Pt(111) electrodes. The adatoms that are able to decorate the steps should also deposit preferentially on the defects, since their electronic properties are very similar. Thus, a very low adatom coverage is able to block completely the defects and thus prevent poison formation. On the other side, selenium adatoms, which cannot decorate the steps, behave as a third body for the poison formation reaction on Pt(111) electrodes.

For stepped surfaces with short (111) terraces, the results are different, since a linear diminution of the poison formation is observed, irrespective of the deposition of the adatom on the step or the terrace (Fig. 4.18) [9, 93]. For surfaces with narrow terraces the interaction between steps and terraces becomes important and the surface cannot be described as a linear combination of terrace and step sites. In fact, all the terrace atoms are affected by the presence of the step and they no longer behave like the atoms in wide terraces, and thus they have different properties toward poison formation. Similar results have been observed with different adatoms.

The electrocatalysis of adatom-modified surfaces has also been studied [95–98]. For bismuth-modified surfaces, the lowest poisoning rate and the highest currents were obtained for surfaces with narrow (111) terraces (two or three atoms wide) in which the step is decorated with bismuth [95, 96]. In this configuration, all the platinum terrace sites are adjacent to a bismuth adatom, indicating that the ensemble bismuth–platinum site provides the highest catalytic activity. Additionally, this ensemble suffers no poisoning from the CO. In the case of the antimony-modified surfaces, the presence of antimony on the stepped surfaces leads to a lower poisoning rate and enhancement of oxidation currents at low potentials [97, 98].

4.5.3 Oxygen Reduction on Stepped Surfaces

Oxygen reduction has also been studied on the different stepped surfaces in both sulfuric and perchloric acid media [99, 100]. The surfaces employed were those having (111) terraces and (100) and (110) monoatomic steps, (100) terraces and (111) steps, and (110) terraces and (111) steps. Since oxygen solubility in water is not very high, the studies were carried out using the hanging-meniscus rotating-disk configuration.

In either sulfuric or perchloric acid solution, the Pt(111) surface showed the lowest catalytic activity among those studied (Fig. 4.19). This was especially the case in sulfuric acid solutions, because of the formation of an ordered sulfate adlayer $(\sqrt{3} \times \sqrt{7})$ R19 on the electrode at high potentials. This ordered structure is able to survive even when the oxygen is being reduced on the surface, since the spike at 0.45 V, which is characteristic of the order–disorder phase transition of this adlayer, is still visible in the presence of oxygen. This spike is very sensitive to the presence of adsorbates and consequently oxygen reduction takes place on a Pt(111) surface almost free of adsorbed intermediates. In fact, its behavior cannot be extrapolated from that of the stepped surfaces having (111) terraces, since on the stepped surfaces the formation of the sulfate adlayer does not takes place. Additionally, the voltammetric response of the Pt(111) electrode depends on the number of defects on the surface.

In contrast, the surfaces with the highest catalytic activity are those with the higher step density, Pt(211), Pt(311), Pt(221), and Pt(331) electrodes. The differences in the catalytic behavior of these surfaces with the basal planes are higher in sulfuric acid solutions, indicating that the adsorbed anions play a significant role in the electrocatalytic activity of the electrodes. However, the difference between the basal planes and the surfaces with the highest electrocatalytic activity is not very high. The presence of oxides on the surface also inhibits the oxygen reduction reaction. In the electrochemical environment, the adsorption process is always competitive in nature, and O_2 adsorption must compete for the adsorption sites with other species present in the aqueous solution. For that reason, anion adsorption hinders oxygen reduction.

The small difference in the catalytic activity for the stepped surfaces in comparison to the basal planes is rather unexpected, since UHV measurements indicate that oxygen is preferentially adsorbed on the step sites [101]. In fact, the dissociative adsorption process of oxygen on Pt(111) and Pt(533) is controlled by the presence of defects or steps [102]. The direct extrapolation of the UHV data would have suggested a significantly higher catalytic activity for the step sites in comparison with the basal planes. As discussed above, owing to the Smoluchowski effect, the step sites will have a lower local pzc than the terrace sites. Therefore, at a given electrode potential anion adsorption will be stronger on the step sites as compared to the terrace. For instance, the much higher catalytic activity of the step sites for the CO oxidation reaction has been associated with the adsorption of OH on these sites [69].

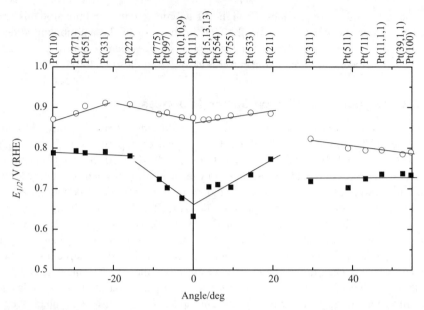

FIGURE 4.19 Half-wave potential for oxygen reduction reaction on different single-crystal electrodes plotted vs. angle of surface with respect to (111) plane in 0.1 M HClO$_4$ (circles) and 0.5 M H$_2$SO$_4$ (squares).

The adsorbed CO molecules on the terrace will diffuse until they reach a step site covered by OH, where it reacts to yield CO$_2$. Therefore, it can be proposed that the step sites are already covered by OH, and oxygen must displace the adsorbed OH to react on these sites. Adsorbed OH is a poison for the oxygen reduction reaction, and the higher adsorption energy for O$_2$ on the step sites is then counterbalanced with a higher adsorption strength of the OH species on these sites, resulting in a catalytic activity lower than the expected value form UHV experiments.

As a conclusion, it can be said that oxygen reduction is a structure-sensitive reaction in both sulfuric and perchloric acid media. Since oxygen must compete for the adsorption sites with the anions and surface oxides, two effects contribute to the structure sensitivity of this reaction. The main effects are the anion adsorption and oxide formation on the electrode, whereas the oxygen adsorption energy on the different sites plays a secondary role in determining the electrocatalytic activity of the electrode.

REFERENCES

1. W. Schmickler, *Interfacial Electrochemistry*, Oxford University Press, Oxford, 1996.
2. J. Clavilier, R. Faure, G. Guinet, and R. Durand, *J. Electroanal. Chem.*, 107 (1980) 205.
3. C. L. Scortichini and C. N. Reilley, *J. Electroanal. Chem.*, 139 (1982) 247.
4. P. N. Ross, *Surf. Sci.*, 102 (1981) 463.
5. S. Motoo and N. Furuya, *J. Electroanal. Chem.*, 172 (1984) 339.

6. K. Itaya, S. Sugawara, K. Sashikata, and N. Furuya, *J. Vac. Sci. Technol.*, A 8 (1990) 515.
7. A. Berná, A. Rodes, J.M. Feliu, F. Illas, A. Gil, A. Clotet, and J.M. Ricart, *J. Phys. Chem. B*, 108 (2004) 17928.
8. N. P. Lebedeva, M. T. M. Koper, J. M. Feliu, and R. A. van Santen, *J. Phys. Chem. B*, 106 (2002) 12938.
9. M. D. Maciá, E. Herrero, J. M. Feliu, and A. Aldaz, *J. Electroanal. Chem.*, 500 (2001) 498.
10. B. Lang, R. W. Joyner, and G. A. Somorjai, *Surf. Sci.*, 30 (1972) 440.
11. M. A. Van Hove and G. A. Somorjai, *Surf. Sci.*, 92 (1980) 489.
12. D. W. Blakely and G. Somorjai, *Surf. Sci.*, 65 (1977) 419.
13. E. Herrero, J. M. Orts, A. Aldaz, and J. M. Feliu, *Surf. Sci.*, 440 (1999) 259.
14. N. Garcia-Araez, V. Climent, E. Herrero, and J. M. Feliu, *Surf. Sci.*, 560 (2004) 269.
15. N. Hoshi, A. Nakahara, M. Nakamura, K. Sumitani, and O. Sakata, *Electrochim. Acta*, 53 (2008) 6070.
16. A. Nakahara, M. Nakamura, K. Sumitani, O. Sakata, and N. Hoshi, *Langmuir*, 23 (2007) 10879.
17. L. A. Kibler, A. Cuesta, M. Kleinert, and D. M. Kolb, *J. Electroanal. Chem.*, 484 (2000) 73.
18. J. Clavilier, K. El-Achi, and A. Rodes, *Chem. Phys.*, 141 (1990) 1.
19. R. Gómez, V. Climent, J. M. Feliu, and M. J. Weaver, *J. Phys. Chem. B*, 104 (2000) 597.
20. D. Strmcnik, D. Tripkovic, D. van der Vliet, V. Stamenkovic, and N. M. Marković, *Electrochem. Commun.*, 10 (2008) 1602.
21. A. M. Funtikov, U. Stimming, and R. Vogel, *J. Electroanal. Chem.*, 428 (1997) 147–153.
22. A. Rodes, K. El Achi, M. A. Zamakhchari, and J. Clavilier, *J. Electrianal. Chem.*, 284 (1990), 245.
23. G. A. Attard, O. Hazzazi, P. B. Wells, V. Climent, E. Herrero, and J. M. Feliu, *J. Electroanal. Chem.*, 568 (2004) 329.
24. K. Domke, E. Herrero, A. Rodes, and J. M. Feliu, *J. Electroanal. Chem.*, 552 (2003) 115–128.
25. P. Rodriguez, E. Herrero, J. Solla-Gullón, F. J. Vidal-Iglesias, A. Aldaz, and J. M. Feliu, *Electrochim. Acta*, 50 (2005) 3111.
26. R. Smoluchowski, *Phys. Rev.*, 60 (1941) 661.
27. K. B. Besocke, B. Krahl-Urban, and H. Wagner, *Surf. Sci.*, 68 (1977) 39.
28. S. Trasatti, *Electrochim. Acta*, 36 (1991) 1659.
29. S. Trasatti, *Surf. Sci.*, 335 (1995) 1.
30. J. Lecoeur, J. Andro, and R. Parsons, *Surf. Sci.*, 114 (1982) 320.
31. G. L. Beltramo, H. Ibach, U. Linke, and M. Giesen, *Electrochim. Acta*, 53 (2008) 6818.
32. G. L. Beltramo, H. Ibach, and M. Giesen, *Surf. Sci.*, 601 (2007) 1876.
33. A. N. Frumkin, O. A. Petrii, and B. B. Damaskin, in *Comprehensive Treatise of Electrochemistry* vol. 1, *Potential of Zero Charge*, J. O. Bockris, B. E. Conway, and E. Yeager (Eds.), Plenum, New York, 1980, pp. 221–289.
34. J. Clavilier, R. Albalat, R. Gómez, J. M. Orts, J. M. Feliu, and A. Aldaz, *J. Electroanal. Chem.*, 330 (1992) 489.
35. J. M. Feliu, J. M. Orts, R. Gómez, A. Aldaz, and J. Clavilier, *J. Electroanal. Chem.*, 372 (1994) 265.
36. J. Clavilier, R. Albalat, R. Gómez, J. M. Orts, and J. M. Feliu, *J. Electroanal Chem.*, 360 (1993) 325.
37. E. Herrero, J. M. Feliu, A. Wieckowski, and J. Clavilier, *Surf. Sci.*, 325 (1995) 131.
38. R. Gómez, J. M. Feliu, A. Aldaz, and M. J. Weaver, *Surf. Sci.*, 410 (1998) 48.

39. V. Climent, R. Gómez, and J. M. Feliu, *Electrochim. Acta*, 45 (1999) 629.
40. R. Gómez, V. Climent, J. M. Feliu, and M. J. Weaver, *J. Phys. Chem. B*, 104 (2000) 597.
41. M. J. Weaver, *Langmuir*, 14 (1998) 3932.
42. A. Cuesta, *Surf. Sci.*, 572 (2004) 11.
43. P. N. Ross, *J. Chim. Phys.*, 88 (1991) 1353.
44. A. Hamelin and J. Lecoeur, *Surf. Sci.*, 57 (1976) 771.
45. G. A. Attard and A. Ahmadi, *J. Electroanal. Chem.*, 389 (1995) 175.
46. V. Climent, G. A. Attard, and J. M. Feliu, *J. Electroanal. Chem.*, 532 (2002) 67.
47. J. F. Jia, K. Inoue, Y. Hasegawa, W. S. Yang, and T. Sakurai, *Phys. Rev. B*, 58 (1998) 1193.
48. K. Wandelt, *Appl. Surf. Sci.*, 111 (1997) 1.
49. V. Climent, B. A. Coles, and R. G. Compton, *J. Phys. Chem. B*, 106 (2002) 5988.
50. V. Climent, B. A. Coles, R. G. Compton, and J. M. Feliu, *J. Electroanal. Chem.*, 561 (2004) 157.
51. V. Climent, N. Garcia-Araez, R. G. Compton, and J. M. Feliu, *J. Phys. Chem. B*, 110 (2006) 21092.
52. N. Garcia-Araez, V. Climent, and J. M. Feliu, *J. Am. Chem. Soc.*, 130 (2008) 3824.
53. N. García-Aráez, V. Climent, and J. M. Feliu, *Electrochim. Acta*, 54 (2009) 966.
54. E. Herrero, V. Climent, and J. M. Feliu, *Electrochem. Comm.*, 2 (2000) 636.
55. V. Climent, E. Herrero, and J. M. Feliu, *Electrochem. Common.*, 3 (2001) 590.
56. N. Garcia-Araez, V. Climent, and J. M. Feliu, *J. Am. Chem. Soc.*, 130 (2008) 3824.
57. C. Nishihara and H. Nozoye, *J. Electroanal. Chem.*, 396 (1995) 139.
58. C. Nishihara and H. Nozoye, *J. Electroanal. Chem.*, 386 (1995) 75.
59. E. A. Abdelmeguid, P. Berenz, and H. Baltruschat, *J. Electroanal. Chem.*, 467 (1999) 50.
60. L. J. Buller, E. Herrero, R. Gómez, J. M. Feliu, and H. D. Abruña, *J. Chem. Soc., Faraday Trans.*, 92 (1996) 3757.
61. L. J. Buller, E. Herrero, R. Gómez, J. M. Feliu, and H. D. Abruña, *J. Phys. Chem. B*, 104 (2000) 5932.
62. R. Francke, V. Climent, H. Baltruschat, and J. M. Feliu, *J. Electroanal. Chem.*, 624 (2008) 228.
63. A. Bewick, M. Fleischmann, and H. R. Thirsks, *Trans. Faraday. Soc.*, 58 (1962) 2200.
64. M. Fleischmann and H. R. Thirsks, in *Advances in Electrochemistry and Electrochemical Engineering*, Vol 3, P. Delehay (Ed.), Willey, New York, 1963, p. 123.
65. M. Bergelin, E. Herrero, J. M. Feliu, and W. Wasberg, *J. Electroanal. Chem.*, 467 (1999) 74.
66. E. Herrero, B. Álvarez, J. M. Feliu, S. Blais, Z. Radovic-Hrapovic, and G. Jerkiewicz, *J. Electroanal. Chem.*, 567 (2004) 139.
67. N. P. Lebedeva, M. T. M. Koper, J. M. Feliu, and R. A. van Santen, *J. Electroanal. Chem.*, 524–525 (2002) 242.
68. N. P. Lebedeva, M. T. M. Koper, J. M. Feliu, and R.A. van Santen, *Electrochem. Comm.*, 2 (2000) 487.
69. N. P. Lebedeva, M. T. M. Koper, E. Herrero, J. M. Feliu, and R. A. van Santen, *J. Electroanal. Chem.*, 487 (2000) 37.
70. M. P. Lebedeva, A. Rodes, J. M. Feliu, M. T. M. Koper, and R. A. van Santen, *J. Phys. Chem. B*, 106 (2002) 9863.
71. M. T. M. Koper, N. P. Lebedeva, and C. G. M. Hermse, *Faraday Discuss.*, 121 (2002) 301.
72. B. Love and J. Lipkowski, *ACS Symp Ser.*, 378 (1988) 484.
73. L. Palaikis, D. Zurawski, M. Hourani, and A. Wieckowski, *Surf. Sci.*, 199 (1988) 183.

74. C. A. Angelucci, E. Herrero, and J. M. Feliu, *J. Solid State Electrochem.*, 11 (2007) 1532.
75. G. Garcia and M. T. M. Koper, *Phys. Chem. Chem. Phys.*, 10 (2008) 3802.
76. R. Parsons and T. VanderNoot, *J. Electroanal. Chem.*, 257 (1988) 9.
77. J. Clavilier, R. Parsons, R. Durand, C. Lamy, and J. M. Leger, *J. Electroanal. Chem.*, 124 (1981) 321.
78. R. R. Adzic, A. V. Tripkovic, and W. O'Grady, *Nature*, 296 (1982) 137.
79. R. R. Adzic, A. V. Tripkovic, and V. B. Vessovic, *J. Electroanal. Chem.*, 204 (1986) 329.
80. S. Motoo and N. Furuya, *Ber. Buns. Gesellsch. Phys. Chem.*, 91 (1987) 457.
81. M. D. Maciá, E. Herrero, and J. M. Feliu, *J. Electroanal. Chem.*, 554–555 (2003) 25.
82. S. G. Sun, Y. Lin, N. H. Li, and J. Q. Mu, *J. Electroanal. Chem.*, 370 (1994) 273.
83. S. G. Sun and Y. Y. Yang, *J. Electroanal. Chem.*, 467 (1999) 121.
84. E. Leiva, T. Iwasita, E. Herrero, and J. M. Feliu, *Langmuir*, 13 (1997) 6287.
85. E. Herrero, A. Fernández-Vega, J. M. Feliu, and A. Aldaz, *J. Electroanal. Chem.*, 350 (1993) 73.
86. E. Herrero, J. M. Feliu, and A. Aldaz, *J. Electroanal. Chem.*, 368 (1994) 101.
87. M. J. Llorca, E. Herrero, J. M. Feliu, and A. Aldaz, *J. Electroanal. Chem.*, 373 (1994) 217.
88. E. Herrero, M. J. Llorca, J. M. Feliu, and A. Aldaz, *J. Electroanal. Chem.*, 383 (1994) 145.
89. E. Herrero, M. J. Llorca, J. M. Feliu, and A. Aldaz, *J. Electroanal. Chem.*, 394 (1995) 161.
90. V. Climent, E. Herrero, and J. M. Feliu, *Electrochim. Acta*, 44 (1998) 1403.
91. P. J. Fiebelman and D. R. Ramman, *Phys Rev. Lett.*, 52 (1984) 61.
92. M. D. Maciá, E. Herrero, J. M. Feliu, and A. Aldaz, *Electrochem. Comm.*, 1 (1999) 87.
93. M. D. Maciá, E. Herrero, and J. M. Feliu, *Electrochim. Acta*, 47 (2002) 3653.
94. S. P. E. Smith, K. F. Ben-Dor, and H. D. Abruña, *Langmuir*, 16 (2000) 787.
95. S. P. E. Smith and H. D. Abruña, *J. Electroanal. Chem.*, 467 (1999) 43.
96. S. P. E. Smith, K. F. Ben-Dor, and H. D. Abruña, *Langmuir*, 15 (1999) 7325.
97. Y. Y. Yang, S. G. Sun, Y. J. Gu, Z. Y. Zhou, and C. H. Zhen, *Electrochim. Acta*, 46 (2001) 4339.
98. Y. Y. Yang and S. G. Sun, *J. Phys. Chem. B*, 106 (2002) 12499.
99. M. D. Maciá, J. M. Campiña, E. Herrero, and J. M. Feliu, *J. Electroanal. Chem.*, 564 (2004) 141.
100. A. Kuzume, E. Herrero, and J. M. Feliu, *J. Electroanal. Chem.*, 599 (2007) 333.
101. P. J. Feibelman, S. Esch, and T. Michely, *Phys. Rev. Lett.*, 77 (1996) 2257.
102. A. T. Gee and B. E. Hayden, *J. Chem. Phys.*, 113 (2000) 10333.

CHAPTER 5

Computational Chemistry Applied to Reactions in Electrocatalysis

AXEL GROSS and SEBASTIAN SCHNUR

Institute of Theoretical Chemistry, Ulm University, D-89069 Ulm, Germany

5.1 INTRODUCTION

In the field of surface science, recent years have witnessed a tremendous progress as far as the microscopic elucidation of structures and processes is concerned [1]. This is not only caused by the development of experimental probes with atomistic resolution, but currently it is also caused to a large extent by advances in theoretical surface science [2]. Due to the ever-improving computer power and the development of efficient algorithms, a reliable theoretical description of complex surface structures and of processes on surfaces based on first-principles electronic structure theory, in particular density functional theory (DFT) [3], has become possible. DFT methods combine computational efficiency with acceptable accuracy. Consequently, theoretical studies are no longer limited to explanatory purposes but have gained predictive power. Thus, theory and experiment can collaborate on the same footing, which has resulted in numerous very fruitful collaborations between theoretical and experimental groups addressing surface science problems. One of the many impressive examples is the design of a successful catalyst for the steam-reforming process based on a close collaboration between fundamental academic research, both experimental and theoretical, and industrial development [4].

In electrochemistry and in electrocatalysis, processes occur not at the solid–gas interface but rather at the solid–liquid interface. Furthermore, external electric fields are applied leading to varying electrode potentials. This adds considerable complexity to the appropriate theoretical treatment of electrochemical processes at the solid–liquid interface. Furthermore, the number of experimental probes with atomic resolution at the solid–liquid interface is limited compared to the solid–vacuum interface. Consequently, even such elementary properties as the exact structure of water at

Catalysis in Electrochemistry: From Fundamentals to Strategies for Fuel Cell Development,
First Edition. Edited by Elizabeth Santos and Wolfgang Schmickler.

the electrode–electrolyte interface are still debated. In addition, for seemingly simple reactions such as the hydrogen evolution/oxidation reaction and the oxygen–reduction reaction in electrocatalysis, relatively little is known about the basic reaction steps [5]. Therefore there is a strong need for theoretical studies leading to a microscopic description and analysis of electrocatalytic reactions.

In fact, there have already been several attempts to model external fields or the electrode potential at the solid–liquid interface within the DFT slab approach [6–10]. In this chapter, we will try to review the current status of the first-principles treatment of reaction in electrocatalysis, including the effect of the electrode potential. We will first give a general introduction into the problems associated with the description of the electrode potential. There are two different approaches to treat charged systems within periodic DFT calculations, either at constant number of electrons or at constant electron chemical potential [6]. Both approaches will be briefly introduced and contrasted. In an electrochemical cell, the potential falls off at the solid–liquid interface within the double layer to a constant value in the electrolyte [11]. The charge on the surface of the electrode is controlled by the potential difference between the metal and the electrolyte and by the capacity of the double layer. Hence it is the potential that is the crucial variable and not the charge so that the calculations should be performed at constant chemical potential rather than constant charge. Yet, calculations at constant charge are usually computationally less demanding.

Furthermore, we will describe DFT studies devoted to the description of the water–metal interaction in the absence of any external fields. Traditionally this was done using parameterized interaction potentials [12], and still the development of water interaction potentials is an active field [13], but nowadays these studies can be performed entirely from first principles [14]. We will in particular focus on the structure of water above late transition metal electrodes.

Finally we will discuss and contrast the different approaches and implementations for modeling external fields and varying electrode potentials. We will give examples of results obtained with the different methods. This is not meant to be an exhaustive overview over recent applications but rather an illustration of the potential of these methods in elucidating microscopic details of processes in electrocatalysis. In fact, these methods are rather powerful as a means of giving insights into fundamental processes at the electrochemical solid–liquid interface. Still there is no commonly accepted method to describe varying electrode potentials in periodic DFT calculations. All the methods used have advantages but also some drawbacks. Consequently, it is certainly fair to say that there is still enough room for improvements in the realistic theoretical description of solid–liquid interfaces in the presence of external fields. Nevertheless, it is anticipated that first-principles calculations in electrochemistry will soon play a similar role as in surface science.

5.2 THEORETICAL FOUNDATIONS

In order to address extended interface systems from first principles, almost exclusively periodic electronic structure calculations based on DFT [3] are used. We will first discuss general aspects of DFT calculations related to periodic DFT calculations at the electrochemical solid–liquid interface and then address specific problems related to the treatment of charged systems and to the determination of the electrode potential.

5.2.1 Periodic DFT Calculations

Periodic DFT calculations using a plane-wave expansion of the one-electron states are computationally rather efficient [15]. Still, at an electrode surface, the three-dimensional periodicity of the bulk system is broken. In order to exploit the efficiency of the periodic DFT codes, a three-dimensional periodicity has to be restored without introducing any artefacts. This is achieved in the so-called supercell approach illustrated in Figure 5.1, in which electrodes are represented by repeated slabs of finite thickness that are infinitely extended in lateral directions. Thus the delocalized nature of the electronic states of metal electrodes can be taken into account. The distance between the slabs has to be large enough so that there is no interaction between the electrodes, and the slabs have to be thick enough to give a good representation of

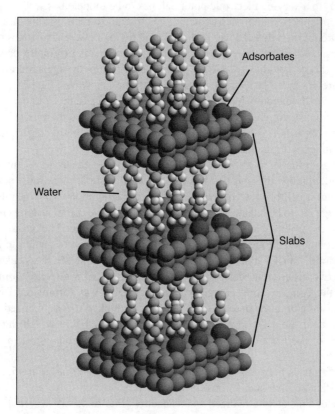

FIGURE 5.1 Illustration of the supercell approach in periodic DFT calculations to describe extended electrode surfaces plus an electrolyte.

the substrates. These conditions can be easily checked by increasing the height of the supercell and the thickness of the slabs, respectively, until convergence of the results of interest with respect to these parameters is reached.

The accuracy of DFT calculations with respect to the quantum many-body effects cannot be easily assessed. These effects are all included in the exchange correlation functional which is unfortunately not known so that approximations are needed. Nowadays the generalized gradient approximation (GGA) [16, 17] is typically used in which the gradient of the electron density is taken into account, but in such a way that important sum rules are obeyed. There are still shortcomings of the GGA, for example, that van der Waals or dispersion forces are not properly reproduced. However, hydrogen bonds are typically well described, which is rather important for DFT calculations applied to electrochemical systems at the solid–liquid interface.

In periodic DFT calculations addressing surface science problems at the solid–gas or solid–vacuum interface, the space between the slabs is kept empty, but the space between the slabs can equally well be filled up with an electrolyte, as indicated in Figure 5.1, so that systems relevant for electrocatalysis can be addressed. However, for a realistic description of the electrochemical system, the presence of external fields, which leads to a varying electrode potential, has to be modeled as well.

A conceptually easy approach to include electric field effects is to create an external field by introducing a dipole layer in the vacuum region between two slabs representing the surface or interface system [18]. This method was originally developed to correct the dipole field in periodic DFT calculations when the slabs representing the substrate are not symmetric along the surface normal. However, the dipole layer can also be used to deliberately introduce an external field [7]. In this approach, the number of electrons is not changed; thus charge neutrality is also maintained. However, within this approach it is not straightforward to relate the applied dipole field to the corresponding resulting electrode potential.

Charging up the slabs also leads to a variation in the electrode potential. Yet, in periodic calculations the unit cell has to be neutral, that is, there must not be a net charge per unit cell because otherwise the electrostatic energy diverges. Hence the excess charge has to be balanced by countercharges. These compensating charges can be realized in various ways, for example as a uniform charge background [9], as a localized counterelectrode [6], or by the explicit introduction of counterions [10].

Furthermore, in DFT calculations describing electrodes with excess charges, there are in fact two different modes to deal with the charges [6], irrespective of the distribution of the compensating charges. Originally, DFT was formulated for systems with a constant number of electrons, N_e. Very quickly it was realized [19] that there is an equivalent grand canonical formulation of DFT in which the chemical potential μ of the electrons instead of the number of electrons is one of the basic quantities.

The differences between these approaches are illustrated in Figure 5.2. Systems with a constant number of electrons correspond to an isolated slab placed in an external electric field. This can be realized as a capacitor [Figure 5.2(a)] where the slab and the counterelectrode carry charges of equal amount but opposite sign. However, the compensating charge does not need to be locally separated from the charged slab

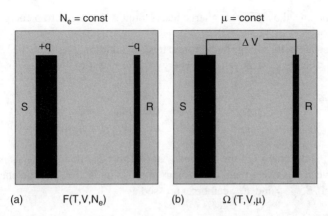

FIGURE 5.2 Illustration of two possible modes to describe charged systems within DFT: (a) constant charge, $N_e = $ const; (b) constant chemical potential, $\mu = $ const. The appropriate thermodynamics potentials to treat a slab together with a reference electrode are the Helmholtz free energy F and the grand potential Ω, respectively.

but can also be distributed uniformly over the unit cell as a compensating charge background [9].

The $\mu = $ const mode, on the other hand, corresponds to a metallic slab that is part of an electric circuit. This mode is not as easy to implement into periodic DFT codes as the $N_e = $ const mode. Therefore most of the calculations of metallic slabs have been performed at a constant number of electrons. There are in fact implementations of periodic DFT calculations that perform self-consistent iterations within the grand canonical formulation of DFT [6, 20], that is, the electron density is calculated in each iteration step as a sum of partial densities over Kohn–Sham orbitals with eigenvalues up to a given chemical potential μ so that the number of electrons is not necessarily conserved. However, such a scheme does exhibit a rather slow convergence, much slower than calculations with a fixed number of electrons. However, calculations in the $N_e = $ const mode can also be related to the $\mu = $ const mode. This is done by performing calculations for different charge states, determining the corresponding potentials, and interpolating the desired quantity as a function of the potential in order to get the correct value for a given arbitrary potential. This will be demonstrated below using oxygen dissociation [21] as an example.

5.2.2 Energy Correction in the Presence of a Constant Charge Background

Introducing a constant charge background to balance the excess charge on a metal slab is relatively easy to implement into periodic DFT codes; in fact, it corresponds to the default procedure in periodic DFT codes to compensate a charged system. However, the charged background interacts with the system under consideration. In order to compare various charged systems at constant potential that will in general be counterbalanced by different charge backgrounds, the interaction energy has to be corrected for.

The presence of the charge background leads to additional terms in the DFT energy E_{DFT} [6, 8, 9]. The derivative of the energy with respect to the charge is then given by

$$\frac{\partial E_{DFT}}{\partial q} = \frac{\partial E_{syst}}{\partial q} + \frac{\partial E_{syst-bg}}{\partial q} + \frac{\partial E_{bg}}{\partial q} \tag{5.1}$$

where E_{syst} is the energy of the charged water–electrode system, $E_{syst-bg}$ is the interaction between the system and the background charge, and E_{bg} is the energy of the background. $E_{syst-bg}$ and E_{bg} can be expressed as

$$E_{syst-bg} = \int \rho_{bg} V_{syst} \, d^3 x \tag{5.2}$$

$$E_{bg} = \int \rho_{bg} V_{bg} \, d^3 x \tag{5.3}$$

where V_{syst} is the electrostatic potential of the charged water–electrode system and V_{bg} is the electrostatic potential of the background charge. Then the derivative (5.1) can be expressed as

$$\frac{\partial E_{DFT}}{\partial q} = \mu - \int \frac{V_{syst} + V_{bg}}{\Omega} d^3 x \tag{5.4}$$

$$= \mu - \int \frac{V_{tot}}{\Omega} d^3 x \tag{5.5}$$

where $\rho_e = -\rho_{bg} = q/\Omega$ was used. The total energy is obtained by integrating the chemical potential μ over the applied charge, that is,

$$E = \int_0^q \mu \, dQ = E_{DFT} + \int_0^q \left[\int \frac{V_{tot}}{\Omega} d^3 x \right] dQ \tag{5.6}$$

5.2.3 Potential in the Presence of a Constant Charge Background

The presence of the charge background does not only affect the total energy of the considered system, it also influences the one-electron potential. Naively one could think that a constant charge background cannot create any potential gradient corresponding to an electric field since it is translationally invariant. However, one has to take into account that the constant charge background is superimposed on the varying charge density of the water–metal system, and the resulting electrostatic potential as a solution of the Poisson equation is a consequence of the whole charge distribution subject to the appropriate boundary conditions. Even in vacuum regions where the charge distribution is entirely given by uniform background charge, this leads to a varying potential. This can easily be seen by inspecting the Poisson equation for a region with a constant charge background:

$$\nabla^2\phi(\mathbf{x}) = 4\pi\rho_0 \tag{5.7}$$

The general solution in Cartesian coordinates is given by

$$\phi(\mathbf{x}) = 4\pi\rho_0 \left(\sum_{i,j=1}^{3} C_{i,j}x_ix_j + \sum_{i} C_i x_i + C_0 \right) \tag{5.8}$$

with $\sum C_{ii} = 1$. Note that there is no proper solution of Eq. (5.7) for an infinitely extended isolated uniform charge background reflecting the fact that the electrostatic energy density diverges for such a system. However, for a finite region of constant charge density the potential follows a quadratic profile according to Eq. (5.8). This is confirmed in periodic DFT calculations as illustrated in Figure 5.3 where

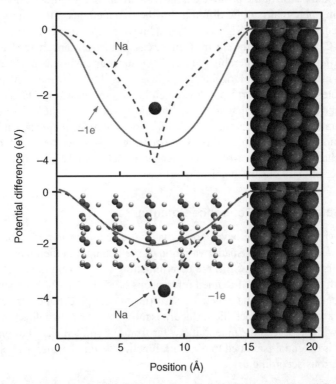

FIGURE 5.3   Calculated potential difference between charged and uncharged Cu(111) slabs with the excess electron density compensated either by a constant charge background (denoted by $-1e$) or a sodium ion pseudopotential (denoted by Na) without (upper panel) and with a water layer (lower panel) in front of the electrode (After Taylor et al. [9].) Note that the plotted atoms are only included as an illustration and do not correspond to the actual positions of the atoms in the calculations.

the upper panel gives the calculated potential difference between a charged and uncharged Cu(111) slab with the excess charge compensated by a constant charge background [9] (indicated by $-1e$). The variation of the electrostatic potential can be understood considering the fact that for positions displaced from the middle of the vacuum region there are unequal amounts of charge in the opposite directions along the surface normal.

It is obvious that this dependence of the potential in the vacuum region can create artefacts. If, for example, the work function of the metal is smaller than the depth of the potential minimum in the middle of the vacuum region, then there will be an unrealistic charge flow from the metal slab to the middle of the vacuum region.

Instead of balancing the excess charge of the slab by a constant charge background, one can also explicitly include counterions at the approximate position of the outer Helmholtz plane as a model for the electrochemical interface. The corresponding potential difference is also included in Figure 5.2 where the counterion is represented by a sodium ion pseudopotential that should lead to the same formal surface charge density as in the case of the positive charge background. As Figure 5.2 demonstrates, the resulting potential is quite different for the two methods. The slope of the two potential curves also differs significantly, leading to an electric field that is more than a factor of 2 larger close to the electrode in the case of the constant charge background.

This difference is considerably reduced in an aqueous environment because of the screening effects of the polarizable water layers, as the lower panel of Figure 5.2 demonstrates, where the corresponding potential differences at a $H_2O/Cu(111)$ interface are shown [9]. In particular, across the inner water layer the resulting electric fields are rather similar. This suggests that the continuum technique might be appropriate to model electrochemical processes occuring in the inner layer. Still one has to be aware that the introduction of a constant charge background can introduce artefacts in the description of electrochemical interfaces.

5.2.4   Selection of a Reference Potential

In the previous section, we have shown that the explicit consideration of the aqueous environment leads to a spatial variation of the potential in the inner water layer that appears to be realistic. Still, the determination of the absolute value of the potential with respect to a well-defined reference is crucial for a true comparison with electrochemical experiments which are performed under potential control.

In the case of a neutral metallic surface, it is most convenient to define the reference potential with respect to the potential ϕ_∞ far from the surface well into the vacuum region. This potential is related to the work function Φ and the chemical potential μ of the electrons according to

$$\phi_\infty = \Phi + \mu \qquad (5.9)$$

There are some subtleties about what "far away" exactly means for finite crystallites [22], but as far as periodic slab calculations are concerned, ϕ_∞ is defined as

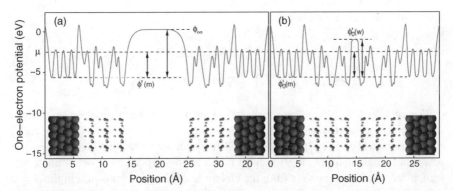

FIGURE 5.4 Schematic illustration of the electrostatic energy profile across the unit cell in a periodic slab calculation for solvated water slabs with (a) and without (b) a vacuum layer in the middle between the metal slabs: μ_e is the chemical potential of the electrons which corresponds to the Fermi energy at $T = 0\,\mathrm{K}$; ϕ_∞ is the vacuum level; and $\phi'(m)$ and $\phi'_0(m)$ are the bulk metal potentials with and without the presence of a vacuum layer in the calculations, respectively. (After Taylor et al. [9].)

the one-electron potential well in the vacuum region between the slabs where it is independent of the position within a certain spatial region.

If, however, the interstitial region between the slabs is filled with water and, in addition, the slab is charged, then it is not so obvious to define a reference potential. Taylor et al. have suggested a so-called double-reference method [9] for the situation in which the charge of the slab is compensated by a uniform background. In a first step, a DFT calculation is performed for a solvated slab without any excess charges and with a vacuum region introduced in the middle of the unit cell between the slabs, as illustrated in Figure 5.4(a). For such a setup, the vacuum level ϕ_∞ and the work function of the metal–water interface are computationally well defined, as in periodic calculations for the metal–vacuum interface.

The potential in the middle of the vacuum layer is used as the first reference by setting $\phi_\infty = 0$. The water layer should be thick enough that the vacuum level is converged with respect to the number of included water layers in Figure 5.4(a). Then it is a reasonable assumption that the electrode potential does not change when the vacuum layer in the water region is omitted [Fig. 5.4(b)]. A point in the interior of the metal slab is selected where the potential variation does not depend on the presence of the vacuum region. The corresponding potential $\phi'_0(m)$ is adjusted according to

$$\phi_0(m) = \phi(m) = \phi'(m) - \phi_\infty \qquad (5.10)$$

where the primed values indicate the unshifted values and the subscript zero denotes the uncharged calculations without a vacuum. All other potentials in the profile are then taken with respect to $\phi_0(m)$:

$$\phi_0(z) = \phi_0'(z) - \phi_0'(m) + \phi_0(m)$$

$$= \phi_0'(z) - \phi_0'(m) + \phi'(m) - \phi_\infty \tag{5.11}$$

For a charged slab, however, we are facing the problem that a variation in the electronic charge q leads to the existence of an electric field at the interface. Consequently, a vacuum reference point cannot be established because there is no region where the potential is flat. Taylor et al. [9] suggest the following procedure to determine a reference potential: A region far from the electrode is fixed at its position in the $q = 0$ calculation and its potential $\phi_0(w)$ is used as the second reference point. The rest of the system is relaxed under the influence of the applied charge, and the potential at all other positions is shifted with respect to the second reference point:

$$\phi_q(z) = \phi_q'(z) - \phi_q'(w) + \phi_0(w) \tag{5.12}$$

Finally, the electrode potential versus the normal hydrogen electrode is obtained by subtracting the work function for the H_2/H^+ couple on Pt in standard conditions,

$$\phi_{\text{NHE}} = -4.85\,\text{eV} - \phi_q \tag{5.13}$$

where $\phi_q = \mu$ is the Fermi potential taken with respect to the vacuum potential. For the unsolvated case it is related to the work function Φ according to $\phi_q = -\Phi$, as can be derived from Eq. (5.9) by setting $\phi_\infty = 0$. It has to be subtracted here since it is defined with respect to the electrons, whereas the electrode potential in electrochemistry is usually taken with respect to a positive probe charge. Note that one has to be cautious about whether the potential plotted in theoretical studies is plotted with respect to a positive or a negative probe charge because it is often not explicitly specified.

In calculational setups in which an external electric field is created either by a localized planar charge distribution [23, 24] of by a dipole layer [7] in the vacuum region, the reference potential is not that easy to determine since there is no field-free space in which a vacuum level ϕ_∞ could be determined. In order to define a reference electrode, the potential at the charged layer in the middle of the vacuum region can be set to zero [6, 25]. Thus results for different systems (e.g., different surface reconstructions) can be compared at the same potential. However, in this approach the position of the charged layer is an essential parameter since the calculated energies such as surface energies depend on the distance between surface and reference electrode [25].

Still, it is possible to define an experimentally meaningful potential in this approach by using the relation

$$dq = C\,d\phi \tag{5.14}$$

between the surface charge q and the electrode potential ϕ, which are connected via the differential capacitance C of the double layer [24]. The determination of the capacitance should in principle also be possible in the framework of DFT calculations, but it is not trivial. Thus, in a DFT study addressing the potential-induced lifting of the Au(100)–(hex) reconstruction, the surface charges were converted to electrode potentials using measured values of the capacitance [24].

5.3 WATER–METAL INTERFACE IN THE ABSENCE OF EXTERNAL FIELDS

The structure of the water–metal interface is naturally of strong interest in electrochemistry. However, the importance of the water–metal interaction in many aspects has also motivated numerous surface science studies (for reviews, see [26, 27]). Likewise, there is also a strong theoretical interest in the structural characterization of the water–metal interface and water–adsorbate interactions at the solid–liquid interface. We will first address, based on a recent excellent review [14], the geometric and electronic structure of water at the water–metal interface according to DFT calculations and then discuss the influence of water on molecule–surface interactions in the absence of external fields.

5.3.1 Structure of the Water–Metal Interface

As far as a single water monomer on metal surfaces is concerned, it binds relatively weakly to the metal atoms with adsorption energies ranging from -0.1 to $-0.4\,\text{eV}$. As for some particular important late transition metals, the interaction strength is ordered according to Au<Ag<Cu<Pd<Pt<Ru<Rh [14]. The water monomers typically bind via their oxygen atom to the top sites of metal surfaces in an almost flat configuration, as illustrated in Figure 5.5, at distances between 2.25 Å (Cu) and 3.02 Å (Au), which are much larger than typical distances of specifically adsorbed or chemisorbed species.

From the electrochemical point of view, the structure of water at a solvated metal electrode is of great interest. Still there are many uncertainties left. In particular, the question of whether water at the solid–liquid interface is crystalline (i.e., icelike) or rather waterlike is not fully answered yet. On late transition metal surfaces, in particular with hexagonal symmetry, it is traditionally assumed that water adsorbs in a bilayer structure [14] similar to that of the densest layer of ice [26]. In this structure, every second water molecule is oriented parallel to the surface in a fashion similar to the water monomer shown in Figure 5.5. For the other water molecules, there are in fact two different possible orientations, namely the so-called H-down and H-up structures with one hydrogen atom either pointing toward or away from the surface. These structures are illustrated in Figures 5.6(a) and (b).

The adsorption energies per water molecule on late transition metals with respect to the free water molecules range between -0.42 and $-0.56\,\text{eV}$ [14, 28]. The H-up structure is energetically favorable on Ni(111), Cu(111), and Ru(0001), whereas on Rh(111), Ag(111), Pt(111), Pd(111) [14], and Pd/Au(111) [28] the H-down structure is more stable. In this context it should be mentioned that these energies are

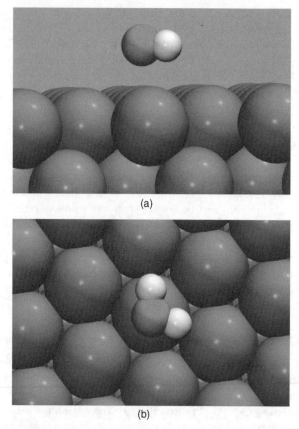

FIGURE 5.5 Side and top views of the typical adsorption configuration of a water monomer on close-packed metal surface.

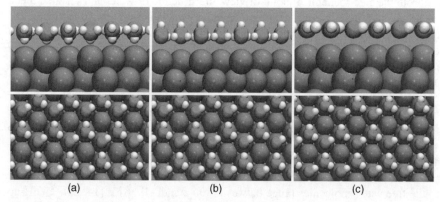

FIGURE 5.6 Side and top views of water bilayer structures: (a) H-down bilayer; (b) H-up water bilayer; (c) half-dissociated water–OH bilayer with the additional hydrogen atoms being removed.

less negative than the calculated sublimation energy of water in a 32-molecule per cell model of ice I_h, $E_{sub} = -0.666$ eV [28]. This means that the considered water adlayers are not thermodynamically stable with respect to conversion to a three-dimensional ice cluster. This might be an indication that the metal–water interaction is underestimated by current DFT functionals.

Therefore it is of interest to decompose the computed adsorption energies in the bilayer structures into the contribution from the water–metal and the water–water interaction. Unfortunately it turns out that there is no unique decomposition of the water adsorption energies into these two contributions [14, 28]. One can assume that the bilayer is first assembled in the gas phase and then deposited on the electrode surface. Alternatively, one can first adsorb the water molecules individually on the surface and then assemble the icelike bilayer structure. The first approach does not take into account the changes of the interwater binding during the adsorption of the water bilayer, whereas the second approach assumes that the water–metal interaction stays constant when the water bilayer is assembled.

However, although there are small quantitative differences in the two energy decomposition schemes outlined above [14, 28], qualitatively they yield rather similar results. The main result is that most of the binding energy in the water bilayers (about 75%) comes from the water–water interaction whereas the water–metal interaction is rather weak with binding energies below 0.2 eV per water molecule. Interestingly enough, the water–water interaction on the considered close-packed surfaces is found to be almost independent of the lattice spacing [14], although the next-nearest-neighbor lattice distance between equivalent water molecules spans values from 4.33 Å for Ni to 5.04 Å for Ag whereas the corresponding calculated equilibrium value for ice, I_h, is 4.50 Å. This is in fact surprising since one would expect that the hydrogen bonding between the water molecules exhibits a stronger dependence on distance.

The water bilayers might in fact not stay intact on the electrode surfaces. On the basis of DFT calculations, Feibelman had suggested that water on Ru(0001) should form a half-dissociated overlayer [29] where every second water molecule is dissociated to OH. Such a H_2O–OH structure with the additional hydrogen atoms being removed is illustrated in Figure 5.6(c). As a matter of fact, Ru is not the only metal where half-dissociated water layers are more stable. In Figure 5.7, the bilayer dissociation energy is plotted which corresponds to the total energy difference between the most stable intact bilayer and the half-dissociated water overlayer per unit cell where the extra hydrogen atom is assumed to be adsorbed at its most favorable threefold hollow site on clean portions of the metal. The negative values for Ru, Rh, and Ni indicate that the half-dissociated water overlayer is more stable whereas on Cu, Pt, and Ag the water bilayers rather stay intact, and Pd is an undecided case. The dissociation energies in Figure 5.7 are plotted against the OH adsorption energies at the on-top sites of the respective surfaces showing a clear correlation. The less noble metals Ru, Rh, and Ni bind OH more strongly than the more noble metals, which apparently provides a driving force for the water dissociation. Still it is not clear whether the water layers indeed dissociate since the formation of the partially dissociated overlayers might be kinetically hindered by the presence of high barriers.

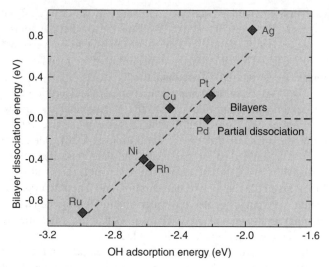

FIGURE 5.7   Bilayer dissociation energy as a function of the OH adsorption energy at atop sites. (After Michaelides [14].)

In order to assess the influence of the presence of water on the interaction of molecules with electrode surfaces, it is interesting to inspect the changes of the electronic structure at the surface upon the adsorption of water. In Figure 5.8, the local density of states (LDOS) of the clean Pt(111) surface is compared with those of water-covered surfaces [30]. For the H-down water bilayer shown in Figure 5.6(a), there are three inequivalent Pt surface atoms per surface unit cell, either noncovered or covered by a water molecule bound via an oxygen atom or a hydrogen atom. The LDOS of these Pt atoms is similar to the one of bare Pt(111). This indicates that there is a rather weak interaction between water in the bilayer structure and metal electrode surfaces with binding energies below 0.2 eV, as discussed above.

An isolated water monomer, on the other hand, is bound by 0.35 eV to Pt(111) [30]. This stronger interaction leads to a more pronounced modification of the LDOS of the Pt atom closest to the water monomer. In particular, the peak at about −4.5 eV is caused by the hybridization of the water $1b_1$ orbital with the Pt d band [31]. Still it should be noted that also in the case of water monomer adsorption the water-induced change in the electronic structure of Pt is small, some peak heights of the LDOS are altered, but the peak positions and their widths remain almost the same.

Although the adsorption energies of the H-up and H-down water bilayers are usually rather similar, there is a strong dependence of the bilayer-induced work function change $\Delta\Phi$ of the metal electrodes on the orientation of the water bilayer resulting in differences in $\Delta\Phi$ of about 2.2 eV on Pd(111) [32] and 2.0 eV on Pt(111) for the two orientations. It has already been noted that these large differences lead to a charge control of the water monolayer–metal interface, that is, the stable water bilayer structure can be tuned by changing the surface charge [32]. It is not surprising that the effect of the H-up and H-down water bilayers on the work function is so different because

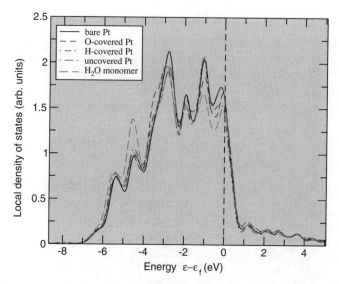

FIGURE 5.8 The LDOS of Pt(111) surface atoms without and in the presence of water. For the H-down water bilayer the LDOS of the three inequivalent Pt atoms within the surface unit cell is shown whereas for the adsorption of the water monomer only the LDOS of the Pt atom directly below the water molecule is plotted. (After Gohda et al. [30].)

their associated dipole moments have different signs. However, it is quite surprising that on Pd(111) [32] as well as on Pt(111) both types of bilayers lower the work function. The H-down water bilayer induced work function change is $\Delta\Phi = -0.23\,\text{eV}$, whereas it is $\Delta\Phi = -2.27\,\text{eV}$ for the H-up water bilayer. Naively one would expect that, because of the oppositely oriented dipole moments, the H-down bilayer would lower the work function while the H-up bilayer would increase the work function.

In order to understand this astonishing result, it is instructive to analyze the charge density difference

$$\Delta\rho = \rho[H_2O/Pt(111)] - \{\rho(H_2O) + \rho[Pt(111)]\} \qquad (5.15)$$

which corresponds to the water adsorption-induced rearrangement of the charge density. In Figure 5.9, the laterally averaged charge density difference as a function of the position perpendicular to the surface for both types of bilayers is plotted. Since oxygen is more electronegative than hydrogen, the free H-up bilayer has a dipole moment that lowers the work function. In addition, due to the interaction of the water layer with the Pt(111) surface, surface electronic charge flows from the water layer to a region close to the Pt atoms. Figure 5.9 also shows the integral of the density difference along the surface normal, and the position of its maximum roughly separates the regions of charge surplus and charge deficiency. This charge rearrangement causes an additional effective dipole layer that further lowers the work function.

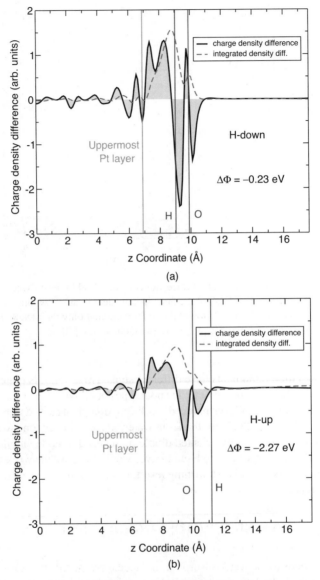

FIGURE 5.9 Laterally averaged charge density difference and its vertical integrated value upon adsorption of an H-down and an H-up water bilayer, respectively, on Pt(111) as a function of the position perpendicular to the substrate. The vertical lines indicate the position of the uppermost Pt atoms, the oxygen atom, and the H-down and H-up atoms, respectively.

In the case of the H-down water bilayer on Pt(111), there is even a much stronger charge flow from the water layer to the region between the metal and water layer. The water layer itself has a dipole moment that would increase the work function of the surface, but the charge rearrangement is so large that the resulting effective dipole moment overcompensates the dipole moment of the water layer leading to a net reduction of the Pt(111) work function by 0.23 eV.

Figure 5.9 also indicates that there is some charge rearrangement within the Pt(111) slab upon the interaction with water, but it is rather small, which is consistent with the minor changes observed in the Pt(111) local density of states in Figure 5.8. Thus the overall lowering of the Pt(111) work function in the presence of both the H-up as well as the H-down water bilayers is mainly a consequence of the high polarizability of water.

The geometric and electronic structure discussed so far are related to static properties of adsorbed water in equilibrium geometries. However, processes in electrochemistry occur at temperatures close to room temperature so that the observed experimental properties correspond to thermal averages. Furthermore, the dynamics of the water layer itself is interesting for an understanding of electrochemical processes on metal surfaces. The structure of the Ag(111)–water interface was addressed by ab initio molecular dynamics (AIMD) simulations [33] based on the Car–Parrinello scheme [34]. In this study, the initial configuration for the AIMD runs was obtained from a classical MD simulation, employing a model chemisorption metal–water potential [12], that was performed for a system of 256 water molecules for 30 ps at a temperature of about 300 K in a $c(8 \times 4)$ surface unit cell. The positions of the 48 water molecules closest to the surface where then used for the initial configuration of the AIMD simulation that was run for 2.1 ps.

AIMD trajectories of the water molecules in the first layer within the surface unit cell are shown in Figure 5.10(a). In these simulations, all water molecules were bound via the oxygen atom to the top sites of the Ag(111) surface. As the trajectories indicate, the water molecules in the first layer remain rather localized above the top sites. The molecules of the second layer seem to be more mobile, as the plotted trajectories of selected molecules in Figure 5.10(b) suggest. Interestingly enough, the configuration of the water molecules in the first layer [Fig. 5.10(a)] does not exhibit any indication of a hexagonal bilayer structure, although the coverage of 0.63 is roughly the same as

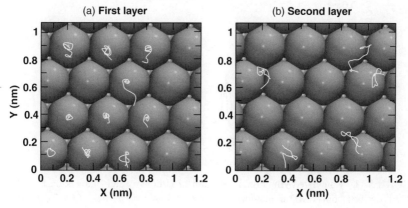

FIGURE 5.10   Trajectories of AIMD runs of water on Ag(111) of water molecules in the first layer (a) and of selected molecules of the second layer (b). (After Izvekov [33].)

in the bilayers. It is not clear whether this is an artefact of the parameterized potential used to determine the initial configuration or whether this is a consequence of the nonzero temperatures.

In order to derive experimentally whether the water structure at the water–electrode interface is crystalline or rather disordered, measured vibrational spectra have been analyzed [35], in particular focusing on the OH stretch vibration in the range between 2800 and 3800 cm^{-1}. Generally, OH stretch vibrations at about 3200 cm^{-1} have been assigned to threefold coordinated water (i.e., water in a disordered liquidlike structure), whereas OH stretch vibrations at about 3400 cm^{-1} have been taken as a signature of fourfold coordinated water (i.e., highly ordered icelike water molecules) [36]. While for Pt(111) two broad peaks at 3200 and 3400 cm^{-1} were obtained [35], the observed vibrational spectra on Au(111) were dominated by a broad peak around 3500 cm^{-1}; however, this peak was attributed to oxide formation. The results were interpreted as being an indication that on Pt(111) ordered and disordered water structures coexist whereas on Au(111) the water is less well ordered.

Vibrational spectra can be evaluated from molecular dynamics runs performing a Fourier transformation of the velocity autocorrelation function [37]. We have carried out AIMD simulations of H-down water bilayers on Pt(111) and Ag(111) for 5 ps with a time step of 1 fs within a $2\sqrt{3} \times 2\sqrt{3}$ surface geometry at a temperature of 300 K based on periodic DFT calculations [15] using the Perdew-Burke-Ernzerhof (PBE) functional [16]. These conditions lead to a spectral resolution of $\delta\omega = 6$ cm^{-1} [37]. The resulting spectra are plotted in Figure 5.11. Two main peaks are obvious, around 3500 cm^{-1} and around 1600 cm^{-1}, which are related to the OH stretch and bending

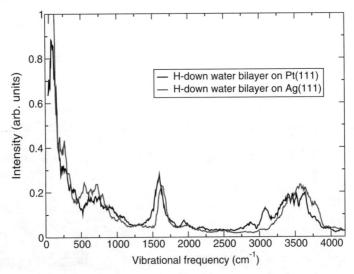

FIGURE 5.11 Calculated vibrational spectrum of H-down water bilayers on Pt(111) and Ag(111) derived from AIMD simulations that have been run for 5 ps in a $2\sqrt{3} \times 2\sqrt{3}$ geometry at a temperature of 300 K.

vibrations, respectively. First of all it is obvious that the water vibrations on Pt(111) are red shifted compared to those on Ag(111), which can be explained by the stronger Pt–water interaction compared to the Ag–water interaction. Interestingly enough the spectra are rather similar to the experimentally measured spectra on Pt(111) and Au(111) [35], in particular as far as the existence of two peaks on Pt(111) is concerned, although no disordered structures evolved on Pt(111) in the AIMD simulations. On Ag(111), on the other hand, the simulations indeed suggest that the water layer might become disordered at room temperature.

5.3.2 Influence of Water on the Molecule–Metal Interaction

The field of electrocatalysis is concerned with reactions of molecules at electrode surfaces in an electrochemical environment. In this context, it is of course interesting how the presence of water influences molecule–metal interactions. In the previous section we have shown that adsorbed water bilayers become strongly polarized but that their presence only weakly influences the electronic structure of the substrate. Thus it is not clear how much the presence of water bilayers modifies adsorption energies in specific adsorption.

This issue was addressed in a DFT study of CO and hydrogen adsorption on a bimetallic PdAu surface consisting of a pseudomorphic Pd overlayer on Au(111). In Table 5.1, the hydrogen and CO adsorption energies on the clean substrate and in the presence of H-down and H-up water bilayers are collected. In the presence of water bilayers, the binding of hydrogen and CO to Pd/Au(111) becomes weaker, but only by less than 10%. For hydrogen, the results are basically independent of the specific form of the water bilayer. For CO, the change in the binding due to the presence of water is somewhat larger than for hydrogen, and also the results for the two orientations of the water bilayer differ by about 60 meV.

In order to understand these trends one should note that the height of the hydrogen atom above the plane of the Pd atoms is only 0.6 Å whereas the carbon atom of the adsorbed CO is located about 1.2 Å above the Pd atoms [38]. The water molecules, on the other hand, are more than 2 Å above the metal atoms, as for example Figure 5.9 indicates. Thus the hydrogen adsorption energies are only weakly influenced by the presence of water because the H atoms are located so close to the surface and water

TABLE 5.1 **Hydrogen and CO Adsorption Energies (eV) at the Most Favorable Adsorption Site on the Pd/Au(111) Overlayer System on the Clean Surface and in the Presence of H-Down and H-Up Water Bilayers**

	Clean Surface	H-Down Bilayer	H-Up Bilayer
H (fcc)	−0.690	−0.661	−0.660
CO (hcp)	−2.043	−1.866	−1.923

Source: From Ref. [7].

Note: The H adsorption energy is taken with respect to the free H_2 molecule.

has only a minor effect on the electronic structure of the metal substrate. The CO molecules are located further away from the surface. Hence they will directly interact with the polarized charge distribution of the water bilayer, which leads to a larger modification of the adsorption energies; furthermore, there is also a dipole–dipole interaction between the adsorbed CO molecules and the water layer that causes the dependence of the CO adsorption energy on the orientation of the water bilayer.

The water bilayers are not only weakly interacting with metal electrodes, their adsorption energy also does not strongly depend on the adsorption site. As Figure 5.6 indicates, in their equilibrium structure the hexagonal rings of the water bilayers are arranged around a metal atom that remains uncovered. Now for both hydrogen and CO the most favorable adsorption site on Pd/Au(111) is the threefold hollow site (see Table 5.1). The fact that the hydrogen and CO binding energies on Pd/Au(111) are reduced in the presence of water demonstrates that there is a repulsive interaction between these adsorbates and water. Since the adsorbed hydrogen atom and the water bilayer are only weakly interacting, the water bilayer is only slightly distorted due to the presence of the hydrogen atoms, as Figure 5.12(a) indicates where a top view of the energy minimum structure is shown. CO and water are interacting more strongly. Furthermore, CO is bound more strongly than the water molecules. As a consequence, the presence of adsorbed CO in threefold hollow positions causes a shift of the water bilayer to maximize the water–CO distance.

These results show that binding energies in specific adsorption are weakly influenced by solvation effects as long as the adsorbates are located close to the electrodes. However, transition state configurations of electrocatalytic reactions are often further away from the surface than the adsorption sites, which might lead to a larger effect of water on the reaction. As a simple example, the influence of water layers on the dissociation of H_2 above Pt(111) was studied [30].

A two-dimensional cut through the potential energy surface of H_2/Pt(111) in the presence of a water double layer is plotted in Figure 5.13 as a function of the H–H distance d and the H_2 center-of-mass distance Z from the surface [30]. Note that on the clean Pt(111) surface H_2 dissociation is only hindered by a barrier of 54 meV. At

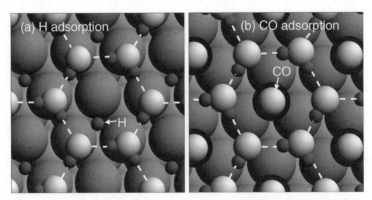

FIGURE 5.12 Relaxed adsorption geometry of the H-down bilayer on the Pd/Au(111) over-layer system together with adsorbed hydrogen (a) and CO (b). (After [7].)

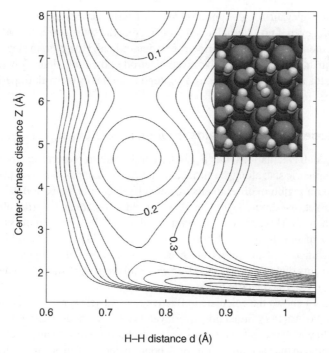

FIGURE 5.13   Two-dimensional cut through the potential energy surface of the interaction of H_2 with Pt(111) covered by two water bilayers. The potential energy is plotted as a function of the H–H distance d and the H_2 center-of-mass distance Z from the surface. The lateral position and orientation of the H_2 molecule correspond to a fcc hollow-top-hcp hollow configuration, as indicated in inset. The contour spacing is 50 meV. (After Gohda et al. [30].)

the solvated surface, this barrier is increased by 167 meV to a value of 221 meV. This barrier is located roughly at the position of the first water bilayer above the surface. It is interesting to note that the increase in the barrier height approximately corresponds to the barrier for H_2 to propagate through a free water bilayer which is very similar to the barrier for the H_2 propagation through the upper water bilayer at a height of about 6 Å above the surface, as shown in Figure 5.13. Thus, to a first approximation, the barrier for the H_2 dissociative adsorption at a Pt(111) surface solvated with icelike water can be regarded as a superposition of the dissociation barrier for the bare surface plus the barrier for propagation through the water bilayer, which indicates that the modification of the barrier is not the result of any water-induced modification of the electronic structure of Pt(111).

5.4   WATER–METAL INTERFACE IN THE PRESENCE OF EXTERNAL FIELDS AND/OR VARYING ELECTRODE POTENTIALS

So far we have only considered the structure of water bilayers on metal surfaces and their interaction with molecules without any external field. These calculations are certainly relevant for the understanding of basic processes in electrocatalysis. Furthermore, even if no external field is considered, the setup corresponds to a certain electrode potential that could in principle be determined. However, in electrocatalysis the basic quantity of interest is the given electrode potential. Here we will describe external fields and/or varying electrode potentials in the framework of periodic DFT calculations.

5.4.1 Modeling Electric Fields through Dipole Layer

In periodic slab calculations, it is a standard procedure to correct the dipole moment of a nonsymmetric slab along the surface normal by a planar dipole layer in the middle of the vacuum region in order to electrostatically decouple the periodically repeated slabs [18]. Yet, the dipole layer can also be used to create an external electric field acting on the slab. This approach was employed in a DFT study to address the stability of the H-down and H-up water bilayers on the Pd/Au(111) pseudomorphic overlayer system as a function of an external electric field [7]. Because of the oppositely oriented dipole moments of the H-down and H-up water bilayers, they should react differently to an applied external electric field.

Figure 5.14 shows the calculated effective one-particle potential through a water layer on a Pd/Au(111) bimetallic overlayer system without and with an external electric field which is created by a dipole layer whose position is indicated. The slope of the potential in the vacuum region corresponds to the applied electric field. Note that inside the metal electrode the one-particle potential is hardly affected demonstrating the good screening properties of metals. In the water layer, the potential is slightly modified by the presence of the electric field.

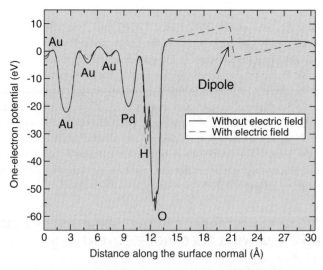

FIGURE 5.14 Effective potential along the surface normal z which passes through the water layer for zero external field and with an electric field of strength $E = -0.7$ V/Å. (After [7].)

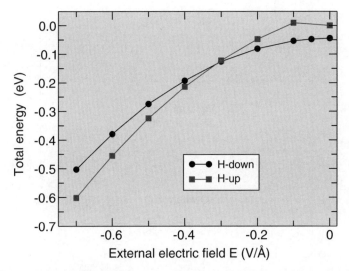

FIGURE 5.15 Change of the total energy of H-down and H-up water bilayers as a function of the external electric field. The energy zero corresponds to the H-up structure without any external electric field. (After [7].)

Surface X-ray scattering experiments had found a surprisingly large inward contraction of the water molecules for positive electrode potential [39, 40]. However, electric fields with a strength of up to 0.5 V/Å only lead to displacements of the atoms in the water bilayer by less than 0.05 Å. Still, the applied electric field has a significant effect on the stability of the water layers, as Figure 5.15 demonstrates. There the total energy of the H-down and H-up water bilayer as a function of the external electric field is plotted. Whereas in the field-free case the H-down structure is more stable, for external electric fields more negative than −0.3 V/Å the H-up bilayer becomes energetically favorable. This is in qualitative agreement with experiments for water on Ag(111) which found a field-induced rotation or flip of adsorbed water molecules [40, 41].

5.4.2 Explicit Consideration of a Counterelectrode

Instead of including a dipole layer, a counterelectrode may be explictly considered. Some authors have used a localized planar charge distribution with a Gaussian shape perpendicular to the surface [23–25] in order to address the reconstruction of charged surfaces. In this approach, the excess surface charge is easy to determine since it corresponds to the charge of the counterelectrode. These computational studies did not consider any explicit water layers at the metal surface. As already discussed in Section 5.2.4, in this setup it is not straightforward to obtain the corresponding electrode potential; it was done either by using relation (5.14) with experimentally derived capacities [24] or by setting the potential at the position of the charged layer to zero [25]. As a general result these calculations showed that to first order in the

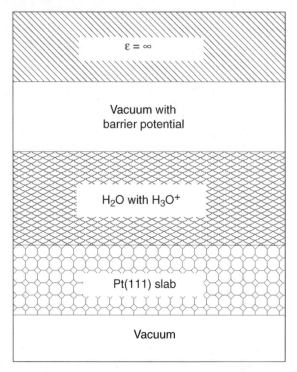

FIGURE 5.16   Computational setup used by Sugino et al. [42, 43] to model the water/Pt(111) interface under bias potential within the effective screening method.

surface charge additional positive charges favor surfaces with smaller work function whereas negative charges favor surfaces with larger work function [25].

The so-called effective screening method (ESM) was suggested by Sugino et al. [42, 44] based on a scheme for the electronic structure calculations that is periodic in the lateral direction but not in the direction perpendicular to the surface. In this method, which is schematically illustrated in Figure 5.16, the laterally periodic slab covered by water and/or adsorbates is placed between two polarizable continua characterized by their dielectric constant ε. In order to model an electrochemical cell, the simplest model is to use vacuum ($\varepsilon = 1$) at the metal side and a perfect conductor ($\varepsilon = \infty$) at the other side. The corresponding electrostatic potential across the system is obtained from the Laplace equation using a Green's function solver [44]. The water layer was modeled by 32 water molecules within a $3 \times 2\sqrt{3}$ unit cell leading to four bilayers. To have acidic conditions, one proton was added to the water layer. The structure of this system was then studied by AIMD simulations. To prevent the water bilayers from desorbing and the electrons from entering the perfect conductor, additional barrier potentials were introduced.

The bias can be varied in this approach by adding excess electrons to the system. These excess charges will induce opposite surface charges in the perfect conductor. The perfect conductor is placed approximately 20 Å above the uppermost Pt layer;

hence the countercharge is located relatively far away from the Pt electrode. Since this setup does not have a three-dimensional periodicity, there is no need for compensating charges. The introduction of the excess electrons leads to the creation of a uniform electric field. During the initial thermalization of the system in the MD run which takes several picoseconds, the double layer is formed and the potential profile becomes almost flat perpendicular to the layer due to the strong polarization of water [43]. Thus a corresponding electrode potential can be derived.

The AIMD production runs have been performed for 4 ps with excess charges of 0, 0.35, and 0.70 e per supercell corresponding to potentials of 0.04, 0.36, and 0.81 V, respectively, with respect to the potential of zero charge [43]. It is found that the density of the first water layer at the electrode, the so-called contact layer, is only slightly increased by raising the potential. The rearrangement within the layers, however, becomes strongly modified for higher potentials leading to a smaller binding between the contact layer and the bulk water above the first layer. It has been speculated whether this is an indication for a hydrophobic water monolayer, as suggested on the basis of experiments [45].

5.4.3 Uniform Compensating Charge Background

Using a uniform charge background to compensate the surface excess charges has been used in a number of studies addressing electrochemical systems at the solid–water interface [8, 9, 21, 32]. Here we will use a study of the dissociation of oxygen on Pt(111) in the presence of water and coadsorbates as a function of the electrode potential [21] as an illustration of this method.

In this work, the O_2 dissociation was studied on Pt(111) within a 3×3 surface unit cell on the clean surface, and on Pt(111) covered with one Na atom per surface unit cell. Water was included in the simulation either as a single molecule or as four icelike bilayers. The nonsolvated slab was kept symmetric, that is, the adsorbates were added to both sides of the slab for a direct evaluation of the work function and in order to avoid spurious dipole–dipole interactions in the case of Na adsorption because of the strongly ionic Na–Pt bond.

The water structure was obtained by adding 24 water molecules to the 3×3 surface unit cell in an icelike structure leading to four bilayers. For the solvated slab, adsorbates were introduced into the system by replacing water molecules with them. For O_2, 2O, or Na adsorption, a single H_2O molecule was replaced, while for the coadsorption of Na with O_2 or 2O two H_2O molecules were replaced, and then the resulting structures were optimized. In addition, the transition state between the configurations for the O_2 and the 2O adsorption was located using automatic transition state search routines [46, 47]. All these calculations were performed for different charge states of the Pt electrode by adding charges corresponding to -1, -0.5, 0, 0.5, and 1 e to the system which were compensated for by a uniform charge background. For the initial and the final states of the O_2 dissociation as well as for the transition state, the potential was derived using the double-reference method [9] described in Section 5.2.4 as a variation of the geometric configuration along the reaction path can also lead to a change in the work function.

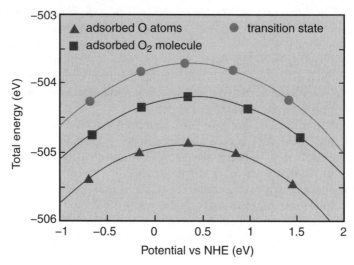

FIGURE 5.17 Total energies versus potential for the initial, transition, and final state of O_2 dissociation on solvated Pt(111). The solids curves are quadratic fits to the results. (After Wasileski and Janik [21].)

Hence these calculations have been performed not for a given specified value of the potential but rather for a given charge state, and the corresponding potential has been derived a posteriori. In order to yield the energies of the considered states for any arbitrary value of the potential, the results for the different charge states can be interpolated as a function of the electrode potential generating a continuous energy-versus-potential curve for each state. This is illustrated in Figure 5.17, where the potential dependence of the initial, transition, and final state energies for the O_2 dissociation on the solvated Pt(111) slab is plotted. The symbols denote the total energies for the different states at the five different charge states considered whereas the solid curves correspond to quadratic fits to the results.

There is in fact a rationale for using a quadratic fit, as it can be regarded as an expansion of the energy as a function of the potential about the potential of zero charge (ϕ_{pzc}):

$$E(\phi) = \tfrac{1}{2}C\left(\phi - \phi_{pzc}\right) + E_{pzc} \tag{5.16}$$

where C is the capacitance and E_{pzc} is the energy of the system at zero charge. The fact that the curvature of the three curves is not identical indicates that the capacitance of the system depends on the actual atomic configuration.

A similar fit as shown in Figure 5.17 was also made for the O_2 dissociation barrier on the solvated slab with an additional Na atom. From a figure such as Figure 5.17, the O_2 dissociation barriers can be read off for constant charge and for constant potential. The energy difference between the symbols corresponds to the dissociation barrier at constant charge, whereas the energy difference of the curves for any given value of

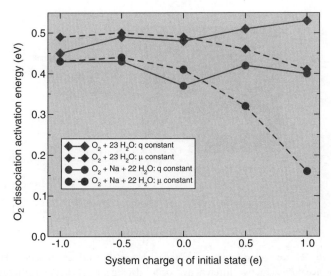

FIGURE 5.18   Dissociation barrier of O_2 on solvated Pt(111) without and with Na coadsorption denoted by O_2 + 23 H_2O and by O_2 + Na + 22 H_2O, respectively, for various system charges q of the system and for constant potential. In the latter mode, the potential of the initial state has been used as the reference and been kept fixed. (After Wasileski and Janik [21].)

the potential corresponds to the dissociation barrier at constant potential. Thus, if the symbols for a particular value of the charge are aligned vertically above each other, then the dissociation barriers are the same in the constant charge and in the constant potential mode. However, an inspection of Figure 5.17 reveals that this is in general not the case.

The O_2 dissociation barriers on solvated Pt(111) without and with Na coadsorption for various system charges are plotted in Figure 5.18 and compared to the values for constant charge where the potential of the initial state has been kept constant. It is obvious that for negative excess charges there is only a small difference between the constant charge and the constant potential mode, whereas for positive charges corresponding to positive potentials there is a significant difference. The Na coadsorption has two effects. First, it reduces the O_2 dissociation barrier at all given charge states and potentials. Second, it makes the potential dependence of the O_2 dissociation barrier much more dramatic. These effects have been attributed to the enhanced polarizability of the system when Na is coadsorbed; however, the exact mechanism is not easy to resolve because of the complexity of the system [21]. Most importantly, the differences between the results in the constant charge and the constant potential mode indicate that it is crucial to consider the electrode potential for the determination of reaction energies in electrochemical systems.

5.4.4   Explicit Consideration of Counterions

We have already seen in the work by Sugino et al. presented in Section 5.4.2 that counterions such as protons can of course be added to the water layer. This approach can be

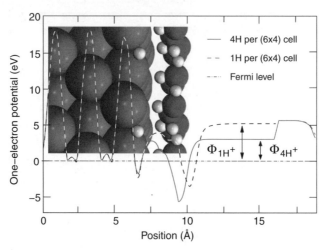

FIGURE 5.19 One-electron potential averaged in lateral direction as a function of the position along the surface normal for two different hydrogen coverages and hence two different electrode potentials. In the vacuum layer, there is a potential drop due to the presence of a dipole layer. (After et al. [10].) The inset illustrates the structure of the interface with additional protons.

taken further by varying the number of electrons/protons in the double layer and thus changing the electrode potential. This can be achieved by introducing coadsorbates such as hydrogen to the system [10, 48]. The added hydrogen atoms become solvated as protons leading to the formation of hydronium ions (H_3O^+), and the electrons move to the metal electrode. By changing the hydrogen concentration, the surface charge and hence the electrochemical potential can then be varied. In this setup, the whole supercell always remains neutral so that no countercharges are needed.

This approach has been used to address the hydrogen evolution reaction in an electrochemical double layer on Pt(111) [10]. In this study, one adsorbed water bilayer on Pt(111) was considered. The potential was varied by adding one to four hydrogen atoms to a 6×4 supercell, one and two hydrogen atoms to 6×2 or 3×4 supercells, and one hydrogen atom to a 3×4 supercell.

The variation of the electrode potential as a function of the adsorbed hydrogen atoms is illustrated in Figure 5.19 where the one-electron potential is shown for two different concentrations of hydrogen atoms, one or four atoms per 6×4 supercell. The underlying inset illustrates the atomic configuration. The corresponding electrode potential can be derived from the work function of the system given by the flat potential in the vacuum region. Note that the potential is averaged in the lateral direction; therefore the potential variations are quite different from, for example, those shown in Figure 5.15 where the potential along one particular line is shown.

This setup has been used to study the elementary processes occuring in the hydrogen evolution reaction on Pt(111), namely the Volmer reaction,

$$H^+ + e^- \rightarrow H_{ad} \tag{5.17}$$

the Tafel reaction,

$$2H_{ad} \rightarrow H_2 \tag{5.18}$$

and the Heyrovsky reaction,

$$H_{ad} + H^+ + e^- \rightarrow H_2. \tag{5.19}$$

Here we will concentrate on the Tafel reaction since it is the reverse reaction of the dissociative adsorption of H_2 on Pt(111) discussed in Section 5.3.2. In order to calculate the barrier of the Tafel reaction as a function of the electrode potential, first the equilibrium hydrogen coverage as a function of the potential was determined which can be derived from the differential adsorption energies as a function of the coverage [10]. Then the barrier for the Tafel reaction was evaluated for different hydrogen coverages and related to the corresponding potentials. This relationship is plotted in Figure 5.20 at the electrode–vacuum interface and in the presence of a water bilayer. The same is plotted also for the Heyrovsky reaction (5.19). Note that the calculated data points for the Heyrovsky reaction at more negative and more positive potentials are not shown.

There is a large gap between the results at negative and at positive potentials. This is caused by the fact that there is a discontinuity in the differential hydrogen adsorption energies once the coverage becomes larger than 1. The gap in Figure 5.20 could only

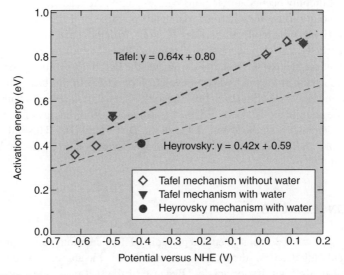

FIGURE 5.20 Calculated activation energy for the Tafel reaction as a function of the potential without (diamonds) and with (triangles) a water bilayer and for the Heyrovsky reaction with water (circles). (After Karlberg et al. [10].) For the Heyrovsky reaction, two further barriers were determined which lie outside the plotted potential range. The dashed lines correspond to linear fits to the data.

be closed if larger unit cells were chosen. Furthermore, according to Figure 5.20, the barrier for the Tafel reaction is not changed significantly by the presence of the water [10]. This seems to be at variance with the results by Gohda et al. [30] in Section 5.3.2, who found that the presence of water leads to an increase of the barrier for the inverse Tafel reaction by about 170 meV. This seeming inconsistency of the results is resolved when one considers that the presence of water also leads to a reduction of the atomic hydrogen-binding energies from 468 meV at the clean Pt(111) surface to 395 meV in the water-covered case. As a consequence, the reduction of the atomic binding energies is compensated by the increase in the barrier height, so that the difference, which is the barrier for hydrogen desorption, remains almost unchanged [30].

The dependence of the barrier for the Tafel reaction on the potential is approximately linear. The slope of the linear fit to the data,

$$\alpha = \frac{dE_a}{d\phi} \tag{5.20}$$

gives the so-called transfer coefficient, which is a measure of the symmetry of the activation barrier. The value of $\alpha = 0.64$ for the Tafel reaction indicates that the barrier location for the Tafel reaction is closer to the initial state at the electrode, whereas $\alpha = 0.42$ for the Heyrovsky mechanism means that in this case the barrier is closer to the outer Helmholtz plane.

At potentials around 0 V, the hydrogen coverage becomes 1, and further hydrogen adsorption only occurs at potentials below -0.5 V. At negative potentials where hydrogen evolution becomes thermodynamically possible, both the Tafel and the Heyrovsky reaction exhibit moderate barriers. Figure 5.20 demonstrates that the calculated barriers for the Heyrovsky mechanism are smaller than for the Tafel mechanism. This suggests that the Heyrovsky reaction dominates the hydrogen evolution. Experimentally, the mechanism for the hydrogen evolution on Pt electrodes has been found to depend on the electrode orientation [49]. As for Pt(111), the exact reaction mechanism could not be unambiguously deduced; the measured activation energy of 0.18 eV [49], however, is smaller than those calculated in the DFT study [10]. This indicates that an even more realistic description of the hydrogen evolution on Pt electrodes might be required.

5.5   CONCLUSIONS

This brief review has demonstrated that periodic DFT calculations can be quite powerful in elucidating details of the metal–water interface in the absence and the presence of external fields from first principles. Because of the still incomplete status of our knowledge about the microscopic nature of structures and processes at the electrochemical solid–liquid interface, there are many open questions remaining which wait to be addressed by theory. However, the theoretical description of the solid–liquid interface in the presence of varying electrode potentials is not trivial. It is certainly fair to say that the first-principles treatment of these systems has not matured yet. Several different theoretical approaches to represent external fields and varying electrode potentials exist, and all have advantages and disadvantages. There is still room for improvements in the realistic theoretical description of electrochemical solid–liquid interfaces. This makes this research field demanding but also exciting.

REFERENCES

1. G. Ertl, *Angew. Chem. Int. Ed.*, 47 (2008) 3524.
2. A. Gross, *Surf. Sci.*, 500 (2002) 347.
3. W. Kohn, *Rev. Mod. Phys.*, 71 (1999) 1253.
4. F. Besenbacher, I. Chorkendorff, B. S. Clausen, B. Hammer, A. M. Molenbroek, J. K. Nørskov, and I. Stensgaard, *Science*, 279 (1998) 1913.
5. N. M. Marković and P. N. Ross, Jr., *Surf. Sci. Rep.*, 45 (2002) 117.
6. A. Y. Lozovoi, A. Alavi, J. Kohanoff, and R. M. Lynden-Bell, *J. Chem. Phys.*, 115 (2001) 1661.
7. A. Roudgar and A. Gross, *Chem. Phys. Lett.*, 409 (2005) 157.
8. J. S. Filhol and M. Neurock, *Angew. Chem. Int. Ed.*, 45 (2006) 402.
9. C. D. Taylor, S. A. Wasileski, J.-S. Filhol, and M. Neurock, *Phys. Rev. B*, 73, (2006) 165402.
10. E. Skúlason, G. S. Karlberg, J. Rossmeisl, T. Bligaard, J. Greeley, H. Jónsson, and J. K. Nørskov, *Phys. Chem. Chem. Phys.*, 9 (2007) 3241.
11. W. Schmickler, *Chem. Rev.*, 96 (1996) 3177.
12. J. I. Siepmann and M. Sprik, *J. Chem. Phys.* 102 (1995) 511.
13. R. Bukowski, K. Szalewicz, G. C. Groenenboom, and A. van der Avoird, *Science*, 315 (2007) 1249.
14. A. Michaelides, *Appl. Phys. A*, 85, (2006) 415.
15. G. Kresse and J. Furthmüller, *Phys. Rev. B*, 54 (1996) 11169.
16. J. P. Perdew, K. Burke, and M. Ernzerhof, *Phys. Rev. Lett.*, 77 (1996) 3865.
17. B. Hammer, L. B. Hansen, and J. K. Nørskov, *Phys. Rev. B*, 59 (1999) 7413.
18. J. Neugebauer and M. Scheffler, *Phys. Rev. B*, 46 (1992) 16067.
19. N. D. Mermin, *Phys. Rev.* 137 (1965) A 1441.
20. A. Alavi, J. Kohanoff, M. Parrinello, and D. Frenkel, *Phys. Rev. Lett.*, 73, (1994) 2599.
21. S. A. Wasileski and M. J. Janik, *Phys. Chem. Chem. Phys.*, 10 (2008) 3613.
22. N. W. Ashcroft and N. D. Mermin, *Solid State Physics*, Saunders College, Philadelphia, 1976.
23. C. L. Fu and K. M. Ho, *Phys. Rev. Lett.*, 63 (1989) 1617.
24. K. P. Bohnen and D. M. Kolb, *Surf. Sci.* 407 (1998) L629 .
25. A. Y. Lozovoi and A. Alavi, *Phys. Rev. B*, 68 (2003) 245416.
26. P. A. Thiel and T. E. Madey, *Surf. Sci. Rep.*, 7 (1987) 211.
27. M. A. Henderson, *Surf. Sci. Rep.*, 46 (2002) 1.
28. A. Roudgar and A. Gross, *Surf. Sci.*, 597 (2005) 42.
29. P. J. Feibelman, *Science*, 295 (2002) 99.
30. Y. Gohda, S. Schnur, and A. Gross, *Faraday Discuss.*, 140 (2009) 233.
31. A. Michaelides, V. A. Ranea, P. L. de Andres, and D. A. King, *Phys. Rev. Lett.*, 90 (2003) 216102.
32. J. S. Filhol and M.-L. Bocquet, *Chem. Phys. Lett.*, 238 (2007) 203.

33. S. Izvekov and G. A. Voth, *J. Chem. Phys.*, 115 (2001) 7196.
34. R. Car and M. Parrinello, *Phys. Rev. Lett.*, 55 (1985) 2471.
35. H. Noguchi, T. Okada, and K. Uosaki, *Faraday Discuss.*, 140 (2009) 125.
36. S. Ye, S. Nihonyanagi, and K. Uosaki, *Phys. Chem. Chem. Phys.*, 3, (2001) 3463.
37. M. Schmitz and P. Tavan, *J. Chem. Phys.*, 121 (2004) 12247 .
38. A. Roudgar and A. Gross, *J. Electroanal. Chem.*, 548 (2003) 121.
39. M. F. Toney et al., *Nature*, 368 (1994) 444.
40. M. F. Toney et al. *Surf. Sci.*, 335 (1995) 326.
41. K. Morgenstern and J. Nieminen, *J. Chem. Phys.*, 120 (2004) 10786.
42. O. Sugino, I. Hamada, M. Otani, Y. Morikawa, T. Ikeshoji, and Y. Okamoto, *Surf. Sci.*, 601 (2007) 5237.
43. M. Otani, I. Hamada, O. Sugino, Y. Morikawa, Y. Okamoto, and T. Ikeshoji, *Phys. Chem. Chem. Phys.*, 10 (2008) 3609.
44. M. Otani and O. Sugino, *Phys. Rev. B*, 73 (2006) 115407.
45. G. A. Kimmel, N. G. Petrik, Z. Dohnálek, and B. D. Kay, *Phys. Rev. Lett.*, 95 (2005) 166102.
46. G. Henkelman and H. Jónsson, *J. Chem. Phys.*, 113 (2000) 9978.
47. G. Henkelman, B. P. Uberuaga, and H. Jónsson, *J. Chem. Phys.*, 113, (2000) 9901.
48. J. Rossmeisl, E. Skúlason, M. J. Björketun, V. Tripkovic, and J. K. Nørskov, *Chem. Phys. Lett.*, 466 (2008) 68.
49. N. M. Marković, B. N. Grgur, and P. N. Ross, *J. Phys. Chem. B*, 101 (1997) 5405.

CHAPTER 6

Catalysis of Electron Transfer at Metal Electrodes

ELIZABETH SANTOS[1,2] and WOLFGANG SCHMICKLER[2]

[1] Instituto de Física Enrique Gaviola (IFE6-CONICET), Facultad de Matemática, Astronomia y Física Universidad Nacional de Córdoba, Córdoba, Argentina
[2] Institute of Theoretical Chemistry, Ulm University, D-89069 Ulm, Germany

6.1 INTRODUCTION

Electrochemical reactions of technological interest, such as the oxygen and the hydrogen reactions, involve several steps, some purely chemical, others involving charge transfer. Nearly all such reactions show a strong dependence on the electrode potential, with transfer coefficients of the order of $\frac{1}{2}$ or larger, indicating that the rate-determining step is a charge transfer. All charge transfer reactions involve the exchange of electrons between the reactant and the electrode, and in this chapter we will consider electron transfer in this wider sense. Thus, we shall include the Volmer reaction $H^+ + e^- \rightarrow H$ under electron transfer, even though it can just as well be called proton transfer.

Electron transfer in solution involves the reorganization of the solvent, and for simple, outer sphere reactions this is the factor that determines the rate. In catalytic reactions, the electrode material plays a decisive role; for example, the rate of the hydrogen evolution varies by six orders of magnitude with the nature of the electrode. However, the role of the solvent remains important, since the ions that participate in the reaction are stabilized by their solvation shell. This effect is particularly important for the proton, which has the highest energy of hydration, about 11 eV, of all ions. Indeed, proper electron transfer, as distinct from the mere formation of a polar bond, cannot take place without the participation of the solvent.

In surface science, quantum-chemical calculations, in particular those based on density functional theory (DFT), have been an invaluable tool in understanding and elucidating reaction mechanisms (see, e.g., refs. 1 and 2). It is natural to apply the same

Catalysis in Electrochemistry: From Fundamentals to Strategies for Fuel Cell Development,
First Edition. Edited by Elizabeth Santos and Wolfgang Schmickler.
© 2011 John Wiley & Sons, Inc. Published 2011 by John Wiley & Sons, Inc.

methods to electrochemical interfaces, and there is presently much activity in this area, and a number of interesting results have already been achieved. However, given the important role of the solvent, it is clear that a theory of electron transfer cannot be based on DFT alone. Realistic calculations would have to include a sizable amount of water molecules and would require proper thermal averages over the solvent configurations. At present, and in the foreseeable future, such calculations are not possible. It could even be argued that such calculations are not really desirable, since merely mimicking a complex reaction on a computer does not provide much intellectual insight. Besides, there is as yet no satisfactory way of including the electrode potential into DFT calculations, and this is a serious shortcoming indeed.

In our work, which we will report below, we have taken a different approach. As our basis, we have taken a model Hamiltonian proposed by Santos, Koper, and Schmickler (SKS) [3, 4], which contains the essential interactions, including the interaction of the reactants with the solvent. The resulting model explains the course of an electron transfer and suggests how a strong interaction of the reactant with a metal d band can catalyze the reaction. For the application to real systems, we have obtained the relevant parameters from DFT calculations as far as possible.

We will start by briefly reviewing the theory of outer sphere electron transfer in order to present the basics. Next we explain how the transfer can be catalyzed by a metal electrode and illustrate this by model calculations both for simple and for bond-breaking electron transfer. Finally, we will apply our theory to the hydrogen reaction and compare our calculations with experimental data. We will keep the presentation as simple as possible, relegating the SKS Hamiltonian, on which much of our work is based, to the appendix.

6.2   ADIABATIC OUTER SPHERE REACTIONS

The simplest kind of electrochemical reaction is an electron exchange of the type

$$A^{m+} \leftrightarrow A^{(m+1)+} + e^- \tag{6.1}$$

When the reactant sits several angstroms away from the electrode surface, so that its interaction is weak, this process is called an *outer sphere* reaction. Typical examples are reactions of complexes such as

$$[Ru(NH_3)_3]^{3+} \leftrightarrow [Ru(NH_3)_3]^{4+} + e^- \tag{6.2}$$

where the ligands and the strong solvation of the highly charged ions keep the reacting center away from the electrode surface. From a practical point of view, they are not an important class of processes. We treat them here as reference systems against which proper catalysis can be contrasted. Also, several basic concepts are best explained in terms of this simplest type of reaction.

The theory of such reactions was founded by Marcus [5] and Hush [6]; it was later taken up by Levich and Dogonadze [7]—the extensive work of the Russian school is well represented in the book by Kuznetsov [8]. Its principal concept is the reorganization of the solvent, which accompanies the electron exchange. Reactant and product carry different charges and therefore have different solvation spheres. Thus, the solvent in the vicinity of the reactant must reorient during the reaction, and this reorganization limits the rate of the reaction, while the rate of the electron transfer as such is very fast over the distance of a few angstroms. The electrons move much faster than the solvent molecules, so that the latter are stationary during the electronic transition (Franck–Condon principle). Therefore, an electron transfer reaction proceeds in the following way: First there is a partial reorganization of the solvent to an intermediate state, in which the reduced and the oxidized states have the same energy; then the electron is transferred, and the solvent relaxes to the final state. If the interaction of the reactant with the metal is sufficiently strong—an interaction energy of the order of 10^{-3}–10^{-2} eV generally suffices—the electron exchange takes place every time the system passes the intermediate state. This is the adiabatic case and is typical for electron transfer at a bare metal electrode, and in the following we restrict ourselves to this case.

In order to describe the state of the solvent, we introduce a generalized solvent coordinate q with the following meaning: If the solvent coordinate has the value q, the solvent is in a state which would be in equilibrium with a charge of $-q$ on the reactant. This simple, one-dimensional model is equivalent to a multidimensional representation if the solvent behaves like a classical bath [9, 10], which we shall assume here. To be specific, we assume a reaction of the form given in Eq. (6.1) with $m = 0$, so that the initial state is uncharged. Then, in the initial state the solvent energy has its minimum at $q = 0$, and in the final state at $q = -1$. In the harmonic approximation, the energy of the solvent can be expressed through a parabola of the form

$$E_{\mathrm{sol}} = \lambda q^2 + 2\lambda q(1 - \langle n \rangle) \tag{6.3}$$

where $\langle n \rangle$ is the occupation probability of the valence orbital that donates the electron to the electrode; thus $\langle n \rangle = 1$ corresponds to the reduced, and $\langle n \rangle = 0$ to the oxidized state, and the term in parentheses gives the charge on the reactant. The *energy of reorganization*, of the solvent λ, characterizes the interaction of the transferring electron with the slow solvent modes. The second terms represent a linear interaction of the solvent with the charge on the reactant, and q has indeed been normalized such that for $\langle n \rangle = 0$ the minimum occurs at $q = -1$.

The valence orbital on the reactant interacts with all the electronic states on the metal electrode. Due to this interaction, it is characterized no longer by a sharp energy level but by a density of states (DOS) $\rho_a(\epsilon)$ which depends on the electronic energy ϵ [11]. In the simplest case the electron interacts uniformly with a wide, structureless band on the metal—typically this would be an sp band. In this so-called wide-band approximation [12], the DOS takes the form of a Lorentzian distribution, which is characterized by the level broadening Δ:

$$\rho_a(\epsilon) = \frac{1}{\pi} \frac{\Delta}{\left[\epsilon - (\epsilon_a - 2\lambda q)\right]^2 + \Delta^2} \tag{6.4}$$

Here, ϵ_a is the energy of the valence orbital in the absence of the interaction with the electrode and with the solvent. The term $-2\lambda q$ in the denominator accounts for

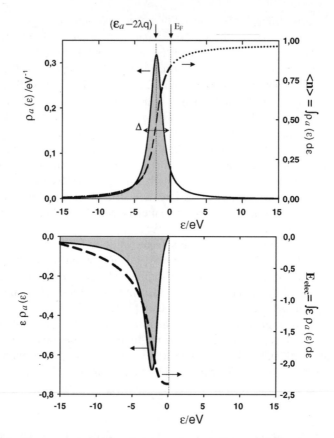

FIGURE 6.1 Upper panel: density of states and occupation probability; lower panel: electronic energy.

the interaction of the electron with the solvent. Thus, the center of the density of states is at $\epsilon_a - 2\lambda q$ and therefore changes with the configuration of the solvent (see Fig. 6.1).

In the adiabatic case the reactant shares its electron with the metal; therefore, its DOS, is filled up to the Fermi level E_F. Ignoring the small width of the Fermi–Dirac distribution, the occupation probability of this orbital is obtained by integrating the DOS up to E_F, which is taken as the energy zero for convenience. Hence,

$$\langle n \rangle = \int_{-\infty}^{0} \rho_a(\epsilon)\, d\epsilon = \frac{1}{\pi} \text{arccot} \frac{\epsilon_a - 2\lambda q}{\Delta} \tag{6.5}$$

Taking the lower limit at $-\infty$ is permissible for metals, which have wide energy bands. The corresponding electronic energy is obtained by multiplying the DOS with the energy ϵ and then integrating up to the Fermi level (see Fig. 6.1):

$$E_{\text{elec}} = \int_{-\infty}^{0} \epsilon \rho_a(\epsilon) \, d\epsilon \tag{6.6}$$

With the Lorentzian form of the DOS, this integral diverges. However, relative energies are well defined, and it is natural to take the initial state, with $q = 0$, as reference. The total energy, written as a function of the solvent coordinate q, is then [13]

$$E(q) = \lambda q^2 + 2\lambda q + (\epsilon_a - 2\lambda q)\langle n \rangle + \frac{\Delta}{2\pi} \ln \frac{\left[\epsilon - (\epsilon_a - 2\lambda q)\right]^2 + \Delta^2}{\epsilon_a^2 + \Delta^2} \tag{6.7}$$

This equation defines the free energy curve for the reaction—it is a free energy rather than an energy because the use of a single solvent coordinate implies that the others have been averaged over. On rearrangement, the first two terms give for the initial state, with $\langle n \rangle = 1$, an energy of $\epsilon_a + \lambda q^2$, which is the energy of the electron plus that of the solvent; for the final state, with $\langle n \rangle = 0$, it gives the energy of the solvent interacting with a positive charge, which can be written as $\lambda(q - 1)^2 - \lambda$. The third term gives the effect of the energy broadening Δ. Inspection shows that for $\epsilon_a = -\lambda$ the system is in equilibrium: There are two minima of equal depth (see Fig. 6.2). The energy ϵ_a of the valence level shifts with the applied electrode potential; with our normalization we have $\epsilon_a = -\lambda + e_0\eta$, where η is the applied overpotential.

For outer sphere reactions the level broadening Δ is of the order 10^{-3}–10^{-2} eV and thus much smaller than the energy of reorganization, which typically lies in the range of $\lambda = 0.5 - 1.0$ eV. In this case, the last term in Eq. (6.7) is negligible, and the occupation probability of Eq. (6.5) becomes a step function: The valence orbital is filled if $\epsilon_a - 2\lambda q$ lies below the Fermi level; otherwise it is empty. Further, the whole free-energy curve is independent of the metal, whose properties enter only into Δ, and

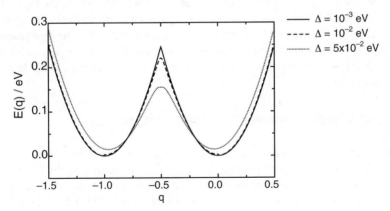

FIGURE 6.2 Adiabatic potential energy $E(q)$ for various values of Δ at equilibrium. The reorganization energy was taken as $\lambda = 1$ eV.

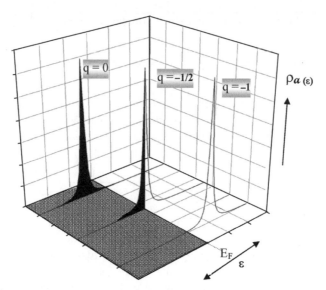

FIGURE 6.3   Density of states for various values of q at equilibrium: $q = 0$ is the initial, $q = -\frac{1}{2}$ the transition, and $q = -1$ the final state.

the rate of electron transfer is also independent of the metal, as has been confirmed experimentally [14, 15].

At equilibrium, the energy of activation is $E_{act} = \lambda/4$ for small Δ, as in the theory of Marcus and Hush. With increasing Δ it decreases, and this can be considered as a catalytic effect. However, strong catalysis, as we shall see below, requires a more detailed theory. At equilibrium, for reasons of symmetry, the solvent coordinate takes the value of $q = -\frac{1}{2}$ at the saddle point and equals the transfer coefficient.

It is instructive to consider the evolution of the density of states $\rho_a(\epsilon)$ during the reaction. In the initial state, for $q = 0$, it lies well below the Fermi level and is fully occupied (see Fig. 6.3). At the transition state, $q = -\frac{1}{2}$, it straddles the Fermi level and is half-filled. In the final state, for $q = -1$, it lies well above the Fermi level and is empty. Thus, the energy of activation can be considered as the energy required to lift the DOS from its initial position to the Fermi level. There, electron transfer takes place, and the solvent relaxes to its new equilibrium position and lifts the DOS higher above the Fermi level.

6.3   CATALYSIS OF SIMPLE ELECTRON TRANSFER

When an electron transfer reaction takes place directly on the electrode surface, as in the Volmer reaction $H^+ + e^- \leftrightarrow H_{ad}$, the details of the metal's electronic structure are important. Most metals contain two kinds of bands of electronic states: sp bands, which are composed of the s and p orbitals of the atoms, and d bands, composed of d orbitals. Since s and p orbitals are extended, the overlap between orbitals on neighboring atoms is large, and they tend to form rather large electronic bands with little structure. In contrast, d orbitals are more localized and the d bands therefore

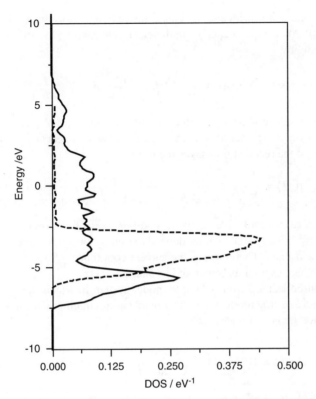

FIGURE 6.4 Densities of state for *sp* band (full line) and *d* band (broken line) at Ag(111) surface. Their integrals have been normalized to unity, and the Fermi level has been taken as the energy zero.

are narrower. As an example, we show in Figure 6.4 the DOS at a Ag(111) surface. The *sp* band shows a peak at lower energies, which is mainly composed of *s* states, the rest is extended and rather structureless. The *d* band is narrow and in this case centered well below the Fermi level.

The interaction of a reactant's valence level with an *sp* band induces a level broadening Δ_{sp}, which to a first approximation can be taken as constant, just like in the case of an outer sphere reaction; however, it is much larger, typically of the order of 0.5 eV. The *d* band induces an additional level broadening, which depends on the electronic energy ϵ. Again to a first approximation, this broadening is proportional to the density of states $\rho_d(\epsilon)$ of the metal *d* band:

$$\Delta_d(\epsilon) = |V|^2 \pi \rho_d(\epsilon) \tag{6.8}$$

where V is the coupling constant between the reactant and the *d* band. Thus, the wide-band approximation, which we employed for outer sphere reactions, can still be used for the interaction with the *sp* band, but not for the *d* band. A consequence of Eq. (6.8) is that $\Delta_d(\epsilon)$ vanishes outside the band.

A strong interaction with a d band leads not only to a broadening of the level but also to a shift $\Lambda(\epsilon)$, which depends on the electronic energy [11, 16]. It is related to the broadening by

$$\Lambda(\epsilon) = \frac{1}{\pi} \, \mathcal{P} \int_{-\infty}^{\infty} \frac{\Delta(\epsilon')}{\epsilon - \epsilon'} \, d\epsilon' \tag{6.9}$$

where $\mathcal{P}$ denotes the principal part. Because the $\Delta(\epsilon)$ and $\Lambda(\epsilon)$ play an important role in the adsorption of a species as well, they are also known as *chemisorption functions*.

The DOS of the reactant now takes the form

$$\rho_a(\epsilon) = \frac{1}{\pi} \frac{\Delta_{sp} + \Delta_d(\epsilon)}{\left[\epsilon - (\epsilon_a + \Lambda(\epsilon) - 2\lambda q)\right]^2 + \left[\Delta_d(\epsilon) + \Delta_{sp}\right]^2} \tag{6.10}$$

At a first glance this looks like a trivial modification; it has, however, far-reaching consequences, since the DOS now depends strongly on its relative position with respect to the d band. Therefore, as the solvent coordinate q fluctuates, the DOS of the reactant changes both its center and its shape.

Much can be learned from a simple model system in which the sp band of the metal is broad and structureless and the d band has a semielliptic form, so that $\Delta_d(\epsilon)$ and $\Lambda(\epsilon)$ have analytical forms [16]:

$$\Delta_d(\epsilon) = |V|^2 \frac{2}{w} \left[1 - \left(\frac{\epsilon - \epsilon_c}{w}\right)^2\right]^{1/2} \theta[w^2 - (\epsilon - \epsilon_c)^2] \tag{6.11}$$

where w is the half-width of the band and ϵ_c its center. The equation for $\Lambda(\epsilon)$ is a little more complex, so we refer to the literature [16]. The form of the two chemisorption function is shown in Figure 6.5. Note that the level shift $\Lambda(\epsilon)$ has a sharp maximum right at the upper band edge and a sharp minimum at the lower one and is antisymmetric.

A strong interaction with a d band greatly modifies the DOS of levels that are close in energy. As an example, the solid curve Figure 6.6 shows the DOS of a level that lies right at the center of a d band. It is strongly broadened and has two peaks near the edge of the band, which can be considered as the bonding and the antibonding orbitals formed with the d band. A level that lies a little below the band is also somewhat broadened, but bonding part of the DOS above the d band is quite small. A level that lies far below the band is hardly affected—this is not shown in the figure.

In an electron transfer reaction, a strong interaction of the reactant with a d band centered near the Fermi level can greatly reduce the energy of activation and hence catalyze the reaction. The underlying mechanism, which was proposed by us [17], is illustrated in Figure 6.7 for the case where the reaction is in equilibrium. In the initial state, with $q = 0$, the reactant's orbital is at $\epsilon = -\lambda$, situated well below the Fermi level, somewhat broadened by the interaction with the sp band, but in the situation depicted hardly influenced by the d band, which lies too far above. During the course of the reaction, a thermal fluctuation of the solvent raises the electronic level of the reactant. The critical phase occurs when this level passes the Fermi level of the metal.

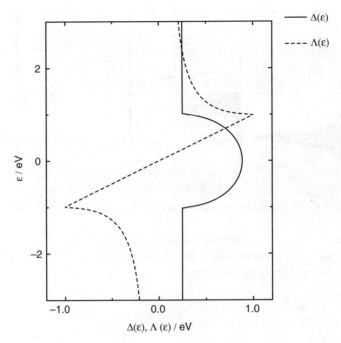

FIGURE 6.5 Chemisorption functions for normalized band with $w = 1$ and $\epsilon_c = 0$.

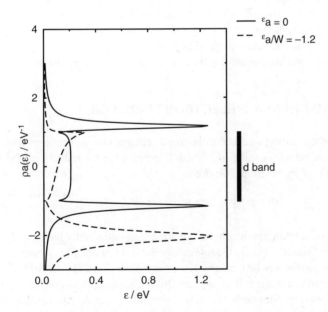

FIGURE 6.6 Density of states of a level with ϵ_a right at the center of a band (full line) or a little below; the solvent coordinate has been set to zero.

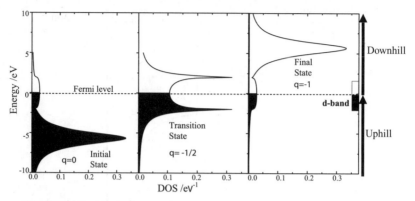

FIGURE 6.7   Mechanism of electrocatalysis by a d band near the Fermi level.

As indicated in the figure, a d band situated close to the Fermi level and interacting strongly with the reactant induces a substantial broadening of the DOS. Since the electronic energy is given by the integral

$$\int_{-\infty}^{0} \epsilon \rho_a(\epsilon)\, d\epsilon \tag{6.12}$$

the part that lies below the Fermi level, situated at $\epsilon = 0$ in our convention, substantially reduces the energy. Thus, a broadening of the DOS as the system passes the Fermi level lowers the energy of activation and thus catalyzes the reaction. This mechanism immediately explains why good metal catalysts, such as platinum or ruthenium, generally have a high density of d states near the Fermi level. However, the mere presence of these states is not enough; they must also interact strongly with the reactant to be effective. Specific examples for this mechanism will be discussed below.

6.4   BOND-BREAKING ELECTRON TRANSFER

A particular interesting case is the electron exchange of a molecule with the electrode with a simultaneous breaking of a bond. Examples are the reduction of chlorine and the oxidation of hydrogen according to

$$Cl_2 + 2e^- \leftrightarrow 2Cl^- \qquad H_2 \leftrightarrow 2H^+ + 2e^- \tag{6.13}$$

The theory of electrochemical bond breaking was initiated by Saveant [18, 19], who extended the Marcus theory. Quantum-mechanical versions have been proposed by German and Kuznetsov [20] and by Koper and Voth [21, 22]. The SKS Hamiltonian, which we shall use in the following (see also the appendix), builds on the latter work.

Even if we consider only the case in which the molecule breaks up into two parts, there are various possibilities such as the subtraction or addition of one or two electrons, homonuclear or heteronuclear molecules. In order to understand the basic principles, we restrict ourselves to the reduction of a homonuclear molecule of the type

$$A_2 + 2e^- \leftrightarrow 2A^- \tag{6.14}$$

and assume that the molecule initially lies flat, so that the two atoms are equivalent [3, 4]. A minimal description of the free-energy surface now requires two coordinates, the solvent coordinate q and the separation r of the reactants; the latter we define such that $r = 0$ for the equilibrium separation of the isolated molecule.

The important electronic states on the molecule are the bonding and the antibonding orbitals. In the isolated molecule, the bonding orbital is filled and the antibonding orbital is empty. Due to the interaction with the surface, each orbital acquires a DOS of the same form as for a simple electron transfer:

$$\rho_x(\epsilon) = \frac{1}{\pi} \frac{\Delta_{sp} + \Delta_d(\epsilon)}{\left[\epsilon - (\epsilon_x(r) + \Lambda(\epsilon) - 2\lambda q)\right]^2 + \left[\Delta_d(\epsilon) + \Delta_{sp}\right]^2} \tag{6.15}$$

where x stands for either the bonding or the antibonding orbital. The electronic energies ϵ_x for the isolated molecules depend on the separation r. It is instructive to consider the binding within the extended Huckel theory augmented by the Wolfsberg–Helmholtz approximation; in this case, the interatomic potential between the two atoms in the molecule takes the form of a Morse potential. When system parameters such as binding energy, interaction with the metal, and energy of reorganization are specified, the free energy surface $V(q, r)$ can be calculated.

A typical example is shown in Figure 6.8 for the case in which the reaction (6.14) is in equilibrium. At $q = 0$, $r = 0$, the surface has a minimum corresponding to the initial state, a stable molecule in front of the metal surface. The final state, two separated anions, is indicated by a valley centered at $q = 2$ which grows deeper with increasing separation r. These two regions are separated by a saddle point which the system has to pass in order for the reaction to proceed. The height of this saddle point determines the energy of activation.

It is instructive to follow the development of the DOS as the reaction proceeds. Initially, the DOS of the molecule has a filled bonding orbital lying well below the Fermi level and an antibonding orbital well above (see Fig. 6.9). In the final state, the bond has been broken; bonding and antibonding orbitals have collapsed into a single orbitals on each molecule that lies below the Fermi level and is filled. For the reaction to proceed, a fluctuation must shift the antibonding orbital below the Fermi level, and the critical phase is when it actually passes E_F. In the situation depicted in the figure, a metal d band centered near the Fermi level interacts strongly with the reactant and induces a splitting of the antibonding orbital, thus lowering the energy of the transition state. This is the same mechanism of catalysis that we discussed before for simple electron transfer (see Figure 6.7), but here it is the broadening of the antibonding orbital that lowers the energy of activation. In contrast, for a reaction of the type $A_2 \leftrightarrow 2 A^+ + 2 e^-$ it is the *bonding* orbital that passes the Fermi level in the critical stage, and its interaction with the d band determines the catalytic behavior. The three types of reactions are compared in Figure 6.10. For a simple

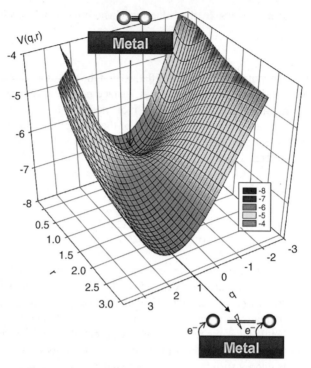

FIGURE 6.8 Typical free-energy surface for equilibrium.

dissociation of the type $A_2 \leftrightarrow 2A$, the distance between the two atoms increases until they separate; at the same time, the interaction between the orbitals is diminished; the separation between bonding and antibonding orbitals becomes smaller until it finally disappears.

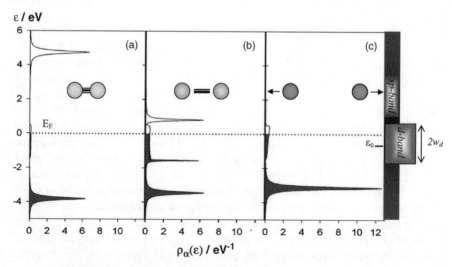

FIGURE 6.9 Development of reactant's density of states during reaction.

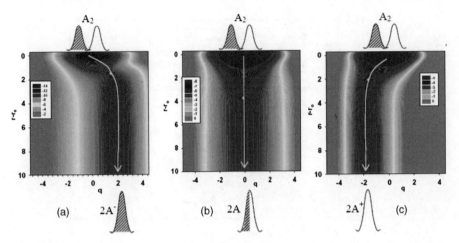

FIGURE 6.10 Free-energy surface for reduction (left), simple dissociation (center), and oxidation (right).

A natural question is: For a given interaction strength, what is the optimum position for the d band in order to catalyze a reaction? For the idealized case of a semielliptic band this question can be answered and is illustrated in Figure 6.11 for the three different kinds of reactions at a constant width of the band. As expected, it lies close to the Fermi level. However, the exact position is a little different for the three types of reactions and lies a little below E_F for the oxidation. For the case of the reduction, Figure 6.12 shows how the energy of activation varies as a function of the position of the band center for various widths. For a band centered near the Fermi level, a narrow

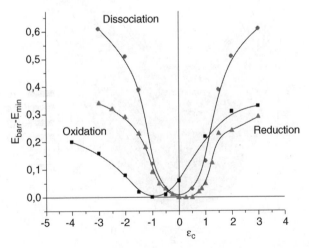

FIGURE 6.11 Energy of activation as function of center of band for three types of reaction. In order to facilitate the comparison for the three kinds of reactions, the minima have been normalized to zero. The coupling strength was set to $|V|^2 = 1$ eV^2 and the width of the d band to 1 eV.

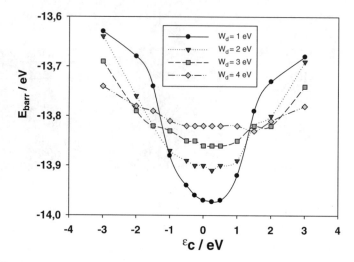

FIGURE 6.12 Energy of activation of reduction reaction as function of center of band for various bandwidths. The coupling strength was set to $|V|^2 = 1$ eV^2.

width is favorable, while for a band far from E_F a wider band is more favorable, which still reaches E_F.

A change of the electrode potential simply changes the electronic levels of the molecule with respect to the Fermi level. Therefore, when the system parameters are known, it is possible to calculate the reaction surfaces for various overpotentials. An

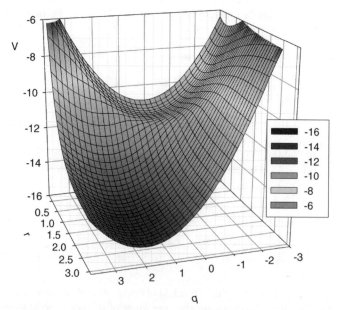

FIGURE 6.13 Free-energy surface for reaction of Figure 6.8, but for overpotential of 0.5 V.

example is shown in Figure 6.13; the overpotential substantially lowers the energy of the final state, which is represented by the valley near $q = 2$. At the same time, the energy of activation is lowered drastically.

6.5 HYDROGEN CATALYSIS

The oxidation of hydrogen is one of the most important electrochemical reactions. Its rate varies by about six orders of magnitude on the various metals [23]; it is therefore a prime example of a reaction that can be catalyzed and an excellent candidate to test the models described above. We shall treat this reaction at two levels: First, we shall apply the SKS formalism (see Appendix), on which the discussion in the previous section was based. This gives a good overall picture, predicts trends, and, above all, provides a basic understanding of the mechanism of catalysis. Second, we shall consider a few particular cases on the basis of a new formalism proposed by us which combines the SKS model with the results of DFT calculations and is better suited for a more detailed and quantitative analysis.

6.5.1 Trends in Exchange Current Densities

As discussed above, the catalysis of an electron transfer is governed by the position of the metal d band and its interaction with the reactant. The former are readily available from DFT calculations, and a list of interaction constants of the metal d band with hydrogen, although approximate, has been published by Hammer and Nørskov [24, 25]. These provide a basis for calculating the dependence of the reaction on the nature of the substrate.

The surface d band of a few selected metals are shown in Figure 6.14. Cadmium is shown as a representative for a typical sp metal whose d band lies well below the Fermi

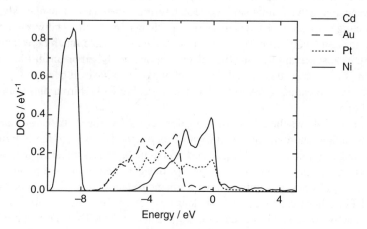

FIGURE 6.14 d-Band density states at (111) surface of selected metals. The integral over the densities has been normalized to unity.

TABLE 6.1 Effective Coupling Constants $|V_{eff}|^2$ to Hydrogen (eV^2) [24, 25]

Re	Ir	Pt	Rh	Au	Ag	Ni	Co	Cu
14.42	10.63	9.44	7.93	8.10	5.52	2.81	3.2	2.42

level and has no catalytic effect. The d band of gold lies much higher, but still well below E_F, while platinum and nickel have d bands that extend well above the Fermi level and are therefore candidates for good catalysts. However, as discussed above, the catalytic properties are determined not only by the position of the d band but also by the interaction with hydrogen. In Table 6.1 we show the coupling constants of a few metals to hydrogen [24, 25]; note that these present average values at a not-well-defined distance. More precise values calculated as a function of the distance, will be presented in the next section. Platinum has a particularly high coupling constant, as do all of the platinum group metals like rhenium and iridium; in contrast, cobalt interacts weakly, just like nickel. In the group of the coin metals the coupling increases down the periodic table, because the orbitals are becoming more extended and the overlap increases. For sp metals, the interaction with hydrogen can be neglected when its valence orbital passes the Fermi level.

Besides the properties of the metal, the energy of reorganization of the solvent is an important quantity that determines the energy of activation. Its value for a proton at the surface is not known and has to be estimated. The energy of solvation of a proton is about 11.5 eV; as a rough guide, the energy of reorganization for a singly charged ion in front of a metal electrode is about one quarter of this value [9]. In the calculations shown below, it has been set to $\lambda = 3$ eV. Since this value is expected to be almost independent of the metal, its exact value is not important if one wants to compare the catalytic activities of various metals.

On the basis of these data the energy of activation for the hydrogen reaction can be calculated within the SKS framework described in the appendix. There is, however, one caveat: While this formalism is good at describing the mechanism of catalysis and at predicting trends, it lacks the correlation–exchange interactions between the metal and the reactant. While this energy is not expected to change much between metals [24, 25], its absence implies that the calculations are not based on an absolute energy scale. In other words, the calculations which we present have all been performed for the same overpotential of unknown absolute value. We will return to this point in the next section.

Figure 6.15 shows the energy of activation for the hydrogen oxidation calculated from this model and compares them with experimental values for the logarithm of the exchange-current density [26]. In addition to the d and sd metals listed, we have also performed calculations for a generic sp metal by setting the coupling constant $|V_{eff}|^2 = 0$ and compared this with experimental values for typical sp metals like Cd and Pb, whose exchange current densities are of the order of 10^{-9} A cm^{-2}. Unfortunately, there is a large scatter of experimental data. This cannot only be due to the fact that many of these values refer to polycrystalline electrodes. The scatter is mostly due to different methods of electrode preparation and to different measuring techniques. As a general rule, the higher rate constants are more likely to be correct,

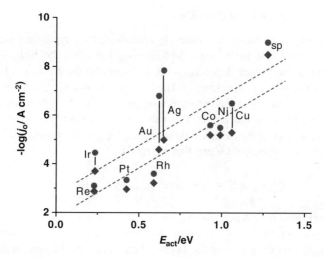

FIGURE 6.15 Experimental data for exchange current density of hydrogen oxidation versus our calculated values for activation energy. The experimental data have been taken from the compilation of Nørskov et al. [27]; in addition values for Ni and Cu have been taken from ref. [28]. Where more than two experimental values existed, the extreme values have been taken in order to indicate the range. For the *sp* metals, we have taken the values for Cd and Pb.

since the most common contaminations tend to inhibit the reaction, and the use of inadequate (i.e., too slow) techniques generally results in too low values. The scatter in the experimental values is largest for gold and silver; however, the lower rates—upper points in the figure—were obtained before the introduction of flame tempering and have probably been obtained from rather imperfect surfaces. Indeed, the higher rate observed for silver is from our own laboratory and was obtained from well-prepared surfaces. Hence there are good reasons to disregard the slow values for Ag and Au.

However that may be, there is a good correlation between the calculated activation energies and the experimental rates. In particular, our calculations explain very well the large difference in the catalytic activity of Pt and Ni: Even though both metals have a high DOS near the Fermi level, Ni is very much less active because its coupling constant is smaller. The behavior of the three *sd* metals can be seen to be more strongly influenced by their coupling constants than by their DOS. The low rate of the *sp* metals is, of course, due to the absence of *d*-band catalysis. It should be noted that Figure 6.15 illustrates the first explanation of the trends in hydrogen catalysis that is based on a theory rather than on a correlation.

In fact, the basic trend of hydrogen catalysis can be understood on the basis of the *d* bands and the coupling constants without calculations: The platinum group metals all have *d* bands that extend to the Fermi level and high coupling constants; hence they are excellent catalysts. Nickel and cobalt have *d* bands at the Fermi level but interact only weakly and therefore do not catalyze well. The *d* bands of the coin metals lie below the Fermi level, and hence they are poor or mediocre catalysts. Finally, the *sp* metals are extremely bad catalysts because their *d* bands lie too low.

6.5.2 Calculations Based on DFT

The theory can be made more quantitative by combining the formalism with the results of DFT calculations. The basic idea is the following: Let $E(r, q)$ be the total electronic energy of the system which depends on the position r of the reactant relative to the substrate and the solvent coordinate q. The case $q = 0$ corresponds to an uncharged system in equilibrium, and the corresponding energy can be calculated from DFT alone; let us call this energy $E_{\text{DFT}}(r)$. For the same configuration, we can calculate the electronic energy from the SKS Hamiltonian for $q = 0$; the difference gives us the exchange and correlations terms that are missing:

$$\Delta E = E_{\text{DFT}}(r) - E_{\text{SKS}}(r, q = 0) \tag{6.16}$$

The question is how ΔE changes with the solvent coordinate. This change will depend on the occupation probabilities of the orbitals involved. For example, in the reaction $H \rightarrow H^+ + e^-$ the occupation of the $1s$ orbital changes with the solvent fluctuations. DFT calculations give the energy for the initial state, the adsorbed H atom. As long as the occupation $\langle n(q) \rangle$ remains unity, the bonding remains unchanged, and hence the correction term does not change either. In the final state, the bond is broken, $\langle n(q) \rangle = 0$, and the correction term vanishes. Hence it is natural to assume that the correction is proportional to the occupation probability [29]:

$$\Delta E(r, q) = \Delta E(r, q = 0) \times \langle n(r, q) \rangle \tag{6.17}$$

This is, of course, an approximation, but all of DFT is based on approximations of this type, and there is no reasons why Eq. (6.17) should be worse than those immanent in DFT. So, as the basis for describing the reaction energetics, we can use

$$E(r, q) = E_{\text{SKS}}(r, q) + \Delta E(r, q = 0) \times \langle n(r, q) \rangle \tag{6.18}$$

The SKS formalism requires as input the DOS of the metal d band, the energy ϵ_a of the reactant's orbital, the coupling constants, and the broadening Δ_{sp} caused by the d band. The d band is easily calculated from DFT; the other parameters can be obtained by fitting the reactant's DOS, as obtained from DFT, to the form of Eq. (6.10). An example is shown in Figure 6.16 for a hydrogen molecule at a Cd(0001) surface. The d band of Cd lies very low, close to the bonding orbital of the hydrogen molecule. The interaction with the band splits the DOS of the binding H_2 orbital in two parts, a bonding and an antibonding part. Since both are filled, no bonding results. The DOS of the molecule can be fitted quite well to the model; the resulting parameters, especially the interaction constants, are expected to be more accurate than the values given in Table 6.1, which are based on simple model wavefunctions.

We consider explicitly the Volmer reaction $H \rightarrow H^+ + e^-$ on a few model surfaces: on Cd(0001) and on the (111) surfaces of Cu, Ag, Au, and Pt. Thus, we have one extremely bad, three mediocre, and one excellent catalysts on our list. In the course of this Volmer reaction the reactant moves from the solution to the surface; therefore we require the interaction of the hydrogen $1s$ orbital with these metals as a function

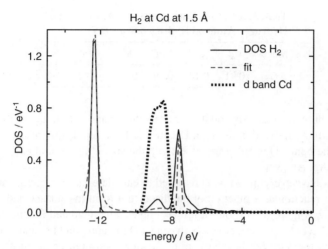

FIGURE 6.16 Density of states of binding orbital of hydrogen molecule at distance of 1.5 Å in front of Cd(0001) electrode.

of the distance. These are shown in Figure 6.17; note that their normalization differs by a factor of π from those in Table 6.1, and refers to the atom; both lists follow the same order.

Before looking at the activation energies, it is instructive to consider the free energies of adsortpion on the standard hydrogen electrode (SHE) scale, that is, the free energy of adsorption of a proton at the equilibrium potential (see Table 6.2); these values can be obtained by DFT alone [27]. On Cd, the d band lies very low (see Fig. 6.16); hence the both bonding and antibonding orbitals formed with the hydrogen $1s$ orbital are occupied, so that the d band plays no role. The bonding with the sp band is weak, with a bonding distance of about 1.3 Å at the hollow site. The other four metals have face-centered-cubic (fcc) structure, and the bonding occurs at the fcc hollow site. In Cu, Ag, and Au, the d bands also lie below the Fermi level (see

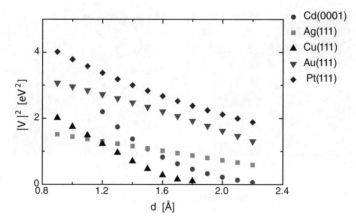

FIGURE 6.17 Interaction constants $|V|^2$ in eV^2 as function of distances for several metals.

TABLE 6.2 Free Energy ΔG_{ad} of Adsorption of Hydrogen Atom on Hydrogen Scale (eV)

Metal	Cd	Cu	Ag	Au	Pt
ΔG_{ad}	0.91	0.10	0.39	0.41	−0.25

Fig. 6.14), but not so far. Also on these metals the d band does not contribute, but the binding with the sp band is stronger. Because of Pauli repulsion, the d band actually weakens the bond. This effect increases with the size of the orbital and hence in the order Cu, Ag, Au [30].

We now have everything that is required to calculate the free-energy surfaces for the Volmer reaction as a function of the distance d from the surface and the solvent coordinate q. Figure 6.18 shows a few examples, all calculated for the equilibrium potential; Ag(111) is intermediate between Cu(111) and Au(111) and is not shown. In all cases the surfaces have two minima: one minimum centered at $q = 0$ and $d \approx 0.9$ Å, which corresponds to the adsorbed hydrogen atom; another minimum is found at $q = -1$ and at large distances corresponding to the solvated proton. The

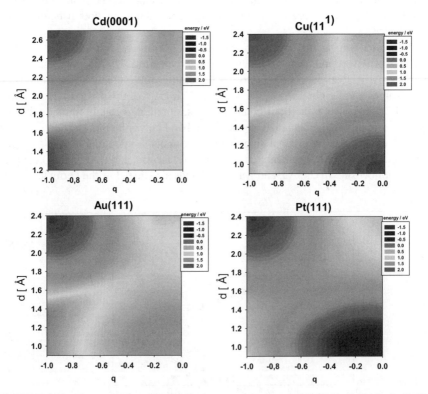

FIGURE 6.18 Free-energy surfaces for hydrogen adsorption at Cd(0001), Cu(111), Au(111), and Pt(111).

TABLE 6.3 Free Energies of Activation for Hydrogen Adsorption on Various Metals

Metal	Cd	Cu	Ag	Au	Pt
λ Variable	0.93	0.71	0.71	0.70	0.30
λ Constant	0.83	0.61	0.57	0.63	0.0

Note: The two sets of values were obtained by slightly different assumptions about the energy of reorganization; for details, see Santos et al. [31].

two minima are always separated by a barrier whose saddle point determines the free energy of activation which are given in Table 6.3.

Figure 6.19 presents a more detailed view for platinum and cadmium. The upper part shows the surfaces for the overall reaction. Clearly, in both cases the Volmer step determines the overall rate. On platinum, there is only a small barrier for the recombination reaction, and on cadmium it is downhill and involves a sizable energy barrier. Further details on the recombination can be found in the literature [29].

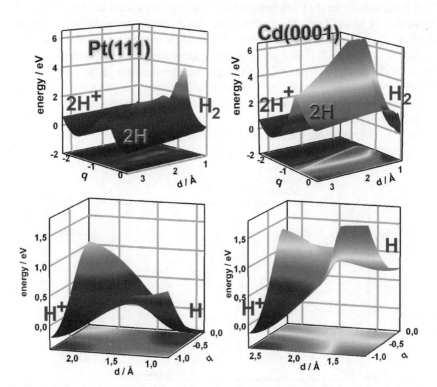

FIGURE 6.19 Above: Free-energy surfaces for the overall reaction, Volmer step and recombination, for Pt(111) and Cd(0001). Below: three-dimensional view of surfaces for Volmer reaction on these metals.

The energy of activation is highest for cadmium, because its d band lies between -10 and -8 eV below the Fermi level and is quite narrow. Therefore, in spite of the relatively high coupling constant, it has no catalytic effect. Indeed, if we switch off the coupling by setting $|V|^2 = 0$, we obtain exactly the same activation energy. Experimentally, hydrogen evolution occurs with roughly the same speed on metals like Cd, Hg, Pb, Tl, and In, all of which have d bands that lie too low to affect the activation energy. Therefore, on these metals the reaction is dominated by the sp band, whose interaction does not vary much between the metals.

On the three coin metals Cu, Ag, Au the reaction has roughly the same energies of activation. This is due to a compensation effect: The interaction $|V|^2$ with the d band increases down the column of the periodic table, which lowers the energy of activation. On the other hand, the energy of the adsorbed hydrogen increases in the same order (see Table 6.2); therefore, the reaction free energy for the adsorption rises, which in turn raises the activation energy.

Of the metals considered, platinum is the only one whose d band extends over the Fermi level. Its interaction with hydrogen is strong, and therefore it has by far the lowest activation energy. At the equilibrium electrode potential, the energy of the adsorbed hydrogen is lower than that of the solvated proton. Therefore adsorption sets in at potentials above the hydrogen evolution, so that one speaks of strongly adsorbed hydrogen, or sometimes of hydrogen deposited at underpotential (H_{upd}). However, there is convincing experimental evidence that this is not the species that takes part in hydrogen evolution [32] but that the intermediate is a more weakly adsorbed species. So our calculations correspond to the deposition of the strongly adsorbed species. Experimentally, this reaction is so fast that it has not been possible to measure its rate. This is in line with the very low energy of activation that we obtained.

In Section 6.3 we explained how a strong interaction with the d band can catalyze a reaction by broadening the reactant's DOS as it passes the Fermi level. Figure 6.20 shows the hydrogen DOS at the saddle point for Cd, Au, and Pt. For Cd, the DOS is

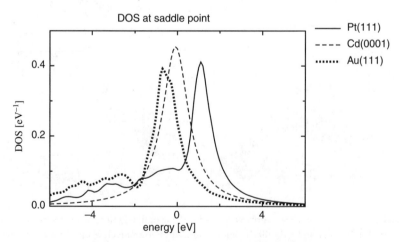

FIGURE 6.20 Hydrogen DOS at saddle point.

TABLE 6.4 $-\log_{10} j_{00}$ **for Hydrogen Evolution**

metal	Cd	Cu	Ag	Au	Pt
this work	9.4 – 11.2	5.6 – 7.3	4.9 – 7.3	6.0 – 7.1	−5 – 0.2
experimental values	10 – 12	5 – 6	5 – 7.8	5.4 – 6.8	—

Note: Upper row: values obtained from Table 6.3; lower row: range of experimental values [26]. The standard exchange current density j_{00} is proportional to the rate constant at the equilibrium potential.

only broadened by the *sp* band, while the DOS at Pt and at Au are widened in the region of their *d* bands, the effect being somewhat larger for Pt. The saddle points occur for different occupancies at the three metals. On Pt, the orbital is about half filled ($\langle n \rangle = 0.55$), while for Au and Cd the occupancy is much higher (for Cd $\langle n_a \rangle = 0.75$ and for Au $\langle n_a \rangle = 0.79$), because the adsorption of the proton is uphill. The significant broadening of the DOS on Au shows that this metal would be quite a good catalyst if the energy of the adsorbed hydrogen were not so high. The overall behavior is completely in line with the qualitative discussion in Section 6.3.

A more quantitative comparison with experimental data can be made by estimating the preexponential factor of the quantity measured by experimentalists, the standard exchange current density j_{00}, which is proportional to the rate constant at equilibrium. The preexponential factor is given by $A = Fc_s^0/\tau \approx 10^5$ A cm^{-2}, where F is Faraday's constant, c_s^0 the surface concentration of the proton (i.e., the concentration in roughly the first 5 Å adjacent to the electrode), and $\tau \approx 10^{11}$ s^{-1} a typical reorientation time for water. As shown in Table 6.4, our calculated free energies of activation have about the correct order of magnitude, and, somewhat surprisingly, the uncertainty in experimental data is about the same as that of our model.

The more exact treatment presented in this section does not invalidate the simpler approach of Section 6.5.1 but adds some details and minor corrections. In particular, the different behavior of the *sp* metals, the coin and the transition metals, is well explained by both methods. However, the DFT-based calculations, which are much more time consuming, establish an absolute scale of energy and potential and are more precise.

6.6 OUTLOOK

In the last few years our understanding of electrocatalysis has progressed rapidly. Much has been learned about the structure of electrode surfaces by the application of DFT, which is very useful for calculating the electronic ground state of a system. In particular, DFT can give the thermodynamics of an electrochemical reaction, not only of the overall reaction but also of individual steps. More can be found in other chapters of this book about the direct application of DFT to electrochemical system.

At the heart of every electrochemical reaction lies an electron transfer step which is strongly coupled to the solvent and therefore cannot be described by DFT alone. It is the virtue of the works described above that they combine the concepts of solvent reorganization with electronic structure theory for metal surfaces. The resulting model provides an attractive picture of how electron transfer takes place as a reactant's orbital

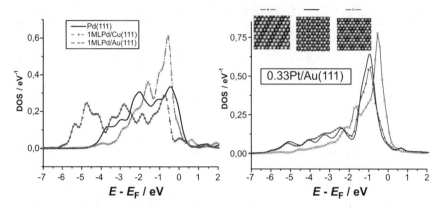

FIGURE 6.21 Left: *d*-band DOS of monolayer of Pd on Au and Cu. Right: *d*-band DOS of Pt for various surface alloys containing one-third Pd atoms.

passes the Fermi level and, most importantly, suggests a mechanism for catalysis by a strong interaction with the metal *d* bands.

Although the solvent reorganization plays a vital role, its details are not known, and in the present theory it enters in the form of one constant, the reorganization energy. In principle, its value could be computed by simulations, but in the absence of reliable calculations it has to be estimated. Still, it is reasonable to assume that the solvent reorganization does not depend much on the electrode material. Therefore, the variation of the reaction rate with the nature of the electrode can be calculated, and this is the principal aim of the theory of electrocatalysis. The first application to the hydrogen reaction has been quite successful, we believe, and it constitutes the first explanation of the its catalysis that is based on a theory and not just on a correlation.

The calculations presented here have all been performed for single crystal surfaces of pure metals. However, real catalysts are usually polycrystalline. Moreover, at present much experimental work is performed on nanostructured electrodes such as surface alloys, clusters, metal overlayers or islands, which hold great promise as catalysts. Naturally, the electronic structure of such surfaces is complicated; nonetheless, its band structure can be elucidated. As examples, Figure 6.21 shows *d* bands of Pd and Pt for various monolayers and surface structures. Evidently, the details of the morphological structure have a significant effect on the electronic structure, and with the methods presented in Section 6.5.2 it should be possible to estimate their relative catalytic activity.

In summary: There has been considerable progress in understanding the relation between electronic structure and catalytic activity, and it now seems realistic to perform calculations for specific systems and estimate their reactivity. For a complete theory, we need to learn more about the role of the solvent.

APPENDIX: SCHMICKLER–KOPER–SANTOS HAMILTONIAN

We consider a diatomic molecule of the form A–B reacting with a metal electrode. The total system contains three parts: the reacting molecule, the metal electrode, and the solvent. Accordingly, we decompose our model Hamiltonian into parts describing the subsystems and their interactions.

We start with the terms for the molecule, which in terms of a tight-binding (or extended Hückel) scheme is written in the form

$$H_{\text{mol}} = \sum_{\sigma} \left(\epsilon_a n_{a,\sigma} + \epsilon_b n_{b,\sigma} + V_{ab} c_{a\sigma}^+ c_{b\sigma} + V_{ab}^* c_{b\sigma}^+ c_{a\sigma} \right)$$
$$+ U_a n_{a,\sigma} n_{a,-\sigma} + U_b n_{b,\sigma} n_{b,-\sigma} \tag{6.19}$$

Here the indices a and b stand for the valence orbitals on the two atoms A and B; n is a number operator that counts the electrons in the indicated state, c^+ and c denote creation and annihilation operators for the electrons, and σ is the spin index. The third and fourth terms in the parentheses effect electron exchange and are responsible for the bonding between the two atoms, while the last terms stand for the Coulomb repulsion between electrons of opposite spin on the same orbital.

This molecule is adsorbed on a metal electrode whose electronic states are labeled by their quasi-momentum k. They interact with the atomic orbitals through exchange terms, so that the metal part of the Hamiltonian becomes

$$H_{\text{met}} = \sum_{k,\sigma} \left[\epsilon_k n_{k,\sigma} + \left(V_{a,k} c_{k,\sigma}^+ c_{a,\sigma} + V_{b,k} c_{k,\sigma}^+ c_{b,\sigma} + \text{h.c.} \right) \right] \tag{6.20}$$

where h.c. stands for the Hermitian conjugate.

Finally we specify the terms for the solvent and any vibrational modes that interact with the electron exchange. They are represented as a phonon bath that interacts linearly with the charge on the molecule:

$$H_{\text{sol}} = \sum_{\nu} \left\{ \tfrac{1}{2} \hbar \omega_\nu (q_\nu^2 + p_\nu^2) + \left[z - \sum_{\sigma} (n_{a,\sigma} + n_{b,\sigma}) \right] \hbar \omega_\nu g_\nu q_\nu \right\} \tag{6.21}$$

Here, ν labels the phonon modes, q_ν and p_ν are dimensionless coordinates and momenta for the modes with frequencies ω_ν, z is the charge number of the molecule when the two orbitals a and b are empty, and the g_ν are the coupling constants describing the interaction between the phonon modes and the molecule. We have assumed that the molecule as a whole interacts with the solvent; obviously, this is no longer true when the bond has been broken and the two atoms are far apart. However, it should be a good assumption at short distances, where the bond is still intact or has just been broken. This is the region of interest for this work.

The total model Hamiltonian is given by the sum

$$H = H_{\text{mol}} + H_{\text{met}} + H_{\text{sol}} \tag{6.22}$$

The Hamiltonian can be solved by Green's function techniques using the Hartree–Fock approximation for the Coulomb repulsion terms. Details are given in the original literature [3, 4].

ACKNOWLEDGMENTS

Financial support by the Deutsche Forschungsgemeinschaft (Schm 344/34-1 and Sa 1770/1-1) and the European Union under COST is gratefully acknowledged. E. S. and W.S. thank CONICET for continued support.

REFERENCES

1. A. Gross, *Theoretical Surface Science—A Microscopic Perspective*, Springer, Berlin, 2002.
2. R. A. van Santen and M. Neurock, *Molecular Heterogeneous Catalysis*, Wiley-VCH, Weinheim. 2006.
3. E. Santos, M. T. M. Koper, and W. Schmickler, *Chem. Phys. Lett.*, 419 (2006) 421.
4. E. Santos, M. T. M. Koper, and W. Schmickler, *Chem. Phys.*, 344 (2008) 195.
5. R. A. Marcus, *J. Chem. Phys.*, 24 (1956) 966.
6. N. S. Hush, *J. Chem. Phys.*, 28 (1958) 962.
7. V. G. Levich, Kinetics of reactions with charge transfer, in *Physical Chemistry, and Advanced Treatise*, Vol. Xb, H. Eyring, D. Henderson, and W. Jost (Eds.), Academic, New York, 1970.
8. A. M. Kuznetsov, *Charge Transfer in Physics, Chemistry and Biology*, Gordon & Breach, Reading, MA, 1995.
9. W. Schmickler, *Chem. Phys. Lett.*, 237 (1995) 152
10. W. Schmickler, *Electrochim. Acta*, 41 (1996) 2329
11. J. P. Muscat and D. N. Newns, *Prog. Surf. Sci.*, 9 (1978) 1.
12. J. W. Gadzuk, in *Surface Physics of Materials*, Vol. II, J. M. Blakely (Ed.), Academic Press, New York, 1975, p. 339.
13. W. Schmickler, *J. Electroanal. Chem.*, 204 (1986) 31.
14. T. Iwasita, W. Schmickler, and J. W. Schultze, *Ber. Bunsenges. Phys. Chem.*, 89 (1985) 138.
15. E. Santos, T. Iwasita, and W. Vielstich, *Electrochim. Acta*, 31 (1986) 431.
16. D. M. Newns, *Phys. Rev.*, 178 (1969) 1123.
17. E. Santos and W. Schmickler, *Chem. Phys.*, 332 (2007) 39.
18. J. M. Savéant, *J. Am. Chem. Soc.*, 109 (1987) 6788.
19. J. M. Savéant, *Acc. Chem. Res.*, 26 (1993) 455.
20. E. D. German and A. M. Kuznetsov, *J. Phys. Chem.*, 98 (1994) 6120.
21. M. T. M. Koper and G. A. Voth, *Chem. Phys. Lett.*, 282 (1998) 100.
22. A. Calhoun, M. T. M. Koper, and G. A. Voth, *J. Phys. Chem. B*, 103 (1999) 3442.
23. S. Trasatti, *J. Electroanal. Chem.*, 111 (1980) 125.
24. B. Hammer and J. K. Nørskov, *Adv. Catal.*, 45 (2000) 71.
25. B. Hammer and J. K. Nørskov, *Surf. Sci.*, 343 (1995) 211.
26. E. Santos and W. Schmickler, *Ang. Chem. Int. Ed.*, 46 (2007) 8262.
27. J. K. Norskøv et al., *J. Electrochem. Soc.*, 152, J23 (2005) 138.
28. B. E. Conway, E. M. Beatty, and P. A. D. DeMaine, *Electrochim. Acta*, 7 (1962) 39.
29. E. Santos, Kay Pötting, and W. Schmickler, *Disc. Farad. Soc.*, 140 (2009) 209.
30. B. Hammer and J. K. Nørskov, *Nature*, 376 (1995) 238.
31. E. Santos, A. Lundin, K. Pötting, P. Quaino, and W. Schmickler, *Phys. Rev. B.*, 79 (2009) 235436.
32. W. Schmickler, *Interfacial Electrochemistry*, Oxford University Press, New York, 1996.

CHAPTER 7

Combining Vibrational Spectroscopy and Density Functional Theory for Probing Electrosorption and Electrocatalytic Reactions

MARC T. M. KOPER

Leiden Institute of Chemistry, Leiden University, 2300 RA Leiden, The Netherlands

The advantages of combining experimental vibrational spectroscopy and computational density functional theory (DFT) for studying and understanding adsorption at electrode surfaces are outlined and applied to a number of model systems, such as cyanide and carbon monoxide adsorption on gold and platinum and formic acid oxidation and ethanol oxidation on gold and platinum, respectively. The nature of the Stark effect in electrochemical adsorption is discussed as well as the potential and pitfalls in using vibrational spectroscopy for the assessment of the strength and nature of the chemisorption bond.

7.1 INTRODUCTION

In situ vibrational spectroscopy has become an indispensable tool in the study of electrochemical interfaces, especially for the identification and characterization of surface-adsorbed species. Experimental techniques such as Fourier transform infrared (FTIR) Spectroscopy, both in internal and external reflection modes, surface-enhanced Raman spectroscopy, and sum frequency generation have the ability to detect species adsorbed on electrode surfaces and are now more or less routinely employed in various interfacial electrochemistry laboratories [1–6]. A particularly attractive though relatively recent strategy in measuring and understanding the vibrational characteristics of chemisorbates is to combine vibrational spectroscopy with first-principles quantum-chemical calculations, typically employing the DFT formalism. Modern

Catalysis in Electrochemistry: From Fundamentals to Strategies for Fuel Cell Development,
First Edition. Edited by Elizabeth Santos and Wolfgang Schmickler.
© 2011 John Wiley & Sons, Inc. Published 2011 by John Wiley & Sons, Inc.

DFT-based quantum chemistry codes have the ability to compute the vibrational features of surface-adsorbed species with a very acceptable level of accuracy such that a direct comparison to experiment becomes meaningful and useful [7]. Moreover, effects of surface coverage, coadsorbates solvent, and electric field or potential may all be included in the computation provided sufficient computing power is available [8]. This affords the possibility to use interfacial vibrational spectroscopy not only as an analytical tool but also as a method for probing chemisorption on a quantum-chemical level.

The aim of this chapter is to provide a succinct selective review of the possibilities and potential of combining vibrational spectroscopy and DFT for examining electrosorption and electrocatalytic reactions. The details of carrying out spectro-electrochemical experiments or ab initio quantum-chemical calculations will not be reviewed, as this has been done extensively by specialists elsewhere in the literature. The chapter will first give some background on theoretical aspects of vibrational spectroscopy of adsorbed molecules, including a general treatment of field or "Stark" effects, as these have been of great interest in electrochemistry ever since the first experiments on vibrational spectroscopy of adsorbed molecules on electrodes. We will then discuss a number of examples, primarily focusing on the adsorption of cyanide and carbon monoxide on gold and platinum electrodes. Finally, we will discuss how the combination of vibrational spectroscopy and DFT may help in elucidating the mechanisms of electrocatalytic reactions, particularly formic acid oxidation on gold and ethanol oxidation on platinum (although in the latter example the role of DFT is very limited).

7.2 THEORY OF VIBRATIONAL SPRECTOSCOPY OF ADSORBED MOLECULES AND ELECTRIC FIELD EFFECTS

It is well known that molecules adsorbed on surfaces change their intramolecular vibrational characteristics due to the interaction with the surface [1–3]. In addition, the bond formed between the adsorbate and the surface leads to new vibrational features. Taken together, these vibrational fingerprints help to identify and characterize the adsorbate. From a more fundamental point of view, the interesting question arises if the vibrational features may also serve as a more quantitative measure of chemisorption bond strength or type of chemisorption bond. Especially the notion of using vibrational fingerprints to draw conclusions on the binding strength of adsorbates would be of great interest to electrochemists, as the limited temperature range that may be applied to electrochemical systems renders the quantitative determination of chemisorption energies problematic. Furthermore, it is commonly observed that vibrational features depend sensitively on the applied potential, commonly referred to as the Stark effect. The exact nature of this effect has been discussed extensively in the literature.

To illustrate how the interaction of a simple diatomic molecule with a metal surface may influence the molecule's intramolecular stretching mode, consider the classic Blyholder model of the interaction of CO with a transition metal surface [9]. One of the main interaction channels of a CO with a metal surface is through hybridization of the CO's $2\pi^*$ lowest unoccupied molecular orbital (LUMO) orbital with the d orbitals of the transition metal, concurrently with an energy broadening and an energy shift due to the interaction with the more delocalized s orbitals. This leads to electron transfer from the metal to the new $2\pi^*$ orbital. Since the $2\pi^*$ orbital is an antibonding orbital for the C−O bond, this typically leads to a red shift (lowering) of the C−O stretching frequency. However, the new $2\pi^*$ orbital is bonding with respect to the metal–CO bond, and there is a relation between the extent of back donation and the binding strength [10]. These observations are sometimes interpreted as a "rule of thumb" that the extent of lowering of the C−O stretching frequency is a measure of the strength of the metal–CO bond. Unfortunately, this rule of thumb hardly ever applies, and in reality one will always have to carry out detailed ab initio calculations of the system under consideration to understand the relation between frequency changes and bonding characteristics.

A good example of the various factors involved in observed changes in C−O vibrational fingerprints may be based on a detailed comparison of the C−O stretching adsorbed on a Pt(111) and a Ru(0001) cluster surface as calculated by DFT (see Table 7.1) [11]. Table 7.1 shows results of a so-called decomposition analysis for CO adsorbed in the atop and hollow site of a 13-atom Pt(111) cluster and at the atop site of a 13-atom Ru(0001) cluster. The idea of this decomposition analysis is to separate out the contributions of steric, donation, and back-donation interactions to the bonding of the CO molecule to the metal surface [11 and references therein]. By applying the same decomposition procedure to energy surfaces instead of energies, one may track the corresponding contributions to changes in a vibrational frequency when a change in bonding takes place. Comparing the frequency decomposition for CO adsorbed atop and hollow on the Pt(111) cluster in Table 7.1 shows that the lower frequency in the hollow site is primarily due to the more negative back donation contribution in the hollow site. Comparing CO adsorbed atop on Pt(111) and Ru(0001) shows that

TABLE 7.1 **Decomposition of Zero-Field Vibrational Frequency Shifts (cm^{-1})**
Compared the Calculated Vacuum Values into Steric and Orbital
Components for CO Chemisorbed in Atop and Hollow Site of Pt(111) and
CO Adsorbed in Atop Site of Ru(0001)^a

System	Steric	Donation	Back Donation	Rest	Final
Pt−CO (atop)	+743	−328(−486)	−569(−411)	+46	1987
Pt−CO (hollow)	+645	−198(−371)	−910(−737)	+82	1714
Ru−CO (atop)	+670	−349(−471)	−562(−440)	+47	1901

^{a}The far-left column is the cluster–adsorbate system. Adjacent three columns (steric/donation/back donation) give the change in frequency due to each contribution, calculated with respect to the uncoordinated DFT frequency of 2095 cm^{-1} for CO. The main entries have been calculated in the order as given in the table, i.e., steric–donation–back donation, whereas the entries within parentheses refer to the order steric–back donation–donation. The column "rest" refers to the residual frequency shift upon adsorption not accounted for by the sum of steric/donation/back donation. The far-right column gives the C−O vibrational frequency of the chemisorbed CO.

the C–O stretch frequency on Ru(0001) is lower due to the smaller steric contribution to the overall frequency change. This is in contrast with a common interpretation in the literature that a lower frequency is related to a stronger back donation and hence a stronger chemisorption bond. A detailed discussion of the absence of a clear-cut relation between the C–O stretching frequency and the chemisorption energy was given by Wasileski and Weaver [12], and we will encounter more examples of the absence of such a relationship further on in this chapter.

The electrode potential is a key control parameter in many electrochemical experiments, and varying the electrode potential often leads to changes in the vibrational signatures of surface-adsorbed species. Before discussing results of ab initio calculations, including the effect of an applied potential or an applied electric field on chemisorbate properties, we summarize a general theory that will be useful in understanding and classifying the relationships found in the calculations [13–15]. Since we are interested in the field dependence of the potential energy curve pertaining to a certain chemical bond, we can write the dependence of the potential energy surface (PES) $E(r,F)$ as an expansion in the bond distance r and the electric field F up to a certain order. Defining $\Delta r = r - r_{eq}$, the formal expression for the PES reads as

$$
\begin{aligned}
E(\Delta r, F) = E_b(\Delta r = 0, F = 0) &+ \frac{\partial E}{\partial \Delta r} \Delta r + \frac{\partial E}{\partial F} F \\
&+ \frac{1}{2} \frac{\partial^2 E}{\partial \Delta r^2} \Delta r^2 + \frac{1}{2} \frac{\partial^2 E}{\partial F^2} F^2 + \frac{\partial^2 E}{\partial \Delta r \partial F} \Delta r \, F \\
&+ \frac{1}{6} \frac{\partial^3 E}{\partial \Delta r^3} \Delta r^3 + \frac{1}{6} \frac{\partial^3 E}{\partial F^3} F^3 + \frac{1}{2} \frac{\partial^3 E}{\partial \Delta r^2 \partial F} \Delta r^2 \, F \\
&+ \frac{1}{2} \frac{\partial^3 E}{\partial \Delta r \partial F^2} \Delta r \, F^2 + O(4)
\end{aligned}
\tag{7.1}
$$

where $O(4)$ signifies all terms higher than fourth order and all partial derivatives are taken in $\Delta r = 0$ and $F = 0$ (making $\partial E / \partial \Delta r = 0$). Equation 7.1 can be rewritten as

$$
\begin{aligned}
E(\Delta r, F) = E_b^0 + \tfrac{1}{2} K_0 \, \Delta r^2 &+ G_0 \, \Delta r^3 - \mu_s(F)F - \mu_D(F)F \, \Delta r \\
&- \alpha_D(F)F \, \Delta r^2 + O(4)
\end{aligned}
\tag{7.2}
$$

with the different parameters defined as

$$
K_0 = \frac{\partial^2 E}{\partial \Delta r^2}
$$

$$
G_0 = \frac{1}{6} \frac{\partial^3 E}{\partial \Delta r^3}
$$

$$\mu_s(F) = -\frac{\partial E}{\partial F} - \frac{1}{2}\frac{\partial^2 E}{\partial F^2}F - \frac{1}{6}\frac{\partial^3 E}{\partial F^3}F^2$$

$$\mu_D(F) = -\frac{\partial^2 E}{\partial \Delta r \, \partial F} - \frac{1}{2}\frac{\partial^3 E}{\partial \Delta r \, \partial F^2}F$$

$$\alpha_D(F) = -\frac{1}{2}\frac{\partial^3 E}{\partial \Delta r^2 \, \partial F}$$

Here, K_0 is the mode's force constant at zero field, G the mode anharmonicity at zero field, $\mu_s(F)$ the field-dependent static dipole moment, $\mu_D(F)$ the field-dependent dynamic dipole moment ($=\partial\mu_s/\partial\Delta r$), and $\alpha_D(F)$ a higher-order polarizability term. Equation (7.2) shows that the static dipole moment μ_s describes, to a first approximation, the field dependence of the bond strength

$$\frac{dE_b(F)}{dF} = -\mu_s(F) \tag{7.3}$$

Wasileski et al. [14] have checked the validity of this relationship by DFT–GGA (=Generalized Gradient Approximation) cluster calculations. We note that the field-dependent μ_s values at real electrode surfaces may depend sensitively on solvation and other effects which may or may not have been included in the DFT calculations.

As binding energies, and hence their field or potential dependence, are difficult to measure experimentally in an electrochemical cell, it is of interest to examine the field dependence of other fundamental binding characteristics, in particular vibrational properties of the chemisorbate. A fundamental equation for the field dependence of a vibrational frequency can be straightforwardly derived from Eq. (7.2). The force constant K_F at field F can be obtained by determining the minimum of Eq. (7.2) at field F and then inserting the resulting value for Δr in the expression for the second derivative of E with respect to Δr. This yields

$$K_F = K_0 + \frac{6G_0\mu_D(0) - 2K_0\alpha_D(0)}{K_0}F \tag{7.4}$$

The harmonic vibrational frequency ν_H is then

$$\nu_H(F) = \frac{1}{2\pi}\sqrt{\frac{K_F}{m*}} \tag{7.5}$$

where $m*$ is the reduced mass of the vibrational mode considered.

From a generalization of these equations, an expression for the slope of the force constant and the vibrational frequency at field F can be derived, that is

$$\frac{dK_F}{dF} = \frac{6G_F\mu_D(F) - 2K_F\alpha_D(F)}{K_F} \tag{7.6}$$

and

$$\frac{dv_H(F)}{dF} = v_H(F)\frac{3G_F\mu_D(F) - K_F\alpha_D(F)}{K_F^2} \tag{7.7}$$

These equations show that the field dependence of the force constant and the vibrational frequency are determined by the field-dependent "dipolar bond stretching parameters" μ_D and α_D. In the common case that the second term in the numerator of Eqs. (7.7) and (7.8) can be neglected with respect to the first, the slope is a direct measure of the dynamic dipole moment of the mode considered. In the case of an adsorbate–substrate bond, the dynamic dipole moment is often a measure of the ionicity of the bond. Hence, we come to the meaningful and interesting conclusion that the field dependence of the adsorbate–surface vibrational frequency is, under certain conditions, a measure of the ionicity or polarity of the surface bond. Note, however, that if the potential energy surface would be harmonic (i.e., $G = 0$) and the dipole moment–distance ($\mu_s - r$) curve is entirely linear (i.e., $\alpha_D = 0$), then the force constant and hence the harmonic vibrational frequency will be field independent, illustrating the importance of the coupling to the higher order anharmonic terms.

Wasileski et al. [14] have validated Eqs. (7.6) and (7.7) numerically by DFT–GGA cluster calculations of simple atomic adsorbates on a 13-atom Pt(111) cluster. The four parameters of interest, K_F, G_F, $\mu_D(F)$, and $-2\alpha_D(F)$, were determined from DFT for four model polar adsorbates, O, Cl, I, and Na, and one model non polar adsorbate, H, at three different fields. This allows the calculation of dK_F/dF as prescribed by Eq. (7.6), and these values were compared with the corresponding quantities extracted directly from the field-dependent DFT-computed potential energy surfaces themselves, labeled as dK_F/dF(DFT). In general, a good approximate correspondence (chiefly to within 10%) was observed between the values for dK_F/dF obtained from Eq. (7.6) and the dK_F/dF(DFT) values. For details of the comparison, the reader is referred to the original paper [14].

The above discussion has focused on the influence of the application of an electric field across the electrode surface on the vibrational properties of chemisorbates. In an electrochemistry experiment, one does, however, not directly apply an electric field but an electrode (or electrochemical) potential with respect to some reference potential. Applying a fixed electrode potential in a quantum-chemical simulation containing a small number of atoms is much less straightforward. The "cleanest" way of applying a potential in a first-principles simulation, at least in this author's view, is to switch to another simulation ensemble in which the potential of the electrons is fixed but not its number [16]. In practice, this grand canonical ensemble method turns out be cumbersome, as it involves an additional iterative loop in an already time-consuming calculation. Therefore, practically all first-principles calculations in the literatute modeling the influence of an electrode potential make use of some kind of "applied" charge distribution in the simulation, either by adding or removing electrons from the metal slab and placing some kind of countercharge somewhere in the simulation cell, or by applying an electric field to the simulation cell. The electrode potential must in such a case always be calculated *after* the calculation has converged or, in the case of an applied electric field, be estimated assuming some effective double-layer thickness. In the latter procedure, the experimental Stark tuning slope may be related to the calculated one by using the relationship

$$\frac{dv}{dF} = d_{dl}\frac{dv}{dE} \tag{7.8}$$

where d_{dl} is the effective thickness of the double layer, typically taken to be $\sim 3\,\text{Å}$ [17]. In experimental terms, such "constant charge" or "constant field" simulations method could be called "coulostatic." For more detailed reviews and discussions of the application of quantum-chemical calculations to electrochemistry, the reader may consult Koper [7] and Janik et al. [8].

7.3 SELECTED EXAMPLES

7.3.1 Cyanide on Gold

Cyanide adsorption on a gold electrode is a good model system for the purposes of this chapter since it has been studied extensively by FTIR, surface-enhanced Raman spectroscopy (SERS), and sum freguency generation (SFG), as well as first-principles DFT calculations including field effects. Figure 7.1 shows a surface-enhanced Raman spectrum of CN^- adsorption on a (roughened) gold electrode, showing the three main

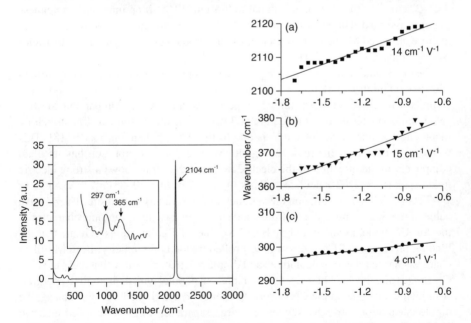

FIGURE 7.1 Left-hand panel: SER spectrum for adsorbed cyanide on gold in 0.01 M KCN and 0.1 M K_2SO_4 at -1.5 V vs. Hg|Hg_2SO_4. Right-hand panel: Peak frequency of C—N stretching mode (a), Au—CN stretching mode (b), and Au—CN bending mode (c) as a function of the electrode potential for adsorbed cyanide on gold in 0.01 M KCN and 0.1 M K_2SO_4 [18].

vibrational features that may be observed in this electrosorption system [18]. There are two low-frequency modes, at $\sim$300 and $\sim$365 cm^{-1}, that have only been observed with SERS due to their low-energy nature [18, 19] and a higher frequency mode at $\sim$2100 cm^{-1} that has been observed with SERS, FTIR, and SFG [18–21].

The vibrational feature observed at $\sim$2100 cm^{-1} is due to the internal C–N stretching mode of the chemisorbed cyanide. This feature has been reported to have a significant Stark tuning slope, although the value of the Stark tuning slope seems to vary between the different spectroscopic methods. Using SERS, Beltramo et al. [18] have observed that the value of the Stark tuning slope may depend on the potential region in which the system is studied and specifically on whether or not the gold dissolves under the influence of the cyanide solution and the applied potential. In the potential region where dissolution may be ruled out, a low Stark tuning slope of $\sim$14 cm^{-1} V^{-1} was observed, in agreement with earlier SERS results. At more positive potentials, where gold (slowly) dissolves, a higher Stark tuning slope of $\sim$40 cm^{-1} V^{-1} was observed, a value closer to that observed with FTIR and SFG experiments [20, 21].

DFT calculations using slab and cluster models of the gold surface found that the atop and bridge binding sites on Au(100) and Au(110) were the most stable adsorption sites for CN on Au [18]. Considering that the atop-bonded CN was calculated to have a C–N stretching frequency close to 2100 cm^{-1} and the bridge-bonded CN a C–N stretching frequency closer to 1900–1950 cm^{-1}, it was concluded that the atop CN is the one observed experimentally. The C–N Stark tuning slope was calculated to be $\sim$24–27 cm^{-1} (V/Å)$^{-1}$, which is equal to 8–9 cm^{-1} V^{-1} if an inner layer thickness of 3 Å is assumed. The Stark tuning slope for the C–N stretch is rather low compared to those observed for the C–O stretch of CO adsorbed on gold or platinum electrodes (see Sections 7.3.3 and 7.3.4).

The low vibrational modes also depend on the applied potential. The feature around 370 cm^{-1} exhibits a Stark tuning slope of 15 cm^{-1} V^{-1}; the feature around 300 cm^{-1} exhibits a much lower Stark tuning slope of $\sim$4 cm^{-1} V^{-1}. Comparison to DFT calculations suggested that the feature at 370 cm^{-1} is due to the Au–CN stretching mode and the feature at 300 cm^{-1} is due to the Au–CN bending mode [18]. This would also explain their Stark tuning slopes. The Au–CN vibration has its main component in the direction of the electric field and therefore shows a strong electric field dependence; DFT calculates a Stark tuning slope of $\sim$50–65 cm^{-1} (V/Å)$^{-1}$, or 17–22 cm^{-1} V^{-1}, depending on the symmetry of the Au adsorption site. The positive value of the Stark tuning slope of this mode also agrees with the anionic character of the Au–CN bond, as suggested by Eq. (7.8). The DFT calculations indicate that the binding preference of CN$^-$ to Au is controlled by the bond ionicity: Both Mulliken charges, charge transfer calculations and dipole calculations show that the Au(110)–CN is most and the Au(111)–CN the least ionic, in parallel with their binding energies. Although the charge on the CN adsorbate cannot be determined unambiguously, the dipole calculations strongly suggest a charge separation of close to a full electron for CN on Au(110) and Au(100). The Au–CN bending mode at 300 cm^{-1} has a negligble Stark tuning slope as its main component is perpendicular to the direction of the electric field.

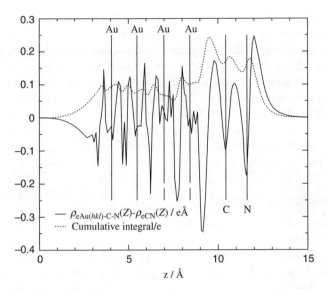

FIGURE 7.2 The solid line represents the charge transfer function for the optimized configuration when CN is adsorbed at the top site of the Au(100) surface. The sign convention is that more positive values indicate more electron rich regions. The dashed line represents the cumulative integral of the charge transfer function from the vacuum at 0 to the plotted z value. The vertical lines indicate the positions of the atoms/slabs as indicated. One can see that an excess of electrons exists between the top Au layer and the C atom, indicating that charge transfer from the metal to the 5σ orbital is responsible for the polar Au–CN bond [18].

As has been shown in various other quantum-chemical studies [22, 23], in the formation of the CN–surface bond, electronic charge is transferred from the metal into the 5σ orbital of CN (or from the 5σ into the metal if cyanide CN$^-$ is taken as the reference state). This is also suggested by Figure 7.2, from which it can be seen from the cumulative integrated charge transfer function that the charge transferred from the metal to the CN is mainly located between the first Au layer and the C atom, which is where the 5σ is located. Compared to carbon monoxide, the role of back donation into the $2\pi^*$ is much less prominent, as the orbital levels (including 5σ and 2π) of the iso-electronic cyanide levels lie $\sim$10 eV higher than that of carbon monoxide. Since back donation into the $2\pi^*$ is responsible for the positive Stark tuning slope of the C–O stretching frequency (see Section 7.3.3), this automatically raises the question of why the C–N stretching frequency also has a smaller but still distinctly positive Stark tuning slope, even if back donation into the $2\pi^*$ does not seem to play a role for chemisorbed cyanide. The interaction with the 5σ orbital probably does not contribute much, as this orbital is nonbonding or only slightly antibonding. However, we believe the positive Stark tuning slope for the C–N stretching frequency can be ascribed tentatively to the contribution of the steric interaction with the surface ("the wall effect" [7, 24]) based on the following argument. Since the adsorbed CN is an anionic adsorbate, the Au–C bond shortens with a more positive applied field.

Typically, in the DFT calculations the Au–C bond distance decreases by about 0.1 Å on all clusters when the field is changed from -0.514 to $0.514 \, V \, Å^{-1}$. The closer the CN is to the metal surface, the stronger the repulsion due to the nonbonding electrons in the metal, leading to a stiffening of the potential energy surface on the carbon side of the C–N vibration. This wall effect is known to lead to an increase in the intramolecular stretching frequency [24]. We believe that this effect, in combination with the only weakly potential dependent back donation, may explain the small Stark tuning slopes of CN adsorbed on Au compared to CO adsorbed on Pt. Bond energy decomposition calculations (such as illustrated in Tables 7.1 and 7.2) have not been carried out for this system but would clearly be worthwhile.

7.3.2 Cyanide on Platinum

Given the results discussed in the previous section, it is interesting the compare cyanide binding on gold to that to on platinum in an electrochemical environment. It is useful to first discuss the SERS experiments for cyanide adsorption on roughened platinum electrodes by Ren et al. [25, 26]. Depending on the cyanide solution concentration, they observed two regions with different behavior. At relatively positive potentials ($>-0.4 \, V$ vs. SCE for $0.1 \, mM \, CN^-$ in $0.1 \, M \, KNO_3$), a region with a clear positive Stark tuning slope of $\sim 65 \, cm^{-1} \, V^{-1}$ for the C–N stretching frequency at $\sim 2100 \, cm^{-1}$ is observed together with a negative Stark tuning slope of $-17 \, cm^{-1} \, V^{-1}$ for the Pt–CN stretching frequency at $\sim 400 \, cm^{-1}$ in the same potential region. The authors postulate binding to the Pt through the C atom by comparison of the frequecies with literature values for Pt–CN complexes. The cyanide adsorption through the C atom was also found to be the favored adsorption mode on Pt(111) in DFT calculations by Ample et al. [22]. Especially the negative Stark tuning slope for the Pt–CN mode is remarkable, as it is opposite to what has been observed on Au (see previous section). According to the theory presented in Section 7.2, such a negative Stark tuning slope for the surface–adsorbate stretch is expected for an electropositive adsorbate, which cyanide is not. Remarkably, below $-0.4 \, V$ vs. SCE the C–N stretching frequency no longer changes with potential, and the Pt–CN band intensity quickly drops to zero. Ren et al. [25] have ascribed this to a flat-lying cyanide on platinum but have also observed the formation of Pt–CN complexes in solution at high potentials and high cyanide concentration, similar to what Beltramo et al. [18] observed for gold in cyanide solution (see Section 7.3.1). Furthermore, although Ren et al. [26] do not mention this, in the potential region where they observe the disappearance of the Pt–CN signal, Pt will catalyze the reduction of nitrate, the electrolye anion in their experiments.

There have also been a number of FTIR experiments of cyanide adsorption, especially on well-defined single-crystal platinum electrodes [27, 28]. On Pt(111), cyanide may be irriversibly adsorbed in an ordered structure of $\sim 0.6 \, ML$. For such a cyanide overlayer, the C–N frequency–potential plot displays two slopes: $\sim 100 \, cm^{-1} \, V^{-1}$ for $E < 0.4 \, V$ versus reversible hydrogen electrode (RHE), and $\sim 30 \, cm^{-1} \, V^{-1}$ for $E > 0.4 \, V$ versus RHE [27]. Tadjeddine et al. [23] have observed a similar change in Stark tuning for CN^- on Pt(111) using SFG and ascribed it to the coadsorption of underpotential deposited hydrogen at the low potentials. Interestingly, in alkaline media there is only a single Stark tuning slope, $\sim 40\,cm^{-1}\,V^{-1}$ [28]. We note that the C–N stretching values in alkaline media are lower than in acidic media. This is in fact also a Stark effect, as the "real" electrochemical potential is more negative in alkaline media than in acidic media. A clear illustration of this effect will be given in Section 7.3.4 when we will discuss CO adsorption on gold electrodes in acidic and alkaline media.

Tadjeddine et al. [23] have performed an extensive study of cyanide adsorption on three basal planes of Pt using SFG and compared the results with DFT cluster calculations. They concluded that cyanide adsorbs to Pt through the C atom, though there was evidence for N adsorption on Pt(110) at high potentials. The DFT calculations indicated that when cyanide adsorbs through the C atom, the bond is partially ionic (as on gold; see Section 7.3.1), whereas when cyanide adsorbs through the N atom, the bond is mainly covalent. No Stark tuning slopes were calculated in their DFT study.

7.3.3 Carbon monoxide on Platinum and Other Transition Metals

Carbon monoxide chemisorption on platinum and platinum group metals has been studied extensively in both ultrahigh vacuum (UHV) and electrochemical (EC) environments. A detailed comparison between the UHV and EC results has been published by Weaver [29] and Weaver et al. [30]. In this section, we will focus primarily on the DFT results and how they have aided in the interpretation of the experimental EC data. One of the objectives of our studies together with the Weaver group was to provide relationships between binding energetics, geometries, and vibrational properties, both of the intramolecular C–O and the metal-adsorbate Pt–CO bond.

Let us first consider the field-dependent binding energetics in the atop and hollow site of a 13-atom Pt(111) cluster [11]. Figure 7.3 shows the total binding energy $E_b(F)$ (upper left panel) and its field-dependent decomposition into steric (lower left panel) and orbital components (right panels), where the latter is further decomposed into contributions of A_1 symmetry (donation contribution, upper right panel) and E symmetry (back-donation contribution, lower right panel). The orbital components have been set to zero for vanishing electric field, as their absolute values are not of interest here. In addition to the calculations in which the C–O bond length was fully relaxed (solid lines), Figure 7.3 also shows the field-dependent binding parameters pertaining to a chemisorbed CO with its C–O bond length fixed at its uncoordinated vacuum value (dashed lines). This bond constraining is especially useful in analyzing the field-dependent steric contribution to the binding energy.

A significant result illustrated in Figure 7.3 is that atop coordination is preferred at positive fields, whereas at negative fields multifold hollow coordination is most favorable. This field-driven site switch on Pt(111) is in qualitative accordance with experiment [29, 30]. It is seen that the back-donation component gets more negative (i.e., more stabilizing) toward the negative field, whereas the opposite is the case for the donation component. These field-dependent donation and back-donation effects are expected on the basis of the Blyholder model [9], in agreement with earlier ab

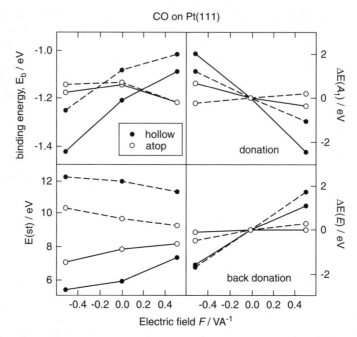

FIGURE 7.3 Field-dependent plots of binding energy E_b and constituent steric $E(\text{st})$ and donation (A_1) and back-donation (E) orbital components for chemisorbed CO in atop (open circles) and hollow (filled circles) sites on Pt(111) surface. Note that the orbital components are plotted as differences with respect to the zero-field values. Dashed plots refer to "bond-constrained" values, as described in the text [11].

initio calculations of Head-Gordon and Tully [31] and Illas and co-workers [32, 33]. The back-donation component is more strongly field dependent for the multifold coordination, that is, more stabilizing toward negative fields. This can be understood in terms of the bonding character of the back-donation interaction, which will prefer to bind to as many surface atoms as possible. By contrast, the donation interaction mainly occurs through an antibonding resonance and should therefore prefer to bind to as few surface atoms as possible. This explains the more destabilizing effect of a more negative field on the back-donation component for multifold coordination as compared to atop coordination. For the C–O "bond-constrained" case, the increase in the back-donation interaction seen toward more negative fields is substantially larger than the opposite field dependence observed for the donation term. Since there is no significant difference in the field dependence of the steric term for atop and hollow coordination (although their absolute values are obviously different), the increased back-donation interaction toward the negative field is the dominant reason for CO to switch site from atop to multifold coordination on Pt(111) under the influence of a more negative electric field. Also note that the steric component becomes more destabilizing toward the negative field, which is due to a combination of the increased electron spill over at negative fields and the shorter Pt–CO bond length. Analyzing the field-dependent components in the case the C–O bond length is allowed to equilibrate (solid lines in Fig. 7.3) is a bit more involved as several effects now play a role. The steric repulsion is now observed to decrease toward the negative field as the C–O bond stretches toward the negative field, leading to a lower *overall* steric repulsion term. Since the field-dependent donation and back-donation components are almost offsetting in the "C–O relaxed" case, the more favorable binding energies for hollow-site binding toward the negative field should now formally be ascribed to the steric repulsion term, changes in which are, as mentioned, mainly due to changes in the internal C–O bond length. This illustrates that the analysis can be quite subtle and may in fact depend on which coordinates are allowed to relax in the calculation. If several coordinates are allowed to relax, it will in general be difficult to disentangle the dominant physical effects in an unequivocal way. For this reason, I prefer the analysis for the "C–O bond-constrained" case as it unequivocally ascribes the site switch from atop to hollow under the influence of the electric field to the enhanced back-donation contribution toward the negative field.

Concerning Stark tuning slopes of C–O stretching frequency, they have been calculated from the slope of the $\nu - F$ curve at $F = 0$ for both CO and NO (nitric oxide) on the four different transition metal surfaces Rh, Ir, Pd, and Pt [11]. A comparison of the DFT-computed Stark tuning slopes with experimental behavior [29, 30] is shown in Figure 7.4. The $d\nu_{DFT}/dF$ values are seen to be in rough accordance with the $d\nu_{exp}/dF$ estimates [by using Eq. 7.8], although the variations of the former

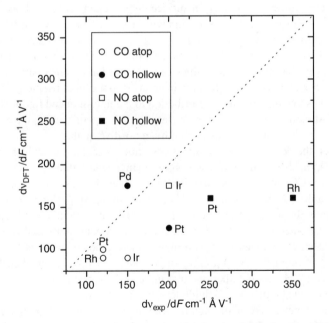

FIGURE 7.4 Comparison of field dependencies of ν_{CO} and ν_{NO} calculated by DFT, $d\nu_{DFT}/dF$, and experimental quantities, $d\nu_{exp}/dF$ as extracted from experimental potential-based values by using Eq. (7.8) [11].

TABLE 7.2 Decomposition of Field-Induced Frequency Shift (cm^{-1}) for CO Adsorbed in Atop and Hollow Site on Pt(111)^a

System	Steric	Donation	Back Donation	Rest	Overall
Pt–CO atop	−131	−7	+264	−16	+110
Pt–CO hollow	−72	−20	+201	+22	+131

^{a}The far-left column is the cluster/binding site configuration. The far-right column is the frequency shift computed for a change in the applied field from −0.514 to +0.514 V Å^{-1}. The middle three columns give the corresponding field-induced frequency shift due to each contribution. The column "rest" refers to the residual field-induced frequency shift not accounted for by the sum of these components.

are smaller. These discrepancies are probably due to the role of coadsorbed solvent in affecting the local electrostatic field. Nevertheless, the larger Stark tuning slopes computed for NO versus CO are in accordance with experimental observations.

The field-dependent changes in the C–O frequencies were analyzed following the decomposition method described in Section 7.2. As reference state, the frequency at the negative field $F = -0.514$ V Å^{-1} was taken, whereas the final state was the system at $F = +0.514$ V Å^{-1}. To the negative field PES the steric, donation, and back-donation components calculated for the positive field were added sequentially and the corresponding frequency changes calculated. Table 7.2 shows the results for CO adsorbed in the atop and hollow site of the Pt(111) cluster. Both the steric and donation contributions are seen be negative. The negative steric contribution likely arises from the lower metal electron surface electron density toward positive fields, diminishing the extent of steric repulsion with the chemisorbate electrons and hence yielding a shallower PES. The negative donation contribution is also readily understandable given that this orbital interaction should lessen toward more positive fields. Note, however, that the absolute values of this contribution are small, indicating an interesting insensitivity of the donation interaction to the electrostatic field. Table 7.2 also illustrates that the dominant contribution to the field-induced frequency changes in the back-donation component, being a large positive term, making the overall Stark tuning slope for the intramolecular vibration a positive quantity.

However, the decomposition analysis does not suggest any overriding trends that could be responsible for the variations observed in the Stark tuning slope as a function of adsorbate (CO or NO), coordination (atop vs. hollow), or metal substrate. For instance, Table 7.2 shows that the higher Stark tuning slope for CO bound in the hollow versus the atop site (a result which itself is in agreement with experiment) cannot be attributed to a higher back-donation component in the hollow site, as one might initially expect, at least not according to this method of analysis. The statement also holds for the higher Stark tuning slopes observed for chemisorbed NO versus CO. The overall picture is rather one that invokes the offsetting influences of two or more components on the overall field–dependent behavior.

The field dependence of the metal–adsorbate ν_{M-CO} was also studied in some detail [34, 35]. Figure 7.5 shows the $\nu_{M-CO} - F$ curves compared to the $\nu_{C-O} - F$ for CO adsorbed at the atop and hollow site on a Pt(111) cluster. For CO in the hollow site and for CO in the atop site at sufficiently positive potential, the $\nu_{M-CO} - F$

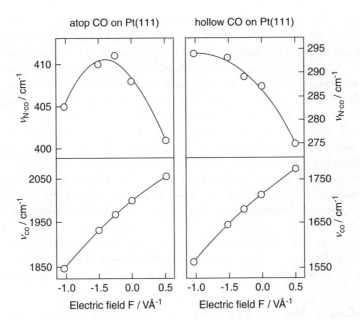

FIGURE 7.5 Stretching frequencies for metal–carbon (ν_{M-CO}) and internal CO (ν_{CO}) bonds plotted against external field F for atop and hollow-site CO on Pt(111) [34].

curve has a negative Stark tuning slope, in agreement with Raman spectroscopic results on polycrystalline Pt [36]. However, the curve for atop CO shows a well-pronounced maximum, as at negative potentials/fields the $\nu_{M-CO} - F$ curve has a positive slope. Such behavior is related, at least qualitatively, to a change in the sign of the dynamic dipole of the M–CO bond as a function of field. For the hollow-site CO, the maximum is not observed. This is most likely related to the much more negative chemisorption bond at negative fields for this geometry (see Fig. 7.5), which favors progressively larger ν_{M-CO} values under these conditions. The combination of the two effects presumably yields the nearly constant ν_{M-CO} values seen at negative fields in the DFT calculations.

The above calculations were all performed using CO adsorption on small 13-atom Pt clusters. More recently, similar calculations have been carried out using the slab geometry. Lozovoi and Alavi [37] and Curulla and Niemantsverdriet [38] have calculatated the field-dependent adsorption of CO on Pt(111). Many of the trends compare well to the previous cluster calculations, such as the positive Stark tuning slope for the C–O stetch, the negative Stark tuning slope for the Pt–CO stretch, and the field-induced site switch from atop at the positive field to multifold at the negative field, but the values obtained especially for the C–O Stark tuning slope differ quantitatively. Whereas on clusters the DFT-calculated Stark tuning slope for the atop-bound C–O stretching frequency on Pt(111) falls in the range of $100–120 \, \text{cm}^{-1}(\text{V/Å})^{-1}$, Lozovoi and Alavi [37] find a value $45 \, \text{cm}^{-1}(\text{V/Å})^{-1}$ for a CO coverage of 0.25 ML. A very similar value [$46–48 \, \text{cm}^{-1}(\text{V/Å})^{-1}$ depending on the adsorption geometry] was found by Curulla and Niemantsverdriet [38] for the same system using a different DFT code. Curulla and Niemantsverdriet also calculated the Stark tuning slope at different coverages to extraploate a value of ~69 cm^{-1}(V/Å)$^{-1}$ in the limit of zero coverage. As discussed by Lozovoi and Alavi, the origin of these significant differences with the cluster calculations is not clear but is most likely due to the limitations of the cluster approach in screening charge transfer effects in the direction along the surface. In favor of the slab calculations, Lozovoi and Alavi note that their value agrees well with the Stark tuning slope of CO adsorbed on Pt(111) obtained under UHV conditions, which at 0.25 ML was estimated to be 56 cm^{-1} (V/Å)$^{-1}$ [39]. Also the coverage dependence of the Stark tuning slope agrees well with the UHV experimental values [38]. This suggests that the higher Stark tuning slope obtained under electrochemical conditions is caused not necessarily by our inaccuracy in estimating fields from electrode potentials using Eq. 7.8 but by a fundamental difference between the two interfaces, presumably primarily reflected in the presence of the solvent. It also suggests that the cluster calculations, though qualitatively useful and insightful, may not yield the correct quantitative results. DFT calculations of the Pt(111)–CO Stark tuning slope in the presence of coadsorbed water have yet be conducted.

7.3.4 Carbon Monoxide on Gold

The adsorption of carbon monoxide on gold electrodes is especially interesting in relation to the unexpected reactivity of gold toward the oxidation of carbon monoxide. A comprehensive FTIR study of CO adsorption to polycrystalline gold electrodes in both acidic and alkaline media was published more than 20 years ago by Kita et al. [40, 41]. A meaningful figure from those papers is reproduced in Figure 7.6, showing

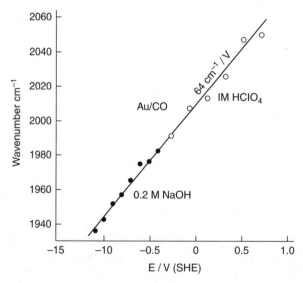

FIGURE 7.6   Potential dependence of C–O stretching frequency of CO adsorbed on gold in acidic and alkaline solutions. Note that potentials are referred to the SHE scale. (From ref. 41.)

the internal C–O stretching frequency on gold in a CO-saturated solution as a function of potential, both in acidic and alkaline media. The corresponding Stark tuning slope is $64 \, cm^{-1} \, V^{-1}$; note the linearity of this plot over a potential range of almost 2 V. Also note that the results obtained in alkaline solution can be nicely extrapolated from acidic solution, showing that on the RHE scale, which is often the preferred potential scale when considering reactivity, CO binding to Au is different in alkaline media from that in acidic media. Kita et al. [40, 41] ascribed the enhanced bonding of CO to gold in alkaline media to the co-adsorption of OH^-. We have recently shown that CO may be adsorbed irreversibly on single-crystalline Au: even Au(111) binds a detectable amount of CO in alkaline media in the absence of CO in solution [42]. In this case, both atop- and bridge-bonded CO can be observed on Au(111), whereas only atop-bonded CO is observed on Au(110).

The presence of irreversibly adsorbed CO has also been claimed by Zhang and Weaver [43] on roughened polycrystalline gold in perchloric acid employing SERS. Their spectra also suggest an enhanced oxygen coadsorption in the presence of adsorbed CO. The low-frequency Au–CO mode at $\sim 465 \, cm^{-1}$ displays a small negative Stark tuning slope of $\sim -5 \, cm^{-1} \, V^{-1}$. However, DFT calculations suggest that the Au–CO stretching mode, both in atop- and bridge-bound modes, should be at a much lower energy, closer to $250 \, cm^{-1}$ (bridge) and $310 \, cm^{-1}$ (atop) [44]. We have not been able to observe a feature at $465 \, cm^{-1}$ in our SERS experiments of CO on Au (in the presence of CO in solution), but we did observe two weak bands at the indicated frequencies [44]. In the C–O stretching region, only atop-bound CO was observed, its stretching frequency increasing with $\sim 40 \, cm^{-1} \, V^{-1}$, in reasonable though not perfect agreement with the earlier results of Kita et al. [40, 41]. We note that the C–O stretching band corresponding to bridge-bonded CO, expected at about $1950 \, cm^{-1}$, was not observed in our experiments. This band was however observed for CO on an Au electrode by Kudelski and Pettinger using SERS [45] but only on certain parts of the electrode. Although the issue of the low-frequency mode may not have been solved yet, we do believe that comparison of experimental spectra to DFT calculations, certainly in the low-energy region accessible to Raman, is a very useful tool in correctly indentifying adsorption modes and thereby adsorbed intermediates.

There have been very few DFT calculations of CO binding to Au surfaces in the presence of an electric field. Some recent results from our own group confirm the large Stark tuning slope for the C–O stretching on both atop and bridge sites on Au(111) slabs [46]. Interestingly, although the C–O stretching red shifts frequency significantly in the presence of a negative applied field, the binding energy hardly changes and adsorption of CO to Au(111) is weak with a ~ 0.2–$0.3 \, eV$ binding energy. This is another example of the lack of a correlation between the binding energy and the C–O stretching frequency.

7.3.5 Formic Acid Oxidation on Gold

To illustrate the potential of combining vibrational spectroscopy and computational chemistry in elucidating intermediates in electrocatalytic reactions, we discuss briefly the gold-catalyzed electrooxidation of formic acid in acidic solution, as studied by

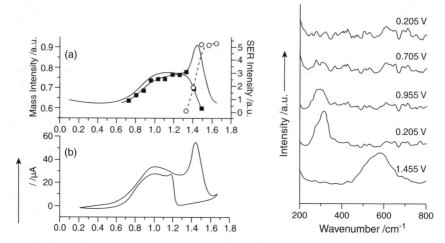

FIGURE 7.7   Left-hand panels: Plot (b) shows a cyclic voltammogram obtained at 2 mV s^{-1} for a gold electrode in 0.05 M HCOOH and 0.10 M HClO$_4$. Plot (a) shows the DEMS intensity corresponding to m/z = 44 (carbon dioxide) and the SERS intensity of the low-frequency Au—formic acid adsorbate mode (■) and the Au—O/Au—OH mode (o) versus potential in the positive-going sweep. Right-hand panel: SER spectrum obtained for a gold electrode in 0.05 M HCOOH and 0.10 m HClO$_4$ at the indicated potentials. [44]

SERS and DFT [44]. We were interested in elucidating whether this dehydrogenation process happens through an adsorbed intermediate, and, if so, if we could determine the nature of this intermediate. The two most likely candidates for such an intermediate would be adsorbed carboxyhydroxyl (COOH$_{ads}$) and formate (HCOO$_{ads}$). Both intermediates have been suggested in the older literature.

Figure 7.7(b) shows the voltammetry of formic acid oxidation on a polycrystalline gold electrode in perchloric acid solution, and Figure 7.7(c) shows the measured SER spectra at the indicated potentials in the voltammogram. At potentials below 1.3 V one low-frequency peak is observed in the Raman spectra, at ~300 cm^{-1}. Above 1.3 V, this peak is replaced by a peak at much higher wavenumbers, ~600 cm^{-1}, as illustrated by the SER spectrum at 1.455 V. By comparison with previous SERS studies, we identify this peak with the formation of gold oxide [43]. The feature at 300 cm^{-1} must be due to a formic acid adsorbate. Figure 7.7(a) plots the intensity of the features at 300 and 600 cm^{-1} as a function of potential, alongside with the signal for carbon dioxide formation as probed by differential electrochemical mass spectrometry (DEMS). Note that the DEMS signal follows closely the current, in agreement with the oxidation of formic acid to CO$_2$. The SERS signal ascribed to the formic acid adsorbate matches the main formic acid oxidation wave, suggesting the following mechanism for the oxidation of formic acid on gold below 1.3 V:

$$HCOOH \leftrightarrows Ads + H^+ + e^-$$
$$Ads \quad\;\; \rightarrow CO_2 + H^+ + e^-$$

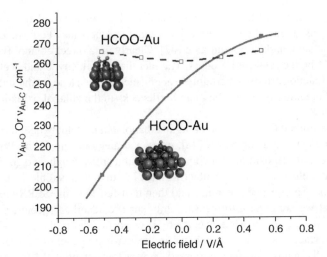

FIGURE 7.8 Field-dependent low-frequency metal–adsorbate stretching frequencies calculated from DFT for COOH and HCOO adsorbates on Au(110) [44].

Above 1.3 V, the SERS signal attributed to the formic acid adsorbate quickly disappears in favor of the gold oxide peak, accompanied by a peak in both current and CO_2 production. We have interpreted this as a second pathway for formic acid oxidation, which takes place above 1.3 V and cleans the gold surface from the formic acid adsorbate:

$$H_2O \leftrightarrows OH_{ads} + H^+ + e^-$$
$$Ads + OH_{ads} \rightarrow CO_2 + H_2O$$

The question of the identity of the formic acid adsorbate "Ads" was solved by comparison with DFT calculations. The experimentally observed feature at $300\,cm^{-1}$ has a substantial Stark tuning slope of $\sim 70\,cm^{-1}\,V^{-1}$. The positive slope suggests that we are dealing with surface-adsorbed anion. From the DFT calculations, the $HCOO_{ads}$ adsorbate is found to be $\sim 1\,eV$ about more stable than the $COOH_{ads}$ adsorbate on Au(110). Moreover, the Stark tuning slope of $HCOO_{ads}$ is sizably positive [$65\,cm^{-1}\,(V/\mathring{A})^{-1}$], whereas that for the $COOH_{ads}$ is close to zero (see Fig. 7.8). Taken together, we proposed that the formic acid adsorbate is the formate *anion*:

$$HCOOH \leftrightarrows HCOO^-_{ads} + H^+$$
$$HCOO^-_{ads} \rightarrow CO_2 + H^+ + 2e^-$$

7.3.6 Ethanol Oxidation on Platinum

The oxidation of ethanol on platinum is another good example of a complex electrocatalytic reaction where the use of in situ FTIR and SERS has proven indispensable in solving the details of the reaction mechanism. The ethanol oxidation is especially challenging as at least three different stable final products may be formed: acetaldehyde, acetic acid, and carbon dioxide [47]. Clearly, in a direct ethanol fuel cell, the latter would be the preferred product. The fact that acetaldehyde and acetic acid are prominent reaction products, certainly in acidic media, often leads to the statement that C–C bond breaking is one of the main bottlenecks in the ethanol oxidation reaction mechanism.

We have studied C–C bond breaking in ethanol and acetaldehyde interacting with a platinum electrode using SERS [48], using a strategy developed by Weaver et al. [3], which entails depositing a thin layer of platinum onto a roughened gold electrode. The platinum electrode was in contact with 0.1 M ethanol in perchloric acid solution at a number of prescribed potentials and then transferred to the SERS measurement cell, so that we are certain to measure only surface-adsorbed species, and there is no interference of any possible acetaldehyde or acetic acid in the layer close to the electrode surface. The results are shown in the left-hand panel of Figure 7.9, which shows four distinguishable features already at very low potential, 0.1 V, which is close to the reversible potential for ethanol oxidation to carbon dioxide. The four peaks come in two pairs: a pair at $\sim$495 cm^{-1} and $\sim$2000 cm^{-1}, which we attribute to the Pt–CO and C–O stretching vibrations of surface-bonded CO, respectively, and a pair at $\sim$426 cm^{-1} and $\sim$2880 cm^{-1}, which we attribute to the Pt–CH$_x$ and C–H stretching vibrations of surface–bonded CH$_x$. The first assignment is based on a comparison to literature SERS results on CO bonding to Pt [43]. For the second assignment,

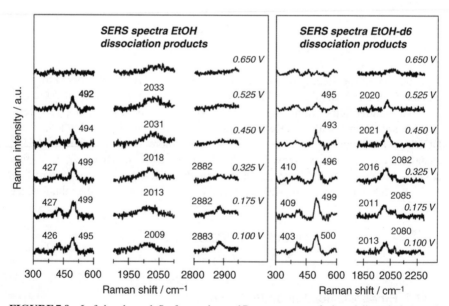

FIGURE 7.9   Left-hand panel: Surface-enhanced Raman spectra of ethanol dissociation products on a Pt electrode in 0.1 M HClO$_4$ recorded at indicated potentials (potential is increasing in the experiment). Right-hand panel: Same experiment but for deuterated ethanol [48].

no literature data are available for comparison. Therefore, we conducted an isotope labelling experiment using deuterated ethanol in D_2O. The results are shown in the right-hand panel of Figure 7.9. The spectra illustrate that the features attributed to surface-bonded CO have not shifted, as expected, but the features attributed to CH_x have both shifted to lower wavenumbers, ~ 405 and $\sim 2080\, cm^{-1}$, respectively, in full agreement with having a CD_x species on the surface. From the magnitude of the isotopic shift and comparison to the surface science and computational literature, we concluded that the most likely species CH_x on the surface must be CH_{ads}. Essentially similar results were obtained for acetaldehyde dissociation, with the main difference that larger amounts of CO and CH were detected on the Pt surface compared to ethanol. This is in good agreement with our earlier voltammetric results on Pt single-crystal electrodes, namely that C–C bond breaking occurs to a much greater extent in acetaldehyde than in ethanol. A second important observation to be made from Figure 7.9 relates to the potential dependence of the intensity of the CO and CH bands. As the potential is made more positive, the CH band loses intensity, whereas the CO band initially gains in intensity (the right-hand panel actually illustrates this the clearest; plotting of the CO integrated band intensity shows that the effect is significant in both light and heavy water). This implies that at these potentials (below ~ 0.4 V) the CH fragment on the surface is oxidized to chemisorbed CO. At higher potentials, this chemisorbed CO is oxidized to CO_2. These results suggest the following mechanism for ethanol oxidation to CO_2 on a platinum electrode. Most likely, ethanol is oxidized first to acetaldehyde, which is a relatively easy reaction. We suggest that acetaldehyde is the precursor to C–C bond breaking, very similar to suggestions made for ethanol dissociation on Pt in UHV [49]. The acetaldehyde dissociates into CH and CO on the surface. This takes place at potentials close to the equilibrium potential for the overall reaction. The chemisorbed CH fragment is oxidized into chemisorbed CO before CO is oxidized to CO_2. This makes CO_{ads} oxidation the potential determining reaction in the formation of CO_2. This mechanism shows that C–C bond breaking need not necessarily be the difficult step in the complete ethanol oxidation to CO_2. In the presence of ethanol in solution, acetaldehyde may be further oxidized to acetic acid, which is inactive on a platinum surface and represents a "dead end" in the oxidation mechanism. In this light, improving the complete oxidation of ethanol to carbon dioxide should focus on favoring the C–C bond breaking in acetaldehyde over oxidizing acetaldehyde to acetic acid.

7.4 CONCLUSIONS

Combining vibrational spectroscopy of adsorbates at electrochemical interfaces with first-principles density functional calculations is a very useful tool in probing and understanding molecular-level details of electrochemical adsorption and reactivity. It may not only help in correctly identyfing adsorbed species on the electrode surface and the nature of their bonding geometry, but detailed scrutiny of the quantum-chemical calculations in relation to the experimental observations, especially as a function of the applied electrode potential or electric field, also aids in understanding the quantum-chemical factors that govern surface bonding. If there is one important take-home message from the brief and selective survey given in this chapter, it is that simple rules of thumb regarding the relationship of vibrational features and binding strength do not always work. In general, one should carefully evaluate quantum-chemical calculations and make use of the various tools that quantum chemists have developed in order to analyze results beyond simply looking at bonding energies and geometries. The future of this fruitful interplay between experiment and theory lies in studying systems with increasing complexity, not only in terms of larger adsorbates, but especially in the more realistic modeling of the electrochemical interface by adding water or solvent molecules to the simulation.

REFERENCES

1. T. Iwasita and F. C. Nart, In situ infrared spectroscopy at electrochemical interfaces, *Prog. Surf. Sci.*, 55 (1997) 271–340.
2. M. Osawa, In-situ surface-enhanced infrared spectroscopy of the electrode/solution interface, in *Advances in Electrochemical Sciences and Engineering*, Vol. 9, R. C. Alkire, D. M. Kolb, J. Lipkowski, and P. N. Ross (Eds.), Wiley, New York, 2006, pp. 269–314.
3. M. J. Weaver, S. Zou, and H. Y. H. Chan, The new interfacial ubiquity of surface-enhanced Raman spectroscopy, *Anal. Chem.*, 72 (2000) 38A–47A.
4. D-Y. Wu, J-F. Li, B. Ren, and Z-Q. Tian, Electrochemical surface-enhanced Raman spectroscopy of nanostructures, *Chem. Soc. Rev.*, 37 (2008) 1025–1041.
5. A. Tadjeddine and A. Peremans, in *Spectroscopy for Surface Science*, R. J. H. Clark and R. E. Hester (Eds.) Wiley, Chichester, 1998, p. 159.
6. G. Q. Lu, A. Lagutchev, T. Takeshita, R. L. Behrens, D. D. Dlott, and A. Wieckowski, Broadband sum frequency generation studies of surface intermediates involved in fuel cell electrocatalysis, in *Fuel Cell Catalysis: A Surface Science Approach*, M. T. M. Koper (Ed.), Wiley, New York, 2009, pp. 375–409.
7. M. T. M. Koper, *Ab initio* quantum-chemical calculations in electrochemistry, in *Modern Aspects of Electrochemistry*, Vol. 36, C. G. Vayenas, B. E. Conway, R. E. White (Eds.), Kluwer Academic/Plenum, New York, pp. 51–130.
8. M. J. Janik, S. A. Wasileski, C. D. Taylor, and M. Neurock. First-principles simulation of the active sites and reaction environment in electrocatalysis, in *Fuel Cell Catalysis: A Surface Science Approach*, M. T. M. Koper (Ed.), Wiley, New York, 2009, pp. 93–128.
9. G. Blyholder, Molecular orbital view of chemisorbed carbon monoxide, *J. Phys. Chem.*, 68 (1968) 2772–2777.
10. B. Hammer, Y. Morikawa, and J. K. Nørskov, CO chemisorption at metal surfaces and overlayers, *Phys. Rev. Lett.* 76 (1996) 2141–2145.
11. M. T. M. Koper, R. A. van Santen, S. A. Wasileski, and M. J. Weaver, Field-dependent chemisorption of carbon monoxide and nitric oxide on platinum-group (111) surfaces: Quantum chemical calculations compared with infrared spectroscopy at electrochemical and vacuum-based interfaces, *J. Chem. Phys.*, 113 (2000) 4392–4407.
12. S. A. Wasileski and M. J. Weaver, What can we learn about electrode–chemisorbate bonding energetics from vibrational spectroscopy? An assessment from density functional theory, *Faraday Discuss.*, 121 (2002) 285–300.

13. D. K. Lambert, Vibrational Stark effect of adsorbates at electrochemical interfaces, *Electrochim. Acta*, 41 (1996) 623–630.

14. S. A. Wasileski, M. T. M. Koper, and M.J. Weaver, Metal electrode-chemisorbate bonding: General influence of surface bond polarization on field-dependent binding energetics and vibrational properties, *J. Chem. Phys.*, 115 (2001) 8193–8203.

15. S. A. Wasileski, M. T. M. Koper, and M. J. Weaver, Field-dependent electrode-chemisorbate bonding: sensitivity of vibrational Stark effect and binding energetics to nature of surface coordination, *J. Am. Chem. Soc.*, 124 (2002) 2796–2805.

16. A. Y. Lozovoi, A. Alavi, J. Kohanoff, and R.M. Lynden-Bell, *Ab initio* simulation of charged slabs at constant chemical potential. *J. Chem. Phys.*, 115 (2000) 1661–1669.

17. M. J. Weaver, Electrostatic-field effects on adsorbate bonding and structure at metal surfaces: Parallels between electrochemical and vacuum systems, *Appl. Surf. Sci.*, 67 (1993) 147–159.

18. G. L. Beltramo, T. E. Shubina, S. J. Mitchell, and M. T. M. Koper, Cyanide adsorption on gold electrodes: A combined surface enhanced Raman spectroscopy and density functional theory study, *J. Electroanal. Chem.*, 563 (2004) 111–120.

19. H. Baltruschat and J. Heitbaum, On the potential dependence of the CN stretch frequency on Au electrodes studied by SERS, *J. Electroanal. Chem.*, 157 (1983) 319–326.

20. K. Kunimatsu, H. Seki, W. G. Golden, J. G. Gordon II, and M. R. Philpott, A study of the gold/cyanide solution interface by in situ polarization-modulated Fourier transform infrared reflection absorption spectroscopy, *Langmuir*, 4 (1988) 337–341.

21. C. Matranga and P. Guyot-Sionnest, Vibrational relaxation of cyanide at the metal/electrolyte interface, *J. Chem. Phys.*, 112 (2000) 7615–7621.

22. F. Ample, A. Clotet, and J. M. Ricart, Structure and bonding mechanism of cyanide adsorbed on Pt(111), *Surf. Sci.*, 558 (2004) 111–121.

23. M. Tadjeddine, J-P. Flament, A. Le Rille, and A. Tadjeddine, SFG experiment and ab initio study of the chemisorption of CN⁻ on low-index platinum surfaces, *Surf. Sci.*, 600 (2006) 2138–2153.

24. P. Bagus and G. Pacchioni, Electric field effects on the surface-adsorbate interaction: Cluster model studies, *Electrochim. Acta*, 36 (1991) 1669–1675.

25. B. Ren, X-Q. Li, D-Y. Wu, J-L. Yao, Y. Xie, and Z-Q. Tian, Orientational behavior of cyanide on a roughened platinum surface investigated by surface enhanced Raman spectroscopy. *Chem. Phys. Lett.*, 322 (2000) 561–566.

26. B. Ren, D-Y. Wu, B-W. Mao, and Z-Q. Tian, Surface-enhanced Raman study of cyanide adsorption at the platinum surface, *J. Phys. Chem. B*, 107 (2003) 2752–2758.

27. F. Huerta, E. Morallon, C. Quijada, J. L. Vazquez, and A. Aldaz, Spectroelectrochemical study on CN⁻ adsorbed at Pt(111) in sulphuric and perchloric media, *Electrochim. Acta*, 44 (1998) 943–948.

28. F. Huerta, F. Montilla, E. Morallon, and J. L. Vazquez, On the vibrational behaviour of cyanide adsorbed at Pt(111) and Pt(100) surfaces in alkaline solutions, *Surf. Sci.* 600 (2006) 1221–1226.

29. M. J. Weaver, Binding sites and vibrational frequencies for dilute carbon monoxide and nitric oxide adlayers in electrochemical versus ultrahigh-vacuum environments: the roles of double-layer solvation, *Surf. Sci.*, 437 (1999) 215–230.

30. M. J. Weaver, S. Zou, and C. Tang, A concerted assessment of potential-dependent vibrational frequencies for nitric oxide and carbon monoxide on low-index platinum-group surfaces in electrochemical compared with ultrahigh vacuum environments: structural and electrostatic implications. *J. Chem. Phys.*, 111 (1999) 368–381.

31. M. Head-Gordon and J. C. Tully, Electric field effects on chemisorption and vibrational relaxation of CO on Cu(100), *Chem. Phys.* 175 (1993) 37–51.
32. F. Illas, F. Mele, D. Curulla, A. Clotet, and J. M. Ricart, Electric field effects on the vibrational frequency and bonding mechanism of CO on Pt(111), *Electrochim. Acta*, 44 (1998) 1213–1220.
33. D. Curulla, A. Clotet, J. M. Ricart, and F. Illas, Ab initio cluster model study of electric field effects for terminal and bridge bonded CO on Pt(100), *Electrochim. Acta*, 45 (1999) 639–644.
34. S. A. Wasileski, M.J. Weaver, and M. T. M. Koper, Potential-dependent chemisorption of carbon monoxide on platinum electrodes: New insight from quantum-chemical calculations combined with vibrational spectroscopy, *J. Electroanal. Chem.*, 500 (2001) 344–355.
35. S. A. Wasileski, M. T. M. Koper, and M. J. Weaver, Field-dependent chemisorption of carbon monoxide on platinum-group (111) electrodes: Relationships between binding energetics, geometries, and vibrational properties as assessed by density functional theory, *J. Phys. Chem. B*, 105 (2001) 3518–3530.
36. P. Gao and M. J. Weaver, Metal-adsorbate vibrational frequencies as a probe of surface bonding: halides and pseudohalides at gold electrodes, *J. Phys. Chem.*, 90 (1986) 4057–4063.
37. A.Y. Lozovoi and A. Alavi, Vibrational frequencies of CO on Pt(111) in electric field: A periodic DFT study, *J. Electroanal. Chem.*, 607 (2007) 140–146.
38. D. Curulla Ferré and J. W. Niemantsverdriet, Vibrational Stark tuning rates from periodic DFT calculations: CO/Pt(111), *Electrochim. Acta*, 53 (2008) 2897–2906.
39. J. S. Luo, R. G. Tobin, and D. K. Lamert, Electric field screening in an adsorbed layer: CO on Pt(111), *Chem. Phys. Lett.*, 204 (1993) 445–450.
40. H. Nakajima, H. Kita, K. Kunimatsu, and A. Aramata, Infrared spectra of carbon monoxide on a smooth gold electrode. Part I. EMIRS spectra in acid and alkaline solutions, *J. Electroanal. Chem.*, 201 (1986) 175–186.
41. K. Kunimatsu, A. Aramata, H. Nakajima, and H. Kita, Infrared spectra of carbon monoxide on a smooth gold electrode. Part II. EMIRS and polarization-modualted IRRAS study of the adsorbed CO layer in acidic and alkaline solutions, *J. Electroanal. Chem.*, 207 (1986) 293–307.
42. P. Rodriguez, J. M. Feliu, and M. T. M. Koper, Unusual adsorption state of carbon monoxide on single-crystalline gold electrodes in alkaline media, *Electrochem Commun*, 11 (2009) 1105–1108.
43. Y. Zhang and M. J. Weaver, The electro-oxidation of carbon monoxide on noble metal catalysts as revisited by real-time surface-enhanced Raman spectroscopy. *J. Electroanal. Chem.*, 354 (1993) 173–188.
44. G. L. Beltramo, T. E. Shubina, and M. T. M. Koper, Oxidation of formic acid and carbon monoxide on gold electrodes studied by surface-enhanced Raman spectroscopy and DFT, *ChemPhysChem*, 6 (2005) 2597–2606.
45. A. Kudelski and B. Pettinger, Fluctuations of surface-enhanced Raman spectra of CO adsorbed on gold substrates, *Chem. Phys. Lett.*, 383 (2004) 76–79.
46. P. Rodriguez, N. Garcia-Araez, A. A. Koverga, S. Frank, M.T.M. Koper, Langmuir 26 (2010) 12425–12432.
47. M. T. M. Koper, S. C. S. Lai, and E. Herrero, Mechanisms of the oxidation of carbon monoxide and small organic molecules on metal electrodes. in *Fuel Cell Catalysis: A Surface Science Approach*, M. T. M. Koper (Ed.), Wiley, New York, 2009, pp. 159–207.

48. S. C. S. Lai, S. E. F. Kleyn, V. Rosca, and M. T. M. Koper, Mechanism of the dissociation and electrooxidation of ethanol and acetaldehyde on platinum as studied by SERS, *J. Phys. Chem. C*, 112 (2008) 19080–19087.

49. Y. Cong, V. van Spaendonk, and R. I. Masel, Low temperature C—C bond scission during ethanol decomposition on Pt(331), *Surf. Sci.*, 385 (1997) 246–258.

CHAPTER 8

Electrochemical Catalysts: From Electrocatalysis to Bioelectrocatalysis

ALEJANDRO JORGE ARVIA, AGUSTÍN EDUARDO BOLZÁN, and
MIGUEL ÁNGEL PASQUALE

Instituto de Investigaciones Fisicoquímicas Teóricas y Aplicadas (INIFTA),
Universidad Nacional de La Plata (UNLP), 1900 La Plata, Argentina

8.1 INTRODUCTORY REMARKS

The technological advance of electrocatalytic processes is sustained by the continuous search for more efficient and longer lifetime performance electrode materials. This goal can be progressively achieved with further knowledge of the basic science lying behind those processes. For this purpose, the study of the structure and composition of solid surfaces at the atomic level, the oxidation state of surface atoms, and the adsorbate bonding capability in different media over a wide range of the applied electric potential, electrolyte composition, pressure (P), and temperature (T) are required. This is possible employing photon, electron, and ion spectroscopies covering wide spatial, time, and energy resolution ranges combined with electrochemical techniques.

The design of high-technology electrocatalysts is based upon a combination of surface science, electrochemical kinetics, and engineering concepts. Electrocatalyst architectures need minimal microporous surface area and maximized electrode surface accessibility utilizing mesoporous (2–50-μm), macroporous, or rough electrodes of some kind. In any case, electrochemical reactions imply that the concentration of electrons as reactants or products scales with the electrified surface area operating in an ionic conductive medium for charge balancing. The mobility of ions into $\approx$1-μm pores is several orders of magnitude slower than in the bulk electrolyte. This means that transport at the surface area in micropores cannot be electrochemically controlled on the same time scale as transport at more easily accessible surfaces.

A variety of materials of electrocatalytic interest are utilized in fuel cell applications, such as single metals, alloys, amorphous materials, oxides, perovskites, carbons,

Catalysis in Electrochemistry: From Fundamentals to Strategies for Fuel Cell Development,
First Edition. Edited by Elizabeth Santos and Wolfgang Schmickler.
© 2011 John Wiley & Sons, Inc. Published 2011 by John Wiley & Sons, Inc.

organometallic complexes, polymers, chemically modified conducting substrates, semiconductors, and biological materials such as enzymes and living organisms. Relatively simple fuel cells are those consisting of inorganic catalysts, either dispersed or single-metal rough anodes fed with either hydrogen or methanol (MeOH) as fuels and oxygen/air as cathode. The overall electrochemical reactions produce energy and water or water and carbon dioxide from hydrogen and MeOH electro-oxidation, respectively. Fuels such as lower order alcohols and alkanes are also frequently used as reformed reactants containing hydrogen to feed the anode compartment.

More complex devices, as those with bioelectrocatalytic electrodes, involving enzymes and living organisms such as bacteria are under continuous research looking for more efficient and environmentally friendly processes. In this respect, to mimic as close as possible natural systems, which have shown high efficiency in energy conversion processes, bioelectrocatalysis is of particular interest. Enzymes in living creatures extract energy from food, such as glucose, to power life. Accordingly, they are excellent candidates for constructing implantable power sources for artificial hearts and devices for monitoring glucose levels in blood and chemicals that signal the onset of a particular disease.

The rapid advance in the development of fuel cell electrodes limits the description of this chapter to some basic aspects related to fuel cell electrochemistry with electrodes of increasing complexity such as nanostructured materials, including some different types of electrodes used at present. Biofuel cells are much further from commercial development than simpler traditional ones and they still face considerable challenges, but significant progress has been achieved in recent times that deserves particular attention. In this chapter, the electrochemical systems are grouped according to their chemical nature as electrocatalytic and bioelectrocatalytic systems. Photoeffects, such as light-powered electrochemistry as energy source, a subject with a promising future [1], are beyond the scope of this chapter.

8.2 SOME TYPICAL ELECTROCATALYSTS

8.2.1 Dispersed and Rough Metal Electrodes

Metal electrodes for energy conversion are usually employed as dispersed particles in a conducting substrate. Electrocatalyst dispersions consist of crystalline metal clusters of average size from nanometers up to micrometers [2]. Patterns of well-defined metal clusters on crystalline substrates can be produced by different procedures: (i) local metal electroplating utilizing the tip of a scanning tunnelling microscope; (ii) molecular paint brushing the microscope tip and subsequent transfer of metal atoms to the substrate by contacting the microscope tip with the surface [3].

In heterogeneous reactions, the influence of the particle size and shape plays a key role in the behavior of the catalyst. This has been well established after it was found that tiny aged gold dots spread on titanium oxide support catalyze a variety of chemical reactions and the catalytic activity increases as dot size shrinks. When this occurs, the number density of gold dots with *potent sites* increases [4]. The catalytic activity of these gold dot dispersions has been tested for the $CO \rightarrow CO_2$

oxidation heterogeneous reaction utilizing gold clusters of different size [Fig. 8.1(a)]. In this case, the rate of CO oxidation shows a maximum for ≈3-nm gold cluster sizes [Fig. 8.1(b)] [4]. The particle size dependence of catalyst performance is consistent with the fact that the surface structure and electronic properties change greatly as the particle size varies from roughly 50 to 1 nm.

As heterogeneous catalysis is an inherently nanoscopic phenomenon, new catalysts with improved performance arise when the multifunctionality combined with the transport of reactants and products is reconsidered in terms of appropriate designs based on nanoscale building blocks, in which deliberate disorder, including voids, become parts of the catalyst structure [5]. Accordingly, the activity or selectivity of catalyst nanoparticles depends on the size, shape (aspect ratio), surface structure, and environmental characteristics [2]. These conclusions are applicable to understanding the behavior of metal electrocatalysts in more complex fuel cell systems.

Theoretical studies of atomic and molecular clusters often seek to explain structure, dynamics, and thermodynamics in terms of an underlying potential energy surface and the type of particle interactions. The overall approach, derived from the global analysis of potential surfaces, indicates that changes in the most favorable cluster morphology can be qualitatively associated with a certain range of particle interaction forces. In this cases, thermodynamic properties can be calculated from a representative sampling of local minima on the potential energy surface and dynamic predictions from the knowledge of the transition states and reaction pathways [6]. In fact, the so-called *compensation effect* (see Chapter 2) suggests that an effective catalyst bonding with reactant species must be tight enough to help them come together but loose enough so that they can react with one another. This conclusion, however, had to be revised after the high electrocatalytic efficiency of gold dots was discovered [4]. In this case, the catalytic efficiency was explained by the interactions of individual molecules with gold dots utilizing powerful tools of surface science and the density functional theory (DFT) and different models for gold atom arrangements packed into a hexagonal geometry in contact with a single-crystal surface substrate [Fig. 8.1(a)]. Thus, the

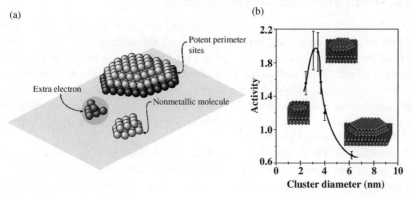

FIGURE 8.1 (a) Scheme of gold dots packed into atomic hexagonal pattern for dot size higher than 3 nm. Gold dots are supported on a conducting TiO_2 crystalline surface. (b) Catalytic activity for heterogeneous oxidation of $CO \rightarrow CO_2$ versus dot average diameter. The maximum catalytic activity occurs for 2–3 nm particle diameter. (Plotted data from Valden and Lai [4].)

few monolayer hexagonal patterns consist of a smooth surface on top surrounded by step edges. Atoms at alternative step edges have fewer neighbor atoms than those at the smooth surface. The number of potent sites that exhibit stronger interactions with reactants and intermediates increases as the nanoparticle shrinks. However, the basic physical properties of the nanoparticle tend to disappear in tiny dots 2 monolayers (ML) thick and 3 nm in diameter. These data provided an insight for understanding possible ways of atom and molecule rearrangements at the catalyst surface in order to increase the catalytic efficiency of a particular process.

A comparable behavior has been reported for rough palladium nanoclusters and nanowires formed on highly ordered graphite (HOPG) by electrodeposition from aqueous acid palladium chloride containing an excess of sodium perchlorate [7, 8] [Figs. 8.2(a,b)]. These 30–100-nm average-size nanoclusters show a 3D core and 2D dendrites merging at 120° [Fig. 8.2(c)]. Their fractal surfaces catalyze the first electron transfer step of the hydrogen evaluation reaction (HER) [Figs. 8.2(d,e)] and the oxygen electrochemical reduction reaction (OERR) in oxygen-saturated aqueous acid solutions [Fig. 8.2(f)].

Fractal surface palladium electrodes can be produced from electrochemical roughening (Chapter 2) by cyclic asymmetric potential reversal techniques [9]. The surface of these electrodes behaves as a self-affine fractal with dimension $D_f = 2.57$. It is made of an irregular columnlike structure. Conventional voltammetry and triangular modulated voltammetry revealed a remarkable enhancement of the H-atom electrosorption processes, the H-atom surface diffusion prevailing over bulk hydrogen diffusion into the metal.

Highly rough platinum electrodes were also prepared by electrochemical roughening. These electrodes also exhibit a self-affine fractal surface [10, 11]. They electrocatalize the electrooxidation of reduced CO_2 adsorbates [12] in aqueous acids. The corresponding electrochemical oxidation charge of reduced CO_2 adsorbate becomes proportional to the electrode roughness ($\mathcal{R}$) (Fig. 8.3) [13].

8.2.2   Dispersed Metal-Carbon Electrodes

The performance of carbon materials for electrochemical applications is determined by their electronic properties, surface cleanliness, and catalyst size. These carbons are most frequently referred to as the sp^2 graphitic allotrope, and their reproducibility depends on their origin and surface preparation procedure. Due to the anisotropy of electronic properties, carbons behave from an essentially metallic material in disordered graphite or glassy carbon to a semimetallic one in HOPG. Variations in electronic properties are reflected in the interfacial capacitance, the adsorption behavior, and the electron transfer kinetics.

The surface chemistry of carbons is rather complex because of the variable number of surface defects and O- and OH-containing functional groups that participate in physisorption or chemisorption processes (Fig. 8.4). Carbon electrodes are of special interest for the fabrication of chemically modified electrodes, as carbon surfaces are relatively inert in electrochemical processes over a wide useful potential range. Physical and electrochemical properties of carbon materials (polycrystalline graphite,

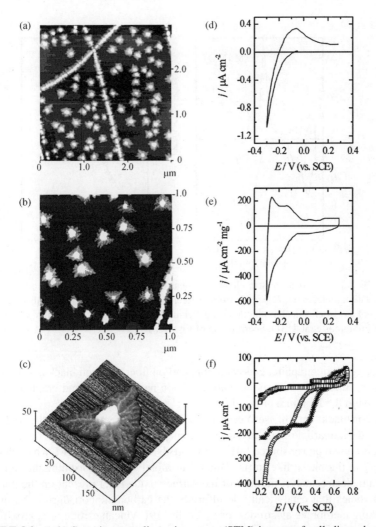

FIGURE 8.2 (a,b) Scanning tunneling microscopy (STM) images of palladium electrode-posited (q_d = 3 mC cm^{-2}) on HOPG from aqueous acidic palladium chloride with excess of sodium perchlorate. (c) Three-dimensional (3D) view of single palladium cluster. (d,e) Voltam-mograms of palladium electrodes in 0.1 M perchloric acid run at v = 0.1 V s^{-1} from E = 0.30 V downward; (d) polycrystalline (e) palladium clusters dispersed on HOPG (q_d = 5 mC cm^{-2}); (f) voltammograms of bare HOPG run from 0.45 to 0.70 V and backward to -0.20 V in O$_2$-saturated 0.05 M perchloric acid, v = 0.05 V s^{-1}, 298 K (■), polycrystalline palladium (○), dispersed palladium clusters on HOPG (q_d = 3 mC cm^{-2}) (X), dispersed palladium deposited on HOPG (q_d = 3 mC cm^{-2}) in deaerated solution (□). (From Gimeno et al. [7, 8] by permission of The American Chemical Society.)

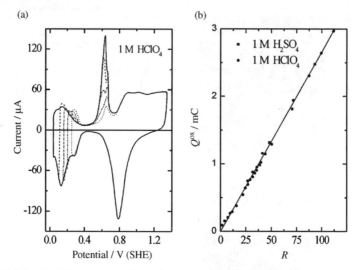

FIGURE 8.3 (a) Voltammetric electrochemical oxidation of reduced CO_2 adsorbates on rough platinum electrode, $\mathcal{R} = 115$, from different cathodic switching potentials up to 1.1 V. (b) Plot of electrochemical oxidation charge versus $\mathcal{R}$ ($1 \leq \mathcal{R} \leq 115$); $v = 5$ mV s^{-1}, $T = 298$ K. (From Marcos et al. [13] by permission of Elsevier.)

HOPG, pyrolytic graphite, glassy carbon, carbon fibers, carbon blacks, metal-doped diamond, fullerenes, carbon nanotubes, carbon nanocapsules, and graphene layers) are given in the literature [14–16].

Carbon blacks are still used because of their physicochemical properties for hindering metal nanoparticle agglomeration, particularly platinum, during operation. One of the well-known power source carbon supports is Vulcan carbon [18], which contains 0.5 wt% of thiophene-like sulfur. This sulfur moiety avoids poisoning the supported platinum electrocatalyst because of the surface oxide layer present on the platinum nanoparticles that catalytically desulfurises the carbon support during the process of making membrane electrodes for fuel cells [19]. Vulcan carbon of approximately 250 m^2 g^{-1} surface area contains micropores in which carbon-supported platinum nanoparticles are placed by means of impregnation synthesis. In this way, the need to maximize platinum surface atom accessibility to fuel can be surpassed.

The electrocatalytic efficiency of dispersed platinum on HOPG for the HER and OERR was compared to those obtained for dispersed platinum on graphite and platinum on AGKTS National graphite, 30% porosity. The latter is a gas diffusion porous electrode structure made of Vulcan XC-72 carbon layer with highly dispersed Teflon sintered at 623 K. In this case, modulated electrodeposition of platinum from 2% H$_2$PtCl$_6$ 6H$_2$O was run at $\eta_a = -0.15$ V and $f = 1$ kHz to avoid dendrite formation [20]. This electrolysis technique favors the growth of facetted platinum crystallites of a rather uniform-size distribution and preferred crystalline orientation on conducting substrates (see Chapter 2). The structure of these electrodeposits depends on f usually in the range $1 \leq f \leq 6$ kHz, and the average applied potential $\langle E \rangle = (E_+ + E_-)/2$,

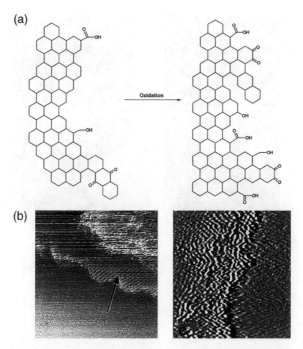

FIGURE 8.4 (a) Scheme of graphite surface before and after thermal oxidation. Before the latter graphite edges show a small number of O-containing species and defects. After thermal treatment a variety of O-containing groups are formed and the number of defects increased. (b) STM image (17×17 nm^2) of typical etched pits on HOPG sample treated at 800°C in air for 30 min. The atomic structure for the HOPG basal plane can be seen far from the pits and terraces. In a region that extends about 2–4 nm from pit edges, a superstructure with the periodicity ($\sqrt{3} \times \sqrt{3}$)a and rotated 30° with respect to the underlying graphite lattice ($a = 0.246$ nm) is observed. (c) 9×9 nm^2 STM region where an interference effect can be seen. The extra contrast at atomic resolution can be assigned to (C–OH, C–OOH) surface heterogeneity. [(b,c) From Klusek [17] by permission of Elsevier.]

where E_+ and E_- are the anodic and cathodic switching potentials, respectively. In this case, platinum electrodeposition is dominated by an activation polarization that strongly depends on the properties of each platinum crystallographic face, that is, E_{pzc}, hydrophobicity, adatom mobility, and the presence of adsorbates [21–26]. For constant f and E_- and adequately chosen values of E_+, a large faceting with a dominating crystallographic orientation can be obtained (Fig. 8.5) [20, 27]. The voltammetric response of these electrodes in aqueous sulfuric acid [Figs. 8.5(c,f)] shows specific distributions of the intensity and location of weakly and strongly bound H-adatom current peaks, which resemble that reported for a stepped single-crystal platinum surfaces with narrow terraces exhibiting either (100) or (111) preferred crystallographic orientation [28]. The stationary polarization curves run for a structure-sensitive reaction such as the OERR in aqueous acid solution utilizing different dispersed facetted platinum electrodes [Fig. 8.5(g)] shows that pyramidal facetted electrodes become

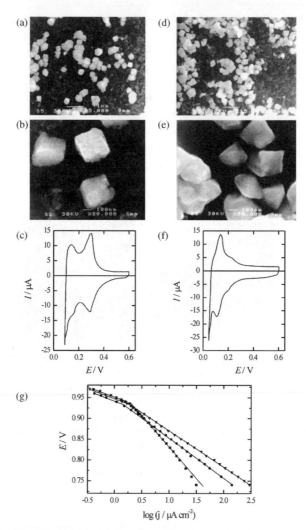

FIGURE 8.5   SEM micrographs of dispersed platinum (a, b, d, e) obtained by pulsating electrolysis on carbon-PTFE substrates in 0.04 M H_2PtCl_6 in 1.1 M HCl aqueous solution at 298 K. (a,b) $E_- = -0.2$ V, $E_+ = 0.9$ V, $f = 2.5$ kHz. Platinum crystallites exhibit a cubic facetted structure; (d,e) $E_- = -0.2$ V, $E_+ = 1.2$ V, $f = 2.5$ kHz. The geometry of crystallites approaches that of facetted pyramids; (c,f) voltammograms run at 0.1 V s^{-1} in aqueous 0.5 M H_2SO_4: (c) platinum cubic crystallites/C-PTFE electrode shown in (a,b), surface area 8.7×10^{-2} cm^2; (f) quasi-pyramidal platinum crystallites/C-PTFE electrode, surface area 4.4×10^{-2} cm^2, depicted in (d,e); (g) Tafel plots for OERR in oxygen-saturated 0.5 M H_2SO_4 at 298 K: (△) pyramidal facetted and (□) cubic facetted platinum crystallites; (○) commercial high-performance carbon E-Tek supported platinum. (From Zubimendi et al. [20] by permission of Elsevier.)

more efficient than high-performance reference commercial E-Tek polycrystalline platinum supported on carbon [20].

The contact between platinum or platinum-based electrocatalyst particles and carbon is extremely important to improve the accessibility of fuel and oxidant to the catalyst. To reach this goal, the incorporation of platinum nanoclusters is a part of the synthesis of porous structures of conductive carbons [29]. However, not only must the carbon support provide more conductivity but also its surface chemistry and morphology must maintain high dispersion of the electrocatalyst over time at the fuel cell operating conditions.

Carbon-supported platinum on tetrafluoroethylene-treated carbon paper followed by passivation in flowing nitrogen at 573 K was used as electrode in the H_2/O_2 fuel cell in phosphoric acid at 463 K. The structure of platinum clusters at different stages of the fuel cell operation was characterized by transmission electron microscopy (TEM) and X-ray absorption spectroscopy. Under cell operation, the platinum–platinum atom coordination number in the clusters increased from 5.5 to 6.1, but no change in the 0.277-nm platinum–platinum distance was observed. The increase in the platinum atom coordination number in the cluster was consistent with the platinum particle average size, which goes from 0.32 to 0.45 nm, accompanied by a wider cluster size distribution function [30].

Electrocatalysis in low-temperature proton exchange membrane fuel cell (PEMFC) dominates the current literature on electrocatalysis for the generation of energy without pollution from byproducts of hydrocarbon combustion [31]. The catalyst required to generate electricity in fuel cells is analogous to the bifunctional heterogeneous catalyst made of nanoparticles highly dispersed on a functional support to achieve the desired reaction rate. In fuel cells, the platinum or platinum-group metal nanoparticles are usually supported on electronically conductive carbons. High-performance fuel cell electrodes require architectures with nanoscopic reaction zones (interphases) assembled from multiple nanoscopic catalytic surface domains whose accessibility for electrons and protons is critical to the kinetics of the electrochemical reactions. These reaction zones are intimately integrated, but they still permit the efficient mass transport of gas–liquid-phase reactants and products to and from the carbon-supported nanoscale electrocatalyst. Many additives, in the form of inks and dispersions, are employed in proton exchange membranes with carbon-supported electrocatalysts in which aggregated carbon nanoparticles mantain the electron paths between the carbon agglomerates.

In recent times, carbon nanotubes (CNTs) were considered as possible support for metal particles, as it was shown, for instance, in the decomposition of hydrazine on iridium nanoparticles supported on CNT. The latter becomes more effective support than alumina. This influence of the supporting material on the electrocatalytic reaction appears of interest for hydrazine-fueled thrusters utilized in space vehicles [32]. The electrocatalytic performance of the OERR to water in alkaline solutions has also been investigated on graphene layer nanocapsules 10–30 nm in size [33].

STM images at the sub-nanometer resolution revealed that the structure of electrodes made of carbon-supported platinum, ionic mixtures, and membrane electrode assemblies consisted of some ordering of particles both on the carbon support and in

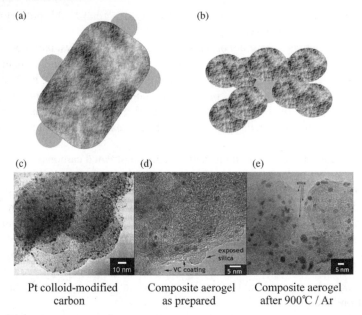

(a)   (b)

(c)   (d)   (e)

| Pt colloid-modified | Composite aerogel | Composite aerogel |
| carbon | as prepared | after 900℃ / Ar |

FIGURE 8.6   Schemes of different architectures of bifunctional catalyst. (a) Traditional impregnated bifunctional catalyst produces metal nanoparticles that are supported on an oxide particle one order of magnitude greater than the metal ones. (b) Catalyst with metal and oxide particles both of similar size. Metal and oxide domains are networked such that junctions with multiple oxide particles exist for each metal nanoparticle. (c) TEM of Vulcan carbon XC-72 modified with preformed 1.8-nm platinum colloids. (d) TEM of platinum-colloid-modified carbon–silica composite aerogel with silica particles disrupting the carbon agglomerates. (e) TEM of platinum-colloid-modified carbon–silica composite aerogel after heating to 900 °C in argon. This structure is resistant to platinum sintering. (From Anderson et al. [35] by permission of The American Chemical Society.)

the membranes. In these cases, the measurement of local reactivity of the electrocatalyst has been made using the scanning tunneling microscope tip as a sensor electrode for the OERR [34].

Platinum-colloid-modified Vulcan carbon nanoglued with silica sol yields a Pt/C-silica composite gel as the basis to produce a continuous 3D mesoporosity for fuel diffusion to the platinum catalyst. This electronically conductive architecture is one order of magnitude more active than the starting colloid modified carbon black for the oxidation of methanol per gram of platinum (Fig. 8.6) [35]. The strategy for designing this electrocatalyst is interesting for the fabrication of multifunctional fuel cell electrodes with maximum activity.

Bifunctional catalyst preparation can be improved by means of silica nanoparticle building blocks that assures a structural continuity of the 3d porous network. The self-aggregation of the carbon primary nanoparticles is clearly described by silica colloidal domains [35]. Silica also appears to contribute to hinder platinum sintering. The 1.8-nm platinum colloids remain unchanged in size after sol–gel and supercritical fluid processing, and the platinum particle size increases to 3–4 nm only after treatment of the composite under argon at 900°C. The sol–gel route to electrocatalytic nanoarchitecture offers flexibility in the design and selection of the composition of nanoscopic building blocks. This confluence of strategies has been pointed out of interest in designing and optimizing multifunctional catalysts [36].

8.2.3 Metal Alloy Electrodes

Metal alloys such as Raney nickel for molecular hydrogen electrooxidation [37], platinum–tin for methanol electrochemical oxidation [38–40], and platinum–chromium for the slow OERR [41] are typical examples of alloy electrocatalysts for low-temperature fuel cells. As recently suggested [42], a key problem to deal with these nanoscale materials is to pin the most active, highest performance face domains, even when exposed to thermodynamic forces that would drive their restructuring, crystallization, or densification and to understand the control of their amorphousness and disorder, in relation to the catalytic activity. Drawing analogies for amorphous or randomly ordered catalytic systems with biocatalysis, in which size and shape selectivity are achieved in the absence of periodicity, will offer new insights for revisiting these hybrid catalytic nanostructures. Data of a number of bimetallic materials of possible use for fuel cell catalysis are presented elsewhere [43, 44].

Platinum-based alloys have been explored for different electrochemical oxidations, for example, the Pt_α–Ru_β for methanol oxidation and for reformated H_2/N_2-containing CO and Pt_α–Cr_β and Pt_α–Co_β for the OERR. These alloys, either as bulk or as nanoparticles, are more efficient electrocatalysts than pure platinum [43]. The interfacial domains of these alloys are often highly disordered even in those with a high surface-to-volume ratio. Mechanistic information of the above processes, run at single crystals and bulk materials, has been extensively described [43], although the validity of this information for processes at electrocatalyst nanoparticles should cautiously be considered. This is the case of the above-mentioned reactions that have been studied from high-order alloys to physicochemical nanocomposites with mixed electron and proton conductivity and their electrosorption characteristics [5]. Furthermore, hydrous nanocomposites such as Pt–$Mo^{V/VII}$–O_x–H_y appear to have a high electrocatalytic activity for the electrochemical oxidation of reformated CO [42].

The activity of the platinum catalyst in the direct methanol fuel cell (DMFC) can be enhanced by adding a second metal such as tin. In this case, methanol electrochemical oxidation in aqueous sulfuric acid undergoes on carbon-supported platinum–tin electrode [45, 46]. For the 3 : 1 Pt–Sn ratio the overall reaction involves two electrons and the rate-determining step (RDS) is similar to that reported for carbon-supported platinum. As the relative amount of tin is increased above 3 : 2, the overall reaction involves a one-electron transfer RDS. This result points out that the presence of tin may assist the charge transfer to platinum via either a local increase in the degree of surface coverage by methanol adsorbates or a decrease in the content of higher valent platinum sites in the catalyst [47].

Transition metal alloy catalysts were also tested as low-cost alternatives for DMFC in aqueous sulfuric acid. Thus, carbon-supported platinum–cobalt catalysts were better than carbon-supported platinum–copper or other transition metals from the VIIIb group [48]. In this case, platinum increases the tolerance to poisons by adsorbed cobalt via the bifunctional mechanism [49]; that is, platinum adsorbs residues from methanol and the second metal activates water decomposition at lower potentials than platinum.

Platinum–ruthenium catalysts supported on glassy carbon were also proposed for methanol electrochemical oxidation. In this case, the reaction order for methanol increases from 0.5 to 0.8 with the ruthenium content in the catalysts, and the apparent activation energy for the alloy anode is about one-half the value for platinum [50]. The influence of ruthenium on the reaction kinetics is explained by the bifunctional reaction mechanism [49, 51]. Accordingly, methanol is adsorbed on platinum sites and oxidized, yielding adsorbed CO species. The overall effect is that the platinum–ruthenium alloy improves the polarization characteristics of methanol electro-oxidation for DMFC operation [52–55].

The functional structure of nanocomposites such as $Pt–Ru^{III/V}–O_x–H_y$ oxide with atomically dispersed platinum [56] and $RuO_2 \cdot xH_2O$ involves mixed–valent hydrous oxides that support electron and proton conductivity. The local and medium-range properties of these nanomaterials were investigated by atomic pair density functional analysis. Results showed that $RuO_2 \cdot xH_2O$, amorphous for X-ray scattering, is an innate nanocomposite involving electron-conducting RuO_2 wires that are networked through hydrous proton-conducting domains, where the volume fractions of the components track the mole fraction of structural water [57]. A mixed-phase electrocatalyst containing platinum and hydrous ruthenium oxides such as $Ru^{III,IV}–O_x–H_y$ exhibits a high electrocatalytic efficiency for methanol oxidation [42]. In this nanocomposite, a phase separation is produced in which a disrupted domain of disordered hydrous ruthenium oxide several layers thick overlaces a platinum-rich core and interconnects multiple $Pt–Ru–O_x–H_y$ particles allowing the full oxidation of methanol to CO_2 and water [42–58].

An interesting material for the OERR electrocatalysis is carbon-supported (about 4%) platinum-doped FeO_x whose activity is comparable to that of 10wt% Vulcan carbon-supported platinum. This electrocatalyst consists of a microporous structure disordered at the nanoscale level with atomically dispersed platinum. When the material crystallizes, water-filled micropores collapse and the electrocatalytic activity disappears [59].

The activity, selectivity, and methanol tolerance of novel supported platinum and Pt_8Me (Me = Ni, Co) catalysts were investigated [60]. The catalysts were prepared by the organometallic route, that is, deposition of preformed platinum and Pt_3Me precursors followed by their decomposition into metal nanoparticles of about 1.7 nm diameter, and evaluated by conventional electrochemical and electrochemical mass spectroscopy. These bimetallic catalysts behave similarly to high-loaded conventional carbon-supported platinum catalysts, the hydrogen peroxide formation becomes negligible, and the methanol tolerance at the cathode improved [60].

Among several selective electrocatalysts for the OERR, $RuSe_x$ is used as part of membrane electrode assemblies and catalyst-coated membranes under methanol-tolerant operating conditions. The DMFC is stable over a 1000-h period, and the amount of hydrogen peroxide formed in the 0.7–0.4-V range is below 1% on carbon-supported Ru and 1–4% on carbon-supported $RuSe_x$ catalysts. The preparation of these cathodes is given elsewhere [61, 62]. For membrane electrode assemblies with a Vulcan X-72 Ru (44wt%)–Se(2.8wt%) catalyst, the highest power output (PO) is about 40% the value obtained with a conventional platinum catalyst at 353 K, although further optimization is still required [63, 65]. The OERR is also catalyzed by WO_3-modified $RuSe_x$ as nanoparticles dispersed in the form of Nafion-containing inks on glassy carbon [65]. The cathodic activation overpotential decreases about 0.1 V as compared to WO_3-free $RuSe_x$.

Co/Pd alloy catalysts to be used as cathodes were synthesized on an XC-72 carbon black and tested for the H_2/O_2 fuel cell with a proton-conducting electrolyte. These cathodes become more efficient than carbon-supported cobalt and more stable to corrosion than carbon-supported palladium. For $E < 0.7$ V, the OERR mechanisms on carbon-supported Co/Pd alloy and carbon-supported platinum are similar, the RDS being the first electron transfer to the adsorbed, previously protonated, oxygen molecule [66].

8.2.4 Chemically Modified Electrodes

The catalytic properties of conducting substrates can be modified by anchoring chemical functional groups to tailor surface structures with a specific electrocatalytic activity. The interactions of functional groups with the substrate imply the formation of electrode surfaces more complex than those of smooth solid surfaces (see Chapter 2). The interaction energy between the anchoring species and the solid surfaces ranges from that of weak adsorption to chemical binding. The structure of the surface layer depends on both the nature of the anchoring species and the specific characteristics of the solid substrate. On the other hand, the binding energy is considerably influenced by the nature of the environment in contact with the solid surface. As the formation energy of the surface layer increases, new specific surface species may be produced. These species may remain homogeneously distributed, or form dimers and clusters, or constitute confined reaction cages at the interface, or generate redox couple centers at the surface. In fact, as the complexity of the interfacial structure increases, the surface layer may tend to acquire a structure closer to biologically anchored species with properties of supramolecular structure moieties, as described further on.

In a simple scheme, these complex surface structures may have electrocatalytic properties assisting the electrochemical conversion of reactants to products by accelerating the electron transfer processes or immobilizing oxidation–reduction couples. Conversely, in the absence of such a mediator, the reactant-to-product conversion, at the naked electrode surface, would undergo a slow electrochemical reaction.

A chemically modified electrode can be obtained by adequately anchoring selected atoms, molecules, or their clusters to the surface of a conducting substrate, utilizing a variety of anchorable species and adequate substrates and appropriate preparation techniques. Substrates that are utilized for making chemically modified electrodes cover from metals to conductive oxides and carbons, each substrate having specific reacting groups for anchoring chemical species. Thus, thermally treated carbon substrates may contain a number of oxidized functional groups and surface defects

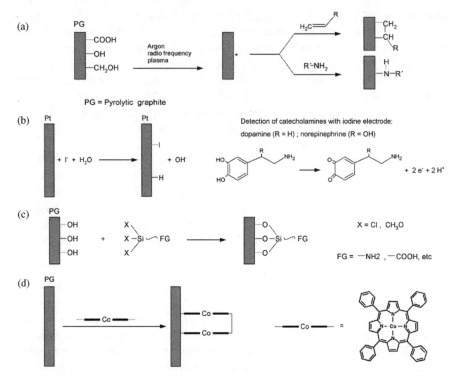

FIGURE 8.7 Schemes of different types of modified electrodes. (a) Covalent modification via activated carbon surface. (b) Iodine-modified (chemisorbed) platinum. (c) Silanization of metal oxide surface. (d) Covalent bonded cobalt porphyrin on carbon. Two adsorbed molecules bridging an oxygen molecule participate in the electrocatalytic reaction.

(Fig. 8.4), which could become useful for producing specific chemically modified surfaces [14, 16, 67]. The concentration and relative distribution of anchored centers at carbon surfaces depend on both the oxidation degree of carbons and the characteristics of anchorable species.

Chemically modified electrodes can be prepared by different conventional procedures (Figs. 8.7 and 8.8) [67–70]:

- *By adsorption:* Substrate/anchored species interaction energies from van der Waals to chemisorption ones. Examples: covalent species produced by atom or molecule binding to graphite that has been activated by thermal, radio frequency plasma, laser, high-energy radiation. Physisorbed adsorbates can be made by direct adsorption, yielding anchored species with a stability lower than that of covalently bonded. Examples: the chemisorption of functionalized vinyl compound on platinum [69, 71, 72]; metal complexes immobilized by a bridge molecule on pyrolytic graphite [69]; iodine-modified platinum electrodes used for catecholamine electrochemical sensors [73].

- *By chemical reactions.* Examples: metal porphyrin film covering an activated carbon electrode in aqueous solution that catalyzes the OERR to water [76];

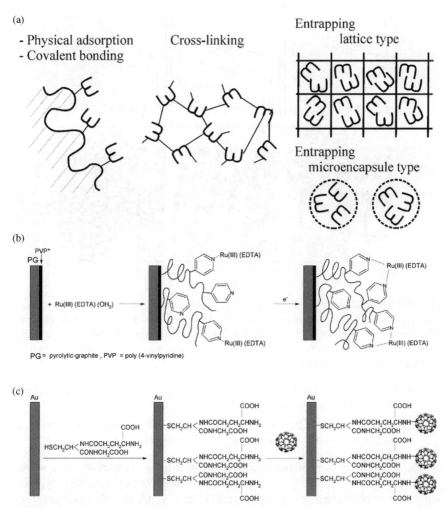

FIGURE 8.8 Schemes of different types of modified electrodes. (a) Schemes of physical and covalent bonding of enzymes; cross-linking bonding and lattice type and microcapsule enzyme entrapping. (b) Polymer-film-modified electrode: immobilized ruthenium complex at pyrolytic graphite (PG) electrode modified with poly(4-vinylpyridine). (c) Ideal assembling of glutathione (GSH) onto gold with covalent bound fullerene onto GSH monolayer. (From Fang and Zhou [77] by permission of Wiley.)

silanization of the SnO_2 electrode surface [75] with dimethyl-dichloride silane yielding an immobilized residual silica and hydrochloric acid in solution, which catalyzes the reaction of OH; silanization of a RuO_2 electrode with $Si(CH_2)_3-NH(CH_2)_2$; oxidized pyrolytic graphite modified by chemically anchoring α-cyclodextrine to produce local cage effects that favor stereoselectivity, for instance, in the electrochemical chlorination of anisole [76].

- *By polymerization:* Substrate–polymer covalent interactions yielding conducting polymer films such as polyacetylene, polyaniline, polypyrrol with functional

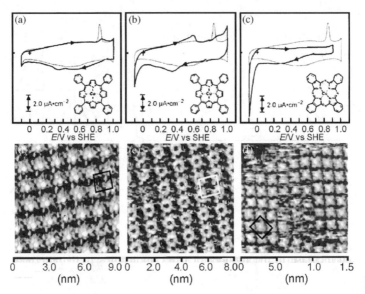

FIGURE 8.9 Voltammetry and STM images of cobalt and copper tetraphenyl porphyrin-modified and cobalt phthalocyanine-modified gold (100) electrodes. Cyclic voltammograms were run in 0.1 M aqueous $HClO_4$ at 0.05 V s^{-1}. STM images taken at 0.17 V reference hydrogen electrode (RHE), tip potential 0.43 V (RHE), and tip current 0.41 nA. (From Suto et al. [79] by permission of The American Chemical Society.)

groups of specific reactivity. Example: pyrolytic graphite modified by polyvinyl-pyridine in which ruthenium complexes are anchored [68].

• Roughly similar procedures can be extended to semiconductor substrates, particularly to be used in photoelectrocatalytic reactions.

More recently, the preparation of chemically modified electrodes has been oriented to the use of nanocomposite materials. This is the case of cobalt porphyrin/polypyrrole template-synthesized nanocomposite for the OERR to water in neutral medium. These nanocomposites were prepared from water-soluble copper and cobalt tetraphenyl porphyrins (TPPs) and cobalt phthalocyanine (Pc) forming aggregates by changing the ionic strength and pH of the medium [78, 79]. These cobalt–TPP aggregates in aqueous neutral solutions can operate as templates to generate different nanostructured cobalt–TPP/polypyrrole (PPY) composites (Fig. 8.9). Polypyrrole was electrochemically synthesized on a fresh gold (100) electrode by potential cycling in the range $0.14 < E < 1.24$ V [vs. normal hydrogen electrode (NHE)] in 0.1 M PY + 1 mM cobalt–TPP + 0.01 M phosphate-buffered saline (PBS) (pH 7.4) solution. The OERR rate on a copolymerized cobalt–TPP/PPY electrode is much higher than that on a PPY-polymerized electrode. This indicates that cobalt–TPP acts not only as a reactant but also as a catalyst for polymerization. Uniform structured nanocomposites can be obtained by application of ultrasonics to the solution. The 3D structured nanocomposite offers a large number of active centers for the reaction when it is prepared by the following three-step method: (i) PPY electrodeposition on the electrode; (ii) electrode immersion in the template solution for 6 h; (iii) immersion in pyrrole solution for PPY electrosynthesis. The resulting electrode showed good electrocatalytic properties for the OERR to water [78, 79].

A variety of new nanomaterials with highly controlled and unique optical, electrical, magnetic, and catalytic properties [80–89] can be used for electrode design and assembling. They comprise a diversity in composition, shape, and readiness of surface functionalizing techniques that permit the fabrication of various functional nanostructured materials, especially for biointerfaces [Figs. 8.8(a,c)]. Nanostructured materials that have found wide applications are gold and platinum nanoparticles, colloidal silver and platinum particles, roughly spherical colloidal semiconductors, quantum dots and wires, carbon nanoparticles, single and multiple-wall-type carbon nanotubes, fullerenes and fibers [90], nanoporous materials with a high specific area, and well-ordered pores with uniform volume and diameter to produce adjustable porous structures. According to the nanoporous size, these materials can be divided into small pore size (below 2 nm), mesoporous (2–50 nm), and macroporous (larger than 50 nm). Some of these nanomaterials exhibit unique optical, electronic, and photophysical properties. The nanostructures can be prepared by different ways, such as physical and chemical adsorption, self-assembled Langmuir–Blodgett (LB) nanofilms, self-assembled monolayers (SAMs), layer-by-layer (LBL) assembled multilayers, sol–gel methods, electrochemical deposition, and electrochemical polymerization. These procedures can be combined with in situ nanoscopies to build functional electrodes (Fig. 8.10).

An interesting modified electrode is the PEDOT [poly(3,4-ethylenedioxythiophene)]/platinum catalyst gas diffusion electrode in air that has been used for DMFC anodes. The enhancement of the electrochemical oxidation of methanol occurs after either overoxidation of PEDOT or a long-time storage of the gas diffusion electrode. Both procedures induce a reorganization and increase in porosity with a better methanol accessibility to platinum active centers. Porosity increase is due to a significant depletion of sulfur and oxygen in the conducting polymer [92].

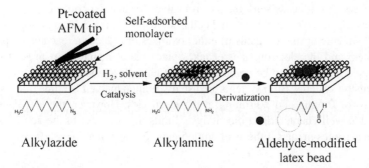

FIGURE 8.10 Stages in formation of functional electrode utilizing atomic force microscope. A self-adsorbed alkylazide monolayer on a single-crystal substrate is reduced to alkylamine by nanocatalytic hydrogenation utilising a platinum-coated microscope tip. The reduced domain is amenable to further derivatization that results in a local cluster of aldehyde-modified latex beads. (From Müller et al. [91] by permission of Science.)

8.3 BIOELECTROCATALYSIS AND BIOELECTRODES

8.3.1 Preliminary Remarks

The history of the relationship between electricity and biological systems started at the end of the eighteenth century when Galvani showed the twitching of severed frog legs by the current of a static electricity generator. But only in 1910 [93] did the idea of producing electricity from a cell driven by a microbiological system such as *Escherichia coli* become a reality. This occurred a relatively long time after the implementation of the first inorganic fuel cell by Groove in 1839 [94]. About 20 years later, in the 1930s, a microbial fuel cell was developed that was able to generate a voltage exceeding 35 V [95]. From 1950 to 1960, different microbial fuel cells were considered, first to explore the possibility of concentration cells driven by the oxygen concentration gradient between the anode and the cathode generated there by the presence of organisms, and later to build waste disposal systems for space flights. Finally, from 1960 onward, the development of biofuel cells based on both living organisms and enzymes was extensively investigated, including cell designs for implantable applications [96–102].

From the standpoint of energy conversion, the study of biological fuel cells is interesting in terms of their efficiency and ecological sustainability as well as in terms of their size for sensor devices. The key components of these systems are organisms and enzymes (purified enzymes) that behave as catalysts with a high selectivity.

Biofuel cells are foreseen as playing an increasing role in years to come for a wide range of ex vivo applications such as (i) power recovery for waste streams with simultaneous bioelectrochemical remediations; (ii) power generation in remote areas or replacement battery packs at electronic goods; (iii) the stomach of a mobile robotic platform for possible application of autonomous robots that can scavenge fuel from their surrounding; (iv) bioremediation on starch plants (bioelectrocatalysis); and (v) biological oxygen demand sensor. Similarly, for in vivo applications: (i) permanent power suppliers to withdraw the fuel from the flow of blood; (ii) devices such as pacemakers; (iii) glucose sensors; and (iv) valves for bladder control.

A scheme of a typical half-biofuel cell is shown in Figure 8.11: a biological component (enzyme or microorganism) either oxidizes or reduces a reactant supplied to the system and the electron transfer takes place via a mediator. At present, large-scale applications tend to be organism-based fuel cells and small-scale applications to enzymatic systems. Both systems are usually limited by their lifetime, for instance, problems of fouling and poisoning at the electrodes. However, the actual number of suitable well-characterized biocatalysts limits the progress in the application of biological routes in the synthesis of compounds for novel materials [103].

8.3.2 Energy Conversion at Biological Cells

Let us consider how efficient and environmentally friendly energy conversion processes take place in the elemental unit of a biological (eukaryote) cell. Thus, in biological cells two coupled processes occur simultaneously: the catabolic one, in which

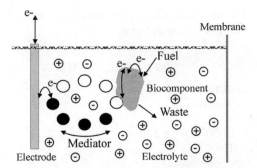

FIGURE 8.11 Scheme of half-biofuel cell. A fuel is oxidized (or oxidant reduced) in the presence of a biological component (enzyme or organism), and electrons are transferred to (from) a mediator, which either diffuses to or is associated with the electrode and it is oxidized (reduced) to its original state. (From Bullien et al. [100] by permission of Elsevier.)

energy is generated, and the anabolic one that leads to energy storage as adenosine $5'$-triphosphate (ATP) to be used to produce proteins, lipids, and nucleic acids required for cell growth and reproduction. The catabolic process comprises three main stages (Fig. 8.12): (i) the fuel (food) is primarily degraded by digestion; (ii) molecules

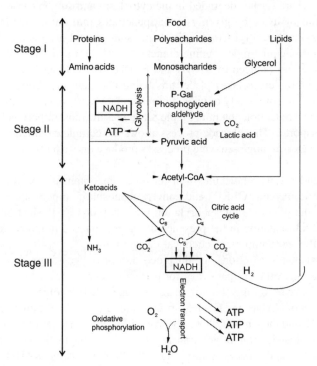

FIGURE 8.12 Scheme of basic processes involved in metabolism of eukaryote cell. For the sake of simplicity, the participation of water in the intermediate processes is omitted.

TABLE 8.1   Redox Half Reactions and Their Standard Potentials (E_0) at pH 7 for Species Involved in Electron Transport Chain

Redox Half Reaction	$E_0(V)$
$2H^+ + 2e^- \rightarrow H_2$	-0.414
$FAD + 2H^+ + 2e^- \rightarrow FADH_2$	-0219
$NAD^+ + 2H^+ + 2e^- \rightarrow NADH$	-0.320
$NADP^+ + 2H^+ + 2e^- \rightarrow$	-0.324
NADP dehydrogenase (FMN) $+ 2H^+ + 2 e^- \rightarrow$	
NADH dehydrogenase (FMNH$_2$)	-0.30
Ubiquinone $+ 2H^+ + 2e^- \rightarrow$ ubiquinol	0.045
Cytochrome $b(Fe^{3+}) + Pe^- \rightarrow$ cytochrome b (Fe^{2+})	0.077
Cytochrome $c1(Fe^{3+}) + e^- \rightarrow$ cytochrome $c1$ (Fe^{2+})	0.22
Cytochrome $c(Fe^{3+}) + e^- \rightarrow$ cytochrome c (Fe^{2+})	0.254
Cytochrome $a(Fe^{3+}) + e^- \rightarrow$ cytochrome a (Fe^{2+})	0.29
Cytochrome $a3$ ($Fe^{3+}) + e^- \rightarrow$ cytochrome $a3$ (Fe^{2+})	0.55
$\frac{1}{2} O_2 + H^+ + 2e^- \rightarrow H_2O$	0.816

The standard potential of these species increases sequentially in going from the initial to the end of the chain, as electrons go from small to high potential regions.

produced in (i) are further degraded in the cytoplasm and simple carbohydrates are converted into pyruvate by glycolysis that penetrates into mitochondria producing acetyl-coenzyme A (acetyl CoA); (iii) at this stage acetyl groups are degraded to CO_2, and both nicotinamide adenine dinucleotide (NAD) and flavin adenine dinucleotide (FAD) are reduced to NADH and FADH$_2$, respectively. At stage (iii), the citric acid (Krebs) cycle takes place, this being the main metabolic pathway that living cells use to completely oxidize biofuels to CO_2 and water. NADH has a high electrochemical reduction potential (Table 8.1) and is the most important high-energy electron transporter. This cascade process allows the complete oxidation of biofuels to CO_2 and water, as single-enzyme processes produce only a partial oxidation of the fuel.

NADH and FADH$_2$ molecules transfer electrons obtained from fuel oxidation to atmospheric oxygen (OERR) and convert adenosine 5′-diphosphate (ADP) into ATP during the oxidative phosphorylation process [104, 105]. Most of the energy taken from fuel oxidation in the presence of water is packed in the nucleoside ATP molecules, the most important chemical energy transporter to those places where specific biosynthetic reactions should proceed. The oxygen atom-dependent reactions undergo via efficient pathways for extracting energy from food (fuel) [106]. This process occurs during the last stage of stage III. It involves the electron and proton transfer of NADH and FADH$_2$ species through a chain of transferring species for the OERR, and the amount of energy given off is utilized to produce ATP and heat (Fig. 8.13). In this electron–proton transfer chain, three enzymatic complexes actuate as proton pumps and electron transporters. The first one is the NADH dehydrogenase, which accepts one electron from each NADH molecule and transfers it first to ubiquinone, which is part of the complex, and then to the cytochrome b–$c1$ complex.

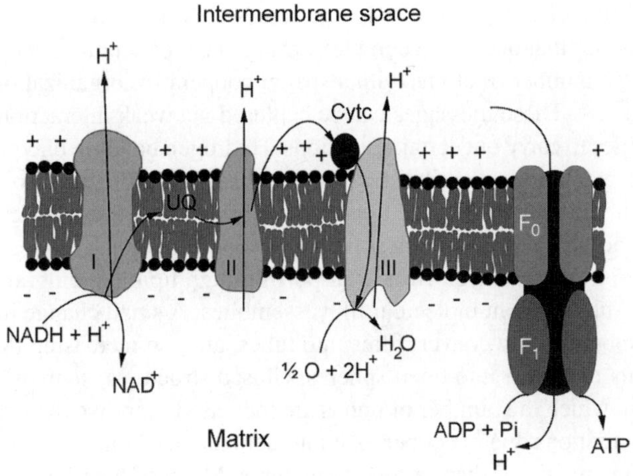

Intermembrane space

FIGURE 8.13 Scheme of third stage involved in respiratory chain. Three main enzymatic complexes participates in the transport of electrons from NADH carriers to be consumed by the OERR: I, NADH dehydrogenase complex; II, cytochrome $bc1$ complex; III, cytochrome oxidase complex. The electron transport generates a H^+ gradient at both sides of the inner membrane of mitochondria.

The latter contributes to the transport of electrons from ubiquinone to cytochrome c (Cyt-c), which subsequently transfers them to the oxidase cytochrome complex (aa3). Eventually, electrons are transferred to an oxygen molecule resulting in a global four-electron and four-proton transfer process yielding two water molecules. The energy obtained from electron transport is used to produce ATP by means of a complex mechanism guided by the enzyme ATP-synthase [104].

The above brief description shows how a tiny biofuel cell actually operates at the level of mitochondria for producing energy in living systems. Natural systems are probably the most efficient energy converters, in which the molecular complexity of the constituents requires special attention. Thus, NADH dehydrogenases have more than 22 polypeptide chains with a mass of ∼800,000 daltons, cytochrome b–$c1$ consists of at least eight different polipeptide chains with a mass of ∼500,000 daltons, cytochrome complex (aa3) is made of a dimer of cytochromes a and $a3$, with a mass of ≈300,000 daltons, and ATP synthase, an enzymatic supramolecular transmembrane complex, contains at least nine polypeptide chains with a mass of ≈500,000 daltons. Summing up: about 40 different proteins are taking part in the electron transport process that allows the very efficient cellular respiration (combustion) process via oxidative phosphorylation to yield about 30 ATP molecules from the combustion of each glucose molecule and 84 ATP molecules from each palmitic acid molecule. Comparing the combustion free-energy differences of lipid and carbohydrate oxidation to carbon dioxide and water to the energy stored in ATP molecules, the efficiency of this process reaches about 40%, that is, a figure exceeding the efficiency of an electric or gasoline engine.

The efficiency of energy conversion at biosystems is the result of a number of coupled processes that occur in a complex arrangement of supramolecular structures made of a large number of atoms with a spatial cooperative organization containing specific messages. These messages can be captured via weak interactions with other macromolecules to carry out certain functions. Hydrogen bonding plays an important role in weak interactions contributing to the cell assembly of supramolecular structures. These interactions allow the free rotation of a part of those structures and the possibility of adopting a quasi-endless number of conformations. Thus, some proteins that specialize in proton or electron transport through lipidic membranes are made of hexagonal subunit monomolecular film assemblies. A small change in the geometry of these subunits may convert films into tubes, and, on increasing the number of subunits, films may turn into open spheres. Closed structures, such as rings, tubes, and spheres in which the number of bonds are increased, improve their stability. Furthermore, fluctuations due to cooperative interactions may induce small changes that either help to maintain or disrupt these structures. Many of these assemblies can be prepared, although others require additional cellular components and the contribution of irreversible enzymatic processes.

The above simplified structural description of the biological system and the energy conversion processes occurring at the mitochondria level will help to understand the problems of design and operation of bioelectrocatalytic electrodes for different purposes. More extended descriptions of metabolism at eukaryotic cells are given in the literature [104, 105].

8.3.3   Biologically Modified Electrodes

Biologically modified electrodes utilize either enzymes or microorganisms as biocatalysts. These electrodes operate in two different ways: (i) the fuel that feeds the cell comes from either a biocatalytic or a metabolic process that occurs in a separate container [Fig. 8.14(a)] or in the electrode compartment separated by a hydrogen permeable membrane [Fig. 8.14(b)]; (ii) the cell consists of single anodic and cathodic electrode compartments separated by a membrane [Fig. 8.14(c,d)]. The biocatalyst at the anode compartment may be directly wired to the anode metallic support [Fig. 8.14(c)] or mediators are used to electrically contact it to the metallic support [Fig. 8.14(d)] [107].

As far as the engineering demands for construction of biosensors is concerned, they are different from those required for biofuel cells [108, 109]. The main limitations of these biocatalysts is that they usually operate in the range 20–40°C and at near neutral pH.

In bioelectrocatalysis either the active components are whole living organisms or a continuous supply of purified enzymes may be free in the electrolyte solution or localized at the electrode surface. For *mediated electron transfer* (MET) devices the charge transfer occurs through a *mediator* that may be either in solution mixed with the active component or located at the electrode surface. For appliances where the active component is located on the electrode, the mediator is usually placed on the supporting conducting surface. On the other hand, mediator-free

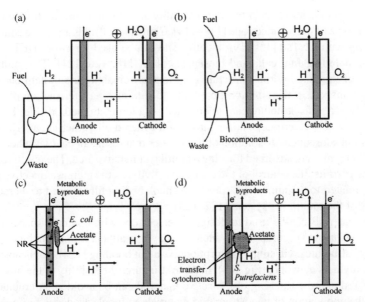

FIGURE 8.14 Schemes of product type biofuel cell systems: (a) external fermenter feeds H_2 to H_2/O_2 inorganic fuel cell; (b) single-compartment fermenter–electrode biofuel cell that utilizes a H_2/H^+ conducting membrane. Schemes of biofuel cell anodes: (c) nondiffusive MET anode made of modified carbon electrode and neutral red (NR), as mediator, to be used in presence of *E. coli*; (d) DET fuel cell anode consisting of unmodified electrode in which electron process occurs via cytochromes in outer membrane of organism (*S. putrefaciens*). In all cases, the OERR is the cathodic reaction. (From Bullien et al. [100] by permission of Elsevier.)

directed electron transfer (DET) devices operate via electron tunneling directly from the active site fixed in the enzyme to the electrode. Recent reviews describing further details of past and present technologies of biofuel cells are given in the literature [100, 110].

The development of bioelectrodes requires materials that, besides retaining the biological activity of the redox biomolecules, impart adequate orientation to the biomolecule to favor electron transfer. An example of this adaptation is the use of modified nanostructured electrodes with a *bridging* system that interconnects the biomolecule with the electrode (CNTs, SAMs, etc.) [111].

Bioelectrodes for fuel cells require a 3D porous structure with minimized resistive losses, full accessibility to the catalyst reactive sites, proton transport and fast charge transfer, and ease of fuel delivery and effluent flowing. For maximizing the power output (PO) of biofuel cells, multidimensional and multidirectional structure 3D electrodes are required.

Multidimensionality provides small pores for enzyme stabilization and large pores for easy mass transport of reactants and products. An example of a multidimensional material is chitosan, a polysaccharide with carboxyl and amine groups utilizable for enzyme immobilization that offers the possibility of controlling the multidimensionality of this type of porous electrode [112–114]. Chitosan scaffolds are made conductive by doping with CNTs [115–119]; a composite can be used for either DET or MET devices, including fuel cells and biosensors [112, 118, 120, 121]. The multidirectionality of the macroporous structure contributes to making the system permeable for fuel transport in the solution phase and to diminishing the fraction of unused solid polymer. This structure can be obtained by a template method (Fig. 8.10), for example, by gold electrodeposition onto a well-ordered micrometric pattern of latex spheres and subsequent dissolution of the spheres that results in cavities where mediator and enzymes are adsorbed forming a multilayer reactive area. The porous ordered structure permits fine control of the catalytic activity, as the increase in active area is linearly related to the number of sphere layers. A Nafion film is used to increase the stability of the multilayer system [122, 123].

Nanomaterials can be used to design bioelectrodes with confined spaces for immobilizing bioactive compounds, such as enzymes attached onto or incorporated into large structures, via simple adsorption, covalent bonding [124], or encapsulation within polymers networks (Fig. 8.8). In these structures the stability of the biocatalyst is increased and the leaching of the active material can be avoided. A crucial problem in the application of nanostructured materials to bioelectrodes is to find simple and reliable assembly routes to build a variety of nanostructured bioelectrochemical interfaces.

The active lifetime of immobilized enzymes, which is between 8 h and 2 days in buffered solutions, can be extended to 7–20 days by using modified electrodes and to more than a year by encapsulation in micellar polymers. These polymers provide a compatible hydrophilic nesting that prevents enzyme deactivation. The following procedure for the preparation of stabilized single-enzyme nanoparticles (SENs) has been proposed [124]: (i) the enzyme is modified by anchoring vinyl groups and solubilized in hexane as an ion pair form; (ii) a polymerization step at the enzyme surface occurs by adding silane monomers containing both vinyl groups and trimethoxysilyl groups to produce linear polymers covalently bound to the enzyme surface; (iii) the treated enzyme is transferred to water where polymer cross-linking takes place; (iv) finally, the resulting SEN can be further immobilized in large-surface-area porous materials. This method provides a stable electrode arrangement without significantly limiting the mass transport characteristics of the substrate to the active enzyme.

8.3.4 Enzyme Electrodes

8.3.4.1 Kinetic Approach to Enzyme Reactions

Considering that the magnitude of the Michaelis constant is commonly used to evaluate the affinity between the substrate (S) and the enzyme (E), it is convenient to briefly analyze the simple enzymatic reaction mechanism

$$E + S <=> [k_1][k_{-1}]ES <=> [k_2][k_{-2}]E + P \tag{8.1}$$

It consists of a first step in which the enzyme E is bound to the substrate S yielding an enzyme–substrate complex (ES) and a second step in which ES decomposes into E and P. In the course of this reaction three regimes can be distinguished. The first one for $t \rightarrow 0$ in which the system essentially consists of pure E and pure S, the concentration of E (c_E) being much lower than that of S (c_S). In the second regime, c_{ES} reaches a constant value because ES is formed and destroyed at the same rate. This part of the process deserves special attention for solving the mechanism of reaction (8.1). Finally, in the third regime the ES decomposition rate becomes greater than its formation rate, and c_{ES} rapidly decreases for $t \rightarrow \infty$.

The second regime results in

$$\frac{dc_{ES}}{dt} = 0 = k_1 c_E c_S - k_{-1} c_{ES} - k_2 c_{ES} + k_{-2} c_E c_P \tag{8.2}$$

where k_1, k_{-1} and k_2, k_{-2} are the rate constants of the first and second steps, respectively. As c_{ES} is small, the rate of the reaction is

$$r = \frac{dc_P}{dt} = k_1(c_{E,0} - c_{ES})c_S - k_{-1} c_{ES} \tag{8.3}$$

where $c_{E,0} = c_E + c_{ES}$ and $c_{E,0}$ is the concentration of E at $t = 0$. Then, the initial reaction rate is

$$r_0 = \frac{k_2 c_{S,0} c_{E,0}}{K_M + c_{S,0}} \tag{8.4}$$

where $K_M = (k_{-1} + k_2)/k_1$ is the Michaelis constant. It represents the value of c_S at which the reaction rate reaches one-half its maximum value ($r_{0,max}$). The reciprocal of Eq. (8.4),

$$\frac{1}{r_0} = \frac{K_M}{c_{S,0} \, r_{0,max}} + \frac{1}{r_{0,max}} \tag{8.5}$$

is a convenient form of Eq. (8.4) for the evaluation of both K_M and $r_{0,max}$ by plotting $1/r_0$ versus $1/c_{S,0}$. Despite the complexity of biofuel electrode reactions and the fact that the first regime (induction period) of the process is neglected, many experimental systems fulfill Eq. (8.5). Anyhow, the above reaction mechanism offers a first quantitative approach to interpreting the kinetics and mechanism involved in enzymatic processes.

8.3.4.2 Enzyme Electrical Connection to Substrate Electrodes

Enzyme-modified electrodes involve an arrangement of enzyme molecules on the conducting substrate due to physical or covalent adsorption, to cross-linking processes, or by lattice or microcapsule entrapping (Fig. 8.7). The electrical connection of the enzyme to the electrode (*enzyme wiring*) depends on the type of enzyme.

As a first approximation, redox enzymes can be divided into three groups according to the location of the redox center of the enzyme protein [111]: (i) enzymes that possess NADH/NAD$^+$ or NADPH/NADP$^+$ redox centers that are often weakly bound to the enzyme protein, so that these centers can diffuse as electron carriers [Fig. 8.15(a)] and the coenzymes behave as natural redox mediators in the biofuel cell; (ii) Enzymes such as peroxidases that have the redox center located near the periphery of the enzyme protein and are able to accept or donate electrons directly from the electrode. Their efficiency depends on their degree of immobilization and appropriate orientation on the surface [Fig. 8.15(b)]; (iii) Enzymes such as glucose oxidase (GOx) that have a redox center strongly bound to a protein or glycoprotein shell. They are unable to communicate directly with the electrode, because they do not give away any species containing the redox center to shuttle electrons to the electrode [Fig. 8.15(c)].

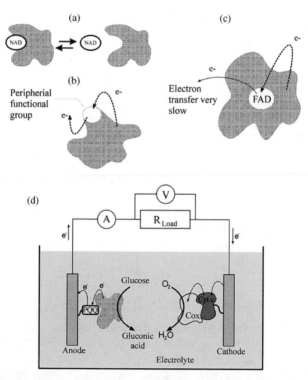

FIGURE 8.15   Classification of enzymes according to location of their active center. (a) The active center can diffuse out of the enzyme and travel to the electrode where electrons are transferred [enzymes with NAD(H) or NADP(H) as active centers]. (b) The active center is located on the periphery of the enzyme shell and is able to transfer (receive) electrons directly to (from) the electrode (a typical case is that of porphyrin derivative active centers). (c) The enzyme has strongly bound redox centers surrounded by a glycoprotein shell (enzyme mediators are required for the electron transfer). (d) A scheme of a single-compartment cell for an enzymatic nondiffusive MET-type electrode. Glucose and oxygen are employed as fuel and oxidant, respectively. (From Bullien et al. [100] by permission of Elsevier.)

In order to favor the anchorage of enzymes to the electrode surface *promoters* are used. They are adsorbed molecules with a substituent (pyridyl, sulfydril, thioether, etc.) and a functional group that transiently interacts with the biomolecule imparting the most favorable orientation for the electron transfer. There are also promoters that act as mediators (ferrocene, hydroquinone, osmium-pyridine complexes, etc.) [125]. Although they have no redox properties, these mediators facilitate the electron transfer between the redox bioactive species and the electrode.

Enzyme electrodes can be classified into diffusion-mediated electron transfer (DMET) electrodes that utilize enzymes of the group (i) and non-diffusion-mediated ones (NDMET) that employ enzymes from groups (ii) and (iii) [100]. DMET electrodes use electron shuttles that diffuse to the electrode. NDMET electrodes with enzymes of group (ii) transfer electrons by direct contact with the peripheral protein redox center of the enzyme. When enzymes of group (iii) are used, the electron transfer occurs either by entrapping them in a conductive polymer or by connecting the active center to a defined structured path or to a highly conductive redox group-modified polymer. The latter electrostatically retains the enzyme that has partially mobile redox groups, and these groups penetrate into redox centers of the enzyme [111]. A scheme of a fuel cell with NDMET electrodes is shown in Figure 8.15(d).

Electrically insulated protein-supporting enzymes can incorporate redox centers with a high density of electrons [111]. When the self-exchange rate of the relays and their density are high, the electron flux through a 1-μm-thick film of a 3D macromolecular network can match or exceed the rate of electron supply to or from the ensemble of enzyme molecules covalently bound to the network. Then, the macromolecular network *wires* the enzyme to the electrode. A practical example of enzyme *wiring* involves the initial formation of a macromolecular complex between GOx and $[Os(Bpy)_2Cl]^+$-containing redox polyamine, cross-linking it with a water-soluble diepoxide, yielding a hydrophilic substrate 1 μm thick.

Redox macromolecules have been used earlier as diffusion mediators for both electron transfer between enzymes and electrodes and enzyme entrapment. For example, GOx can be entrapped in a polypyrrole network, a degenerated semiconductor polymer with traces of platinum that offers a high surface area on which water peroxide from the one-electron molecular oxygen reduction is produced with the participation of $FADH_2$. Other polymers such as polyaniline [126, 127], poly(o-phenylendiamine) [128], polyindole [129], and ferrocene-modified polypirrole were also used for the same purpose [130, 131].

On the other hand, the stability of GOx is enhanced by assembling it on a silicon electrode with a semi-open spatial confined structure created on the unmodified silicon surface without losing the catalytic activity of enzymes under destabilizing conditions [132]. For the electrocatalytic oxidation of glucose, the mediator 4-carboxy-2,5,7-trinitro-9-fluorenylidine malonitrile, the coenzyme NADH and $CaCl_2$, and the enzyme glucose dehydrogenase are adsorbed on this electrode surface forming a multilayer reactive system [111, 122, 123].

Diffusion mediators can be obviated when the electrically insulated redox centers of the enzyme are electrically connected to the electrode directly. In this case, electrons transferred from the substrate to the enzyme are relayed to the external circuit.

Unless the maximum turnover rate of the enzyme is approached, the current increases monotonically with the substrate flux and concentration. In general, the *wire* is a redox macromolecule adequately designed to complex the enzyme protein that electrically connects the enzyme redox center to the electrode. In this way, the enzyme physically attached to the electrode surface remains active [111].

Molecular *wiring* can be improved considerably by electrostatic self-assembling, as in the case of electrodes covered by supramolecular multilayers of multiwalled CNT modified with a ferrocene-derivatized poly(allylamine) redox polymer and glucose. In this case, using indium as substrate, the bioelectrocatalytic activity increases linearly with the number of deposited bilayers. The increase in the glucose oxidation current, the bimolecular rate constant of $FADH_2$ oxidation, and the electron diffusion coefficient indicate an enhancement of electron transport among the enzyme redox active sites, the redox polymer, and the CNT [133]. The latter were also used in hybrid systems as support for coassembled hemoproteins, such as horseradish peroxidase, hemoglobin, myoglobin or Cyt-*c*, and a surfactant such as sodium dodecylsulfate or *N,N*-dimethylformamide. In this case, the enhancement of the electrochemical response makes the use of these systems interesting for biofuel cells [134].

Improving wiring of an enzyme active site to conducting polymers results in a more stable electrode response, as occurs when bilirubin oxidase (BOx) from *Myrothecium verrucaria* is replaced by BOx from *Trachyderma tsunodae* wired to the redox polymer by the stronger new adduct/polymer electrostatic bonding [135]. The replacement of methinine by phenylamine in the axial position of the copper center of BOx shifts the potential of the enzyme positively and consequently the anodic overpotential decreases [135, 136].

The redox couple $NADH/NAD^+$ is one of the redox couples that was studied on platinum modified with polymer-bound dopamine. Dopamine with poly(methacryol chloride) results in a modified polymer with hydroquinone functionalities that in buffered 0.1 M aqueous sodium chloride participates in the reversible redox reaction [77]

$$(\text{Polymer})\text{quinone} + e^- \longrightarrow (\text{polymer})\text{hydroquinone} \qquad (8.6)$$

and the reduced polymer-catalyzed NADH oxidation

$$NADH \rightarrow [-e^-]NADH^{\cdot+} \rightarrow [-H^+]NAD^{\cdot+} \rightarrow [-e^-]NAD^+ \qquad (8.7)$$

Modified flavin-functionalized SAMs [137] have also been successfully used for obtaining biomimetic systems for the oxidation of NADH. These electrodes have been prepared via the SAM technique to immobilize enzymes to increase their stability and prevent their denaturing. The oxidation of NADH in thin biofilms has been extensively reported [138], as, for example, the catalytic activity of flavin-functionalized SAMs [137, 139–144]. Flavins play a key role in the shuttling of electrons in a number of biological redox reactions. Thus, flavin sites buried inside the protein structure of enzymes such as glutathione reductase are active for the very efficient oxidation of NADH [77, 145]. The ideal scheme of the modified fullerene–self-assembled glu-tathion film–gold electrode is shown in Figure 8.8(c). A new flavin-modified electrode based on a self-assembled monolayer of flavin analogue tethered to a gold surface by N-10 linkage has been reported [146]. In this case, to obtain the flavin-tethered SAM, thiol-carrying reactive NH_2 groups (cysteamine) in water are previously ad-sorbed on the gold surface, and subsequently methylformylisoalloxazine is chemically attached. The resulting modified electrode showed good catalytic activity toward β-NADH oxidation.

8.3.4.3 Electron Transport at Bioelectrodes

The electron transport at bioelectrodes depends on the strategy of wiring, which de-termines the nature and location of the enzyme redox center in the protein matrix. Usually, the electron transport through redox centers in polymers is evaluated by deter-mining an apparent electron diffusion coefficient D_{app}^e by electrochemical impedance [147–152].

Electron transport can occur through different mechanisms, such as percolation between immobile redox centers, collision of mobile reduced and oxidized redox centers, and conduction by either electrons or holes through a conjugated backbone [153]. In the absence of the latter, percolation between immobile redox centers is the most frequently found mechanism. In this case, the larger the mobility of re-dox centers tethered to the polymer, the faster the diffusion of electrons. Mobility can be increased by polymer solvation and decreased by cross-linking. This is the case of a PVP-[Os-(4,4'-dimethyl-2,2'-bipyridine)$_2$(4-aminomethyl-4'-methyl-2,2'-bipyridine)Cl]$^{2+/3+}$ hydrogel redox polymer used for laccase oxygen reduction en-zyme wiring [154]. This polymer exhibits a greater value of D_{app}^e and an onset potential for the OERR to water lower than for cross-linked redox polymers. Values of D_{app}^e for several hydrogel polymers range between 10^{-12} and 10^{-6} s cm^{-2} [154].

For enzymes, where redox centers are rather strongly bound to proteins entrapped into a conducting polymer modified with redox groups, the latter being able to inter-act with the redox centers of the enzyme. Likewise, redox groups are conveniently connected to the backbone of the polymer by spacers of different chemical properties and chain length. These spacers determine both the flexibility and hydrophilicity of the supramolecular system. In this case, after some simplifying assumptions, it has been demonstrated that, in this case, the value of D_{app}^e scales with the square root of the spacer chain length [155, 156].

Hydrogels based on polyacrylamide (PAM) and chitosan have been used to im-mobilize hemoglobin (Hb) [157]. The 3D hydrophilic polymer network of hydrogels favors swelling, which confers a flexibility to the polymer comparable to that of nat-ural tissues. Voltammograms of glassy carbon (GC) covered by those polymers in buffer solutions, pH 7.0, show a reversible pair of current peaks related to the Hb Fe^{3+}/Fe^{2+} redox couple, indicating a direct electron transfer from the enzyme redox center to the electrode [157].

The voltammetric characteristics of hydroquinone immobilized in poly(acrylic acid) spacers with and without GOx in 0.1 M phosphate buffer suggest that hydroquinone moiety aggregation into the hydrophobic spacers might impede hydroquinone contact with the electrode [158]. However, when hydrophilic di(ethylene oxide) chain is used as spacer, the hydroquinone oxidation/reduction voltammetric current peaks are observed. Another example is GOx immobilized in a 3D carbon black substrate [159]. When these electrodes, which are obtained by graft polymerization, contain hydrophilic spacers, they exhibit the best electrochemical response for glucose oxidation [160, 161].

8.3.4.4   The OERR at Enzyme Bioelectrodes

After discovering the possible mechanisms of the OERR to water at multi-copper oxidases [162], these enzymes became of interest for molecular oxygen cathodes in biofuel cell fabrication. Enzymes responsible for the OERR to water such as laccase, ceruloplasmin, ascorbate oxidase, and bilirubin oxidase contain a trinuclear copper active site with an additional mononuclear copper site at a 120-nm distance. The trinuclear site is the minimal structural subunit required for biological molecular oxygen bond cleavage and reduction to water. In this process, the three Cu(I) sites are oxidized via a four-electron process to produce a mixed-valence cluster with bound oxide ligands to stabilize one out of the three copper sites in the much lesser stable trivalent oxidation state [163]. Three types of copper atoms can be distinguished: types I and II, which accept electrons from electron-donating substrates, and types II and III, which serve as electron-donating sites for the OERR to water.

A H_2/O_2 biofuel cell, where a cathode and an anode compartment are separated by a Nafion membrane, utilizes a platinum gauze coated with platinum black anode and either a GC or a platinum foil cathode and 0.2 M acetate buffer electrolyte at pH 4. Laccase was first used as a biocatalyst incorporated into the catholyte solution for the OERR to water. The efficiency of the reaction is improved by mediators in solution such as 2,2′-azinobis(3-ethylbenzothiazoline-6-sulfonate) (ABTS), whose formal potential is very close to that of type I Cu(II) site of fungal laccase (from *Pyricularia oryzae*). ABTS exhibits a large stability in acid solution where laccase is more active. Laccase and ABTS mediator were poured into the cathode compartment and kept under stirring to diminish the concentration polarization, and the cathodic overpotential was smaller than that observed for pure platinum and GC cathodes. At the cathode voltage 0.720 V (vs. NHE), $j = 25$ μA cm^{-2}. The cathode potential is much higher than the 0.04 and -0.29 V that result for platinum and GC cathodes, respectively, and the same value of j and the maximum PO = 42 μW cm^{-2} under a 1000Ω load were obtained.

A high-operational-stability laccase electrode for the direct OERR to water was obtained by attaching the enzyme on graphite modified with phenyl derivatives, which produces an enzyme molecule adequately oriented for a DET process. In the absence of redox mediator, j is about 0.5 mA cm^{-2}, suggesting a preferred orientation of type 1 copper centers of laccase toward the electrode. Laccase activity is inhibited by the presence of fluoride ions [164].

The glucose/O_2 biocell involves the anodic oxidation of glucose to gluconic acid and the cathodic OERR to water. A fuel cell, which gets fuel (glucose) from the body and processes it with enzymes (laccase), behaves to some extent as a sort of small power plant in a living system [165]. A DET mechanism is obeyed by several enzyme systems as in the case of oxygen laccase [166, 167], bilirubin oxidase [168] and bioanodes/glucose oxidase [169]. A laccase-based composite electrode was constructed by wiring laccase, from *Coriolus hirsutus* through the redox polymer, PVI-Os(terpyridine)(4,4′-dimethyl-2,2′-bipyridine)$^{2+/3+}$ (PVI stands for poly-N-vinyl imidazole) complexed with Os(tpy)(dme-bpy)$^{2+/3+}$), to the fibers of a hydrophilic carbon cloth. Molecular oxygen was reduced to water at $j = 5$ mA cm^{-2} and 0.7 V (vs. NHE) in citrate buffer [170]. Laccase-based electrodes lose activity almost completely as the solution pH approaches 7 and the chloride ion concentration reaches 100 mM. Laccase enzymes of different origin wired on carbon electrodes show a distinct behavior with both the variation of pH and chloride concentration [171].

A non-compartmentalized, miniature glucose/O$_2$ biofuel cell was constructed employing two 2-cm-long 7-μm-diameter carbon fibers coated with different redox polymers for entrapping and wiring laccase at the cathode and glucose oxidase at the anode. The current density reaches 268 μA cm^{-2} at 0.78 V (vs. NHE) under air and 10% power loss per day functioning continuously for a week [172]. This configuration was implanted in a grape, demonstrating the fuel cell operation in a living plant [173].

Employing the same design based on microcarbon fiber substrate electrodes, a noncompartmentalized glucose/O$_2$ biofuel cell with a cathode made of wired BOx was reported [174]. At the cathode, the bioelectrocatalyst was wired using the electrostatic adduct of BOx (a polyanion at physiological pH) and the polycationic redox copolymer of polyacrylamide and poly(N-vinylimidazole) complexed with Os(4,4′-dichloro-2,2′-bipyridine)$_2$ Cl$^{+/2+}$ [175]. At the anode, the electrocatalyst was the electrostatic adduct of glucose oxidase (GOx) (a polyanion under the working conditions) with the redox polymer poly(N-vinylimidazole) partially quaternized with 2-bromoethylamine and complexed, in part, with [Os(4,4′-diamino-2,2′-bipyridine)$_2$ Cl$^{+/2+}$. Using these polymers the operation potential of the cell was extended. At 0.5 V, PO = 50 μW cm^{-2}, although this figure depended on the sodium chloride concentration. The overall biofuel cell process is limited by the glucose electrochemical oxidation and remains almost independent of the molecular oxygen pressure.

Cyt-c oxidase (COx) is a heme-copper oxidase that catalyses the OERR to water. Molecular oxygen binds to the reduced active site Fe(II)–Cu(I) of COx to form a heme-superoxide complex (Fe(III)O$_2$–Cu(I)) [176].

Biocatalyzed oxidation employing a membraneless glucose/O$_2$ fuel cell based on BOx and glucose dehydrogenase (sGDH) involves a BOx wiring to the cathode by poly(4-vinylpyridine) complexed with Os(2,2′-bipyridine)$_2$Cl and quaternized with bromoethylamine [177]. The sGDH was wired by poly(1-vinylimidazole) complexed with Os(4,4′-dimethyl-2,2′-bipyridine)$_2$Cl to oxidize glucose to gluconate. The electrolyte was buffered 55 mM glucose pH 7 aerated solution containing 3 mM CaCl$_2$ to ensure sGDH activity. The maximum value of PO is 58 μW cm^{-2} at 0.19 V. BOx is immobilized as a multiple layer in a cationic poly-L-lysine matrix on a plastic formed carbon electrode (PFCE) by electrostatic entrapment. The irreversible voltammetric behavior of the anodic process under rotation has been

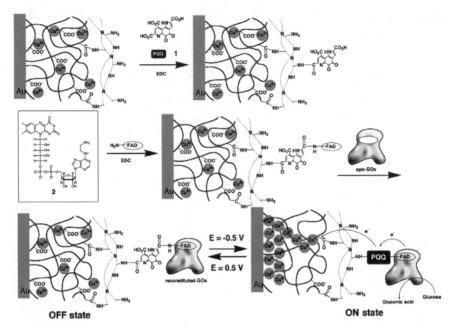

FIGURE 8.16   Preparation and operation of biocatalytic anode. Stepwise covalent binding of PQQ (1) and N-(2-aminoethyl)-flavin adenine dinucleotide (2) to polymer-functionalized electrode followed by reconstitution of apo-glucose oxidase and reversible activation and deactivation of biocatalytic anode by electrochemical reduction of Cu^{2+}–polymer film and oxidation of Cu^{0}–polymer film, respectively. (From Katz and Willner [179] by permission of The American Chemical Society.)

interpreted by a reaction layer model in which BOx diffuses in the immobilized layer [178].

A noncompartmentalized biofuel cell based on the biocatalyzed oxidation of glucose and OERR to water with switchable and tunable PO (Figs. 8.16 and 8.17) was constructed [179]. It consists of two 50-μm-thick evaporated-gold glass plates as anode and cathode substrates separated by a 2-mm-thick O-ring that is penetrated by two metallic needles. These needles are utilized for the electrolyte circulation that is driven by a peristaltic pump and occasionally used as counter- and reference electrodes in the cell. The anode consists of Cu^{2+}-poly(acrylic acid) film in which the redox relay pyrroloquinoline quinone (PQQ) and the FAD cofactor are covalently linked. Apo-GOx is reconstituted on the FAD sites to build the GOx-functionalized electrode. The cathode is made of the same Cu^{2+}-containing film that provides the functional interface for the covalent linkage of Cyt-c that is further linked to COx. The reduction of the Cu^{2+} film at -0.258 V (vs. NHE) at both the anode and the cathode, using the needle as electrode, yields Cu^{0}-poly(acrylic acid) matrices that electrically contact the GOx and Cox/Cyt c electrodes, respectively. The short-circuit current density and open-circuit potential are 550 μA cm^{-2} and 0.120 V, respectively. The maximum PO is 4.3 μW cm^{-2} for 1 Ω external loading. The oxidation of the

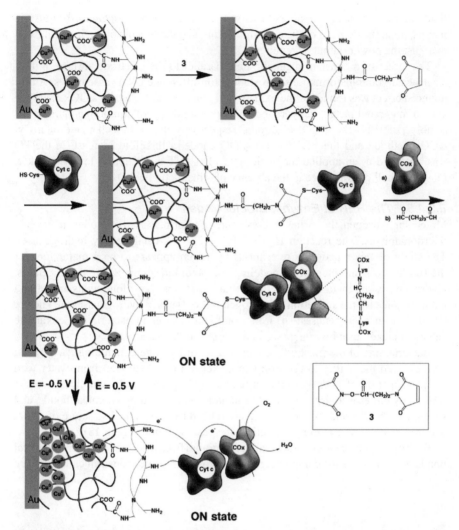

FIGURE 8.17 Preparation and operation of biocatalytic cathode. Covalent attachment of iso-2-Cyt-*c* to polymer-functionalized electrode surface using *N*-succinimidyl-3-maleimidopropionate (3) as heterobifunctional linker followed by affinity binding of COx and cross-link of protein complex layer and reversible activation and deactivation of biocatalytic cathode by electrochemical reduction of Cu²⁺–polymer film and oxidation of Cu⁰–polymer film, respectively. (From Katz and Willner [179] by permission of The American Chemical Society.)

Cu⁰ polymer film at 0.5 V turns it into the nonconductive Cu²⁺-containing film, and biofuel cell operation is blocked. By adequate electrochemical reduction and oxidation cycling the biocell is reversibly switched between *on* and *off* states. This type of biocell may find possible future applications as implantable devices in physiological fluids, as electric energy suppliers to activate machines such as pacemakers or insulin pumps, upon the demand of the consuming unit, and the tunable operation of the cell to adjust the power output as required.

The effect of a constant magnetic field parallel to the electrode surface on bio-electrocatalytic transformations was studied for different enzyme assemblies. A pronounced effect was found for the oxidation of glucose and lactate by a GOx electrode and an integrated lactate dehydrogenase/nicotinamide/pyrroloquinoline–quinine assembly (LDH/NAD$^+$–PQQ) electrode, respectively. For a biocell made of a Cyt c/COx cathode and the LDH/NAD$^+$–PQQ anode, a threefold increase in the PO was observed at an applied magnetic field of 0.92 T and external load of 1.2 kΩ, relative to cell performance in the absence of a magnetic field [180].

8.3.4.5 The HER at Enzyme Bioelectrodes

Molecular hydrogen, the earliest energy source on Earth, is used or produced by many microorganisms. The reaction $H_2 \leftrightarrow 2\ H^+ + 2\ e^-$ is catalyzed by hydrogenases. The study of this reaction is relevant as hydrogen appears to be a major fuel for the future and is an interesting modeling compound to mimic active metal centers. Hydrogenases are metal-enzymes that contain first-row transition metals and offer a natural choice for the activation of hydrogen. Three types of hydrogenases are known: (i) single-iron hydrogenases; (ii) iron–iron hydrogenases in bacteria and eukaryote cells; (ii) nickel–iron hydrogenases in bacteria and Archaea.

Bacteria, one of the most efficient producers of hydrogen, including *Escherichia coli*, evolved the ability to use iron and nickel to make hydrogen from water with the participation of hydrogenases. In natural hydrogenases a bimetallic active site formed by either iron–iron or nickel–iron atoms connected by chemical bonds, and surrounded by CN and CO groups that are related to the physical structure of nickel and iron proteins surrounding the active sites, can be found (Fig. 8.18).

Both hydrogenases contain several sulfur atoms (from cysteine) that form chemical bonds with metal atoms comparable to those found in some industrial catalysts. The

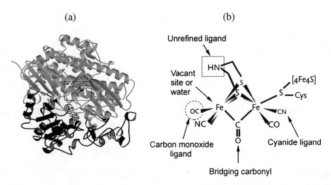

FIGURE 8.18   (a) Model of typical hydrogenase with bimetallic Fe–Fe active site. (b) The Fe atoms are surrounded by S atoms, carbon monoxide, and cyanide ligands, working together to help move electrons from one metal atom to protons.

crystal structure of Fe hydrogenase shows a mononuclear iron coordinated by a sulfur atom of cysteine, 2 CO, and the sp^2-hybridized nitrogen atom of 2-pyrodinol with back-bonding properties similar to those of cyanide. The 3D arrangement of the ligands in binuclear nickel–iron and iron–iron hydrogenases is similar to that of thiolate, CO, and CN ligated to the low-spin iron atom in these hydrogenases. The related iron atom ligation pattern of hydrogenases presumably plays an essential role of activation in the hydrogen molecule [181].

The oxidation of molecular hydrogen involves a first coordination of a molecule to a metal [182], and subsequently a metal bond hydride and a proton are formed. The latter is rapidly removed by a nearly bi-base and the hydride is oxidized, yielding another proton that is also removed. Likely, molecular hydrogen production operates in the reversed way.

For iron hydrogenase [181] the likely catalytic mechanism would involve a hydride generated by splitting the hydrogen molecule that is directly transferred to methenyl-tetrahydromethanepterin. The transfer of this hydride is analogous to that observed for flavin oxidoreductases and nicotinamide adeninedinucleotide phosphate–dependent oxidoreductases. Only one metal is required for catalysis without any formal change in its oxidation state [101].

In general, hydrogenase activity is sensitive to molecular oxygen pressure, a fact that explains the need to perform electrochemical studies under anaerobic conditions. Therefore, molecular oxygen tolerance is a requisite for hydrogenases to be used as anodes in membraneless biofuel cells. Recently, it was found that iron–nickel hydrogenase, which appears bound to the cell membrane of *Knallgas* bacteria, retains a substantial molecular hydrogen oxidation activity under atmospheric oxygen [183]. An oxygen-tolerant membrane-bound nickel–iron hydrogenase from *Ralstonia* spp. catalyzes hydrogen production in addition to hydrogen oxidation. The former process proceeds with a minimal overpotential [184]

Spectroscopic studies showed that hydrogenases produce molecular hydrogen in a totally different way from the way a platinum catalyst does. Hydrogenases stick both electrons on the same proton making a H^- ion that then reacts with H^+ yielding molecular hydrogen. It appears that ligands that surround the metal atoms in the active site work together to help move electrons from one metal atom to protons. For iron hydrogenase the active site would stabilize a *torque geometry*. The twisted iron–iron bond has one more electron than its partner at the terminal iron atom. Then, both the electron and hydrogen tunnel to the active site. The unusual ligand-coordinated iron plays a key role in the formation of a mixed oxidation state and the transfer of the necessary electron and protons to the metal atoms. The CO group appears to be moved between iron atoms as the reaction proceeds. This flip-flop creates a transient chemical intermediate in which the terminal iron is bonded to one CO. The chemical arrangement puts the iron atom in a high-energy state that facilitates the transfer of 2 e^- to a single H^+ yielding hydroxide. The above atom and ligand rearrangement would be different for nickel–iron hydrogenases. Advances are being made to develop catalysts that mimic what the enzyme active sites do to produce molecular hydrogen. Similarly, catalysts based on nickel–iron hydrogenases could improve the hydrogen-oxidizing process that is central to hydrogen fuel cells. It has been shown that nickel–iron hydrogenase (purified) adsorbed onto a graphite electrode becomes a hydrogen-splitting catalyst as good as platinum [185, 186].

Desulfovibrio desulfuricans catalyzes the metabolic conversion of sulfate to sulfide [187]. In this case, the iron–iron and iron–nickel hydrogenases convert protons into molecular hydrogen. The fuel is oxidized at the bioanode, the OERR to water being the common cathodic reaction at the biocell.

The electroreduction of H^+ to molecular hydrogen undergoes better with iron–iron hydrogenases, but their activity is rapidly spoiled by atmospheric oxygen. Thus, for biofuel cells, the more oxygen-tolerant iron–nickel hydrogenases appear to be more suitable for anodes [185]. A pyrolytic graphite edge electrode (PGE) modified by direct adsorption of *Allochromatium vinosum* nickel–iron hydrogenase yields $j \approx 3 \text{-} 4$ mA cm^{-2} at 45°C and overpotentials between 0.4 and 0.6 V, that is, a response similar to that of dispersed carbon-supported platinum electrodes, but it exhibits poor stability [185, 188]. By electropolymerization of a pyrrol layer on a carbon filament and subsequent immobilization of *Thiocapsa roseopersicina* nickel–iron hydrogenase, the electrode stability is improved, yielding j values of 0.5–1.5 mA cm^{-2} and retaining about 50% of its activity after 6 months when it is stored at 4°C [189]. A larger stability was obtained by covalently bonding *Desulfovibrio gigas* nickel–iron hydrogenase to PG and CNT electrodes using a carbodiimide compound [190]. This bond involves the protein dipole originated by the high concentration of glutamate groups surrounding the 4Fe4S cluster and the amine groups on the electrode surface. The value $j \approx 1$ mA cm^{-2}, which persists for a month, was obtained [190].

A fuel cell comprising a graphite anode with the hydrogenase of *Ralstonia metallidurans* CH34, resistant to oxygen deactivation and a graphite cathode modified by a *fungal laccase* enzyme in aqueous electrolyte, exposed to an atmosphere of about 3% hydrogen in air (Fig. 8.19), yields $j = 11$ μA cm^{-2} at 0.5 V and a maximum PO = 5.2 μW cm^{-2} at 47 kΩ.

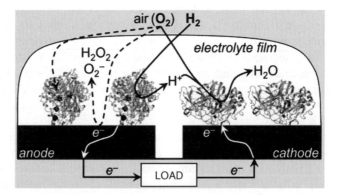

FIGURE 8.19   Fuel cell comprising graphite anode modified with O_2-tolerant hydrogenase of *R. metallidurans* CH$_{34}$ and graphite cathode modified with high-potential *fungal laccase* in aqueous electrolyte under 3% H_2 in air. (From Vicent et al. [191] by permission of The Royal Society of Chemistry.)

8.3.4.6 *Multienzyme Systems*

Single-enzyme biofuel cells may have low efficiency because of a partial oxidation of the fuel. Enzymatic catalysis utilizes isolated and purified enzymes for specific reactions, as these materials are much better defined than microbial catalysts. Conversely, living cells can completely oxidize food (fuel) to carbon dioxide and water combining the action of several enzymes to complete oxidation of fuels via a cascade process.

The first biofuel cell oxidation of alcohols [192] was made with alcohol dehydrogenase (ADH), aldehyde dehydrogenase (AldH) and formate dehydrogenase (FDH). As the effect of hydrogenases depends upon the presence of NAD^+, the enzyme diaforase is added to regenerate NAD^+ by reducing the mediator benzyl viologen.

For the complete oxidation of methanol, the cell consists of anode and cathode compartments separated by a membrane [192]. The immobilization of enzymes onto carbon electrodes using a Nafion matrix made it possible to operate a membraneless-methanol/oxygen biofuel cell. The hydrogenases that catalyze methanol/oxygen biofuel cell reactions give rise to an open-circuit potential of 0.8 V and PO = 0.68 μW cm^{-2}. The complete oxidation of methanol is a NAD-mediated process, which can be expressed by the following sequence of reactions:

$$CH_3OH + 2\,NAD \rightarrow [ADH]CH_2O + 2\,NADH \tag{8.8}$$

$$CH_2O + H_2O + 2\,NAD^+ \rightarrow [AldDH]HCOOH + 2\,NADH^+ \tag{8.9}$$

$$HCOOH + 2N\,ADH^+ \rightarrow [FDH]CO_2 + 2\,NADH^+ \tag{8.10}$$

Then, the reaction at the cathode is

$$3\,NADH + 6\,mediator(ox) \rightarrow [diforase]2\,NAD^+ + 6\,mediator(red) \tag{8.11}$$

$$6\,mediator(red) \rightarrow 6\,mediator(ox) + 6\,e^- \tag{8.12}$$

$$3O_2 + 3H_2O + 6e^- \rightarrow 6OH^- \tag{8.13}$$

The working voltage for this biocell becomes smaller than 0.5 V due to kinetic limitations of the enzymatically catalyzed process [193].

The bioelectrochemical oxidation of a food mixture by glucose oxidase and lactate dehydrogenase [180] and a two-step ethanol oxidation to acetate using alcohol dehydrogenase and aldehyde dehydrogenase [194] have also been studied, although in both cases rather low PO value were obtained. In practical biofuel cells, the complete oxidation of a fuel mixture requires immobilizing a cascade of enzymes either at an electrode surface or in a 3D entrapment polymer to mimic natural pathways [119, 195, 196]. Improved regeneration systems from NAD^+ for NAD^+-dependent dehydrogenase mediators can be achieved through a better matching in potential to increase cell voltage by the presence of benzoylviologen (BV), or more effectively by coimmobilization of both mediators and enzymes.

The different steps of ethanol electrochemical biooxidation were independently studied with both a single enzyme and a mixture of enzymes [194, 197] on a

TABLE 8.2 Polarization Data of 100 mM Ethanol/Air Biofuel Cells Employing from Single Dehydrogenase to Six-Dehydrogenase Cascade for Complete Oxidation of Ethanol

Dehydrogenase Cascades	j ($A\,cm^{-2}$), $\times 10^4$	PO ($W\,cm^{-2}$), $\times 10^4$
ADH	3.77 ±1.97	1.16 ±0.88
ADH, AldDH	5.30 ±2.50	1.30 ±0.52
ADH, AldDH, IDH	8.02 ±1.98	1.63 ±0.37
ADH, AldDH, IDH, KDH	9.84 ±6.47	2.02 ±0.83
ADH, AldDH, IDH, KDH, SDH	16.2 ±0.35	2.86 ±0.36
ADH, AldDH, IDH, KDH, SDH, MDH	36.0±0.23	10.1 ±0.01

Source: From ref. 202.

polymethylene-coated gas chromatograph [198]. This polymer coating behaves as an electrocatalyst for NADH [199, 200]. In this cell, where the fuel oxidation occurs in the presence of NAD^+-dependent dehydrogenase enzymes, the coenzyme NAD^+ is reduced to NADH and regenerated by oxidation of the latter according to reactions (8.11)–(8.13).

More recently [201, 202], an ethanol/air biocell consisting of an anode made of carbon paper and a gas diffusion hydrogen cathode made of carbon-supported 20% platinum with an amount of Nafion ion exchange resin and Teflon wet proofing was investigated with different dehydrogenase cascades. The working electrodes were separated by Nafion membrane modified by tetrabutylammonium bromide (TBAB). Table 8.2 shows polarization data for 100 mM ethanol/air fuel cells with different dehydrogenase cascades that have been tested. Single hydrogenase data are used to compare the behavior of multiple hydrogenase cascades for complete oxidation of ethanol. Both j and PO exhibit maximum values for the enzymatic mixture containing ADH, AldDH, IDH (isocitric dehydrogenase), KDH (alpha-ketoglutarate dehydrogenase), MDH (malic dehydrogenase) and SDH (succinate dehydrogenase) [202]. A scheme of this biofuel cell is shown in Figure 8.20.

The citric acid cycle at the electrode surface of the above biocell can be mimicked considering a TBAB-modified Nafion membrane that permits the immobilization of multienzyme systems including all enzyme/cofactors/coenzymes of the citric acid cycle. In this case, methylene green polymer behaves as an efficient electrocatalyst for the complete oxidation of ethanol. The addition of more electron-generating enzymes results in an increase of PO as compared to the single-enzyme system. The value of PO for the six-dehydrogenase enzymatic cascade attains a maximum at pH 7.5 in different phosphate buffers. The change in pH from 6.5 to 7.5 for the six-dehydrogenase enzymatic cascade produces a 25.8-fold increase in the PO value [202].

8.3.5 Microbiological Fuel Cells

Microbially catalyzed biofuel utilizes whole living organisms as a continuous source of catalysts as they have highly efficient electron transfer and control systems through

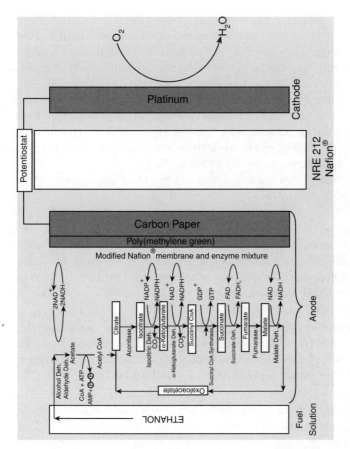

FIGURE 8.20   Scheme of complete biofuel cell. Ethanol is oxidized serving as fuel source at the anode (dark lettering represents dehydrogenase enzymes, whereas light lettering represents other non-energy-producing enzymes). Oxygen is reduced to water at 20% platinum on carbon gas diffusion electrode (GDE) cathode. (From Sokic-Lazic and Minteer [202] by permission of Elsevier.)

multiple enzymatic processes. The complete enzymatic pathway operates on variable feedstocks and is usually capable of oxidizing the substrate to carbon dioxide and water. As is the case of electrochemical systems with purified enzymes, microbially catalyzed fuel cells may have the active component and mediators either in solution or located at the electrode surface. The same classification in MET and DET processes stands for microbially catalysed biofuel cells, although considering their characteristics, whole organisms tend to operate as an indirect electron transfer system with free diffusing mediators (DMET). The electron transfer operates across the cell membrane in both directions, and the exchange of electrons between the metabolic system inside cell membranes and the electrode becomes easier. On the other hand, when the system tends to behave as a nondiffusive MET, the contact of the membrane and the exchange of electrons occur without penetrating the cell membrane. For both DMET and MET the closer the reaction potential of the mediator to that of the biological component, the higher the efficiency of the biofuel cell. In some organisms the DET process occurs between the surface of the microorganism and the electrode assisted by the specific electron transfer capacity within the cell membrane.

Microbial organisms also have the possibility of releasing a secondary fuel product that is used to feed an electrochemical converter. Schemes of these product-type systems are shown in Figure 8.14. Practical operation of this type of cell involves the possibility of coupling bioelectrocatalysis with other biochemical processes such as fermentation, where a secondary fuel is produced [203].

Some examples of microbial biofuel cells involving either anodic or cathodic reactions are:

- Glucose (fuel) oxidation by *Proteus vulgaris* (MET biofuel cell). In this cell the microorganism is immobilized on an anaerobic graphite felt anode surface by an amide link. An air cathode at platinized carbon paper. The cell constituents are 2-hydroxy-1,4-naphthoquinone as diffusive mediator, potassium phosphate buffer At pH 7 in anode, and potassium ferricyanide in the cathode. At 0.7 V, $j = 3.2 \ \mu A \ cm^{-2}$ and PO $= 3.2 \ \mu W \ cm^2$ [204].

- Glucose/O_2 (MET anaerobic, single-compartment biofuel cell) utilizing suspended *Escherichia coli* from sewage sludge. Woven doped graphite electrodes are utilized; the electrolyte is made of phosphate buffer, sodium lactate, peptone, and yeast extract at pH 7; $100 \le j \le 180 \ \mu A \ cm^{-2}$, $75 \le PO \le 85 \ \mu W \ cm^{-2}$ [205].

- Glucose/ferricyanide (anaerobic DET biofuel cell) utilizes *Rhodoferax ferrireducens*. Graphite (rod, foam or felt) electrodes are used. N_2/CO_2 is flushed through the glucose solution at the anode and a potassium ferricyanide buffer solution (pH 7) at the cathode. At 0.265 V with a load of 1000 Ω, $j = 3.1 \ \mu A \ cm^{-2}$, PO $= 0.8 \ \mu W \ cm^{-2}$, coulombic efficiency 81% [206].

- Product biofuel cell (starch, glucose, molases)/ H_2/ferricyanide. This cell works with *Clostridium butyricum* or *Clostridium beijerinckii*. A platinized woven graphite cloth modified with poly(tetrafluoroaniline) and a woven graphite are utilized as anode and cathode, respectively. The N_2-purged anode solution contains the substrate plus yeast extract, peptone, L-cysteine–HCl, magnesium sulfate, potassium bisulfate, sodium chloride, sodium bicarbonate, resazurine. The cathode solution contains potassium ferricyanide in phosphate buffer at pH 7; open-circuit potential $= 0.895$ V, $j = 2.6 \ mA \ cm^{-2}$, PO $= 1.2 \ mW \ cm^{-2}$ [207].

- Lactate/O_2 (anaerobic DET biofuel cell) containing a suspension of *Shewanella putrefaciens*, graphite felt anode and cathode, and phosphate buffer with sodium chloride at pH 7; cell voltage ≈ 0.5 V, j decays from 0.7 to 0.4 $\mu A \ cm^{-2}$, and PO varies between 0.02 and 0.021 $\mu W \ cm^{-2}$ [206].

- H_2(MET)/*Desulfovibrio vulgaris* suspension/O_2(MET) biofuel cell. A carbon felt in an H_2-saturated solution anode and a carbon felt cathode in an O_2-saturated

BOx solution are used. The electrolyte is phosphate buffer pH 7 with potassium chloride. Methyl viologen antraquinone sulfonate is used as mediator at the anode and ABTS as mediator at the cathode. The open-circuit potential is 1.17 V, $j = 176$ μA cm^{-2}, PO ≈ 100 μW cm^{-2}. The current is limited by the BOx concentration [209].

- DET/MET cell for marine sediment and seawater constituents. This is a complete single-compartment fuel cell that employs mixed natural bacteria, sea water, and drilled graphite electrodes. The anode is placed below the sediment surface and the cathode above it; open-circuit voltage is 0.75 V, $j = 7$ μA cm^{-2}, PO ≈ 3 μW cm^{-2} [210].

An extended review of microbial bioelectrochemical devices is given elsewhere [100].

8.3.6 Conclusions

The increasing complexity of electrode designs for fuel cells in going from single metals to cells in which enzymes and living organisms are involved shows that an interdisciplinary approach is required for tackling this matter, particularly when one attempts to improve the cell electrocatalytic efficiency and operation continuity attempting to approach the behavior found in nature. In doing so, a framework based in terms of traditional scientific disciplines has to be bypassed in favor of an interdisciplinary approach such as that given in the functional supramolecular systems scheme that links life and material sciences that has been proposed by Ringsdorf [72] (Figure 8.21). Functional supramolecular systems point out that scientific efforts to mimic natural molecular assemblies have to focus on molecular performances of self-organised molecular aggregates where, for handling the properties of matter, "the whole is more than the sum of its parts."

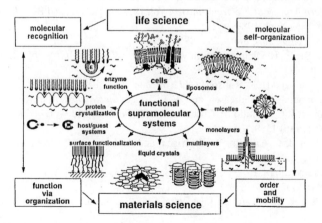

FIGURE 8.21 Functional supramolecular system-connecting links between life science and materials science. (From Ringsdorf [72] by permission of Wiley.)

ACKNOWLEDGMENT

Financial support from the Consejo Nacional de Investigaciones Científicas y Técnicas (CONICET), Agencia Nacional de Promoción científica y Tecnológica (ANPCYT, PICT 34530), and Comisión de Investigaciones Científicas de la Provincia de Buenos Aires (CICPBA) of Argentina is acknowledged.

REFERENCES

1. E. Reisner, J. C. Fontecilla-Camps, and F. A. Armstrong, *Chem. Commun.*, (5) (2009) 550.
2. A. P. Alivatos, *Science*, 271 (1996) 933.
3. C. G. Lugmair, A. T. Bell, and T. D. Tilley, *J. Am. Chem. Soc.*, 124 (2002) 13194.
4. M. Valden, X. Lai, and D. W. Goodman, *Science*, 281 (1998) 1647.
5. D. R. Rolison, *Science*, 299 (2003) 1698.
6. D. Walis, *Science*, 271 (1996) 925.
7. J. A. H. Creus et al., *J. Phys. Chem. B.*, 108 (2004) 10785.
8. Y. Gimeno, A. H. Creus, S. González, R. C. Salvarezza, and A. J. Arvia, *Chem. Mater.*, 13 (2001) 1857.
9. T. Kessler et al., *Langmuir*, 12 (1996) 6587.
10. L. Vásquez et al., *J. Am. Chem. Soc.*, 109 (1987) 1730.
11. A. J. Arvia, R. C. Salvarezza, and W. E. Triaca, *Electrochim. Acta*, 34 (1989) 1071.
12. J. Giner, *Electrochim. Acta*, 9 (1964) 63.
13. M. L. Marcos, J. M. Vara, J. G. Velasco, and A. J. Arvia, *J. Electroanal. Chem.*, 224 (1987) 189.
14. R. L. M. Creery, in *Electroanalytical Chemistry*, Vol. 17, A. J. Bard (Ed.), Dekker, New York, 1994.
15. K. Kinoshita, *Carbon; Electrochemistry and Physicochemical Properties*, Wiley, New York, 1998.
16. M. R. Tarasevich and E. I. Khrushcheva, Electrocatalytic properties of carbon materials, in *Modern Aspects of Electrochemistry*, Vol. 19, J. O. Bockris and B. E. Conway (Eds.), Plenum, New York, 1989, p. 295.
17. Z. Klusek, *Appl. Surf. Sci.*, 125 (1998) 339.
18. J. M. McBreen, H. Olender, S. Srinivasan, and K. V. Kordesch, *J. Appl. Electrochem.*, 11 (1981) 787.
19. K. E. Swider and D. R. Rolison, *J. Electrochem. Soc.*, 143 (1986) 813.
20. J. L. Zubimendi, G. Andreasen, and W. E. Triaca, *Electrochim. Acta*, 40 (1995) 1305.
21. A. J. Arvia, J. C. Canullo, E. Custidiano, C. L. Perdriel, and W. E. Triaca, *Electrochim. Acta*, 31 (1986) 1359.
22. A. Visintin, J. C. Canullo, W. E. Triaca, and A. J. Arvia, *J. Electroanal. Chem.*, 239 (1988) 67.
23. J. L. Zubimendi, M. E. Vela, R. C. Salvarezza, J. M. Vara, and A. J. Arvia, *Phys. Rev. E*, 50 (1994) 1367.
24. C. Alonso, R. C. Salvarezza, J. M. Vara, and A. J. Arvia, *Electrochim. Acta*, 35 (1990) 1331.
25. A. R. Despic and K. I. Popov, *J. Appl. Electrochem.*, 1 (1971) 275.
26. O. M. Magnussen, *Chem. Rev.*, 102 (2002) 679.

27. A. J. Arvia, R. C. Salvarezza, and W. E. Triaca, *J. New Mater. Electrochem. Syst.*, 7 (2004) 133.

28. N. Furuya and S. Koide, *Surf. Sci.*, 220 (1989) 18.

29. N. L. Pocard, D. C. Alzmeyer, R. L. Mackriri, T. X. Neenan, and M. R. Callstrom, *J. Mater. Chem.*, 2 (1992) 771.

30. J.-R. Chang, J.-F. Lee, S. D. Lin, and A. S. Lin, *J. Phys. Chem.*, 99 (1995) 14798.

31. J. Larminie and A. Dick, *Fuel Cells Systems Explained*, Wiley, Chichester, 2000.

32. K. P. D. Jong, and J. W. Geus, *Catal. Rev. Sci. Eng.*, 42 (2000) 481.

33. Y. M. Lin et al., *J. Appl. Electrochem.*, 38 (2008) 507.

34. R. Hiesgen, D. Eberhardt, E. Aleksandrova, and K. A. Friedrich, *J. Appl. Electrochem.* 37 (2007) 1495.

35. M. L. Anderson, R. M. Stroud, and D. R. Rolison, *Nanoletters*, 2 (2002) 235.

36. Z.-L. Lu, E. Linder, and H. A. Mayer, *Chem. Rev.*, 102 (2002) 3543.

37. M. Cooper and G. G. Botte, *J. Electrochem. Soc.*, 153 (2006) A1894.

38. T. Frelink, W. Visscher, and J. A. R. van Veen, *Electrochim.*, Acta 39 (1994) 1871.

39. X. H. Xia and T. Iwasita, *J. Electrochem. Soc.*, 140 (1993) 2559.

40. Y. Du, B. Su, N. Zhang, and C. Wang, *Appl. Surf. Sci.*, 255 (2008) 2641.

41. S. Mukejee, *J. Appl. Electrochem.* 20 (1990) 537.

42. J. W. Long, R. M. Stroud, K. E. Swider-Lyons, and D. R. Rolison, *J. Phys. Chem. B*, 104 (2000) 9772.

43. N. M. Markovic and J. P. N. Ross, *Surf. Sci. Rep.*, 45 (2002) 117.

44. A. A. Wragg and C. H. Hartnig, A. (Eds.), Efficient oxygen reduction for electrochemical energy conversion, *J. Appl. Electrochem.*, 37 (2007) 1395.

45. T. Iwasita, F. C. Nart, and W. Vielstich, *Ber. Bunsenges. Phys. Chem.*, 94 (1990) 1030.

46. A. Hamnett, B. J. Kennedy, and F. E. Wagner, *J. Catal.*, 124 (1990) 30.

47. A. S. Aricó et al., *J. Power Sources*, 50 (1994) 295.

48. T. Page, R. Johnson, J. Hormes, S. Noding, and R. Rambahn, *J. Electroanal. Chem.*, 485 (2000) 34.

49. M. Watanabe and S. Motoo, *J. Electroanal. Chem.*, 60 (1975) 367.

50. J. M. Sieben, M. M. E. Duarte, and C. E. Mayer, *J. Appl. Electrochem.* 38 (2008) 483.

51. V. S. Bagotzky and Y. B. Vassilyev, *Electrochim. Acta*, 12 (1967) 1323.

52. N. Watanabe, M. Uchida, and S. Motoo, *J. Electroanal. Chem.*, 199 (1986) 311.

53. N. Watanabe, M. Uchida, and S. Motoo, *J. Electroanal. Chem.*, 229 (1987) 395.

54. S. Surampudi et al., *J. Power Sources*, 47 (1994) 377.

55. H. A. Gasteiger, N. Marcovic, P. N. Ross, and E. Cairns, *J. Electrochem. Soc.*, 261 (1989) 375.

56. S. Mukerjee and R. C. Urian, *Electrochim. Acta*, 47 (2002) 3219.

57. D. Dmowski, T. Egami, K. E. Swider-Lyons, C. T. Love, and D. R. Rolison, *J. Phys. Chem. B*, 106 (2002) 12677.

58. R. M. Stroud, J. W. Long, K. E. Swider-Lyons, and D. R. Rolison, *Microsc. Microanal.*, 8 (2002) 50.

59. K. E. Swider-Lyons, G. B. Cotton, J. A. Stanley, W. Domowski, and T. Egami, Annual progress report for the hydrogen fuel cells and infrastructure technologies program, Tech. rep., U.S. Department of Energy, Washington DC, 2002.

60. L. Colmenares et al., *J. Appl. Electrochem.*, 37 (2007) 1413.

61. K. Wippermann et al., *J. Appl. Electrochem.*, 37 (2007) 1399.

62. G. Zehl, P. Bogdanoff, I. D. S. Flechter, K. Wippermann, and C. Hartnig, *J. Appl. Electrochem.*, 37 (2007) 1475.

63. C. Christenn, G. Steinhilber, M. Schultze, and K. A. Friedrich, *J. Appl. Electrochem.*, 37 (2007) 1463.
64. A. Racz, P. Bele, C. Cremers, and U. Stimming, *J. Appl. Electrochem.*, 37 (2007) 1555.
65. P. J. Kulesza et al., *J. Appl. Electrochem.*, 37 (2007) 1439.
66. M. R. Tarasevich et al., *J. Appl. Electrochem.*, 37 (2007) 1503.
67. J. Schreurs and E. Barendrecht, *Recl. Trav. Chim. Pays-Bas*, 103 (1984) 205.
68. F. Murray, *Electroanalytical Chemistry*, Vol. 13, Dekker, New York, 1999.
69. E. Laviron, *J. Electroanal. Chem.*, 112 (1980) 1.
70. C. W. B. Bezerra et al., *Electrochim. Acta*, 53 (2008) 4937.
71. K. L. Wolf, H. Frahm, H. Harms, *Z. Phys. Chem. B*, (1937) 237.
72. H. Ringsdorf, *Angew. Chem. Int Ed.*, 29 (1990) 1269.
73. R. F. Lane, A. T. Hubbard, K. Fukunaga, and R. J. Blanchard, *Brain Res.*, 114 (1976) 346.
74. J. Zagal, R. K. Sen, and E. Yeager, *J. Electroanal. Chem.*, 83 (1977) 207.
75. R. W. Murray, *Sililated Surfaces*, Gordon and Breach, New York, 1980.
76. T. Matsue, M. Fujihira, and T. Osa, *J. Electrochem. Soc.*, 129 (1979) 500.
77. C. Fang and X. Zhou, *Electroanalysis*, 13 (2001) 949.
78. S. Yoshimoto, A. Tada, K. Souto, S.-L. Yau, and K. Itaya, *Langmuir*, 20 (2004) 3159.
79. K. Suto, S. Yoshimoto, and K. Itaya, *Langmuir*, 22 (2006) 10766.
80. M. Daniel and D. Astruc, *Chem. Rev.*, 104 (2004) 293.
81. K. Kelly, E. Coronado, L. Zhao, and G. Schatz, *J. Phys. Chem B*, 107 (2003) 668.
82. M. Bruchez Jr., M. Moronne, P. Gin, S. Weiss, and A. P. Alivisatos, *Science*, 281 (1998) 2013.
83. M. Alvarez, J. Khoury, T. Schaff, M. Shafigullin, I. Veamar, and R. Whetten, *J. Phys. Chem. B*, 101 (1997) 3706.
84. J. Hicks, F. Zamborini, A. Osisek, and R. Murray, *J. Am. Chem. Soc.*, 123 (2001) 7048.
85. J. Hicks, D. Miles, and R. Murray, *J. Am. Chem. Soc.*, 124 (2002) 13322.
86. S. Chen and R. Murray, *J. Phys. Chem. B*, 103 (1999) 9996.
87. T. Trindade, P. O´Brien, and N. Pickett, *Chem. Mater.*, 13 (2001) 3843.
88. P. Schwerdfeger, *Angew. Chem. Int Ed.*, 42 (2003) 1892.
89. R. Gonpadhyay and A. De, *Chem. Mater.*, 12 (2000) 608.
90. Y. Zhang, J. Li, Y. Shen, M. Wang, and J. Li, *J. Phys. Chem. B*, 108 (2004) 15343.
91. W. T. Müller, D. L. Klein, T. Lee, J. Clarke, P. L. McEuen, and P. G. Schultz, *Science*, 268 (1995).
92. J. Drillet, R. Dittmeyer, and K. Jüttner, *J. Appl. Electrochem.*, 37 (2007) 1219.
93. M. C. Potter, *Proc. Royal Soc. B*, 84 (1910) 260.
94. W. Groove, *Philos. Mag.*, 14 (1839) 297.
95. B. Cohen, *J. Bacteriol.*, 21 (1931) 18.
96. H. A. Videla and A. J. Arvia, *Experientia*, 18 (1971) 667.
97. H. A. Videla and A. J. Arvia, *Biotechnol. Bioeng.*, 17 (1975) 1529.
98. E. A. Disalvo, H. A. Videla, and A. J. Arvia, *Biotechnol. Bioeng.*, 6 (1979) 493.
99. M. R. Tarasevich, A. Sadkowski, and E. Yeager, in *Comprehensive Treatise of Electro-chemistry*, Vol. 7, B. E. Conway, J. O. Bockris, E. Yeager, S. U. Khan, and R. E. White (Eds.), Plenum, New York, 1983.
100. R. A. Bullien, T. C. Arnot, J. B. Lakeman, and F. C. Walsh, *Biosens. Bioelectron.*, 21 (2006) 2015.
101. F. A. Armstrong and J. C. Fontecilla-Camps, *Science*, 321 (2008) 498.
102. G. C. Hill et al., *Biosens. Biolelectron.*, 18 (2003) 327.
103. A. Bommarus and K. Polizzi, *Chem. Eng. Sci.*, 61 (2006) 1004.

104. D. L. Nelson and M. M. Cox, *Lehninger: Principles of Biochemistry*, W. H. Freeman, New York, 2004.

105. H. Lodish et al., *Molecular Cell Biology*, W. H. Freeman, New York, 2003.

106. G. Badcock and M. Wikstrom, *Nature*, 356 (1982) 301.

107. P. Barlett, P. Tebbut, and R. Whitaker, *Proc. React. Kinet.*, 16 (1991) 55.

108. I. Willner, E. Katz, and B. Willner, *Electroanalysis*, 9 (1997) 465.

109. L. Habermuller, M. Mosbach, and W. Schuhmann, F. J., *Anal. Chem.*, 366 (2000) 560.

110. F. Davis and S. P. J. Higson, *Biosens. Bioelectron.*, 22 (2007) 1224.

111. A. Heller, *J. Phys. Chem.*, 96 (1992) 3579.

112. Y. Liu, M. Wang, F. Zhao, Z. Xu, and S. Dong, *Biosens. Bioelectron.*, 21 (2005) 984.

113. X. Wel, J. Cruz, and W. Goski, *Anal. Chem.*, 74 (2002) 5039.

114. B. Falk, S. Garramone, and S. Shivkumar, *Mater Lett.*, 58 (2004) 3261.

115. Y. Liu, X. Qu, H. Guo, H. Chen, B. Liu, and S. Dong, *Biosens. Bioelectron.*, 21 (2006) 2195.

116. S. Rul, F. Lefevre-chlick, E. Capria, C. Laurent, and A. Peigney, *Acta Materialia*, 52 (2004) 1061.

117. J. K. W. Sandler, J. E. Kirk, I. A. Kinloch, M. S. P. Shaffer, and A. H. Windle, *Polymer*, 44 (2003) 5893.

118. B. E. Kilbride et al., *J. Appl. Phys.*, 92 (2002) 4024.

119. S. D. Minteer, B. Y. Liaw, and M. J. Cooney, *Curr. Opin. Biotechnol.*, 18 (2007) 228.

120. M. Zhang, A. Smith, and W. Gorski, *Anal. Chem.*, 76 (2004) 5045.

121. X. Tan, M. Li, P. Cai, L. Luo, and X. Zou, *Anal. Biochem.*, 337 (2005) 111.

122. S. Ben-Ali et al., *Electrochem. Commun.*, 5 (2003) 747.

123. S. Ben-Ali, D. Cook, N. Bartlett, and A. Kuhn, *J. Electroanal. Chem.*, 579 (2005) 181.

124. J. Kim, J. W. Grate, and P. Wang, *Chem. Eng. Sci.*, 61 (2006) 1017.

125. D. Chen, G. Wang, and J. Li, *J. Phys. Chem. C*, 111 (2007) 2351.

126. M. Shinohara, T. Chiva, and M. Aizawa, *Sens. Actuators*, 13 (1987) 79.

127. M. Shaolin, H. Huaiguo, and Q. Bidong, *J. Electroanal. Chem.*, 304 (1991) 7.

128. C. Malitesta, F. Palmisano, L. Torsi, and P. G. Zambonin, *Anal. Chem.*, 62 (1990) 2735.

129. P. C. Pandey, *J. Chem. Soc. Faraday Trans. I*, 84 (1988) 2259.

130. N. C. Founds and C. R. Lowe, *Anal. Chem.*, 60 (1988) 2473.

131. J. M. Dicks, S. Hattori, I. Karube, A. P. F. Turner, and T. Yokozawa, *Ann. Biol. Clin.*, 47 (1989) 607.

132. G. Wang and S.-T. Yau, *J. Phys. Chem. C*, 111 (2007) 11921.

133. L. Deng et al., *Biomacromolecules*, 8 (2007) 2063.

134. Y. Yan, W. Zheng, L. Wang, L. Su, and L. Mao, *Langmuir*, 21 (1989) 6560.

135. N. Mano, H.-H. Kim, and A. Heller, *J. Phys. Chem. B*, 106 (2002) 8842.

136. N. Mano, H.-H. Kim, Y. Zhang, and A. Heller, *J. Am. Chem. Soc.*, 124 (2002) 6480.

137. G. Cooke, F. Duclairoir, P. John, N. Polwart, and V. Rotello, *Chem. Commun.*, (2003) 2468.

138. E. Simon and P. Bartlett, in *Modified Electrodes for the Oxidation of NADH in Biomolecular Films: Design, Function and Applications*, J. Rusling (Ed.), Marcel Dekker, New York, 2003, Chapter 11.

139. T. Edwards, V. Cunnae, R. Parsons, and D. Gani, *J. Chem. Soc., Chem Commun.*, (15) (1989) 1041.

140. L. Wingard, *Bioelectrochem. Bioeng.*, 9 (1982) 307.

141. B. Mallik and D. Gani, *J. Electroanal. Chem.*, 326 (1992) 37.

142. S.-W. Tam-Chang, J. Mason, I. Iverson, K.-O. Hwang, and C. Leonard, *Chem. Commun.*, (1) (1999) 65.

143. R. Blonder, I. Willner, and A. Bueckmann, *J. Am. Chem Soc.*, 120 (1998) 9335.

144. J. Liu, N. Paddon-Row, and J. Gooding, *J. Phys. Chem. B*, 108 (2004) 8460.

145. E. Pai and G. E. Schulz, *J. Biol. Chem.*, 258 (1983) 1752.

146. E. Calvo et al., *Langmuir*, 7907.

147. C. Cameron and P. Pickup, *J. Am. Chem. Soc.*, 121 (1999) 11773.

148. C. Cameron, T. Pittman, and P. Pickup, *J. Phys. Chem. B*, 105 (2001) 8838.

149. B. Gregg and A. Heller, *J. Phys. Chem.*, 95 (1991) 5970.

150. R. Rajagopalan, A. Aoki, and A. Heller, *J. Phys. Soc.*, 100 (1996) 3719.

151. R. Foster and J. Voss, *J. Electrochem. Soc.*, 139 (1992) 1503.

152. A. Aoki, R. Rajagopalan, and A. Heller, *J. Phys. Chem.*, 99 (1995) 5102.

153. A. Aoki and A. Heller, *J. Phys. Chem.*, 97 (1993) 11014.

154. N. Mano, V. Soukharev, and A. Heller, *J. Phys. Chem. B*, 110 (2006) 11180.

155. D. Blauch and J.-M. Saveant, *J. Am. Chem. Soc.*, 114 (1992) 3323.

156. D. Blauch and J.-M. Saveant, *J. Phys. Chem.*, 97 (1993) 6444.

157. X. Zeng, W. Wei, L. Xuefang, Z. Jinxiang, and L. Wu, *Bioelectrochemistry*, 71 (2007) 135.

158. T. Tamaki, T. Ito, and T. Yamaguchi, *J. Phys. Chem. B*, 111 (2007) 10312.

159. T. Tamaki, T. Ito, and T. Yamaguchi, *Ind. Eng. Chem. Res.*, 45 (2006) 3050.

160. N. Tsubokawa, *Prog. Polym. Sci.*, 17 (1992) 417.

161. H. Mizuhata, S. Nakao, and T. Yamaguchi, *J. Power Sources*, 138 (2004) 25.

162. W. Shin et al., *J. Am. Chem. Soc.*, 118 (1996) 3202.

163. A. Cole, D. Root, P. Mukherjee, E. Solomon, and T. Stack, *Science*, 273 (1996) 1848.

164. C. V. Dominguez et al., *Biosens. Bioelectron.*, 24 (2008) 531.

165. T. Chen, S. C. Barton, G. Binyamin, Z. Gao, Y. Zhang, H.-H. Kim, and A. Heller, *J. Am. Chem. Soc.*, 123 (2001) 8630.

166. Y. Yan, W. Zhen, and L. Su, *Adv. Mater.*, 18 (2006) 2639.

167. W. Zheng, Q. Li, L. Su, Y. Ya, J. Zhang, and L. Mao, *Electroanalysis*, 18 (2006) 2639.

168. R. Duma and S. D. Minteer, *Polym. Mater. Sci. Eng.*, 94 (2006) 592.

169. D. Ivnitski, B. Branch, P. Atanassov, and C. Apblett, *Electrochem. Commun.*, 8 (2006) 1204.

170. S. C. Barton, H. H. Kim, G. Binyamin, Y. Zhang, and A. Heller, *J. Phys. Chem. B*, 105 (2001) 11917.

171. S. C. Barton, M. Pickard, R. Vázquez-Duhalt, and A. Heller, *Biosens. Bioelectron.*, 17 (2002) 1071.

172. N. Mano, F. Mao, W. Shin, T. Chen, and A. Heller, *Chem. Commun.*, 4 (2003) 518.

173. N. Mano, F. Mao, and A. Heller, *J. Am. Chem. Soc.*, 125 (2003) 6588.

174. H.-H. Kim, N. Mano, Y. Zhang, and A. Heller, *J. Electrochem. Soc.*, 150 (2003) A209.

175. N. Mano, F. Mao, and A. Heller, *J. Am. Chem. Soc.*, 124 (2002) 12962.

176. S. Ferguson-Miller and G. T. Babcock, *Chem. Rev.*, 96 (1996) 2889.

177. S. Tsujimura, K. Kano, and T. Ikeda, *Electrochemistry*, 70 (2002) 940.

178. S. Tsujimura, K. Kano, and T. Ikeda, *J. Electroanal. Chem.*, 576 (2005) 113.

179. E. Katz and I. Willner, *J. Am. Chem. Soc.*, 125 (2003) 6803.

180. E. Katz, O. Lioubashevski, and I. Willner, *J. Am. Chem. Soc.*, 127 (2005) 3979.

181. S. Shima et al., *Science*, 321 (2008) 572.

182. G. J. Kubas, *Chem. Rev.*, 107 (2007) 4152.

183. K. A. Vincent, J. A. Cracknell, O. Lenz, I. Zebger, B. Friedrich, and F. A. Armstrong, *Proc. Natl. Acad. Sci. U. S. A.*, 102 (2005) 16951.

184. G. Goldet et al., *J. Am. Chem. Soc.*, 130 (2008) 11106.

185. K. A. Vincent, A. Parkin, and F. A. Armstrong, *Chem. Rev.*, 107 (2007) 4366.

186. M. Y. Darensbourg, E. J. Lyon, and J. J. Smee, *Coord. Chem. Rev.*, 206/207 (2000) 533.
187. W. Habermam and E.-H. Pommer, *Biotechnology*, 35 (1991) 128.
188. A. K. Jones, E. Sillery, S. P. J. Albracht, and F. A. Armstrong, *Chem. Commun.*, (8) (2002) 866.
189. A. A. Karyakin, S. Morozov, E. E. Karyakina, S. D. Varfolomeyev, N. A. Zorin, and S. Cosnier, *Electrochem. Commun.*, 4 (2002) 417.
190. M. A. Alonso-Lomillo et al., *Nano Lett.*, 7 (2007) 1603.
191. K. Vicent et al., *Chem. Commun.*, (48) (2006) 5033.
192. G. T. R. Palmore, H. Bertschy, S. H. Bergens, and G. M. Whitesides, *J. Electroanal. Chem.*, 443 (1998) 1809.
193. G. Palmore, H. Bertschy, S. Bergens, and G. Whitesides, *J. Electroanal. Chem.*, 443 (1998) 155.
194. N. L. Akers, C. M. Moore, and S. D. Minteer, *Electrochim. Acta*, 50 (2005) 2521.
195. M. D. Arning, B. L. Treu, and S. D. Minteer, *Polymer. Mater. Sci. Eng.*, 90 (2004) 566.
196. M. C. Beike and S. D. Minteer, *Polym. Materials Sci. Eng.*, 94 (2006) 556.
197. T. S. Thomas, K. E. Ponnusamy, N. M. Chang, K. Galmore, and S. D. Minteer, *J. Membr. Sci.*, 213 (2003) 55.
198. W. J. Blaedel and R. A. Jenkins, *Anal. Chem.*, 47 (1975) 1337.
199. D. Zhou, H. Fang, H. Chen, V. Ju, and Y. Wang, *Anal. Chim. Acta*, 329 (1996) 41.
200. C. M. Moore, N. L. Akers, A. D. Hill, Z. C. Johson, and S. D. Minteer, *Biomacromolecules*, 5 (2004) 1241.
201. R. L. Arechederra, B. L. Treu, and S. D. Minteer, *Power Sources*, 173 (2007) 156.
202. D. Sokic-Lazic and S. D. Minteer, *Biosens. Bioelectron.*, 24 (2008) 939.
203. D. Das and T. N. Veziroglu, *Int. J. Hydrogen Energy*, 26 (2001) 13.
204. R. Allen and H. Bennetto, *Appl. Biochem. Biotechnol.*, 39 (1993) 27.
205. D. Park and J. Zeikus, *Biotechnol. Bioeng.*, 81 (2003) 348.
206. S. Chaudhuri and D. Lovley, *Nat. Biotechnol.*, 21 (2003) 1229.
207. J. Niessen, U. Schroeder, and F. Schloz, *Electrochem. Commun.*, 6 (2004) 955.
208. H. J. Kim, H. S. Park, M. S. Hyun, I. S. Chang, M. Kim, and B. H. Kim, *Enz. Microb. Technol.*, 30 (2002) 145.
209. S. Tsujimura, M. Fujita, H. Tatsumi, K. Kano, and T. Ikeda, *Phys. Chem. Chem. Phys.*, 3 (2001) 1331.
210. L. M. Tender et al., *Nat. Biotechnol.*, 20 (2002) 821.

CHAPTER 9

Electrocatalysis at Bimetallic Surfaces Obtained by Surface Decoration

HELMUT BALTRUSCHAT, SIEGFRIED ERNST, and NICKY BOGOLOWSKI

Abteilung Electrochemic, Universität Bonn, D-53117 Bonn, Germany

9.1 INTRODUCTION: WAYS OF ACTION OF COCATALYSTS

The acceleration of electrocatalytic reactions by the combined action of two (or more) metallic elements has been known for a long time [1] and raises the interest for the application of such catalysts (e.g., in fuel cells).

One distinguishes three different ways of action of the cocatalysts: a geometric (ensemble) effect, an electronic (ligand) effect, and an effect according to the bifunctional mechanism. These effects will be exemplified for the adsorption and oxidation of CO and methanol in Figure 9.1.

Wherever an ensemble of more than one active atom is necessary for the adsorption process, the geometric effect will occur. In the case of methanol (reacting to adsorbed CO) it is generally accepted that three to four Pt atoms are necessary for the accommodation of the methanol molecule [2]. This is the reason for the inactivity of PtSn surfaces for methanol oxidation and also for the fact that Pt Ru alloys with a low Ru content are best for methanol oxidation.

In the case of a ligand effect the electronic properties of the substrate are modified by the second component such that either the adsorption energy (strength of interaction with the first component) is modified or the activation energy is changed (usually decreased). Such an electronic effect may have two origins: a lateral strain exerted by the second metal on the primary active metal, causing a broadening and lowering (or narrowing and increase) of the energy of the surface d-band of the primary metal (Pt) [3]; or the second component has a direct influence on the electronic states of the first metal (true electronic effect) [4, 5].

An enhancement of the oxidation rate according to the bifunctional effect is believed to occur for the oxidation of adsorbed CO on Ru-modified Pt surfaces: CO

Catalysis in Electrochemistry: From Fundamentals to Strategies for Fuel Cell Development,
First Edition. Edited by Elizabeth Santos and Wolfgang Schmickler.
© 2011 John Wiley & Sons, Inc. Published 2011 by John Wiley & Sons, Inc.

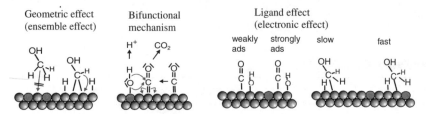

FIGURE 9.1   Sketch of ways of action for cocatalysts (dark spheres).

reacts with an adsorbed oxygen species, which is more abundant on Ru (or adsorbed at lower potentials on Ru) than on Pt.

In the following, we shall briefly discuss the most often used techniques for the study of bimetallic catalysts; then, preparation of model systems with an emphasis on step decoration will be reviewed. Finally, as model reactions, hydrogen evolution and oxidation and the oxidation of CO, methanol, and ethanol on such bimetallic surfaces will be treated.

9.2   EXPERIMENTAL TECHNIQUES

Experimental techniques involve those needed for the characterization of the catalyst itself, those for the identification of adsorbates and intermediates, and those for the identification of products. A detailed discussion of all methods is, of course, not possible here, and the reader is referred to the references given.

9.2.1   Catalyst Characterization of Single Crystals

The ultimate techniques for the characterization of model electrode surfaces certainly are those typically used in ultrahigh voltage (UHV) [6–8]. There, electron spectroscopic methods like Auger electron spectroscopy (AES), X-ray photoelectron spectroscopy (XPS), and ultraviolet photoelectron spectroscopy (UPS) can be used. AES is a comparatively simple technique yielding information on the atomic composition of the catalyst; furthermore it is very sensitive for the detection of surface contaminations; XPS, in addition, allows for a better quantification of the atomic composition of the top layer(s) but also a determination of the oxidation state. UPS, on the other hand, determines the binding energy of electrons in the conduction band and therefore is sensitive to electronic changes.

Due to the limited mean free path of the escaping electrons, these techniques are largely surface sensitive. However, spectra obtained still correspond to several atomic layers, for which, using tabulated atomic sensitivity factors, average concentrations are obtained. In order to obtain true surface concentrations, a special calibration procedure has to be used [9, 10, 11]. A more advanced technique for the determination of the surface composition is LEIS (low-energy ion-scattering spectroscopy, also called ISS—ion, scattering spectroscopy) [2, 12]. Here noble gas ions such as He^+ or alkaline ions are used; they collide with the surface atoms and loose part of their impact energy. The amount of this energy and impact loss depends on the atomic mass at the surface atom, thus allowing its identification.

For single-crystalline model surfaces, also the atomic structure has to be known, and in UHV this is obtained by LEED (low-energy electron diffraction) or RHEED (reflection high-energy electron diffraction) [13].

When using UHV techniques for the characterization of electrode surfaces, the surface is prepared in UHV (which involves cleaning by Ar^+ sputtering and annealing), characterized for cleanliness and surface order, then transferred into an electrochemical cell for surface modification and electrocatalytic experiments, and then transferred back to UHV to control surface modification, structure, and possibly molecular adsorbates formed during the electrocatalytic reactions [14–18]. Several problems are to be considered:

(a) A special transfer system has to be used which ensures that the transfer occurs without any contamination. At a first glance, this seems not to be difficult in light of the fact that electrodes, for example, prepared according to the Clavilier method in ambient atmosphere, can be easily transferred between different cells through air without contamination. Experience, however, shows that upon transfer of an electrode into the UHV system contamination of the surface is much more difficult to avoid, in particular for catalytically active surfaces. One reason seems to be that upon evacuation of the transfer chamber the protecting droplet of electrolyte evaporates and the surface is dry. When such a clean transfer, however, is ensured, one has the additional advantage that, due to preparation (annealing) of the surface in UHV, its state before the experiment is much better known, which is particularly important for alloys.

(b) Is the surface after emersion and transfer identical to that before in the electrolyte? After emersion, a considerable amount of electrolyte usually stays on the surface except in those cases when the surface is hydrophobic. After drying, this leads to high surface concentrations of those species on the surface which are not volatile. The surface concentration therefore does not correspond to their concentration in the double layer. Lu et al. [19] estimated that the thickness of the emersed electrolyte is $10 \, \mu m$ and, therefore, in order to avoid such effects, the electrolyte concentration must not exceed $10^{-4} \, M$. The emersed electrolyte layer also may lead to further effects: The chemical potential during drying drastically changes, and therefore chemical reactions might occur in that layer. In particular, oxo-anions such as perchlorate or sulfate might oxidize organic compounds or less noble metallic components of bimetallic surfaces.

The standard method for the in situ determination of the surface and adsorbate structure is STM (scanning tunnelling microscopy) [20–22]. It has been applied with great success for the determination of the structure of adsorbate layers, including metallic adlayers with a cocatalytic effect such as Ru and Sn [23, 24]. Different elements, however, can be distinguished only in very special cases, such as Pd and Au in an AuPd alloy surface [25]. Surface X-ray scattering (SXS) also has been used for the determination of the surface of bimetallic model catalysts [26–28].

9.2.2 Catalyst Characterization: Nanoparticles

In the context of fuel-cell-related research, nanoparticles are usually supported on active carbon. Also in model studies, for example, those involving rotating-disc electrodes (RDEs) or DEMS (differential electrochemical massspectroscopy), such carbon-supported catalysts were used as a thin catalyst layer on an inert substrate such as glassy carbon [29, 30]. Transmission electron microscopy (TEM) is used for the determination of their size distribution and a possible agglomeration of the nanoparticles, for example, during the electrochemical experiments. In optimum cases, atomic resolution is obtained, revealing the single-crystalline structure of these particles. Elemental analysis is possible by TEM or SEM (scanning electron microscopy) using electron energy loss spectroscopy (EELS) or energy-dispersive X-ray spectroscopy (EDX) or classical elementary analysis. An average size is also obtained from XRD measurements from the line width (FWHM, fixed width at half maximum). A big problem, however, is the distribution of the elements within a nanoparticle. The method of choice here is EXAFS (extended X-ray absorption fine structure), but because of the extensive fitting procedures, results are not always unequivocal [31–34].

9.2.3 Adsorbate, Intermediate, and Product Characterization

A determination of the electrochemical reactivity without knowing what the product is is not really helpful in the development of better catalysts. In the test of fuel cell catalysts for ethanol oxidation, for example, it has to be ascertained that the catalyst yields the final product CO_2 and not only acetaldehyde or acetic acid. Furthermore, a method for the online product determination also should allow working under convection or high transport conditions. Although in principle determination of products by chemical means is also possible after reaction in a batch reactor for some time, this is disadvantageous for two reasons: A fast determination in parallel to cyclic voltammetry helps in the interpretation; in order to determine the kinetics of a reaction, in a batch reactor, an analysis of products would have to be done as a function of time.

Most often used for the insitu *online* determination of electrochemical products are DEMS and infrared spectroscopy.

For DEMS, coupling to a mass spectrometer via a porous Teflon membrane (Gore Tex[TM]) can be achieved with a variety of electrochemical cells, discussed in detail in elsewhere [35]. Best suited for desorption experiments at smooth electrodes is the thin-layer cell shown in Figure 9.2(a) [36, 37]. A massive electrode with a diameter of 1 cm is separated from the porous hydrophobic Teflon membrane, which is mechanically supported by a steel frit, via a 50–100-μm-thick Teflon ring spacer. Species produced at the electrode diffuse to the Teflon membrane through the thin electrolyte layer

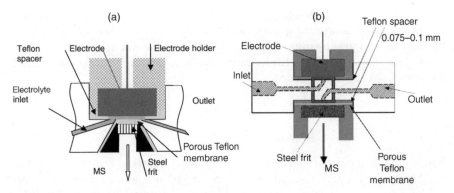

FIGURE 9.2   (a) One-compartment thin-layer cell and (b) dual-compartment thin-layer cell for experiments under constant flow-through conditions.

within 1 or 2 s. In this arrangement all volatile species enter the mass spectrometer and the collection efficiency N is nearly 1.

Faradaic reactions, in which the reactant is continuously consumed (so-called bulk reactions), should be better performed under a continuous flow of electrolyte The "dual thin-layer cell" shown in Figure 9.2(b) is better suited in this case [38]. Here, the complete cell consists of two separate compartments, the upper one for the electrochemistry and the opposite, lower one for the mass spectrometric detection. The electrolyte first enters the upper thin-layer compartment in front of the electrode ("wall-jet" geometry) and then flows through one of four (or six) capillaries into the lower compartment ("detection cell"), where volatile products can cross the Teflon membrane. Despite the geometry, the convective–diffusion behavior in this cell is largely governed by the thin layer and the channel type with a dependence of the diffusion-limited current an the flow rate u given by $u^{1/3}$ [39]. This cell type can also be coupled with other methods such as quartz crystal microbalance or even insitu detection of adsorbates by surface-enhanced infrared spectroscopy as recently described by Behm and co-workers [40, 41].

Coupling electrodes in the hanging-meniscus arrangement with a mass spectrometer has become possible with the help of pinhole inlets [42, 43]; however, a disadvantage with this arrangement is the more complicated diffusion behavior. Local resolution of electrode activity may be achieved using a scanning DEMS (S-DEMS) [44, 45].

Quantification can be achieved by calibration of the mass spectrometer with either the pure substances or a defined mixture (gas mixtures or mixtures of volatile species) of the investigated substances. The relation between the detected ion current and the flux of the species is

$$I_i = K^0 \frac{dn_i}{dt} \tag{9.1}$$

where $dn_i/dt = J_i$ is the rate of species i entering the mass spectrometer and K^0 is the calibration factor containing all settings of the mass spectrometer and the ionization probability of the corresponding species. Due to electron impact ionization (EI), fragmentation of the ionized species usually occurs. Therefore possible interference by fragments originating from different products has to be corrected for.

When the species is produced electrochemically, I_i is given by the faradaic current I_F corresponding to that process:

$$\frac{dni}{dt} = \frac{NI_F}{zF} \tag{9.2}$$

where z is the number of electrons, F is the Faraday constant, and N is the overall collection efficiency, that is, the ratio of the amount of species entering the mass spectrometer and the total amount of species produced electrochemically. The value of N may be less than 1 because a fraction of the produced species does not enter the mass spectrometer. Thus, the detected ion current is finally linked to the faradaic current in the electrochemical cell by the calibration constant K^*:

$$I_i = \left(\frac{K^*}{z}\right) I_F \quad K^* = \frac{K^0 N}{F} \tag{9.3}$$

Whereas in the one-compartment cell N is close to 1 because of the stagnant electrolyte, in the dual thin-layer flow-through cell it is dependent not only on the flow rate but also on the volatility of the compounds to be detected. If diffusion of the product within the capillaries at the outlet is fast and flow is in the range of $1\ \mu L\ s^{-1}$, it is identical to the thin-layer collection efficiency f_2 of the second compartment. However, usually diffusion is too slow and the flow rate is higher. Therefore, the product concentration is different at the upper part (electrode) of the thin layer and the lower one. Because of the laminar flow, this difference persists also at the outer and inner side of the capillary and the lower (Teflon membrane) and upper part of the detection compartment (incomplete mixing) [39]. Careful calibration therefore is necessary. For CO_2, typical experimental values for f_2 and N are $N=f_2=0.9$ at $1\ \mu L\ s^{-1}$ and $N=0.6, f_2=0.37$ at $10\ \mu L\ s.^{-1}$

The usual approach for performing IR spectroscopy at electrode surfaces involves reflection of the IR beam at the surfaces from the solution side and measurement of the absorption bands (infrared reflection absorption spectroscopy, IRRAS) nowadays using FTIR instruments [46]. During the measurement, the electrode is positioned onto a window, typically a CaF_2 prism. Only a very thin electrolyte layer of a few micrometers ensures potential control. Due to the strong absorbance of water, difference spectra (of spectra obtained at two different potentials) have to be calculated. Contributions from the electrolyte and from the electrolyte can in part be distinguished by using polarized light—only p-polarized light has a component of the electric field normal to the surface which does not vanish. Components parallel to the surface vanish in its proximity at a distance of $\lambda/4$. This technique therefore is not truly surface sensitive. Adsorbates are usually distinguished from solution species from the potential dependence of their vibration frequencies. Only species with a component of the dynamic dipole moment normal to the surface are IR active.

The extremely thin electrolyte layer is a severe disadvantage. Measurements under reaction conditions, that is, with current flow, are not possible because of the large electrolyte resistance and IR drop in the thin layer and the complete depletion of the reactants. Only the latter problem can be circumvented by drilling a hole into the IR prism, which allows a continuous flow of electrolyte [47].

The alternative to performing IR measurements at electrode surfaces is the ATR (attenuated total reflection) configuration. Here, the IR beam is internally reflected by a thin metallic layer deposited onto an IR prism. The intensity of the electromagnetic field at the opposite side of the metal film, that is, the electrode surface, is sufficiently high so that adsorbed molecules can be vibrationally excited. Single crystals cannot be used, but the thin film can have a preferential "quasi-single crystalline" orientation. Some roughness of the electrode increases sensitivity due to an electromagnetic enhancement. (SEIRAS—surface-enhanced IR absorption spectroscopy). This method can nicely be coupled to DEMS [40, 41].

Due to the low cross section of the Raman process, the electromagnetic enhancement in surface Raman spectroscopy is even more important (SERS—surface-enhanced Raman spectroscopy). It can only be used for metals in which the conduction electrons behave like free electrons, that is, Au, Ag, or Cu. But it can be used for d-metals as well if a thin metallic overlayer is deposited on a roughened Au surface [48, 49].

9.3 PREPARATION OF BIMETALLIC MODEL SURFACES

A prerequisite in fundamental mechanistic studies is to know the atomic composition of the surface. *Ex-situ* techniques are often employed to determine the surface composition of Pt–Ru electrodes [50, 51]. For a more detailed elucidation of the mechanism of action of bimetallic surfaces also the atomic distribution has to be known.

In UHV, ordered bimetallic surfaces formed from ordered alloy superlattices help in the understanding of the role of geometric and electronic effects. Examples are ordered intermetallic compounds such as Ni_3Al [52] and Pt_3Sn [53–55]. Preparation and control of the atomic surface distribution, however, are difficult and limited to certain combinations of elements. In electrochemistry, this approach was used for ordered Pt–Sn alloys for studying methanol and CO oxidation [56–59].

An alternative is to use small amounts of a second component so that most of the atoms of this second component are monomers. Such surfaces can be obtained by either depositing a monolayer of the alloy from a corresponding solution (e.g., Au and Pd) [25] on the pure substrate or by incorporating small amounts of a second metal into the surface by chemical deposition followed by slight annealing, for example, Pd or Rh into Pt [60, 61]. A statistical analysis using STM data may also be correlated to coverage or catalytic data provided that a chemical contrast exists between the different metals [25].

Quite another possibility is the deposition of a complete monolayer of a second metal, for example, by UPD (underpotential deposition, i.e., the deposition at potentials positive of its equilibration potential for bulk deposition) or the galvanic displacement or surface-limited reduction reaction (SLRR) of a less noble UPD metal like Cu or Pb by another, for example, the replacement of a Cu UPD monolayer by Pt [62], which is used in electrochemical atomic layer epitaxy (ECALE). The catalytic properties of such a surface may differ considerably from those of the corresponding bulk metal, either because of a direct electronic influence of the substrate or the different lattice parameters: That of the monolayer is often that of the substrate; this leads to a shift of the d-band center as compared to the bulk metal [63] For Pd on Au and other surfaces, the effect on hydrogen adsorption and evolution has been thoroughly studied by Kibler et al. [64–67].

As an alternative for obtaining ordered bimetallic surfaces, step decoration of regularly stepped single - crystal surfaces vicinal to low - indexed planes by a second metal can be used. Equidistant rows of the second element alternating with the substrate element are thus obtained. It has been shown that such regularly stepped surfaces can be prepared not only in UHV [68–70] but also under ambient conditions by flame annealing and subsequent cooling in a hydrogen-containing Ar atmosphere [71–73]. By using such a reducing atmosphere, facetting, which occurs in an atmosphere containing traces of oxygen, is avoided [6, 74]. Stepped surfaces vicinal to the Pt(111) plane with (111)- and (100)-oriented steps and vicinal to the Pt(100) plane with (111)- and (110)-oriented steps thus have been prepared. In the case of the Pt electrodes, the quality of the preparation is most easily controlled by recording a cyclic voltammogram in sulfuric acid. (In perchloric acid, peaks in cyclic voltammograms often appear less sharp although — or better because — perchlorate ions are not adsorbed.)

Decoration of steps on electrode surfaces has been reported for many systems [75]. Step decoration was established by Clavilier et al. for Bi on Pt(332) [76]. In the case of Pt single-crystal electrodes, preferential deposition at step sites leads to the suppression of the characteristic hydrogen adsorption peaks corresponding to the step sites [77, 78]. Nishihara et al. examined in detail the underpotential deposition of Cu on stepped Pt(111) surfaces with steps of local (100) orientation; they prepared the decorated steps by successive dissolution of a full monolayer of Cu after electrolyte exchange in a Cu-free solution [79]. Abruna and co-workers [80] deposited submonolayer amounts of Cu (just enough for a complete step decoration) from very diluted Cu^{2+} solutions (10^{-5} M) on Pt(111) stepped surfaces with (111)-oriented steps, (Pt(S)-[n(111) × (111)]). From these results and others (e.g., for Pt(332)/Bi [78], Pt(S)-[n(111) × (100)]/Cu [79], Pt(S)-[n(111) × (111)]/Cu [80], Au(111)/Cu [81], Au(111)/Pb [82], Au(100)/Pb and Ag(100)/Pb [83], Sn, Ru, and Mo on Pt(332), Pt(665), and Pt(755)) [84, 85], it is obvious that preferential adsorption of UPD metals at step sites is a common phenomenon on stepped surfaces. Our own preliminary results, however, indicate that on stepped Pt surfaces with (100) terraces this might not be true in general.

As an example, Cu UPD on Pt(332) will be discussed in detail (Figure 9.3). The inset shows the UPD peaks on a Pt(332) electrode observed during cyclic voltammetry [86]. The sweep rate had to be lowered to 1 mV s^{-1} due to the slowness of the Cu adsorption/desorption process. It is tempting to identify the most positive of these peaks with Cu desorption from step sites, the larger at 0.55 V with Cu UPD at the (111) terraces and that at 0.95 V, which is close to the start of bulk deposition, to Cu

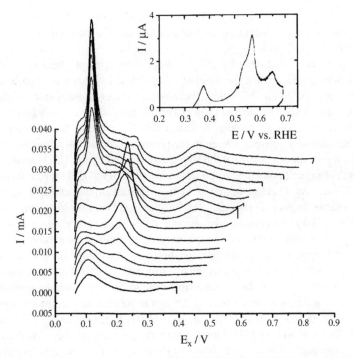

FIGURE 9.3 Dissolution of monolayer of Cu on Pt(332) in 0.5 M H$_2$SO$_4$ by potential sweeps with increasing anodic of potential limit; $v_{scan} = 20$ mV s^{-1}. Inset: Cyclic voltammogram of Pt(332) in 0.5 M H$_2$SO$_4$, 2 × 10^{-4} M CuSO$_4$; $v_{scan} = 1$ mV s^{-1}; $T = 31$ °C. (From ref. 86.)

at the outermost row. This has been verified in experiments in which a monolayer of Cu was dissolved in a solution free of Cu^{2+} ions by sweeping to increasing anodic potential. When the anodic potential limits reach 550 and 600 mV, changes become visible in the hydrogen region. These potentials correspond to the peaks in the cyclic voltammogram of this surface (cf. inset), where due to the presence of Cu^{2+} the peak potentials are approximately 50 mV higher. When Cu is desorbed from the outermost row (at 550 mV), a peak appears at 230 mV, which must not be confused with the hydrogen adsorption peak at steps with a local (100) orientation. The maximum of the charge corresponding to this additional peak (II) almost has the same magnitude as that of the peak (I) (at 120 mV) due to hydrogen adsorbed at the steps. It has been ascribed either to induced anion or hydrogen adsorption at Pt terrace atoms adjacent to the Cu steps [80, 86, 87]. A similar peak also appears in perchlorate solutions or in the presence of chloride ions [88]. Also for other UPD metals, like Ag [89, 90] or even Sn [86], this peak is observed at the same potential.

Buller et al. [80] report on the decoration of the steps of a variety of Pt single crystals with an n (111) × (111) orientation with Cu-UPD, and in ref. [86], where the Cu coverage was varied, it seems that at least two rows of Pt atoms of the terrace have to be uncovered by the UPD metal in order for this extra peak to occur. The number of free Pt rows above that minimum of two rows does not seem to play a role for the charge. Interestingly, the peak potential is not very dependent on the anion [88] (for 110 steps) and completely independent of the decorating metal. An induced adsorption of hydrogen is improbable because Cu should rather decrease the interaction with hydrogen.

Displacement of hydrogen by anion adsorption is ruled out because in the case of the nondecorated surface the amount of adsorbed hydrogen is already close to zero at the peak potential of 230 mV. Therefore, one has to assume that the charge of this peak is determined by the adsorption of anions not displacing adsorbed hydrogen. The potential, however, seems to be determined by the hydrogen adsorption isotherm: Only when during a potential sweep in the positive direction the terraces are free of hydrogen do the anions adsorb at the step site decorated by the UPD metal.

Whereas the peak potential largely is independent of the UPD metal, its charge is not: For Cu and Ag it is close to that of the hydrogen adsorption at steps, but for other metals, notably Te [91], Tl [88], Bi [91], Sn [86], and Ge, it is only a small fraction.

In a typical UPD process the adsorption and desorption process is reversible; that is, in the limit of sufficiently low sweep rates the potentials of the anodic and cathodic process are identical. This is the case for the above-mentioned metals. However, in many other cases, metals and in particular semimetals adsorb at underpotentials, and often [mostly on Pt (111)] show a reversible peak in the cyclic voltammogram which is due not to adsorption/desorption of the metal but to a surface redox process or adsorption of OH. This is true for Se, Bi, Te [91, 92], Ge [93], Sn [23, 84, 94], Mo [95], W [96], and As [97]. Other metals are also irreversibly adsorbed but do not show any surface redox peak, such as Ru and Os. They are also preferentially adsorbed at steps, but the peak for the redox process is only observed for the (metal) species adsorbed at (111) terraces, not for that at the step site.

Whereas for vicinal Pt(111) surfaces step decoration seems to occur for all sorts of UPD systems [not much is known for Au(111)], for stepped surfaces vicinal to Pt(100) step decoration might even be impossible. A first study for Cu UPD on vicinally stepped Pt(100) showed that Cu is preferentially adsorbed not at step sites but rather at terrace sites. A reason could be that the difference in the number of coordinating atoms at the surface is smaller: 3 at terraces versus 5 at step edges at Pt(111) vicinal surfaces and 4 at terraces versus 5 at Pt(100) vicinal surfaces with (111) step orientation. Another reason could be a strong anion adsorption at these steps. A similar result was found for Au deposited at Pt [98].

Even if on these surfaces deposition occurs preferentially on terraces they might be used for the generation of ordered bimetallic surfaces. Such surfaces could even be particularly interesting because the reactive Pt step sites are still available.

Step decoration on other metal substrates is less obvious. Certainly STM can be (and often has been) used for the examination of metal deposition of submonolayer amounts of a second metal. However, the resolution within a terrace is limited due to the finite size of the tip. Whereas in UHV step decoration of vicinally stepped surfaces has been observed [99], the authors are not aware of a similar report for regularly stepped electrode surfaces.

But step decoration is better visible at steps of nominally flat surfaces: Here deposition often is starting at such steps, and this may be taken as a strong indication that also step decoration of regularly stepped surfaces occurs. An example is given in Figure 9.4 for Pd deposition on Au(111).

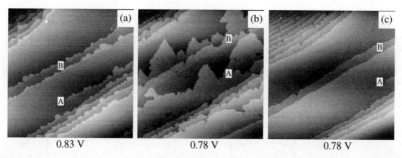

FIGURE 9.4 STM images of Pd deposition at Au(111) in 0.1 M H_2SO_4 + 0.1 mM $PdSO_4$. Image area $250 \times 250\,nm^2$, $E_{tip} = 1.18$ V. All potentials vs. Pt/PtO. (a) Start of deposition (step decoration); (b) further growth of Pd layer; (c) full monolayer of Pd. The dislocation A and the small terrace B are marks to keep track of the positions of the sample regardless of drift (from) [100].

Such a result, however, does not necessarily mean that also a step decoration by a single row of the second component is possible or that the absence of a visible step decoration means that a decoration by a single row is not occurring. First, step decoration might depend on step orientation, which usually is not known for steps on flat surfaces; also a step decoration might only be stable if more than one row of atoms is deposited. Second, a single row of atoms might be decorating a step, although a smooth 2D deposition is not initialized at the steps.

An example for the latter case is Ru. Ru deposition at Pt(111) surfaces has been thoroughly examined by STM. Ru is deposited in 2–3-nm large islands, most of which, but not all, are one monolayer high [101, 102]. The reason for this behavior could be that, due to the lattice mismatch, larger islands would involve attachment of Ru to low-coordinated Pt sites and would show a Moiré-pattern, which seems to be unfavorable in this case. Rather, further Ru is deposited in a second layer.

With the STM, no indication of a preferential deposition or indication of the deposition at steps is observed. Nevertheless, cyclic voltammetry clearly shows that hydrogen adsorption at steps (peak for step sites at 270 mV) is suppressed at vicinally stepped Pt(111) surfaces after deposition of small amounts of Ru (Figure 9.5) [85]. Here, Ru deposition was achieved from a deaerated solution of 0.005 M ruthenium in 0.5 M H_2SO_4 for 5 min at a potential of 570 mV following the procedure of Cramm et al. [103]. Deposition under these conditions should lead to a Ru coverage of around 20%.

Whereas on stepped Pt single-crystal electrodes the occurrence of step decoration is unequivocally revealed from the suppression of the corresponding hydrogen peaks in cyclic voltammetry, step decoration on vicinal surfaces other than Pt has to be confirmed independently.

In the case of Pd decorating the steps of surfaces vicinal to Au(111) this confirmation was performed in a series of experiments:

Deposition of submonolayer amounts of Pd was achieved by sweeping the potential to values close to the bulk deposition potential. Due to the slow irreversible deposition of Pd, a clear UPD peak is not visible in those cyclic voltammograms. Deposition

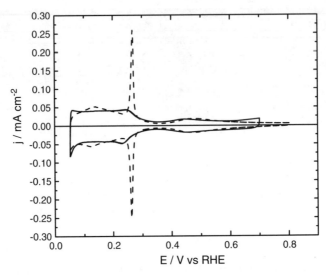

FIGURE 9.5   CV of Pt(755) in 0.5 M H₂SO₄ before (- - - - -) and after (——) Ru deposition; $dE/dt = 50$ mV/s. (From ref. 85.)

was stopped after deposition of the desired amount by emersion of the electrode and transfer to an electrolyte not containing Pd.

Figure 9.6 shows the cyclic voltammograms for a Au(332) surface onto which increasing amounts of Pd were deposited [100].

Interestingly, no hydrogen adsorption peak is observed when the amount of Pd roughly corresponds to only one row of atoms. Only when the amount corresponds to

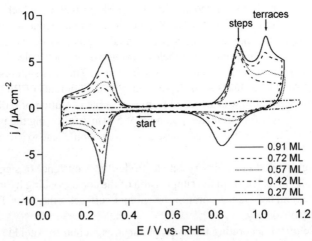

FIGURE 9.6   First voltammetric cycle for Au(332)/Pd in 0.1 M H₂SO₄ with different Pd coverage after five cycles between 0.5 and 0.1 V. Starting potential 0.5 V in the cathodic direction, scan rate $v = 5$ mV s⁻¹. Pd was deposited from PdSO₄ solution. (After ref. 104.)

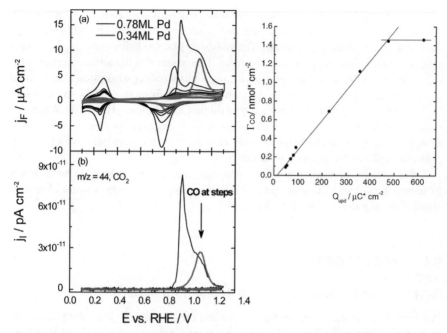

FIGURE 9.7 (a) Stripping voltammograms for adsorbed CO at Pd modified Au(332). ($E_{ad} = 0.35$ V, $\nu = 10$ mV s^{-1}, 0.10 M H$_2$SO$_4$). (b) Corresponding mass spectrometric ion current for CO$_2$. (c) CO coverage (obtained from mass spectrometric ion charge) as function of Pd coverage.

two atomic rows decorating the steps does an adsorption peak for hydrogen become visible, which increases with increasing amount of Pd. Similarly, a peak for oxygen adsorption only appears when two rows of Pd are formed. With an increasing amount of Pd deposited, this peak does not further increase; instead, a second peak for oxygen adsorption evolves. Obviously, the first peak corresponds to step sites, whereas the second peak corresponds to oxygen adsorption at terraces sites. The absence of hydrogen adsorption for a monoatomic row is different from the observation of Maroun et al., who find that for hydrogen adsorption on Pd embedded in a Au(111) surface already a Pd dimer is sufficient, while a single Pd atom is sufficient for CO adsorption [25].

How do we know that in the first CV in Figure 9.6 a row of Pd is really decorating the steps? Figure 9.7 shows an experiment in which CO was adsorbed on electrodes modified by different amounts of Pd [104,105].

The amount of CO adsorbed cannot be simply elucidated from the cyclic voltammograms because in parallel to CO oxidation oxygen is adsorbed. Instead, DEMS was used for the elucidation of the CO coverage [Figure 9.7(b)]. Figure 9.7(c) shows that the CO coverage increases linearly with the Pd coverage and is zero at zero Pd coverage. On the contrary, hydrogen adsorption for an identically prepared surface only starts at Pd coverages above that corresponding to step decoration.

Interestingly, CO at steps is oxidized at more positive potentials than CO at terraces, although Kibler et al. [106] had shown that at surfaces free of steps CO oxidation is shifted to more positive potentials. The reason is that oxidation only occurs at step sites, to which CO molecules have to diffuse. This means that the activation energy for CO oxidation is decreased at step sites despite of its stronger adsorption as compared to terrace sites. [The situation is somewhat different from stepped Pt surfaces like Pt(755), in particular when adsorbed from dilute CO solutions [107–110]. When CO is oxidized on such a surface, in IR measurements the corresponding band for CO at steps disappears first; obviously oxidation starts at the steps where CO is most probably replaced by an oxygen species.]

Other species, in particular ethene, propene, and benzene, also have been shown to adsorb at such monoatomic rows on which hydrogen adsorption is not observed, confirming step decoration [105, 111].

9.4   CASE STUDIES

9.4.1   Hydrogen Evolution at Pd-Modified Au Surfaces

Whereas for practical purposes there is no real need to replace Pt as the catalyst for hydrogen oxidation in fuel cells, for fundamental purposes the study of hydrogen evolution on other bimetallic surfaces is important: Since it is the simplest electro-catalytic reaction, from this reaction much might be learned which is valid also for other reactions.

Stimming and co-workers [112] observed that single Pd clusters deposited on Au(111) substrates displayed an enormous increase in electrochemical activity for H_2 evolution. This rate was determined by using the tip of the scanning tunnel-ing microscope for the detection of the amount of H_2 evolved employing a careful numerical analysis. The increase by several orders of magnitude as compared to bulk Pd was ascribed to a spillover of hydrogen atoms formed at Pd onto the Au part of the surface, where they would recombine.

Because of our interest in hydrogen adsorption on Au surfaces which also seems to be important in the case of formaldehyde oxidation [113, 114], we used step decoration of vicinally stepped Au(111) electrodes to elucidate the role of the 1D interface between Au and Pd for the hydrogen evolution rate [104, 115, 116].

The activity of such surfaces was measured by recording the current as a function of low overpotentials, where diffusion limitation can be neglected (Figure 9.8).

Whereas no reaction is occurring at such low overpotentials at the bare Au elec-trode, step decoration with 0.2-monolayer Pd leads to an activity which no more increases with increasing coverage of Pd up to 1 monolayer. Instead, at a higher coverage, the activity decreases again. Obviously, the boundary between Au and Pd, which is determined by the step length and does not vary when the coverage is increased, is necessary for such a high rate; this seems to support the assumption of a rate-limiting hydrogen spillover from Pd onto Au. Another interpretation, however, is also possible: Strong hydrogen adsorption at monoatomic rows does not occur, as was shown above. According to the Sabatier principle [117], this might be the reason for

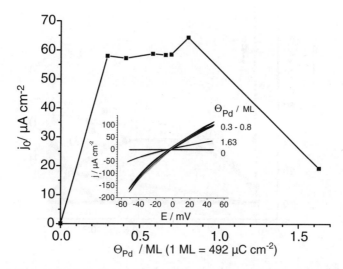

FIGURE 9.8 Exchange current densities for H_2 evolution on Pd/Au(332) as function of Pd coverage. Inset: Current–voltage curves in 0.1 M H_2SO_4 saturated with H_2, 25 mV s^{-1} (no effect of scan rate was found) [104, 116].

the enhanced activity for hydrogen evolution: When an intermediate is too strongly adsorbed, its desorption is slow and limits the overall rate. In the case of very weak adsorption, the first step, the adsorption of the reactant, is too slow and limiting the overall reaction. In this respect, Pd atoms at step sites in proximity to Au might just have an optimum adsorption energy.

Similar results were obtained by Kibler [65]: Exchange current densities for the hydrogen evolution reaction over submonolayers of Pd on Au(111) were reported which were essentially independent of Pd coverage above $\Theta_{Pd} = 0.2$ (Figure 9.9). He ascribed this to the varying coverage-dependent contributions of the free surfaces of Au and Pd and (at Pd coverages close to 1) of Pd–Pd steps. The major contribution, however, was assumed to be due to the high catalytic activity of the border line between Pd islands and the Au substrate. Astonishingly, he also found a very high and sharp maximum of the exchange current density for Pd coverages below 0.1, which is not yet well understood.

9.4.2 The Oxidation of Adsorbed CO on Sn-Modified Pt Surfaces

The promoting effect of Sn on CO oxidation has long been known [118]. The large pseudocapacitive currents observed particularly for Sn adsorbed on Pt(111) [94] and $Pt_3Sn(111)$ [57, 119] and which are caused by surface oxidation of Sn, make a direct examination of this effect by cyclic voltammetry somewhat difficult. However, with the help of DEMS these oxidation charges can well be separated from the oxidation charge from adsorbed CO.

Figure 9.10 shows that on Pt(111) adsorbed Sn hardly has an influence on the oxidation of coadsorbed CO. Also on Pt (332) the main oxidation peak is hardly shifted, but the usual prepeak, which corresponds to the oxidation of so-called weakly

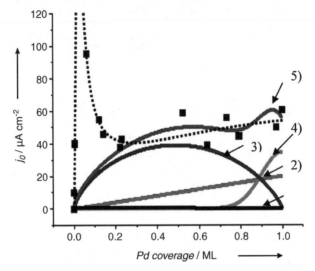

FIGURE 9.9 Exchange current density j_0 for hydrogen evolution over Pd submonolayer deposits on Au(111) as a function of coverage. Contributions from various active sites in dependence of the Pd coverage are included: (1) uncovered Au(111) surface, (2) Pd islands), (3) Pd/Au sites at the perimeter of the Pd islands, (4) Pd/Pd step sites, and (5) sum of the above contributions . Experimentally obtained data points (squares) are connected with a dotted line as guide to the eye. (From ref. 65.)

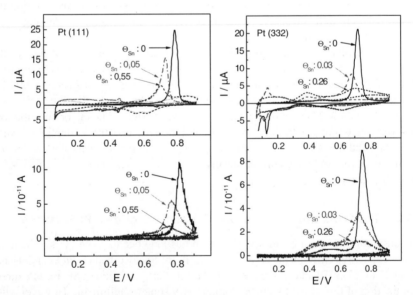

FIGURE 9.10 Top: Cyclic voltammograms of CO oxidation on Sn-modified Pt(111) (left) and (facetted) Pt(332)$_{fac}$ (right) in a thin-layer cell: 0.5 M H$_2$SO$_4$; 10 mV s^{-1}. CO was adsorbed at 70 mV (RHE). Sn coverages as indicated. Bottom: ion currents for CO$_2$ (corrected for different sensitivities of the DEMS system). (Data from refs. 86, 94, and 120.)

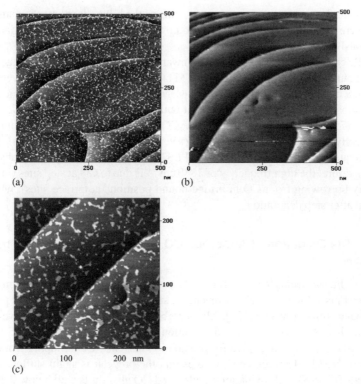

FIGURE 9.11 STM images of CO adsorption on Sn-covered Pt(111) surface in CO-saturated 0.05 M H_2SO_4, Θ_{Sn}: (a) adsorption of CO at 0.2 V, tunneling current 1 nA, bias voltage 50 mV (apparent height of islands 0.22 nm); (b) as (a) but after oxidation of CO, tunneling current 1.1 nA, bias voltage 50 mV. (c) detail of (a). [From ref. 23.]

adsorbed CO on pure Pt, is not only shifted from 550 to 450 mV but also largely increased due to coadsorbed Sn. Up to 50% of CO_{ad} is forced into the weakly adsorbed state. This suggests that the action of Sn is by an electronic effect, which destabilizes adsorbed CO. Similar oxidation in two peaks was also observed on Pt(111)–(2 × 2) Sn but ascribed to the surface oxidation of Sn (bifunctional mechanism) [121]. However, particularly on such an ordered alloy, one would expect oxidation of all CO as soon as the surface OH species becomes available, that is, in one peak, if the bifunctional mechanism would dominate.

The small effect of Sn on Pt(111) is explained by the formation of Sn islands and the large distance from the CO molecules to the Sn domains [23]. These Sn domains are visible in STM only after coadsorption of CO, which is a further indication of the repulsion between CO and Sn (cf. Figure 9.11). After oxidation of CO, the islands disappear again, although Sn remains on the surface. On the CO free surface, Sn atoms are not visible because they are too mobile.

The large effect on Pt(332) is due to the fact that Sn decorates the steps and is thus forced into a more homogeneous distribution over the surface. In the literature, it is commonly believed that Sn acts according to the bifunctional mechanism. If this really were the case, all adsorbed CO should be oxidized at a lower potential, which is obviously not the case.

In IR, coadsorbed Sn leads to a downward shift of the stretching frequency for on-top bound CO by $5\,cm^{-1}$ on both Pt(111) and Pt(332). For low Sn coverages on Pt(111), a band splitting indicates the presence of domains which are influenced by Sn and those which are not, in agreement with the above STM results [107, 109].

On Pt(755), a band for bridge-bonded CO at $1880\,cm^{-1}$ (which is astonishingly high) nearly disappears and is replaced by a very weak band around $1830\,cm^{-1}$, which is the frequency observed at Pt(111) and Pt(332) (in both the presence and absence of Sn). Therefore, the high frequency of $1880\,cm^{-1}$ is due to CO at step sites (including intensity borrowing from CO in bridge-bound positions at terrace sites), which are blocked after step decoration

9.4.3   The Oxidation of Adsorbed CO on Mo and W-Modified Pt Surfaces

At first sight, Mo seems to act as a cocatalyst very similar to Sn. CO oxidation on Mo-modified Pt is sufficiently fast to prevent the surface from being completely poisoned in methanol reformate gas [122]. Mo adsorbed on Pt is oxidized at an even lower potential than Sn, giving rise to an oxidation peak at 0.4 V. Again, DEMS helps in distinguishing pseudocapacitive from oxidation currents. CO stripping from a Mo-modified Pt(332) electrode demonstrates that the prepeak is again shifted to lower potentials (Figure 9.12); oxidation of adsorbed CO also starts well below 0.2 V.

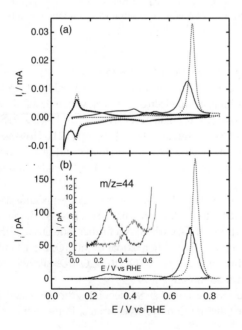

FIGURE   9.12 Simultaneously measured CV (a) and MSCV (b) $m/z = 44$, for oxidation of preadsorbed CO on Pt(332) ($\cdots\cdots$) and Mo-modified Pt(332) ($\Theta_{Mo} = 0.17$)(———) in 0.5 M H_2SO_4 solution in the dual thin-layer cell. Potential scan rate: $10\,mV\,s^{-1}$, electrolyte flow rate $5\,\mu L\,s^{-1}$ (inset: expanded current scale for low-potential region). (Data from ref. 95.)

The position of the main peak is hardly shifted. In contrast to the case of Sn, however, the amount of CO oxidized in the prepeak is not increased but stays below 10% for all Mo coverages and at different surfaces [polycrystalline Pt, Pt(111)], simultaneously sputter-deposited Pt–Mo) [84, 95]. We therefore believe that Mo does act according to the bifunctional effect, but this is only true at potentials at which Mo is not yet oxidized, that is, below 0.5 V. The effect is not large enough to shift the main oxidation peak to a value below 0.5 V, and therefore it is only effective for the weakly adsorbed CO. A similar effect was observed for W on polycrystalline Pt [96].

9.4.4   Oxidation of Adsorbed CO on Ru-modified Pt Surfaces

The possibility of step decoration by Ru was demonstrated above for Pt(665) and Pt(755); it leads to the suppression of the hydrogen adsorption peak for step sites (at 125 or 270 mV, respectively) in the cyclic voltammogram [85].

Oxidation charges are much larger on surfaces partially covered by Ru than on pure Pt surfaces. By measurements with an electrochemical quartz crystal microbalance combined with DEMS, we have shown before that a PtRu (50:50) alloy electrode is covered by an anionic species even at 100 mV [38]. This species, probably oxygen or hydroxide, is displaced upon adsorption of CO. It is, however, readsorbed when the CO is oxidized, leading to an additional charge for oxidation even after background subtraction. For a lower integration limit of 0.28 mV, an additional charge of 40% was thus obtained for the PtRu alloy electrode, whereas for a pure polycrystalline electrode only 20% of the oxidation charge (after background subtraction) is due to this double-layer charging.

Oxidation of CO adsorbed at these surfaces occurs in oxidation peaks which are shifted negative by 200 mV with respect to the corresponding Ru-free surface, as observed with polycrystalline PtRu alloys, Ru adlayers on polycrystalline Pt [51, 123, 124], and Pt(111) electrodes modified by Ru [101–103, 125–127] (Figure 9.13).

The slow rise of the current above 0.4 V overlaps with the prepeak observed on the Ru-free surfaces and is also paralleled by CO_2 formation, as was revealed by DEMS for faceted Pt(332) [84]. The peak is much narrower than without Ru, but a second peak is visible around 0.6 V on the Ru-covered Pt(665) electrode. This has also been observed for Pt(111) partially covered by small amounts of Ru. Also for Pt(775), the oxidation peak is much narrower after step decoration by Ru; however, no clear shoulder is observable at its high-potential side.

We did not observe this shoulder for the Ru-modified Pt(332) surface cooled in Ar/H_2; only when the Pt(332) crystal was cooled in pure Ar (leading to facetting) did the CV look similar to that of Pt(665) cooled in Ar/H_2, in accordance with the similar step width.

Using electrochemical mass spectrometry we showed that both peaks at (facetted) Pt(332)$_{fac}$ and Pt(111) correspond to the oxidation of CO_{ad} to CO_2 [84]. We also showed (as many others afterwards [128–130]) that the CO corresponding to the first peak can be oxidized separately from that in the second peak when, during cyclic voltammetry, the potential is stopped at the onset of the first peak for several minutes: The second peak becomes visible in the subsequent continuation of the

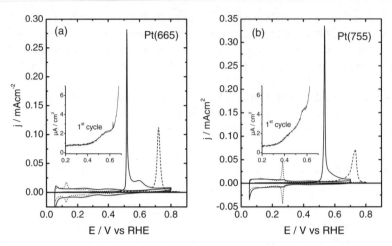

FIGURE 9.13 CV of oxidation of adsorbed CO in CO-free electrolyte at Ru step-decorated electrodes (——), (a) Pt(665) and (b) Pt(755). For comparison: anodic stripping of CO from the Ru-free electrode (- - - - -) (1st cycle) and (·····) (2nd cycle); $0.5\,M\ H_2SO_4$, $dE/dt = 10\,mV\,s^{-1}$. Inset: 1st cycle for CO oxidation at Ru-free electrode. (Data from ref. 85.)

anodic potential sweep. The existence of two adsorbate states for CO was confirmed in galvanostatic experiments.

Our explanation for these two different adsorption sites is the following: Due to the electronic influence of Ru, the adsorption enthalpy in the neighborhood of Ru is increased (i.e., its absolute value is decreased) also as suggested by thermodesorption experiments in UHV [131] and by DFT calculations [132]. (cf. Figure 9.14). This influence extends over all the terrace of the Pt(332) crystal, that is, over at least four rows of atoms. In the case of Pt(665) [and Pt(332)$_{fac}$ cooled in Ar], about one-third of the adsorption sites are not influenced. The two peaks in the cyclic voltammogram correspond to these different adsorption sites. Because of its lower stability, CO

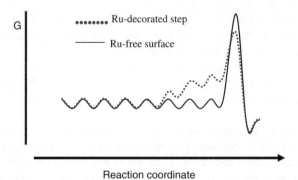

FIGURE 9.14 Influence of Ru step decoration on the adsorption enthalpy of CO$_{ad}$ with respect to the reaction coordinate (schematically): (·····) for Ru-decorated step; (——) for Ru-free surface.

adsorbed in the neighborhood of Ru is oxidized in the first peak. CO adsorbed at a greater distance from Ru only becomes unstable (with respect to its oxidation to CO_2) at potentials of the second peak. These molecules then diffuse to Ru (where OH species are adsorbed) via the higher energy sites (as an intermediate state). Oxidation of these CO molecules adsorbed at sites not influenced by Ru still occurs at somewhat lower potentials than at Ru-free electrodes because the activation barrier is decreased due to the increased availability of OH on Ru according to the bifunctional mechanism.

Figure 9.15 summarizes our view of the cocatalytic effect of Mo, Sn, and Ru for CO oxidation [95]. Here, the free adsorption enthalpies of CO in the weakly adsorbed and strongly adsorbed state are considered as being independent of potential, whereas the free enthalpy of the product ($CO_2 + 2H^+ + 2e^-$) is. The (apparent) free enthalpy of the activated state is potential dependent, due to the electron transfer involved in the reaction and/or due to the potential dependence of the coverage with the reaction partner OH_{ad}. (Please note that this refers to the overall reaction $CO_{ad} + H_2O \rightarrow CO_2 + 2H^+ + 2e^-$, and thus the free enthalpy of adsorption of the reaction intermediate OH_{ad} is part of the free enthalpy of activation.) The reaction in a CV will start when the free enthalpy of activation becomes lower than some critical value G_{crit}, that is, when the lines for the difference $G^{\neq} - G_{crit}$ and the line for the free enthalpy of the weakly or strongly adsorbed CO intersect.

For Mo, only in the low-potential range is the activation barrier decreased due to the oxygen spillover effect. This decrease of the activation barrier therefore is only active for the weakly adsorbed state with its higher free enthalpy. In the case of Sn, the population of the weakly adsorbed state is increased due to a repulsion between Sn and CO. The above STM study showed, that due to this repulsion, Sn adsorbed at Pt(111) is forced into islands only when coadsorbed with CO. (In addition, a relatively small decrease of the activation energy due to the spillover effect may be possible).

For Ru, a short-range electronic effect is effective (only in the neighbourhood of Ru) in addition to the oxygen spillover effect (which lowers the free activation enthalpy for all adsorbed CO). The spillover effect leads to a decrease of the free activation energy, but only above a certain potential. The oxidation potential for the weakly adsorbed state is not lowered: Note that below 0.45 V, that is, at the onset of the prepeak (oxidation of the weakly adsorbed CO), the curves for the bare Pt(332) and the Pt(332) surface partially covered by Ru overlap. Maybe, an OH adsorption at Ru only starts at this potential (0.45 V) and therefore the free activation enthalpy is only decreased above this potential. Another possibility is the following: Before OH adsorption on Ru can set in, some of the CO molecules adsorbed on Ru have to be oxidized. As long as a weakly adsorbed CO exists on Pt, there is a competition between weakly adsorbed CO (on Pt) and OH from solution for Ru sites. Only when the surface pressure of CO from Pt is reduced due to possible oxidation in the prepeak can OH adsorb on Ru. (This effect may only be clearly visible at low surface contents of Ru.)

9.4.5 The Oxidation of Adsorbed CO on Epitaxial Pt–Rh Alloy Surfaces on Pt(100)

Finally, we present an example for CO oxidation on a surface alloy obtained by forced deposition in a reducing hydrogen atmosphere followed by annealing. In the

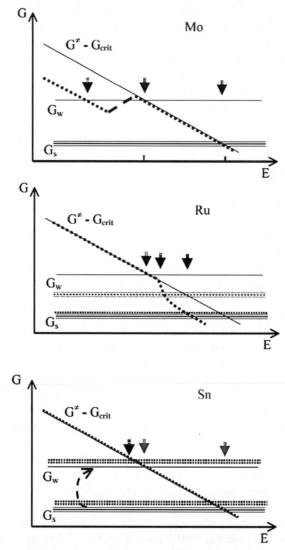

FIGURE 9.15 Schematic sketch of cocatalytic effect of Mo (top), Ru (middle), and Sn (bottom) on electrochemical CO oxidation. Grey lines refer to pure Pt. The free enthalpy of the activated state minus the critical enthalpy of activation is, for simplicity, assumed to decrease linearly with the electrode potential. The free enthalpies of weakly and strong adsorbed CO, are given as G_w and G_S respectively. The short arrows indicate the intersection of this line with the free enthalpy of adsorbed CO; this is the potential at which the activation enthalpy is identical to G_{crit} and therefore the potential at which the oxidation of CO starts in the CV. The larger, broken arrow in the bottom figure indicates the change of population from the strongly adsorbed state to the weakly adsorbed state due to Sn.

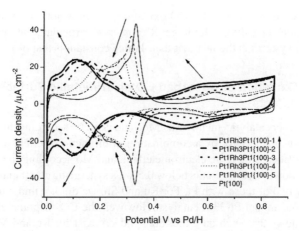

FIGURE 9.16 Formation of PtRh{100} alloys on Pt{100} by flame annealing Pt1Rh3Pt1 sandwich layers. Test solution: 0.1 M H$_2$SO$_4$. Sweep rate: 20 mV s^{-1}. Arrows indicate changes associated with increasing rhodium amount at the surface. Integers indicate the number of flame-annealing steps undertaken. (From ref. 61.)

example of Figure 9.16, Rh was deposited three times on Pt(100) from a 10^{-5} M RhCl$_3$ solution; then Pt was deposited followed by slight annealing [61]. This procedure leads to smooth, epitaxial alloy layers on Pt. Additional annealing steps result in a decrease of the Rh content in the surface layer, as seen from the continuous change in the hydrogen adsorption peaks. As can be seen from Figure 9.17, this leads to a large

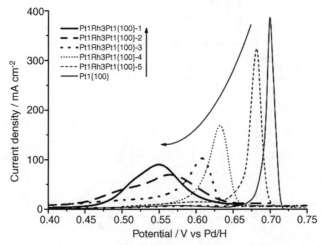

FIGURE 9.17 Voltammograms of CO stripping on Pt–Rh surface alloys formed by flame annealing of Pt1Rh3Pt1 sandwich adlayers on Pt{100}. Test solution: 0.1 M H$_2$SO$_4$. Sweep rate: 20 mV s^{-1}. Arrows indicate changes associated with increasing rhodium amount at the surface. Integers indicate the number of times the surface has been flame annealed. (From ref. 61.)

decrease of the onset potential for CO oxidation, which is not observed when Rh is deposited without annealing or for larger Rh contents. Particularly interesting in these epitaxial alloy systems is the fact that their lattice constant is that of Pt.

9.4.6   The Oxidation of Adsorbed CO on Trimetallic Surfaces: Synergistic Effects

Since the three cocatalytic metals Sn, Mo, and Ru act according to three different mechanisms, namely Ru by a combination of bifunctional effect plus electronic effect, Sn only according to an electronic effect, and Mo according to a bifunctional effect (which, however, only works below 0.5 V), a synergistic effect may be expected on combining two of these with Pt. For Ru and Sn, we did not find a true synergistic effect for our model Pt(332) catalyst. However, the CO-stripping experiment in Figure 9.18 shows that such an effect does indeed exist for Ru and Mo: Here, the steps of the Pt(332) crystal were first decorated by Ru; then Mo was deposited to

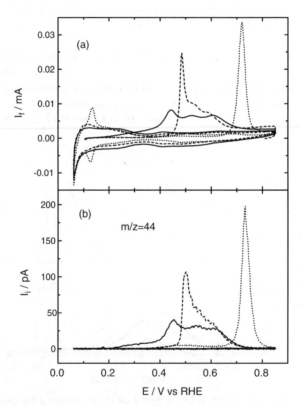

FIGURE 9.18   Simultaneously measured CV (a) and MSCV (b), $m/z = 44$, for oxidation of preadsorbed CO on Pt(332) ($\cdots\cdots$), Ru-modified Pt(332) ($\Theta_{Ru} = 0.2$) (——), and Ru- and Mo-modified Pt(332) ($\Theta_{Ru} = 0.2$, $\Theta_{Mo} = 0.07$) (——) in 0.5 M H$_2$SO$_4$ solution in dual thin-layer cell. Potential scan rate: 10 mV s^{-1}, electrolyte flow rate: 5 μL s^{-1}. (Data from ref. 95.)

achieve a surface coverage of $\Theta_{Mo} = 7\%$. Clearly not only is the onset of CO oxidation decreased by 0.2 V [with respect to both the unmodified Pt(332) and the Ru modified Pt(332)], but also the potential of the main peak is somewhat decreased (by 50 mV) with respect to the surfaces modified by Ru. This is understandable on the basis of the above-mentioned mechanisms: the main peak on the Ru-modified surface is at 0.5 V, a potential at which Mo is not yet completely oxidized. Even if the bifunctional effect of Mo may be added to only the electronic effect of Ru, this leads to the observed effect: For simplicity we assume that the potential changes due to the different actions can be added to give the total change in oxidation potential. Then we can conclude from Figure 9.18 that on Ru-modified Pt(332) the total change consists of the bifunctional effect of 130 mV (peak at 0.6 V) and the electronic effect of 100 mV (peak at 0.5 V). In the presence of Mo, the bifunctional effect amounts to about 200 mV (taken from the shift of the prepeak in Fig. 9.18), a combination of this with the electronic effect of Ru gives a total of 300 mV, and the main peak therefore should be shifted from 730 down to 430 mV, in close agreement with the observation. Interestingly, this synergistic effect could be observed also on Ru- and Mo-modified nanoparticles supported on active carbon [95].

9.4.7 The Oxidation of Adsorbed CO on Nanoparticles

Two oxidation peaks were also observed when Pt nanoparticles were decorated with Ru [129]. Again, CO corresponding to the first peak can be separately oxidized, and the more stable CO in the second peak "survives". In contrast to the previous case where Pt single-crystal surfaces with (111)-oriented terraces were decorated by Ru, the second peak coincides with that for CO oxidation at pure Pt (see Fig. 9.19) [133]. We therefore assume that the origin of the peak splitting is different: The second peak corresponds to CO adsorbed at pure Pt faces of the nanoparticles, whereas the first is due to CO oxidation at faces modified by Ru. At the low average Ru coverage of about 20%, Ru deposition does not occur homogeneously; rather some faces or even complete particles remain completely unmodified.

9.4.8 Methanol Oxidation at Ru-Modified Pt Single Crystal Electrodes

Before discussing the effect of Ru as a cocatalyst for methanol oxidation, the reaction mechanism will briefly be summarized (see Fig. 9.20): According to the generally accepted parallel-path reaction mechanism, one reaction path leads to adsorbed CO, which is then further oxidized to CO at sufficiently high potentials but acts as a catalyst poison at lower potentials. The other reaction path leads to formaldehyde and formic acid. In the absence of fast convection, these are also further oxidized to CO_2. When convection is fast, though, they are transported into the electrolyte without further reaction. This results in low overall current efficiencies for CO_2 between 20 and 50%, depending on surface structure, composition, and concentration but not on a further increase of the convection, that is, the electrolyte flow rate (actually, this low, but

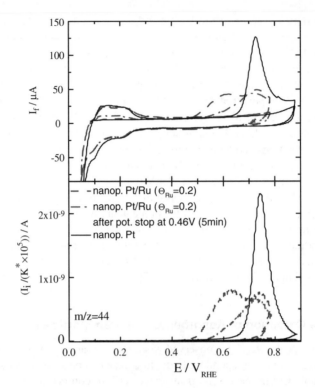

FIGURE 9.19 Simultaneously measured faradaic currents and corresponding ion currents of $^{12}CO_2$, $m/z = 44$, during oxidation osf ^{12}CO adsorbed on (a) Ru-modified polycrystalline Pt, (b) Ru-modified Pt nanoparticles, and (c) Ru- and Mo-modified Pt nanoparticles in 0.5 M H_2SO_4 (dashed line). Scan rate: 5 mV s^{-1}. Electrolyte flow rate: 5 µl s^{-1}. Dash-dotted line: after oxidation of the first peak for 5 min, then going back to 100 mV. Full line: for comparison; oxidation of ^{12}CO adsorbed on pure Pt surfaces. (From ref. 133.)

finite current efficiency could be taken as a proof that both reaction paths really play a role) [30, 134, 135].

Ruthenium is not only a good cocatalyst for CO oxidation, but so far it also seems to be the best cocatalyst for methanol oxidation. Cyclic voltammograms on Ru-modified Pt(111) and Pt(332) are shown in Figure 9.21. The main current peak is not changed

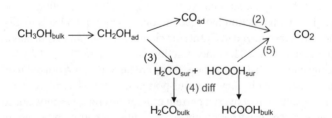

FIGURE 9.20   Sketch of reaction pathways for methanol oxidation.

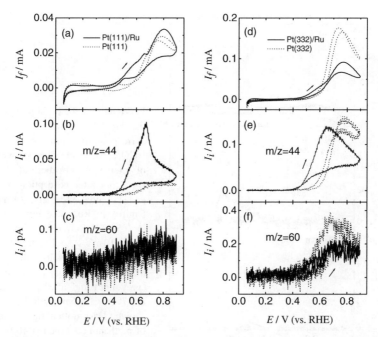

FIGURE 9.21 Simultaneously recorded CVs (a, d), MSCVs of $m/z = 44$ (i.e., carbon dioxide) (b, e) and MSCV of $m/z = 60$ (i.e., methylformate) (c, f) on Pt(111), Pt(332), Pt(111)/Ru, and Pt(332)/Ru electrodes in 0.1 M methanol + 0.5 M H_2SO_4 solution; $v = 10 \, mV \, s^{-1}$. Electrolyte flow rate: 5 $\mu L \, s^{-1}$. Arrows: direction of potential sweep. (From ref. 134.)

on Pt(111) due to Ru deposition and even is reduced on Pt(332). This is because the Ru sites are blocked by OH at the more positive potentials and therefore are not active. More important, however, is that a shoulder appears below 0.6 V. The ion current for CO_2 shows that at Ru-modified Pt the current efficiency for the formation of CO_2 largely increases [to 1 at Pt(111)/Ru and to 0.6 at Pt(332)/Ru]. Since Ru is deposited as 2D islands on Pt(111), the effect is more pronounced due to the introduction of steps. On the one hand, the higher current efficiency for CO_2 is due to the higher CO oxidation rate. On the other hand, we have already shown that Ru also enhances the methanol adsorption rate (formation of adsorbed CO) [135]; both effects lead to a preference of the reaction path via adsorbed CO. Whereas the presence of steps increases the rate of the first common reaction step and therefore the rate of both reaction paths, the reaction path via dissolved intermediates is not increased by Ru.

In the previous section it was shown that CO oxidation is influenced by an electronic effect (leading to destabilization of adsorbed CO) and the bifunctional mechanism (decreasing the activation barrier). Since the first is active only in close vicinity of Ru whereas the second, due to surface diffusion of CO, is a long-range effect, two oxidation peaks were observed. How is the oxidation of the methanol adsorption product influenced by Ru? Figure 9.22 shows a stripping experiment after adsorption of methanol for 2 min at 0.4 V and electrolyte exchange under continuous flow

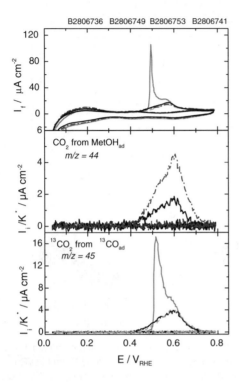

FIGURE 9.22 Simultaneously measured faradaic currents and corresponding ion currents of $^{12}CO_2$, $m/z = 44$, during oxidation of methanol adsorbed at 400 mV for 2 min after adsorption of ^{13}CO and potential stop at 370 mV for 5 min on $Ru_{0.2}$-decorated Pt(665) in 0.5 M H_2SO_4 (black line). Scan rate: 10 mV s^{-1}. Electrolyte flow rate: 5 μL s^{-1}. Additionally, $^{13}CO_{ad}$ oxidation without potential stop (grey line) and after potential stop at 370 mV for 5 min (dashed line) is shown. Dash-dotted line: oxidation of methanol adsorbed at 400 mV for 2 min. The current is referred to the true Pt surface area. (From ref. 133.)

(dash-dotted line). In another experiment, ^{13}CO was adsorbed at 0.05 V, partially oxidized at 0.37 V, and, after electrolyte exchange to ^{12}C methanol solution at 0.05 V, methanol was adsorbed at 0.4 V for 2 min. After a further electrolyte exchange for supporting electrolyte at 0.05 V, the resulting adsorbate was stripped (black line). The ion current for $m/z = 44$ shows that some but not much methanol was coadsorbed, but the ion current for $m/z = 45$ also shows that the amount of CO in the second peak is hardly influenced. (The dashed line represents the corresponding experiment without coadsorption of methanol after oxidation of the first peak, the grey line the CO stripping experiment of the complete CO monolayer.) This result shows that the methanol adsorbate only occupies the more stable adsorption sites, that is, those at some distance to Ru. Longer adsorption times or lower adsorption potentials did not lead to an increase in adsorbate coverage. In the literature it was sometimes reported that adsorbate coverages (on polycrystalline Pt) obtained from methanol were as high as those obtained from CO. This, however, may be due to the fact that such experiments usually are performed without any forced convection, and adsorption can occur from formaldehyde or formic acid which are formed as intermediates and are accumulated in the electrolyte.

Oxidative stripping of the methanol adsorption product at Ru-modified Pt nanoparticles is shown in Figure 9.23. Not only the onset potential but also the peak potential shifts in the anodic direction with decreasing CO (i.e., methanol adsorbate) coverage. Also on these surfaces it can be assumed that the different adsorption sites have

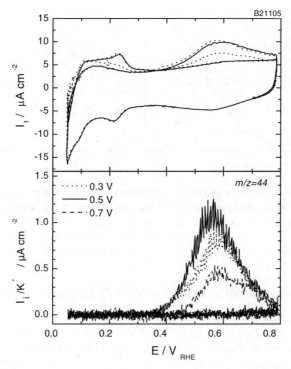

FIGURE 9.23 Simultaneously recorded faradaic current and corresponding ion current of CO_2, $m/z = 44$, during oxidation of methanol adsorbate on Ru ($\theta \cong 0.2$)–modified reduced Pt oxide particles in 0.5 H_2SO_4 solution. Scan rate: 10 mV s^{-1}. Flow rate of electrolyte: 5 μL s^{-1}. Methanol adsorption for minutes at potential: 500 mV (solid line), 300 mV (dotted line), and 700 mV (dashed line). Catalyst loading: 10 μgPt cm^{-2}. The current is referred to the true Pt surface area determined by CO adsorption. (From ref. 133.)

different adsorption energies and Pt sites are occupied preferentially, whereas those in close vicinity to or on Ru are occupied only at high coverage but oxidized first upon anodic stripping. Note that these differences are clear from the mass spectrometric cyclic voltammograms but hardly visible from cyclic voltammetry alone due to high background currents of the carbon support.

Figure 9.24 demonstrates that the performance of such nanoparticles is better than that of commercial PtRu alloy nanoparticles. The main effect of Ru is the increase in current efficiency; at high potentials the current is hardly increased compared to pure Pt, but the onset is decreased similar to what was seen for the Ru-modified Pt(332) surface.

A positive synergistic effect of Mo and Ru (compared to the bimetallic PtRu nanoparticles) was also found in the case of methanol oxidation using DEMS by Behm et al. [136] with cyclic voltammetry by Lin et al. [137] and also on sequentially electrodeposited PtRuMo particles on HOPG surfaces by Cabrera et al. [138].

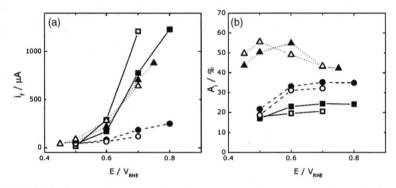

FIGURE 9.24 Dependence of (a) faradaic currents and (b) current efficiencies for CO_2 on potential during potentiodynamic (filled) and potentiostatic (open) methanol oxidation on reduced commercial Pt oxide nanoparticles (squares, surface area $A_{exp} = 4.4 \, cm^2$), PtRu alloy nanoparticles (circles, $A_{exp} = 6.1 \, cm^2$), and Ru ($\theta \cong 0.2$)–modified reduced Pt oxide particles (triangles, $A_{exp} = 4.64 \, cm^2$) in 0.1 M methanol + 0.5 M H_2SO_4 solution. (From ref. 133.)

9.4.9 Ethanol Oxidation at Ru-Modified Pt Single-Crystal Electrodes

Ethanol is even more interesting as a fuel in fuel cells because of its possible production from plants and because the infrastructure for proliferation would not need to be changed much. Its oxidation, on the other hand, at moderate temperatures is even more difficult than that of methanol: In addition to the C–H bond, the C–C bond has to be cleaved. Similar to methanol, one oxidation pathway via adsorbed intermediates and one via dissolved intermediates exist. Here the final product of the route via acetaldehyde is acetic acid, which itself is (nearly) completely unreactive, whereas in the case of methanol the corresponding product (formic acid) is oxidized further to CO_2 if the residence time in proximity of the surface is high enough. Therefore the route via adsorbed intermediates is even more important.

What are the adsorbed intermediates? In the DEMS experiments shown in Figure 9.25, ethanol was allowed to adsorb from 0.1 M solution in sulfuric acid at 0.3 V for a few minutes; after an electrolyte exchange, the potential was swept in the negative or positive direction. During a negative sweep, CH_4 is evolved, demonstrating the presence of an adsorbed CH_x species [139, 140]. Such an adsorbed CH fragment was recently also identified by SERS on Pt electrodeposited on Au electrodes [49]. A decomposition of an adsorbate with an intact C–C bond only during the negative sweep, however, cannot be excluded, similar to what was discussed for adsorbed ethane before [140–142]. A corresponding acetyl species was observed by ATR-SEIRAS [143]. In the subsequent sweep in the positive direction, CO_2 is formed at a potential where usually CO_{ad} is oxidized. In the experiments in Figure 9.25(b), isotopically labeled ethanol was used, and a sweep was started in the positive direction. The peak around 0.8 V again is typical for CO_{ad}, which obviously is formed from both C atoms. At more positive potentials, where typically hydrocarbons are oxidized, the formation of only $^{12}CO_2$ shows that only the carbon atom from the methyl group is oxidized in this potential region, in accordance with the above-mentioned CH_x

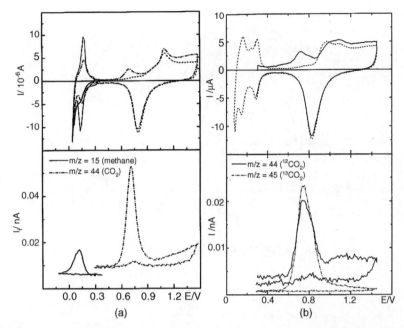

FIGURE 9.25 (a) Cathodic desorption of preadsorbed ethanol (0.1 M) from Pt(110) and subsequent oxidation of remaining adsorbate: $E_{ad} = 0.3$ V, $v = 0.0125$ V s^{-1}; 0.1 M H$_2$SO$_4$. (Top) CV: (——) first sweep in cathodic direction; (–·—·–) subsequent oxidation; (·····) Pt(110) in supporting electrolyte; (– – –) CV after a few cycles between 0.06 and 1.45 V. (Bottom) MSCV. (b) Oxidation of preadsorbed ethanol-(1-^{13}C, 10^{-2} M) on polycrystalline Pt: $v = 0.0125$ V s^{-1}, 0.1 M H$_2$SO$_4$. (Top) CV: (—) first sweep; (- - - - -) CV in supporting electrolyte. (Bottom) MSCV. (From refs. 139 and 140.)

species. Important is the fact that some of the methyl groups are oxidized to adsorbed CO. This is probably a slow follow-up reaction from CH$_x$ to CO, as indicated in the reaction scheme

$$CH_3CH_2OH \xrightarrow{K_{ad}} CO_{ad} + CH_{x,ad} \begin{array}{c} \xrightarrow{k_{CH}} \\ \xrightarrow{k_{CO}} CO_2 \end{array}$$

Crucial is the current efficiency for CO$_2$ during bulk oxidation of ethanol. Much differing values have been reported in the literature. Unfortunately, both CO$_2$ and acetaldehyde give a signal on $m/z = 44$. In principle, by measuring the ion current for $m/z = 29$ (the HCO fragment), the contribution of acetaldehyde on the $m/z = 44$ signal can be calculated, albeit with a relatively large error. Better is the use of deuterated ethanol, the corresponding ^{4}d-acetaldehyde has its molecular peak at $m/z = 48$ (instead of 44). Such an experiment is shown in Figure 9.26.

The CV has the usual shape with an oxidation peak around 0.7 V. After the first sweep, the surface is roughened, and the current in the subsequent sweeps increases

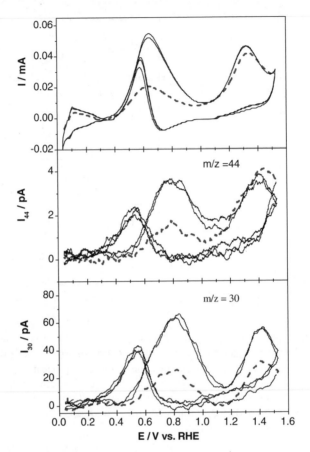

FIGURE 9.26 Electrooxidation of 10^{-2} M C_2D_5OD on Pt(332) in 0.1 M $H_2SO_4 + 0.5$ M $HClO_4$ at 10 mV s^{-1} and flow rate of 1.6 μL s^{-1}. Dashed line: first sweep, $m/z = 44$ corresponds to CO_2 and $C_2D_2O^+$ fragment, $m/z = 30$ to acetaldehyde (DCO^+ fragment).

by a factor of 2. In the cathodic sweep, the current only increases after reduction of adsorbed oxygen. The ion currents for both $m/z = 44$ and $m/z = 30$ closely follow the faradaic current. However, in the oxygen region, the ion current for acetaldehyde is lower than one would expect from the faradaic current, which is twice as large as that at 0.6 V. This points to an additional formation of acetic acid. A quantitative analysis shows that the ion current for $m/z = 44$ (CO_2 and the $C_2D_2O^+$ fragment) is $\approx 7\%$ of that for acetaldehyde, irrespective of surface structure [Pt(332), Pt(331), polycrystalline Pt, both in potential step and potential sweep experiments] and can completely be ascribed to the $C_2D_2O^+$ fragment of acetaldehyde; the amount of CO_2 therefore is negligible.

The amount of acetic acid cannot directly be determined by DEMS, but it can be estimated from the amount of acetaldehyde. The current efficiency for the formation of acetaldehyde on polycrystalline Pt is roughly 100%, that is, the amount of acetic

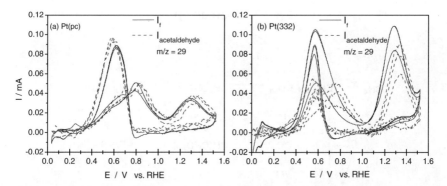

FIGURE 9.27 Comparison of faradaic current with corresponding ion current for acetaldehyde after conversion to faradaic current units on polycrystalline Pt (left) and Pt(332) (right). Conditions as in Figure 9.26, but at a flow rate of $5 \, \mu L \, s^{-1}$. (From ref. 144.)

acid is negligible under convective conditions in the flow-through cell. Figure 9.27 shows the ion currents for $m/z = 29$ after conversion to faradaic current values plotted together with the faradaic current.

On polycrystalline Pt, both curves closely overlap. On Pt(332), however, where the same conversion factor was used, the current due to the formation of acetaldehyde is lower than the ion current. Only at around 0.8 V are both currents identical. (Although there is some incertitude in the calibration factor, it is clear that the current efficiency cannot exceed 100%. The difference to the above result with deuterated ethanol is due to the fact that a higher electrolyte flow rate was used here, resulting in a lower time constant for detection, and also due to the kinetic isotope effect.) Obviously, in the current peak another product is formed. Since the ion current for $m/z = 44$ closely follows that of acetaldehyde and is mainly determined by acetaldehyde, CO_2 cannot be the origin of this discrepancy. Rather, acetic acid is formed in the current peak, but no longer at higher potentials. Certainly, it is somewhat astonishing that the product with the higher oxidation state is formed at a lower potential than that with the lower oxidation state. But it has to be kept in mind that also at higher potentials acetaldehyde can be oxidized further in a follow-up reaction if there is no strong convection that leads to a fast transport away from the electrode. Therefore, the formation of acetic acid seems to be due to a direct reaction at the electrode surface without acetaldehyde as an intermediate. (A similar conclusion has been obtained from IR measurements on Pt(111) without convective flow [145].) Whereas the reaction leading to acetaldehyde is a simple dehydrogenation, the formation of acetic acid involves also a reaction with oxygen or OH^- from water. Although one might assume that the availability of surface-bound OH^- or activated water increases with potential, (bi)sulfate adsorption, which is completed between 0.6 and 0.7V at (111) terraces of Pt, and the adsorption of the produced acetate might impede the adsorption of an activated water or OH^- species. Above this potential only dehydrogenation may take place at step sites. Below this potential, dehydrogenation also takes place at step sites, and a further reaction to acetic acid is possible after an (admittedly speculative) spillover to the terraces sites.

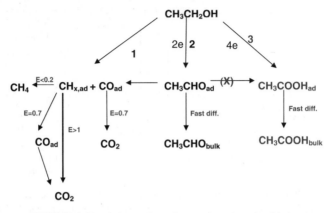

FIGURE 9.28   Schematic pathways for ethanol oxidation.

These results are summarized in the reaction scheme shown in Figure 9.28.

Can the current efficiency for CO_2 be increased by a second metallic component? Figure 9.29 shows that the main effect of step decoration by Ru is a decrease of the overall current. As soon as the steps are completely covered by Ru, the surface is hardly active at all.

This has been studied in detail by the Alicante group [146]. They also decorated the steps by Ru and found an increased current for a coverage of the steps by Ru

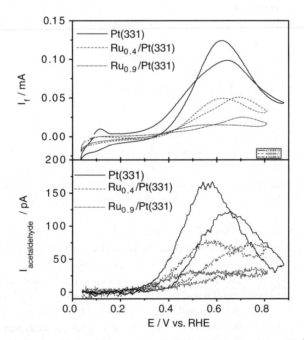

FIGURE 9.29   Electrooxidation of ethanol on Pt(331) and Ru/Pt(331) in 10^{-2} M ethanol in 0.1 M H_2SO_4 + 0.5 M $HClO_4$ at 10 mV s^{-1}. Flow rate 5 µL s^{-1}. (From ref. 144.)

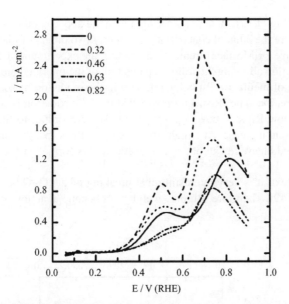

FIGURE 9.30 Positive-going scan for Pt(775) electrode in 0.5 M C_2H_5OH + 0.1 M H_2SO_4 with different ruthenium coverage on the step. Scan rate: 50 mV s^{-1}. (From ref. 146.)

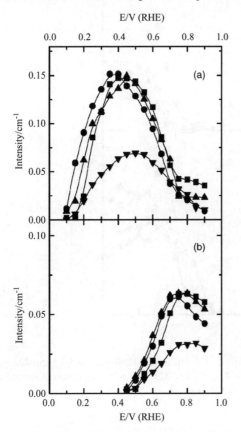

FIGURE 9.31 Integrated band intensity for (a) adsorbed CO_{linear} at 2040 cm^{-1} and (b) CO_2 at 2340 cm^{-1} as a function of the potential on the Pt(332) electrode with different ruthenium coverages on the step: 0.00; 0.10; 0.55; 0.85. (From ref. 146.)

below 0.5 (cf. Fig. 9.30). When comparing this to the above results, it has to be kept in mind that here the ethanol concentration is nearly two orders of magnitude higher than in the above DEMS measurements; moreover, the experiments are performed in a stagnant solution allowing the follow-up reaction of the intermediate acetaldehyde. As in the case of methanol, Ru helps not only in the oxidation of adsorbed CO but also in its formation: The IR spectra reveal that the CO oxidation is shifted to lower potentials for low Ru step coverage (cf. Fig. 9.31). According to the authors, this causes the current increase. At a high Ru coverage at the steps, they are blocked, no more C–C bond cleavage can occur, and acetic acid is formed at the (111) terraces (cf. Fig. 9.32).

In this context, it is very interesting that breaking of the C–C bond is easier for acetaldehyde [49]. Therefore, Ru may well have a beneficial action on acetaldehyde

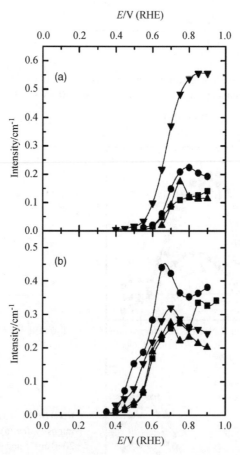

FIGURE 9.32 Integrated band intensity for (a) acetic acid at 1290 cm^{-1} and (b) acetaldehyde at 1113 cm^{-1} as a function of the potential on the Pt(332) electrode with different ruthenium coverages on the step: 0.00; 0.10; 0.55; 0.85. (From ref. 146.)

oxidation rather than on ethanol oxidation. This could explain discrepancies in the literature. Acetic acid, on the other hand, is completely unreactive. The task for the future is to find a catalyst which not only is active for C—C bond splitting in ethanol and/or acetaldehyde but also completely unreactive for the oxidation to acetic acid, that is, a very selective catalyst.

REFERENCES

1. J. O. M. Bockris and H. Wroblowa, *J. Electroanal. Chem.*, 7 (1964) 428–451.
2. H. A. Gasteiger, N. Markovic, P. N. Ross, and E. J. Cairns, *J. Phys. Chem.*, 97 (1993) 12020–12029.
3. M. Mavrikakis, B. Hammer, and J. K. Norskov, *Phys. Rev. Lett.*, 81 (1998) 2819–2822.
4. J. R. Kitchin, J. K. Norskov, M. A. Barteau, and J. G. Chen, *Phys. Rev. Lett.*, 93 (2004) 156801-1–156801-4.
5. J. R. Kitchin, J. K. Norskov, M. A. Barteau, and J. G. Chen, *J. Chem. Phys.*, 120 (2004) 10240–10246.
6. G. Somorjai, *Chemistry in Two Dimensions: Surfaces*, Cornell University Press, Ithaca, NY, 1981.
7. M. Henzler and W. Göpel, *Oberflächenphysik des Festkörpers*, Teubner-Studienbücher: Physik, Stuttgart, 1994.
8. H. Ibach, *Physics of Surfaces and Interfaces*, Springer, Berlin, 2006.
9. D. Briggs and M. P. Seah, *Practical Surface Analysis by Auger and X-ray Photoelectron Spectroscopy*, Wiley, New York, 1983.
10. C. D. Wagner, W. M. Riggs, L. E. Davis, J. F. Moulder, and G. E. Muilensberg, *Handbook of X-ray Photoelectron Spectroscopy*, Physical Electronics Industries, Eden Prairie, 1979.
11. E. D. Davis and N. C. MacDonald, *Handbook of Auger Electron Spectroscopy*, Physical Electronics Industries, Eden Prairie, 1976.
12. H. A. Gasteiger, P. N. Ross, and E. J. Cairns, *Surface Sci.*, 293 (1993) 67–80.
13. G. Ertl and J. Küppers, *Low Energy Electrons and Surface Chemistry*, VCH Verlagsgesellschaft, Weinheim, 1985.
14. A. T. Hubbard, J. L. Stickney, S. D. Rosasco, D. Song, and M. P. Soriaga, *Surface Sci.*, 130 (1983) 326–347.
15. D. M. Kolb, D. L. Rath, R. Wille, and W. N. Hansen, *Berichte Der Bunsen-Gesellschaft-Phys. Chem. Chem. Phys.*, 87 (1983) 1108–1113.
16. F. T. Wagner and P. N. Ross, *J. Electroanal. Chem.*, 150 (1983) 141–164.
17. A. T. Hubbard, et al., *J. Electroanal. Chem.*, 168 (1984) 43–66.
18. H. Hoster and H. Gasteiger, in *Handbook of Fuel Cells—Fundamentals, Technology and Applications*, Vol. 2, H.A. Gasteiger, W. Vielstich, A. Lamm (Eds.) Wiley, Chichester, 2003, pp. 236–265.
19. F. Lu, G. N. Salaita, H. Baltruschat, and A. T. Hubbard, *J. Electroanal. Chem.*, 222 (1987) 305–320.
20. B. C. Schardt, S. L. Yau, and F. Rinaldi, *Science*, 243 (1989) 1050–1053.
21. O. M. Magnussen, J. Hotlos, R. J. Nichols, D. M. Kolb, and R. J. Behm, *Phys. Rev. Lett.*, 64 (1991) 2929.
22. H. Baltruschat, U. Bringemeier, and R. Vogel, *Faraday Discuss.*, 94 (1992) 317–327.
23. X. Xiao, S. Tillmann, and H. Baltruschat, *Phys. Chem. Chem. Phys.*, 4 (2002) 4044–4050.

24. K. A. Friedrich, A. Marmann, U. Stimming, W. Unkauf, and R. Vogel, *Fres. J. Anal. Chem.*, 358 (1997) 163–165.

25. F. Maroun, F. Ozanam, O. M. Magnussen, and R. J. Behm, *Science*, 293 (2001) 1811–1814.

26. C. A. Lucas, N. M. Markovic, B. N. Grgur, and P. N. Ross, *Surface Sci.*, 448 (2000) 65–76.

27. C. A. Lucas, N. M. Markovic, and P. N. Ross, *Surface Sci.*, 448 (2000) 77–86.

28. N. M. Markovic and P. N. Ross, *Surface Sci. Repts.*, 45 (2002) 117–229.

29. T. J. Schmidt, M. Noeske, H. A. Gasteiger, R. J. Behm, P. Britz, and H. Bonnemann, *J. Electrochem. Soc.*, 145 (1998) 925–931.

30. H. S. Wang, C. Wingender, H. Baltruschat, M. Lopez, and M. T. Reetz, *J. Electroanal. Chem.* 509 (2001) 163–169.

31. T. Page, R. Johnson, J. Hormes, S. Noding, and B. Rambabu, *J. Electroanal. Chem.*, 485 (2000) 34–41.

32. H. Modrow, G. Kohl, J. Hormes, H. Bonnemann, U. Endruschat, and R. Mortel, *Phys. Scripta*, T115 (2005) 671–674.

33. I. V. Malakhov, S. G. Nikitenko, E. R. Savinova, D. I. Kochubey, and N. Alonso-Vante, *J. Phys. Chem. B*, 106 (2002) 1670–1676.

34. V. I. Zaikovskii, et al., *J. Phys. Chem. B*, 110 (2006) 6881–6890.

35. H. Baltruschat, *J. Am. Soc. Mass Spectrom.*, 15 (2004) 1693–1706.

36. H. Baltruschat and U. Schmiemann, *Ber. Bunsenges. Phys. Chem.*, 97 (1993) 452–460.

37. T. Hartung and H. Baltruschat, *Langmuir*, 6 (1990) 953–957.

38. Z. Jusys, H. Massong, and H. Baltruschat, *J. Electrochem. Soc.*, 146 (1999) 1093.

39. J. Fuhrmann, A. Linke, H. Langmach, and H. Baltruschat, *Electrochim. Acta*, 55 (2009) 430–438.

40. Y. X. Chen, M. Heinen, Z. Jusys, and R. J. Behm, *Ang. Chemie*, 118 (2006) 995–1000.

41. T. Smolinka, M. Heinen, Y. X. Chen, Z. Jusys, W. Lehnert, and R. J. Behm, *Electrochim. Acta*, 50 (2005) 5189–5199.

42. Y. Gao, H. Tsuji, H. Hattori, and H. Kita, *J. Electroanal. Chem.*, 372 (1994) 195–200.

43. A. H. Wonders, T. H. M. Housmans, V. Rosca, and M. T. M. Koper, *J. Appl. Electrochem.*, 36 (2006) 1215–1221.

44. K. Jambunathan and A. C. Hillier, *J. Electrochem. Soc.*, 150 (2003) E312–E320.

45. K. Jambunathan, S. Jayaraman, and A. C. Hillier, *Langmuir*, 20 (2004) 1856–1863.

46. F. M. Hoffmann, *Surface Sci. Repts.*, 3 (1983) 107.

47. J. D. Roth and M. J. Weaver, *J. Electroanal. Chem.*, 307 (1991) 119–137.

48. M. F. Mrozek and M. J. Weaver, *J. Phys. Chem. B*, 105 (2001) 8931–8937.

49. S. C. S. Lai, S. E. F. Kleyn, V. Rosca, and M. T. M. Koper, *J. Phys. Chem. C*, 112 (2008) 19080–19087.

50. N. M. Markovic, H. A. Gasteiger, C. A. Lucas, I. M. Tidswell, and P. N. Ross, *Surface Sci.*, 335 (1995) 91–100.

51. F. Richarz, B. Wohlmann, U. Vogel, H. Hoffschulz, and K. Wandelt, *Surface Sci.*, 335 (1995) 361–371.

52. H. Niehus, W. Raunau, K. Besocke, R. Spitzl, and G. Comsa, *Surface Sci.*, 225 (1990) L8–L14.

53. M. T. Paffett, S. C. Gebhard, R. G. Windham, and B. E. Koel, *Surface Sci.*, 223 (1989) 449–464.

54. M. T. Paffett, S. C. Gebhard, R. G. Windham, and B. E. Koel, *J. Phys. Chem.*, 94 (1990) 6831–6839.

55. C. Becker, T. Pelster, M. Tanemura, J. Breitbach, and K. Wandelt, *Surface Sci.*, 435 (1999) 822–826.

56. K. L. Wang, H. A. Gasteiger, N. A. Markovic, and P. N. J. Ross, *Electrochim. Acta*, 41 (1996) 2587–2593.

57. A. N. Haner and P. N. Ross, *J. Phys. Chem.*, 95 (1991) 3740–3746.

58. V. Stamenkovic, M. Arenz, B. B. Blizanac, K. J. J. Mayrhofer, P. N. Ross, and N. M. Markovic, *Surface Sci.*, 576 (2005) 145–157.

59. B. E. Hayden, M. E. Rendall, and O. South, *J. Am. Chem. Soc.*, 125 (2003) 7738–7742.

60. F. J. Vidal-Iglesias, A. Al-Akl, D. J. Watson, and G. A. Attard, *Electrochem. Communi.*, 8 (2006) 1147–1150.

61. L. Fang, F. J. Vidal-Iglesias, S. E. Huxter, and G. A. Attard, *J. Electroanal. Chem.*, 622 (2008) 73–78.

62. S. R. Brankovic, J. X. Wang, and R. R. Adzic, *Surface Sci.*, 474 (2001) L173–L179.

63. J. Greeley, J. K. Norskov, L. A. Kibler, A. M. El-Aziz, and D. M. Kolb, *ChemPhysChem*, 7 (2006) 1032–1035.

64. L. A. Kibler, A. M. El-Aziz, R. Hoyer, and D. M. Kolb, *Ang. Chemie*, 117 (2005) 2116–2120.

65. L. A. Kibler, *ChemPhysChem*, 7 (2006) 985–991.

66. L. A. Kibler, *Electrochim. Acta*, 53 (2008) 6824–6828.

67. Y. Pluntke, L. A. Kibler, and D. M. Kolb, *Phys. Chem. Chem. Phys.*, 10 (2008) 3684–3688.

68. G. Somorjai and D. W. Blakely, *Surface Sci.*, 65 (1977) 419.

69. H. Baltruschat, et al., *J. Electroanal. Chem.*, 217 (1987) 111.

70. K. Kuhnke and K. Kern, *J. Phys.-Condens. Matter*, 15 (2003) S3311–S3335.

71. J. Clavilier, R. Faure, G. Guinet, and R. Durand, *J. Electroanal. Chem.*, 107 (1980) 205–209.

72. A. Rodes, K. El Achi, M. A. Zamakchardi, and J. Clavilier, *J. Electroanal. Chem.*, 284 (1990) 245–253.

73. J. Clavilier, D. Armand, S. G. Sun, and M. Petit, *J. Electroanal. Chem.*, 205 (1986) 267–277.

74. E. Herrero, J. M. Orts, A. Aldaz, and J. M. Feliu, *Surface Sci.*, 440 (1999) 259–270.

75. E. Budevski, G. Staikov, and W. J. Lorenz, *Electrochemical Phase Formation and Growth*, VCH, Weinheim, 1996.

76. J. Clavilier, J. M. Feliu, and A. Aldaz, *J. Electroanal. Chem.*, 243 (1988) 419–433.

77. J. Clavilier, K. El Achi, and A. Rodes, *Chem. Phys.*, 141 (1990) 1–14.

78. A. Rodes, K. E. Achi, M. A. Zamakhchari, and J. Clavilier, *J. Electroanal. Chem.*, 284 (1990) 245–253.

79. C. Nishihara and H. Nozoye, *J. Electroanal. Chem.*, 396 (1995) 139–142.

80. L. J. Buller, E. Herrero, R. Gomez, J. M. Feliu, and H. D. Abruna, *J. Chem. Soc. Farad. Trans.*, 92 (1996) 3757–3762.

81. M. H. Hölzle, V. Zwing, and D. M. Kolb, *Electrochim. Acta*, 40 (1995) 1237–1247.

82. M. P. Green, K. J. Hanson, R. Carr, and I. Lindau, *J. Electrochem. Soc.*, 137 (1990) 3493–3498.

83. U. Schmidt, S. Vinzelberg, and G. Staikov, *Surface Sci.*, 348 (1996) 261–279.

84. H. Massong, H. S. Wang, G. Samjeske, and H. Baltruschat, *Electrochim. Acta*, 46 (2000) 701–707.

85. G. Samjeské, X.-Y. Xiao, and H. Baltruschat, *Langmuir*, 18 (2002) 4659–4666.

86. P. Berenz, S. Tillmann, H. Massong, and H. Baltruschat, *Electrochim. Acta*, 43 (1998) 3035–3043.

87. L. J. Buller, E. Herrero, R. Gomez, J. M. Feliu, and H. D. Abruna, *J. Phys. Chem. B*, 104 (2000) 5932–5939.

88. E. A. Abd El Meguid, P. Berenz, and H. Baltruschat, *J. Electroanal. Chem.*, 467 (1999) 50–59.

89. R. Bussar, PhD Thesis, Universität Bonn, Bonn, 2002.

90. K. F. Domke, X.-Y. Xiao, and H. Baltruschat, *Electrochim. Acta*, 54 (2009) 4829–4836.

91. E. Herrero, V. Climent, and J. M. Feliu, *Electrochem. Communi.*, 2 (2000) 636–640.

92. J. M. Feliu, R. Gomez, M. J. Llorca, and A. Aldaz, *Surface Sci.*, 289 (1993) 152–162.

93. P. Rodriguez, E. Herrero, J. Solla-Gullon, F. J. Vidal-Iglesias, A. Aldaz, and J. M. Feliu, *Electrochim. Acta*, 50 (2005) 4308–4317.

94. H. Massong, S. Tillmann, T. Langkau, E. A. Abd El Meguid, and H. Baltruschat, *Electrochim. Acta*, 44 (1998) 1379–1388.

95. G. Samjeské, H. Wang, T. Löffler, and H. Baltruschat, *Electrochim. Acta*, 47 (2002) 3681–3692.

96. T. Nagel, N. Bogolowski, G. Samjeske, and H. Baltruschat, *J. Solid State Electrochem.*, 7 (2003) 614–618.

97. X. Y. Xiao and H. Baltruschat, *Langmuir*, 19 (2003) 7436–7444.

98. O. A. Hazzazi, G. A. Attard, P. B. Wells, F. J. Vidal-Iglesias, and M. Casadesus, *J. Electroanal. Chem.*, 625 (2009) 123–130.

99. P. Gambardella, M. Blanc, H. Brune, K. Kuhnke, and K. Kern, *Phys. Rev. B*, 61 (2000) 2254–2262.

100. F. Hernandez and H. Baltruschat, *Langmuir*, 22 (2006) 4877–4884.

101. K. A. Friedrich, K. P. Geyzers, A. Marmann, U. Stimming, and R. Vogel, *Zeits. Phys. Chem.-Int. J. Res. Phys. Chem. Chem. Phys.*, 208 (1999) 137–150.

102. E. Herrero, J. M. Feliu, and A. Wieckowski, *Langmuir*, 15 (1999) 4944–4948.

103. S. Cramm, K. A. Friedrich, K. P. Geyzers, U. Stimming, and R. Vogel, *Fres. J. Anal. Chem.*, 358 (1997) 189–192.

104. F. Hernandez, J. Sanabria-Chinchilla, M. P. Soriaga, and H. Baltruschat, in *Electrode Processes*, Vol. 7, V. I. Birss, M. Josowicz, D. Evans, and M. Osawa (Eds.), Electrochemical Society, Pennington, NJ, p. 15.

105. J. Steidtner, F. Hernandez, and H. Baltruschat, *J. Phys. Chem. C*, 111 (2007) 12320–12327.

106. A. M. El-Aziz and L. A. Kibler, *J. Electroanal. Chem.*, 534 (2002) 107–114.

107. S. Tillmann, G. Samjeske, A. Friedrich, and H. Baltruschat, *Electrochim. Acta*, 49 (2003) 73–83.

108. S. Tillmann, Ph.D. Dissertation, University of Bonn, Bonn, 1999.

109. S. Tillmann, A. Friedrich, H. Massong, and H. Baltruschat, in *The 195th Meeting of the Electrochemical Society*, The Electrochemical Society, Seattle, WA, 1999, Vol. Abstracts 99-1, p. 1097.

110. H. Baltruschat, R. Bussar, S. Ernst, and F. Hernandez-Ramirez, in *In-situ Spectroscopic Studies of Adsorption at the Electrode and Electrocatalysis*, S.-G. Sun, P. A. Christensen, and A. Wieckowski (Eds.), Elsevier, Amsterdam 2007, pp. 471–537.

111. J. Sanabria-Chinchilla, J. H. Baricuatro, M. P. Soriaga, F. Hernandez, and H. Baltruschat, *J. Colloid Interface Sci.*, 314 (2007) 152–159.

112. M. Del Popolo, E. Leiva, H. Kleine, J. Meier, U. Stimming, M. Mariscal, and W. Schmickler, *Appl. Phys. Lett.*, 81 (2002) 2635–2637.

113. H. Baltruschat, N. A. Anastasijevic, M. Beltowska-Brzezinska, G. Hambitzer, and J. Heitbaum, *Ber. Bunsenges. Phys. Chem.*, 94 (1990) 996–1000.

114. R. Stadler, Z. Jusys, and H. Baltruschat, *Electrochim. Acta*, 47 (2002) 4485–4500.

115. F. Hernandez, Ph.D. Thesis, Rheinische-Friedrich-Wilhelms-Universität, Bonn, 2006.

116. F. Hernandez and H. Baltruschat, *J. Solid State Electrochem.*, 11 (2007) 877–885.

117. P. Sabatier, *Berichte der deutschen chemischen Gesellschaft*, 44 (1911) 1984–2001.

118. S. Motoo, M. Shibata, and M. Watanabe, *J. Electroanal. Chem.*, 110 (1980) 103–109.

119. V. R. Stamenkovic, M. Arenz, C. A. Lucas, M. E. Gallagher, P. N. Ross, and N. M. Markovic, *J. Am. Chem. Soc.*, 125 (2003) 2736–2745.

120. H. Massong, Ph.D. Thesis, Bonn University, Bonn, 2004.

121. B. E. Hayden, M. E. Rendall, and O. South, *J. Mol. Catal. Chem.*, 228 (2005) 55–65.

122. B. Grgur, N. Markovic, and P. Ross, *J. Phys. Chem. B*, 102 (1998) 2494.

123. M. Watanabe and S. Motoo, *J. Electroanal. Chem.*, 60 (1975) 275–283.

124. H. A. Gasteiger, N. A. Markovic, N. Philip, J. Ross, and E. J. Cairns, *J. Phys. Chem.*, 98 (1994) 617–625.

125. A. V. Petukhov, W. Akemann, K. A. Friedrich, and U. Stimming, *Surface Sci.*, 402–404, (1998) 182–186.

126. W. Chrzanowski and A. Wieckowski, *Langmuir*, 13 (1997) 5974–5978.

127. K. A. Friedrich, K.-P. Geyzers, U. Linke, U. Stimming, and J. Stumper, *J. Electroanal. Chem.*, 402 (1996) 123–128.

128. J. C. Davies, B. E. Hayden, D. J. Pegg, and M. E. Rendall, *Surface Sci.*, 496 (2002) 110–120.

129. F. Maillard, G. Q. Lu, A. Wieckowski, and U. Stimming, *J. Phys. Chem. B*, 109 (2005) 16230–16243.

130. G. Q. Lu, P. Waszczuk, and A. Wieckowski, *J. Electroanal. Chem.*, 532 (2002) 49–55.

131. F. B. de Mongeot, M. Scherer, B. Gleich, E. Kopatzki, and R. J. Behm, *Surface Sci.*, 411 (1998) 249–262.

132. M. T. M. Koper, T. E. Shubina, and R. A. van Santen, *J. Phys. Chem. B*, 106 (2002) 686–692.

133. B. Lanova, Ph.D. Thesis, Rheinische Friedrich-Wilhelms Universität, Bonn, Germany, 2009.

134. H. Wang, T. Löffler, and H. Baltruschat, *J. Appl. Electrochem.*, 31 (2001) 759–765.

135. H. Wang, and H. Baltruschat, in *DMFC Symposium, Meeting of the Electrochemical Society 2001*, S. R. Narayanan (Ed.), The Electrochemical Society, Washington, DC, 2001.

136. Z. Jusys, T. J. Schmidt, L. Dubau, K. Lasch, L. Jorissen, J. Garche, and R. J. Behm, *J. Power Sources*, 105 (2002) 297–304.

137. Z. B. Wang, G. P. Yin, and Y. G. Lin, *J. Power Sources*, 170 (2007) 242–250.

138. T. Y. Morante-Catacora, Y. Ishikawa, and C. R. Cabrera, *J. Electroanal. Chem.*, 621 (2008) 103–112.

139. T. Iwasita and E. Pastor, *Electrochim. Acta*, 39 (1994) 531–537.

140. U. Schmiemann, U. Müller, and H. Baltruschat, *Electrochim. Acta*, 40 (1995) 99–107.

141. T. Löffler and H. Baltruschat, *J. Electroanal. Chem.*, 554–555 (2003) 333–344.

142. T. Löffler, R. Bussar, X. Xiao, S. Ernst, and H. Baltruschat, *J. Electroanal. Chem.*, 629 (2009) 1–14.

143. M. H. Shao and R. R. Adzic, *Electrochim. Acta*, 50 (2005) 2415–2422.

144. A. A. Abd-El-Latif, E. Mostafa, S. Huxter, G. Attard, H. Baltruschat, *Electrochimica Acta* 2010, 55, 7951.

145. M. J. Giz and G. A. Camara, *J. Electroanal. Chem.*, 625 (2009) 117–122.

146. V. Del Colle, A. Berna, G. Tremiliosi-Filho, E. Herrero, H., and J. M. Feliu, *Phys. Chem. Chem. Phys.*, 10 (2008) 3766–3773.

CHAPTER 10

CO Adsorption on Platinum Electrodes

ÁNGEL CUESTA and CLAUDIO GUTIÉRREZ

Instituto de Química Física "Rocasolano," E-28006 Madrid, Spain

10.1 INTRODUCTION

The study of carbon monoxide, a deceptively simple molecule, is arguably well worth a lifetime, especially when its adsorption on platinum is concerned. So, it is widely known that CO is a biological poison, because the binding affinity of myoglobin and hemoglobin for CO is about 200 times as large as that for oxygen. It is far less known that CO is produced in the degradation of erithrocytes (life span 120 days) and that therefore vertebrates exist because this difference of binding affinities is not too large, so that only a high concentration of CO causes death. However, an isolated heme group in solution binds CO about 25,000 times more strongly than O_2, this binding increase being two orders of magnitude higher than that of 200 for the above proteins because in them the C–O bond is at an angle (because of steric hindrance of distal histidine) with respect to the Fe–N bond, while CO binds to the heme group forming a linear configuration, yielding a stronger binding, of the N–Fe–C–O atoms [1]. Therefore, vertebrates, both small and large, exist thanks to a small tilt of a very small molecule, CO, which dramatically reduces its binding affinity for myoglobin and hemoglobin.

As will be discussed below, the Pt–CO–electrolyte system is unique in the sense that, as far as we know, it is the only one in which controlling the potential at which the substrate, CO, is introduced in the cell dramatically affects the catalytic activity of the electrode for the electrooooxidation of the substrate. This surprising influence of the substrate admission potential was discovered by accident in 1988 and lay dormant until its rediscovery in 1991, which started an intensive search for its origin that has not abated yet.

The amount of work done on the adsorption of CO on Pt electrodes is so large that a comprehensive review of this subject would require a book rather than a chapter.

Catalysis in Electrochemistry: From Fundamentals to Strategies for Fuel Cell Development,
First Edition. Edited by Elizabeth Santos and Wolfgang Schmickler.
© 2011 John Wiley & Sons, Inc. Published 2011 by John Wiley & Sons, Inc.

For this reason, after a brief historical perspective we will focus here on the work performed by our group during the last 20 years.

10.2   BRIEF HISTORICAL PERSPECTIVE

It is perhaps not widely known that as long ago as 1890 Ludwig Mond tried to convert the hydrogen produced from his gas plant into electricity using a fuel cell and that it was found in his London laboratory that a carbon monoxide impurity in the hydrogen rapidly poisoned the Pt black catalyst [2]. Six score years later a viable solution to this problem has not yet been found, in spite of very intensive research made possible by a cornucopia of public and private funding. Consequently, the amount of publications on the CO–Pt–electrolyte system is so large that an account of the evolution of research in this field must be very selective.

It could be argued that significant work on CO adsorption on a Pt electrode started in the 1960s in the Research Laboratories of General Electric in Schenectady, New York, with Gilman and Breiter, the former under a contract with the Advanced Research Projects Agency, the present DARPA. Gilman used a length of annealed Pt wire with an estimated roughness coefficient of 1.5 and several then state-of-the-art electrochemical techniques (multipulse sequences, oscilloscopic recording of currents, linear scans at rates of up to $1 \, \text{kV s}^{-1}$). He proposed [3] that the electrooxidation of CO took place by what he called a "reactant pair" mechanism, in which the activated complex in the loss of the first electron from CO is an adsorbed CO molecule adjacent to an adsorbed water molecule, in what is in effect a Langmuir–Hinshelwood mechanism.

As far as we know, Breiter was the first to detect, by the galvanostatic method, that both on platinized platinum [4] and smooth Pt [5] electrodes there were two types of chemisorbed CO: a more weakly adsorbed CO which became oxidized between 0.2 and 0.4 V versus the reference hydrogen electrode (RHE) and a more strongly adsorbed CO which was oxidized at higher potentials.

It is worth pointing out that, until scanning tunneling microscopy (STM) unequivocally showed how CO molecules were arranged on single-crystal surfaces, it was universally believed that a CO molecule in a bridge position blocked two Pt atoms, on which no other CO molecule could be adsorbed, so that a Pt surface fully covered with bridge CO would have a coverage of only 0.5 CO molecules per Pt atom. This assumption, once routinely used to calculate the coverage of linear and bridge CO from the ratio of the CO stripping charge to the hydrogen under potential deposition (UPD) charge ("electrons per site"), is now known to be false. For example, in the (2×2)–3CO ($\theta_{CO} = 0.75$) and $(v19 \times v19)23.4°$–13CO ($\theta_{CO} = 0.684$) structures of CO adsorbed on Pt(111), if we assume that every CO adsorbed on-top blocks only 1 Pt atom, the threefold-hollow COs of the former structure and the twofold-hollow COs of the latter one would occupy at most 1.5 Pt atoms.

As said in Section 10.1, a phenomenon which is unique to the CO–Pt system was reported in 1988 by Kita and co-workers [6]. They serendipitously found, while studying the efect of a Sn(IV) pretreatment of Pt on the CO–Pt system, that, if a polycrystalline Pt electrode was held at a potential in the hydrogen region while CO was first introduced in the cell, dissolved CO was electrooxidized at a much lower potential than that of 0.9 V versus RHE universally reported until then, since the usual procedure was to bubble CO at open circuit, that is, at potentials in the double-layer region. This unique case of influence of the admission potential on the activity of an electrode lay dormant until it was rediscovered in 1991 [7] (actually the editor of the journal had to be convinced that the phenomenon was real). The charge of the prepeak in CO-stripping cyclic voltammograms (CVs) recorded in parallel experiments in the absence of solution CO was about 15% of the total charge of adsorbed CO, and it was independently concluded by Wieckowski et al. [8] and by Kita et al. [9] that the low-potential electrooxidation of dissolved CO took place only on these CO-free Pt sites. It was later confirmed [10] that, in spite of changes in the experimental conditions (quiescent or stirrred solution, length of holding time) that produced changes in the potential and intensity of the peak of dissolved CO electrooxidation at low potentials, this prepeak always preceded the main peak of electrooxidation of adsorbed CO. The nature of this prepeak has been the subject of intensive research [7–9, 11–13] and will be discussed in Section 10.5.

10.3 FUNDAMENTAL ASPECTS OF CO ADSORPTION ON Pt ELECTRODES

10.3.1 Ultraviolet Electrolyte Electroreflectance and Differential Reflectance Studies of CO Adsorbed on Pt Electrodes

In the 1970s much effort was devoted to the study of the electrochemical interface by means of ultraviolet–visible spectroscopies, as exemplified by a symposium of the Faraday Society exclusively devoted to optical studies of adsorbed layers at interfaces [14] and by a *Faraday Discussion of the Chemical Society* on intermediates in electrochemical reactions [15]. However, after the pioneering work of Alan Bewick, who first reported infrared spectra of an electrochemical interface, opening up a still expanding field of research, optical spectroscopies were nearly completely abandoned in favor of infrared ones. In particular, the technique of electrolyte electroreflectance (EER), first reported by Cardona in 1965 for the study of the band structure of semiconductors [16] and which had been successfully applied by Bewick [17] and by Kolb [18, 19] to electrochemical problems, became scarcely used.

In spite of the above, in our laboratory we serendipitously found, while trying to detect by EER adsorbed acetate presumably formed in the electrooxidation of ethanol on polycrystalline Pt, that chemisorbed CO, a poisoning reaction product, gave rise in the EER spectrum to a peak at 270 nm. This finding was very surprising, since no transition at this wavelength had been reported in any of the vast array of publications on chemisorbed CO carried out with the full panoply of powerful spectroscopies available to surface scientists. Therefore, given the uniqueness of this peak, we carried out a series of tests in order to confirm that it was due to chemisorbed CO, that is, that it was not an experimental artifact. The first test was to check that the intensity of the

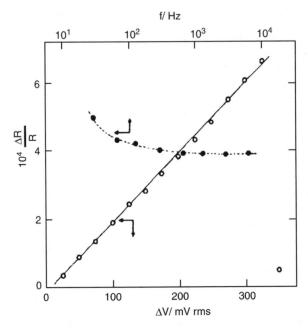

FIGURE 10.1   Dependence of intensity of EER maximum of CO-covered Pt at 280 nm on modulation frequency using modulation amplitude of 200 mV rms (dashed line) and on modulation amplitude using frequency of 65 Hz (solid line). Potential: +0.47 V vs. RHE. (Reprinted from ref. 7. Copyright 1991, with permission from Elsevier.)

EER signal at 270 nm was proportional to the modulation amplitude, as was indeed the case (Fig. 10.1), and that the signal intensity was independent of the modulation frequency, which was also the case (Fig. 10.1), except at the lower frequencies, perhaps because of the flicker noise, which is proportional to the reciprocal of the frequency.

Another test was carried out by means of reflectograms, in which the EER signal intensity at a given wavelength of polycrystalline (Pt(poly)) in CO-saturated 0.5 M $HClO_4$ was recorded as a function of the potential during a cyclic voltammogram at $2 \, mV \, s^{-1}$. As can be seen in Figure 10.2, with a CO admission potential of 0.07 V the signal intensity at the wavelength of the EER maximum remained constant up to 0.60 V and then decreased abruptly and changed sign, as was to be expected, since at this potential chemisorbed CO is electrooxidized. Furthermore, in the negative sweep the EER signal intensity below 0.7 V was nearly the same as in the positive sweep, indicating that CO had adsorbed again on Pt. The reflectogram in the second CV further confirmed that the EER signal is due to chemisorbed CO, since the behavior is the same as in the first CV, but with the crucial difference that the abrupt decrease of the signal intensity now occurs at 0.81 V, in full agreement with the higher potential at which CO is electrooxidized in the second as compared with the first CV.

Yet a further test was effected by recording the reflectograms at 270 nm of (poly) during stripping CVs of chemisorbed CO, obtained after eliminating the dissolved CO by N_2 bubbling. The signal intensity decreased abruptly at 0.47 and 0.57 V for admission potentials of 0.07 and 0.27 V, respectively, again in perfect agreement with

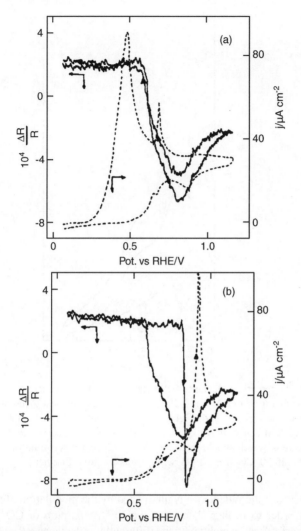

FIGURE 10.2 Voltammograms (dashed lines) and reflectograms (solid lines) at $2\,mV\,s^{-1}$ of CO-covered Pt in CO-saturated 0.5 M $HClO_4$, in first and second sweeps; $E_{adm} = + 0.07$ V vs. RHE. The modulation amplitude and frequency were 100 mV rms and 65 Hz, respectively. (a) First sweep. (b) Second sweep. (Reprinted from ref. 7. Copyright 1991, with permission from Elsevier.)

the potentials at which chemisorbed CO was electrooxidized, and which are lower than the corresponding potentials in CO-saturated solution.

Finally, CO chemisorbed on Ru, Rh, and Pd also gave rise to a peak in the EER spectrum at about the same wavelength as the peak observed with Pt. In the case of Au the EER peak appeared at a longer wavelength, about 330 nm, and blue shifted with increasing potential from 360 to 315 nm. Actually the wavelength maximum of an EER peak should shift with changes in the steady-state potential, since no EER

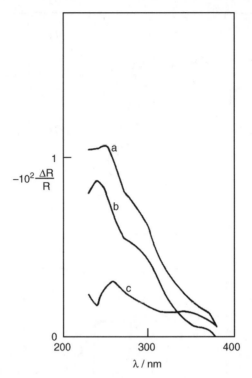

FIGURE 10.3 Differential reflectance (DR) spectra of CO-covered Pt obtained from reflectograms such as that in Figure 1 of ref. 20 from reflectance values at given potential in first
positive sweep (CO-covered Pt) and in second positive sweep (CO-free Pt). Curve *a*, DR spectrum at 0.43 V vs. RHE; curve *b*, DR spectrum at 0.71 V vs. RHE; curve *c*, difference between
curve *a* and curve *b*, which should yield the electroreflectance spectrum of CO-covered Pt.
(Reprinted from ref. 20, Copyright 1995, with permission from Elsevier.)

signal is observed if the reflectance is unaffected by the modulation of the potential.
Therefore, in order to further check that the EER maximum of CO chemisorbed
on Pt was not an artifact, we obtained the differential reflectance, $(R_{CO/Pt} - R_{Pt})/R_{Pt}$,
spectra of CO-covered Pt in CO-free 0.5 M HClO$_4$ by means of reflectograms recorded
every 10 nm for two electrode potentials, 0.43 and 0.71 V, and taking as reference
the reflectance of Pt in the second positive sweep, in which CO had already been
stripped off the Pt surface [20] (Fig. 10.3). The differential reflectance spectrum of
CO-covered Pt at 0.43 V had a larger intensity than that at 0.71 V, and the difference
spectrum between these two showed a maximum at 260 nm, in good agreement with
the maximum at 270 nm found by EER, which confirms its assignment to an electronic
transition of the CO–Pt system.

The high intensity, about 1%, of the differential reflectance spectrum at 250 nm,
both at 0.43 and 0.71 V, brought about by only one monolayer of chemisorbed CO,
clearly indicates that a charge transfer (CT) transition is involved, since CT transitions,
being not parity forbidden, have a high oscillator strength. The question remains
of which energy level of chemisorbed CO is involved, since the lowest electronic

transition of chemisorbed CO detected by electron-energy loss spectroscopy (EELS) appears at a higher energy, 6–7 eV, as compared with 4.6 eV for the EER maximum.

10.3.2   Charge Density and pzc of CO-Covered Pt Electrodes

One of the major contributions to Interfacial Electrochemistry during the past 10 years was the development by the Alicante Group of the CO charge displacement method (see ref. 21 and references therein). The method was based on the assumption that the charge density of the CO-covered electrode (which is usually, although not always, a single-crystal platinum electrode) is negligible, and, hence, the charge flowing during the potentiostatic adsorption of CO corresponds to the negative of the charge initially present on the electrode surface. The CO charge displacement method allowed the determination, with unprecedented accuracy, of the coverage at a given potential of Pt single-crystal electrodes by hydrogen and specifically adsorbed anions, of the potential of zero total charge (pztc) of platinum single-crystal electrodes, and of the coverage of platinum electrodes by adsorbed CO (see Section 10.3.3). However, the assumption that the charge density on the CO-covered electrode is negligible, although reasonable, is not exact, as first noted by Weaver [22], who emphasized that the presence of a finite, potential-dependent charge density on CO-covered platinum electrodes introduced a significant error in the value of the pztc determined from CO charge displacement measurements.

Our group has recently addressed this problem for the case of Pt(111) electrodes [23]. We determined the charge density residing on the CO-covered Pt(111) electrode in 0.1 M HClO₄, which allowed us to obtain a quantitative determination of the error incurred when estimating the pztc by the CO charge displacement method and to determine the point of zero charge (pzc) of the CO-covered Pt(111) electrode.

Our experimental approach was based on the definition of the pztc as the potential at which no charge flows through the interface upon formation of the electrochem-ical double layer. If the measurement is performed in a solution where no specific adsorption occurs, the pztc will coincide with the potential of zero free charge, pzfc. Accordingly, if an initially uncharged surface is immersed at a controlled potential in a solution, the charge flowing upon immersion corresponds to the charge density of the metal at the immersion potential. Such an experiment is illustrated in path 1 of Scheme 10.1. The CO charge displacement method, illustrated in path 2 of Scheme 10.1, can also be considered to be based on the definition of the pztc given above, but, instead of measuring the charge flowing when the double layer is created (Q_1 in Scheme 10.1), it attempts to measure the charge flowing when the double layer is quenched.

The immersion tactic has been successfully applied to the determination of the pztc and pzfc of Au(111) electrodes [24]. Any other metal surface would immediately react, during the transfer from ultra high vacuum (UHV) to the solution, with any oxygen traces present in the atmosphere above the working solution, or even with water vapor molecules, which would result in the formation on the metal surface of an oxide or OH adlayer, this certainly being the reason why this tactic yielded a too positive value for the pztc and pzfc of Pt(111) electrodes in the work of Hamm et al. [24]. In the case of CO-covered Pt electrodes, the tactic can be applied if CO is adsorbed on a freshly annealed, clean platinum surface from the gas phase. Under such

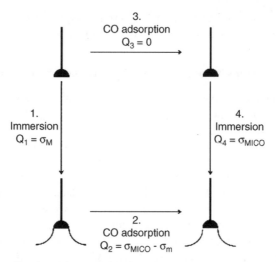

SCHEME 10.1   (Reprinted from ref. 23, Copyright 2004 with permission from Elsevier.)

conditions, no charge can be injected on the electrode surface, whose charge density hence remains nil. Recently the same experimental procedure has been successfully applied to the determination of the pzc of Au(111) electrodes modified by thiol self-assembled monolayers (SAMs) [25].

Current transients recorded during immersion of a dry, CO-covered Pt(111) electrode in a CO-saturated 0.1 M $HClO_4$ solution at potentials between 0.04 and 0.44 V versus the standard hydrogen electrode (SHE) are shown in Figure 10.4, and the corresponding charge densities obtained by integration of the transients are given in Table 10.1. As can be observed, the charge flowing across the interface is always negative, indicating that the pzc of the CO-covered Pt(111) electrode must be more positive than 0.44 V versus SHE. Figure 10.5 shows the cyclic voltammogram of a CO-covered Pt(111) electrode in CO-saturated 0.1 M $HClO_4$ between 0 and 0.44V versus SHE and the charge density–potential curve obtained by integration of the current in the positive sweep using the charge density measured in the immersion transient at 0.04 V versus SHE as the integration constant. Extrapolation of the charge density–potential curve yields a pzc for the Pt(111)/CO–aqueous solution interface of 1.10 ± 0.04 V versus SHE [23], in excellent agreement with previous estimates [22] based on the experimental value of 5.6 eV for the work function of a Pt(111) surface covered by 0.65 ML of CO and by 2 ML of D_2O in UHV [26].

The experimental measurement of the charge density residing on CO-covered Pt(111) electrodes allowed us to determine the error incurred when estimating the pztc of Pt(111) electrodes by the CO charge displacement method, which is -0.05 V in 0.1 M $HClO_4$ and -0.09 V in 0.1 M $LiClO_4 + 10^{-3}$ M $HClO_4$ [23], and to estimate the pzfc of Pt(111) as 0.23 V versus SHE.

The pzc is related to the work function by the relationship [27]

$$E_{\text{pzc}} = \frac{F}{e} + \left[d\chi_0^{\text{M}} - g^{\text{H}_2\text{O}}(\text{dip})_0 \right] - E_k \qquad (10.1)$$

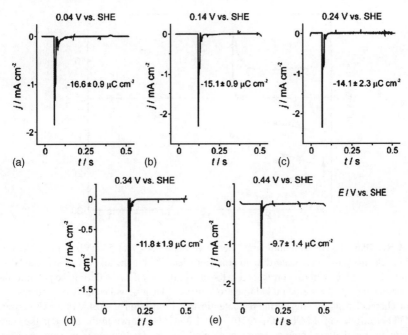

FIGURE 10.4 Current transients measured during potentiostatic immersion of CO-covered Pt(111) electrode prepared according to procedure described in the text in CO-saturated 0.1 M HClO₄; $E_{imm} = 0.04$ V (a); 0.14 V (b); 0.24 V (c); 0.34 V (d); 0.44 V (e). (Reprinted from ref. 23. Copyright 2004, with permission from Elsevier.)

where Φ is the work function, $d\chi_0^M$ the perturbation of the metal surface dipole by the neighboring solvent at the pzc, $g^{H_2O}(dip)_0$ the surface potential of the solvent at the pzc, and E_k the so-called *absolute* potential of the SHE, estimations of which range from 4.4 to 4.8 eV [28]. The values obtained for the pzc of the clean Pt(111) and the CO-covered Pt(111) surfaces (0.23 and 1.10 V vs. SHE, respectively) can

TABLE 10.1 Charge Densities Obtained by Integration of Current Transients Measured During Immersion at Controlled Potential of CO-Covered Pt(111) Electrode

E, V vs. SHE	Q, $\mu C\ cm^{-2}$
0.04	-16.6 ± 0.9
0.14	-15.1 ± 0.9
0.24	-14.1 ± 2.3
0.34	-11.8 ± 1.9
0.44	-9.7 ± 1.4

Source: From ref. 23.

Note: Prepared according to the procedure described in the text in 0.1 M HClO₄.

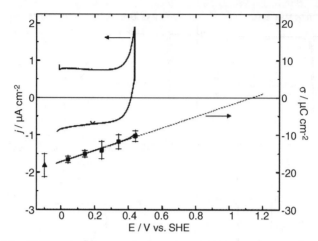

FIGURE 10.5 Cyclic voltammogram between 0 and 0.44 V at 50 mV s⁻¹ of CO-covered Pt(111) electrode prepared according to procedure described in the text in CO-saturated 0.1 M HClO₄. The square symbols correspond to the charge density flowing during the potentiostatic immersion of the CO-covered Pt(111) electrode in 0.1 M HClO₄, while the triangle corresponds to the charge density flowing during the potentiostatic immersion at −0.1 V of the CO-covered Pt(111) electrode in 0.1 M LiClO₄ + 10⁻³ M HClO₄. Also shown in the figure is the charge density–potential curve obtained by integration of the cyclic voltammogram using the charge density flowing during immersion at 0.04 V as integration constant. (Reprinted from ref. 23. Copyright 2004, with permission from Elsevier.)

be introduced in Eq. 10.1 in order to calculate $\left[d\chi_0^M - g^{H_2O}(\text{dip})_0\right]$, which can be interpreted as the contribution of the interfacial water layer to the potential drop at the interface. Using a simple model, in which water molecules are considered to be arranged in a square lattice with a lattice constant of 3 Å [29], the so-calculated contribution of the interfacial water layer to the potential drop at the interface can be used to determine the average angle between the water dipole moment and the electrode surface according to

$$\sin\theta = \frac{\varepsilon_0 \left[d\chi_0^M - g^{H_2O}(\text{dip})_0\right]}{N\mu} \qquad (10.2)$$

where N is the number of water molecules per unit area, μ the dipole moment of a water molecule (6.24×10^{-30} C m), ε_0 the permittivity of vacuum, and θ the angle between the dipole moment and the surface.

The values calculated for $\left[d\chi_0^M - g^{H_2O}(\text{dip})_0\right]$ (using a value of 5.9 eV for the work function of both clean and CO-covered Pt(111) [30] and a value of 4.6 eV for the absolute potential of the SHE) and the corresponding values of θ are given in Table 10.2. The average angle between the water dipole moment and the surface at the pzc decreases from 7.9° at the clean Pt(111) surface to 1.5° at the CO-covered Pt(111) surface, suggesting an important change in the structure of the interfacial water layer upon CO adsorption on Pt. This structural change, which has been deduced

TABLE 10.2 Contribution of Interfacial Water Layer to Potential Drop at Interface, Average Angle between Water Dipole Moment and Direction Parallel to Electrode Surface, and Temperature Coefficient of Surface Dipole

	$[\delta\chi_0^M - g^{H_2O}(dip)_0]_{r.t.}$, V vs. SHE	$[\delta\chi_0^M - g^{H_2O}(dip)_0]_{95K}$, V vs. SHE	$\theta_{r.t.}$	θ_{95K}	$\frac{\Delta[\delta\chi_0^M - g^{H_2O}(dip)_0]}{\Delta T}$, mV K^{-1}
Clean Pt(111)	−1.07	−0.7/−1.0	7.48°	5.13°/7.34°	−0.35/−1.85
CO-covered Pt(111)	−0.2	−0.3/−0.6	1.46°	2.20°/4.39°	0.5/2

Source: From ref. 23.

Note: $[\delta\chi_0^M - g^{H_2O}(dip)_0]_{r.t.}$: contribution of interfacial water layer to potential drop at interface at room temperature calculated with Eq. (10.1) using values of pzfc of clean Pt(111) (0.23 V) and CO-covered Pt(111) (1.10 V) determined in present work, a value of 5.9 eV for work function of both clean and CO-covered Pt(111) [30], and value of 4.6 eV for absolute potential of SHE. $[\delta\chi_0^M - g^{H_2O}(dip)_0]_{95K}$: contribution of interfacial water layer to potential drop at interface at 95 K as deduced from decrease in corresponding work function upon dosing clean or CO-covered Pt(111) surface with D_2O in UHV as given in ref. 26. $\theta_{r.t.}$: average angle between dipole moment of water molecules and electrode surface at room temperature calculated using Eq. (11.2) and estimated value of $[\delta\chi_0^M - g^{H_2O}(dip)_0]_{r.t.}$. θ_{95K}: average angle between dipole moment of water molecules and electrode surface at 95 K calculated using Eq. (11.2) and value deduced for $[\delta\chi_0^M - g^{H_2O}(dip)_0]_{95K}$.

here exclusively on the basis of purely electrochemical measurements, is in perfect agreement with the surface-enhanced infrared absorption spectroscopy in the attenuated total reflection configuration (ATR-SEIRAS) results of Miki et al. [31], who found that, upon CO adsorption, hydrogen-bonded water molecules (broad band around $3500 \, cm^{-1}$) are squeezed out of the interface and substituted by water molecules free from hydrogen bonding (sharp band at $3660 \, cm^{-1}$).

10.3.3 Accurate Determination of CO Coverage of Pt Electrodes

The determination of the coverage of electrodes by adsorbed species, either by reductive or oxidative stripping, appears, in principle, to be a simple task, since Faraday's law provides a direct link between the charge crossing the interface and the number of atoms or molecules adsorbed. However, double-layer charging, which must be subtracted from the total charge measured in order to obtain the net faradaic charge, often renders the accurate determination of coverages difficult, if not impossible.

During the early 1990s, the maximum CO coverage attainable on Pt electrodes was still a matter of discussion. Based on measurements of the charge flowing during CO stripping on platinum single-crystal electrodes, Clavilier and the Alicante Group claimed that the CO coverage is higher in sulfuric acid than in perchloric acid (0.85 vs. 0.65 ML, although the latter depended strongly on the procedure used for double-layer correction) [32], while the groups of Weaver and Wieckowski reached the conclusion, based on UHV data, in situ infrared spectroscopy, and in situ STM, that there is no difference in the CO coverage in different electrolytes and that, in the case of Pt(111), the maximum possible coverage is 0.75 ML in the presence of CO in the solution and between 0.60 and 0.70 ML in its absence [33].

This controversy was solved by the development by the Alicante Group of the CO charge displacement method and its application to the calculation of double-layer corrections in CO-stripping voltammograms (for a review, see ref. 21). As far as we know, Wolter and Heitbaum were the first to record the charge flowing during the potentiostatic adsorption of CO [34]. They argued correctly that the change in pzc and double-layer capacity upon formation of the CO adlayer should be included in the calculation of the double-layer correction, but, apparently, they did not realize that these changes are precisely responsible for the charge measured during CO adsorption, which hence provides the key to the determination of the double-layer correction.

As noted in Section 10.3.2, the procedure developed by the Alicante Group was based on the assumption that the charge density on the CO-covered electrode is negligible, and, hence, the charge flowing during the potentiostatic adsorption of CO corresponds to the negative of the charge initially present on the electrode surface. This provides the constant necessary for the determination of the charge density present at any potential on the CO-free platinum electrode by simple integration of the corresponding cyclic voltammogram between the potential at which CO was adsorbed (i.e., at which the charge density on the CO-free electrode surface has been determined) and the desired potential. Subtraction of the charge on the CO-free metal from the CO-stripping charge yielded the CO coverage. With this procedure, they could show that the CO coverage on single-crystal platinum electrodes is independent

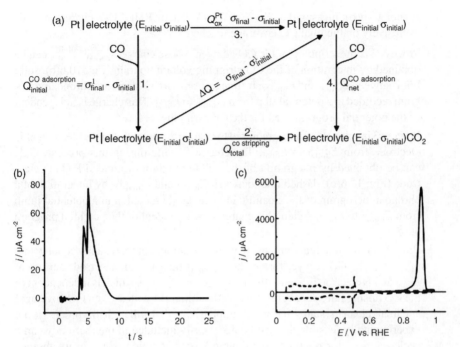

FIGURE 10.6 (a) Thermodynamic cycle from which true, exact thermodynamic double-layer correction necessary for determination of CO coverages from CO-stripping voltammograms can be deduced. (b) CO charge–displacement measurement at $E_{initial} = 0.10$ V. RHE in 0.1 M H_2SO_4 from which $Q_{initial}^{CO\ adsorption}$ can be determined. (c) CO-stripping voltammogram (solid line) and subsequent cyclic voltammogram of clean Pt(111) electrode (dashed line) at 500 mV s^{-1} in 0.1 M H_2SO_4 from which $Q_{total}^{CO\ stripping}$ and Q_{ox}^{Pt}, respectively, can be determined. (From ref. 36. Copyright Wiley-VCH Verlag GmbH & Co. KGaA. Reproduced with permission.)

of the electrolyte used and that for Pt(111) in CO-free solutions, it is always below 0.70 ML [21].

As noted in Section 10.3.2, the assumption that the charge density on the CO-covered electrode is negligible is not exact and, as we have recently shown, unnecessary, since for systems like CO–Pt the charge density is a state function whose value depends only on the state of the system, defined exclusively by the potential and the CO coverage (the pressure, temperature, and electrolyte remain constant in the experiment), and not on how the state has been reached. Therefore, the charge displaced during the potentiostatic CO adsorption can be included in a thermodynamic cycle of charge density, with which the double-layer correction can be calculated [35, 36]. The procedure is illustrated in Figure 10.6(a), which is a thermodynamic cycle composed of the following processes:

- Process 1 is the potentiostatic adsorption of CO, whose charge, $Q_{initial}^{CO\ adsorption}$, is that flowing during a CO charge displacement measurement like that in

Figure 10.6(b), measured at $E_{initial} = 0.10$ V versus RHE. This charge depends exclusively on the final CO coverage reached.

- Process 2 is the stripping of the CO adlayer, whose charge, $Q_{total}^{CO\ stripping}$, can be obtained by integration of the CO-stripping voltammogram [Fig. 10.6(c), solid line] between $E_{initial}$ and E_{final} or by integration of a CO-stripping chronoampero-gram recorded in a potential step from $E_{initial}$ to E_{final}. This charge is independent of the potential program used for the CO stripping.

- Process 3 involves the charge required to change the potential of the clean Pt(111) electrode from $E_{initial}$ to E_{final}. The charge flowing during this process, Q_{ox}^{Pt}, can be obtained by integration of the voltammogram of a clean Pt(111) elec-trode [Fig. 10.6(c), dashed line] between $E_{initial}$ and E_{final} or by integration of a chronoamperogram of the clean Pt(111) electrode recorded in a potential jump from $E_{initial}$ to E_{final}. Again, this charge is independent of the potential program used.

- Process 4 is an imaginary process in which an amount of CO exactly equal to that present in the CO adlayer formed at $E_{initial}$ is dosed to the electrode surface at E_{final} and is electrooxidized at this potential, in a process which is independent of the CO dosing rate, depending only on the initial (Pt with a given CO coverage) and final (clean Pt) states. It is worth emphasizing that although this process is not experimentally feasible, it must be necessarily included in the thermodynamic cycle required for obtaining the double-layer correction, since in its absence it would be impossible to formulate a true cycle. The charge flowing during this process is the sought-after charge density corresponding exclusively to the faradaic oxidation of the CO adlayer, $Q_{net}^{CO\ oxidation}$, and, although not accessible experimentally, it can be calculated, as easily deduced from the thermodynamic cycle, as

$$Q_{net}^{CO\ oxidation} = Q_{total}^{CO\ stripping} - Q_{ox}^{Pt} + Q_{initial}^{CO\ adsorption} \tag{10.3}$$

Therefore, the double-layer correction is simply

$$\Delta Q = Q_{ox}^{Pt} - Q_{initial}^{CO\ adsorption} \tag{10.4}$$

and corresponds to the difference between the surface charge density of the CO-free Pt(111) electrode at E_{final}, σ_{final}^{Pt}, and the surface charge density of the CO-covered Pt(111) electrode at $E_{initial}$, $\sigma_{initial}^{Pt|CO}$.

Obviously, the so-calculated double-layer correction is intrinsically exact, and the error incurred will depend only on the accuracy with which the different charges involved in the calculation can be determined. For this reason, we have called it the *true, exact thermodynamic double-layer correction* [37], and, as illustrated below, we have used this procedure to accurately determine the CO coverage on Pt(111) [37], Pt(100) [38], and Pt(poly) [39] electrodes as a function of potential in CO-free solutions and on Pt(111) electrodes as a function of the CO partial pressure in the gas in equilibrium with the electrolyte [36].

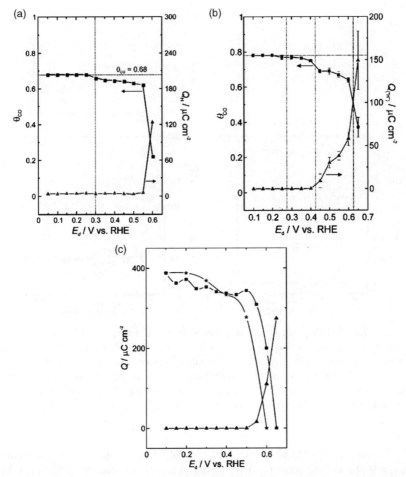

FIGURE 10.7 Plot of CO coverage in CO-free 0.1 M H_2SO_4 as function of dosing potential on Pt(111) (a), Pt(100) (b), and Pt(poly) (c) surfaces. The total charge, without double–layer correction, in the hydrogen region of the cyclic voltammograms (triangles) is also included. The horizontal lines in (a) and (b) correspond to the maximum CO coverage in CO-free solutions. The vertical line in (a) indicates the dosing potential at which a decrease in the CO coverage is detected. The vertical lines in (b) separate the regions giving rise to different types of CO-stripping voltammograms. The stars in (c) correspond to CO–coverage data obtained from the integration of the CO_2 band in FTIR spectra after stripping of the CO adlayer. (Adapted from refs. 37–39. Copyright 2005, 2006 with permission from Elsevier.)

Figure 10.7 shows plots of the true CO coverage obtained with the thermodynamic cycle on Pt(111), Pt(100), and Pt(poly) electrodes in CO-free 0.1 M H_2SO_4 as a function of the dosing potential (E_d). In the case of the Pt(poly) electrode, the coverage as obtained from the integration of the CO_2 band in Fourier transform infrared (FTIR) spectra has also been included.

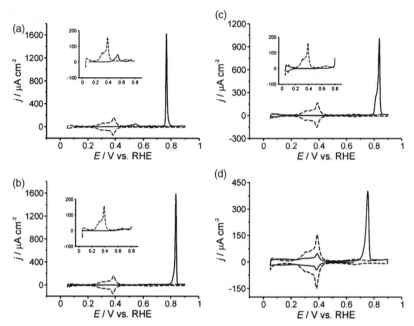

FIGURE 10.8 CO-stripping voltammograms at $50\,mV\ s^{-1}$ (solid line) of Pt(100) electrode covered by CO adlayer formed at $E_d = 0.10\,V$ (a), $E_d = 0.30\,V$ (b), $E_d = 0.55\,V$ (c), and $E_d = 0.65\,V$ (d) in $0.1\,M\ H_2SO_4$. The dashed line corresponds to the cyclic voltammogram at $50\,mV\ s^{-1}$ of a clean and well-ordered Pt(100) electrode in the same solution. The insets in (a), (b), and (c) show the potential region between 0.05 and 0.8 V in an expanded scale. (Reprinted from ref. 38. Copyright 2006, with permission from Elsevier.)

In the case of Pt(111) electrodes, the highest CO coverage in CO-free solution is $\theta_{CO} = 0.68$, and the most positive dosing potential at which the Pt(111) surface is completely blocked for hydrogen adsorption is $E_d = 0.50\,V$ versus RHE, corresponding to $\theta_{CO} = 0.63$, in good agreement with the value of $\theta_{CO} \geq 0.65$ reported by Lebedeva et al. [40].

In the case of Pt(100) electrodes, we observed the formation of four different kinds of CO adlayers as a function of E_d, as illustrated in Figure 10.8:

(i) CO adlayers formed when $E_d \leq 0.40\,V$ versus RHE [Figs. 10.8(a) and (b)], characterized by a complete blocking of hydrogen adsorption, two different types being possible:

   (a) CO adlayers formed when $E_d \leq 0.25\,V$ versus RHE [Fig. 10.8(a)], characterized by a complete blocking of hydrogen adsorption, by a pre peak around 0.55 V versus RHE, and a main CO-stripping peak at 0.77 V versus RHE.

   (b) CO adlayers formed when $0.30\,V \leq E_d \leq 0.40\,V$ versus RHE [Fig. 10.8(b)], characterized by a complete blocking of hydrogen adsorption and a main CO-stripping peak at 0.84 V versus RHE (preceded by a small hump extending from 0.60 to 0.80 V vs. RHE).

(ii) CO adlayers formed when $0.45 \text{ V} \leq E_d \leq 0.60 \text{ V}$ versus RHE [Fig. 10.8(c)], with the hydrogen adsorption states appearing at $E > 0.20 \text{ V}$ versus RHE completely blocked, while those occurring at $E < 0.20 \text{ V}$ are free. CO stripping occurs in two peaks at 0.84 and $\sim 0.80 \text{ V}$ versus RHE.

(iii) Unstable CO adlayers, formed when $E_d = 0.65 \text{ V}$ versus RHE, which completely disappear after 25–30 min in the absence of CO in the solution because this potential already lies at the foot of the CO-stripping peak. They are characterized by the presence of free hydrogen adsorption states both below and above 0.20 V versus RHE and by a single CO-stripping peak at 0.75 V versus RHE [Fig. 10.8(d)].

As can be seen in Figure 10.7(b), the highest CO coverage on Pt(100) electrodes in CO-free solutions is $\theta_{CO} = 0.78$, which can be achieved only when $E_d = 0.20 \text{ V}$ versus RHE, nearly coinciding with the condition necessary for the formation of CO adlayers of type (ia). Between $E_d = 0.25 \text{ V}$ versus RHE and $E_d = 0.40 \text{ V}$ versus RHE the CO coverage decreases slightly, from 0.77 to 0.75 ML, this region nearly exactly coinciding with that where CO adlayers of type (ib) are formed. The value $E_d = 0.40 \text{ V}$ versus RHE is the most positive potential at which the Pt(100) surface is completely blocked for hydrogen adsorption, corresponding to $\theta_{CO} = 0.75$. A steep decrease in the CO coverage occurs for $E_d = 0.45 \text{ V}$ versus RHE, for which $\theta_{CO} = 0.69$. Between E_d values of 0.45 and 0.60 V versus RHE, exactly coinciding with the potential region where CO adlayers of type (ii) are formed, the CO coverage decreases moderately, from 0.69 to 0.64 ML.

In the case of Pt(poly) electrodes, a monotonic decrease of the CO coverage when E_d increases from 0.10 to 0.50 V versus RHE can be observed [Fig. 10.7(c)], completely hydrogen blocking CO adlayers being still formed at $E_d = 0.50 \text{ V}$ versus RHE.

The saturation θ_{CO} value on Pt(111) in CO-free solutions of 0.68 ML found by us is clearly lower than that of $\theta_{CO} = 0.75$ of the (2×2)–3CO adlayer, which, as it is now well established, is the highest possible CO coverage of a Pt(111) electrode [41–43]. This result suggests that removing CO from the solution results in a decrease of θ_{CO} through molecular desorption of some of the CO molecules initially on the surface. Experimental evidence of this partial desorption was obtained by FTIR spectroscopy, the process being completely reversible, with a rapid exchange of CO molecules between the surface and the solution. The experiment, illustrated in Figure 10.9, was performed as follows: After flame annealing and after having checked the cleanliness and [in the case of Pt(111) and Pt(100)] the order of the electrode surface, the potential was set at $E_d = 0.05 \text{ V}$ versus RHE and CO was bubbled through the solution for 15 min, after which the electrode was pressed against the fluorite window and a reference spectrum was taken. Then the CO flow was stopped and we started to bubble N_2 through the solution, taking a spectrum every 5 min while keeping the electrode potential at 0.05 V during the whole experiment. After 60 min, no more changes were observed in the differential spectra. At this point, we recorded a new reference spectrum, stopped the N_2 flow, and started to bubble CO through

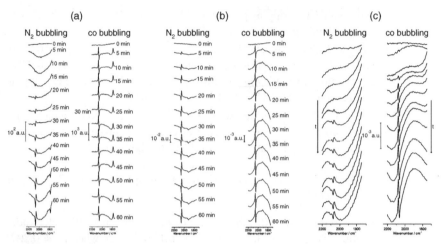

FIGURE 10.9 Time-dependent FTIR spectra of the CO adlayer on Pt(111) (a), Pt(100) (b), and Pt(poly) (c) in 0.1 M H_2SO_4 recorded at constant potential $E = 0.05$ V. The spectra on the left side of each panel were recorded with a 5-min time interval while the initially CO-saturated solution was being purged by bubbling N_2 through it. The spectra on the right side of each panel were recorded immediately afterward, during saturation of the solution with CO. (Adapted from refs. 37–39. Copyright 2005, 2006, with permission from Elsevier.)

the solution, taking a spectrum every 5 min, always without changing the potential, which was kept at 0.05 V. As shown in Figure 10.9 (and as would be expected), the adsorption–desorption process occurs irrespective of the structure of the Pt surface.

Coming back to the calculation of the double-layer correction and taking into account the paragraph above, it must be noted that, due to the thermodynamic nature of the cycle in Figure 10.6(a), the calculated net charge corresponding exclusively to the faradaic oxidation of CO, $Q_{net}^{CO\ oxidation}$, will be strictly correct only if the CO coverage of the adlayer formed during the potentiostatic adsorption of CO and that of the adlayer present on the surface of the Pt electrode at the beginning of the CO-stripping voltammogram are exactly the same, or, from a practical point of view, if the variation of the charge density of the CO-covered electrode between these two CO coverages is negligibly small. When pure CO is blown into the hanging meniscus during the potentiostatic adsorption of CO, the local CO concentration is high, and consequently the CO coverage will be that corresponding to a CO-saturated solution. This coverage will decrease during equilibration of the solution with the purging gas down to that of an irreversibly adsorbed CO adlayer (i.e., a CO adlayer in contact with a CO-free solution). The change in the surface charge density due to the change in CO coverage can be measured if, after a saturated CO adlayer has been formed at $E = 0.10$ V versus RHE on the Pt electrode, and after CO has been removed from the solution by purging with pure N_2 during at least 30 min, CO is again blown into the hanging meniscus (while still keeping the potential at $E = 0.10$ V) in order to increase the CO coverage from that in the CO-free solution to that in the CO-saturated solution, and the current flowing during the process is monitored. Such an experiment is illustrated

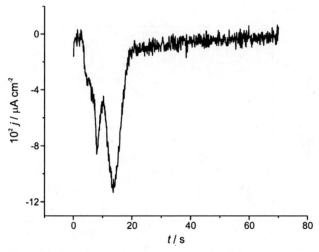

FIGURE 10.10   Current transient measured when CO coverage of Pt(111) electrode in CO-free 0.1 M H_2SO_4 is increased from 0.68 to 0.75 by blowing pure CO into the meniscus at 0.10 V vs. RHE. (From ref. 36. Copyright Wiley-VCH Verlag GmbH & Co. KGaA. Reproduced with permission.)

in Figure 10.10 for the case of Pt(111), from which it was concluded that the charge density of a CO-covered Pt(111) electrode with $\theta_{CO} = 0.75$ is only $1.2 \pm 0.1 \, \mu C$ cm^{-2} more negative than that of a CO-covered Pt(111) electrode with $\theta_{CO} = 0.68$, which is much smaller than the experimental error ($\pm 3 \, \mu C \, cm^{-2}$ for a charge of $140 \, \mu C \, cm^{-2}$ measured at 0.10 V vs. RHE) typical of CO charge displacement experiments. This small change in the charge density demonstrates that the coverage data reported by us are correct and, taking into account that the capacitance ($\sim 15 \, \mu F$ cm^{-2}) of the CO-covered Pt(111) does not vary appreciably between CO-saturated and CO-free solutions, corresponds to a negative shift of $\sim 80 \, mV$ in the potential of zero charge (pzc, which has been determined to be 1.10 ± 0.10 V vs. RHE for $\theta_{CO} = 0.75$ [23]) when the CO coverage decreases from 0.75 to 0.68. Assuming that the structure of the interfacial water layer and hence the potential drop across it do not change appreciably due to the decrease in CO coverage, the shift in the pzc must be due to an equivalent variation in the work function of the CO-covered Pt(111) electrode. We note in passing that a variation of 80 mV in the work function would hardly be measurable in UHV, where typical errors in work function measurements are of the order of 200 mV.

We have plotted in Figure 10.11 the charge of the negative transient recorded during the potentiostatic adsorption of CO on Pt(111) at 0.10 V versus RHE as a function of the CO coverage reached. The charge displaced by CO adsorption at 0.10 V versus RHE increases linearly with increasing CO coverage, but only up to $\theta_{CO} = 0.64$, at which it is $148 \, \mu C \, cm^{-2}$, remaining constant at this value for higher CO coverages. This was to be expected, since hydrogen desorption constitutes the major contribution to the current measured during the potentiostatic adsorption of CO

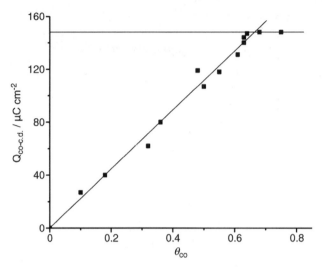

FIGURE 10.11 Plot of charge density displaced during potentiostatic adsorption of CO at 0.10 V vs. RHE as function of CO coverage achieved. The horizontal line indicates the highest measurable charge density, corresponding to $148 \pm 3 \ \mu C \ cm^{-2}$. (From ref. 36. Copyright Wiley-VCH Verlag GmbH & Co. KGaA. Reproduced with permission.)

at 0.10 V, and, as has been shown [37], the hydrogen coverage decreases linearly with increasing CO coverage, but only up to θ_{CO} <0.63–0.65. Additionally, Figure 10.11 indicates that, if the charge displaced during the formation of a saturated CO adlayer is used to determine CO coverages for θ_{CO} <0.64, instead of using the smaller charge displaced during the formation of the subsaturated adlayer, the final coverage will be in error, the error being the larger the lower the coverage.

The CO adlayer with a coverage $\theta_{CO} = 0.68$ (most likely with the $(\sqrt{19} \times \sqrt{19})R23.4°-13CO$ structure) must be especially stable, since it remains on the Pt(111) surface even in CO-free solution, it being necessary to increase the CO partial pressure (P_{CO}) above a certain threshold for the CO coverage to increase above this value [37]. For this reason, we decided to determine quantitatively, using CO-stripping voltammetry, the dependence of the equilibrium CO coverage on Pt(111) electrodes as a function of P_{CO} in the gas in equilibrium with the electrolyte [36]. As a qualitative check of the validity of the coverage data obtained from CO-stripping voltammetry, we used the FTIR fingerprint spectra of the CO adlayers formed on Pt(111) in the presence of different CO concentrations. The high scan rate at which the CO-stripping voltammograms were recorded ($500 \ mV \ s^{-1}$), together with the low CO concentrations used (between 50 ppm and 0.8%), ensured that the amount of dissolved CO oxidized during the potential sweep was negligible compared to the amount of CO adsorbed onto the electrode surface.

The values of the total CO-stripping charges, $Q_{total}^{CO \ stripping}$, measured with different CO concentrations, the corresponding calculated net faradaic charges, $Q_{net}^{CO \ oxidation}$, and the resulting CO coverages are given in Table 10.3, while Figure 10.12 shows the

TABLE 10.3 Experimentally Measured Total CO-Stripping Charge, $Q_{total}^{CO\,stripping}$,[a] Calculated Net Faradaic Charges for CO Stripping, $Q_{net}^{CO\,oxidation}$,[b] and CO Coverages, θ_{CO},[c] in 0.1 M H$_2$SO$_4$ Solutions in Equilibrium with CO–N$_2$ Mixtures of Different CO Concentrations

P_{CO}	$Q_{total}^{CO\,stripping}/\mu C\ cm^{-2}$	$Q_{net}^{CO\,oxidation}/\mu C\ cm^{-2}$	θ_{CO}
0.005%	466 ± 3	327 ± 6	0.68 ± 0.01
0.01%	467 ± 2	328 ± 5	0.68 ± 0.01
0.1%	472 ± 2	333 ± 5	0.69 ± 0.01
0.2%	481 ± 1	342 ± 5	0.71 ± 0.01
0.3%	484 ± 3	345 ± 6	0.72 ± 0.01
0.4%	494 ± 3	355 ± 6	0.74 ± 0.01
0.6%	498 ± 3	359 ± 6	0.75 ± 0.01
0.7%	499 ± 4	360 ± 6	0.75 ± 0.01
0.8%	502 ± 9	363 ± 10	0.76 ± 0.02

Soruce: From ref. 36.

[a]Obtained by integration of CO-stripping voltammograms at 500 mV s^{-1} between 0.10 and 1.00 V.

[b]Calculated using charge density displaced during potentiostatic adsorption of CO at 0.10 V, $Q_{initial}^{CO\,adsorption} = 148 \pm 3\ \mu C\ cm^{-2}$, and charge density obtained by integration of voltammogram of CO-free Pt(111) between 0.10 and 1.00 V, $Q_{ox}^{Pt} = 287 \pm 4\ \mu C\ cm^{-2}$, to determine true, exact thermodynamic double-layer correction.

[c]Calculated as $\theta_{CO} = Q_{net}^{CO\,oxidation} / \left(2Q_{Pt(111)}\right)$, where $Q_{Pt(111)} = 240\ \mu C\ cm^{-2}$ is charge corresponding to a process in which one electron per Pt atom on Pt(111) surface flows across electrode–electrolyte interface.

spectra at 0.05 V of CO adsorbed on Pt(111) electrodes in 0.1 M H$_2$SO$_4$ solutions in equilibrium with N$_2$ + CO gas mixtures with different CO concentrations. For P_{CO} below 0.01%, the CO coverage, $\theta_{CO} = 0.68 \pm 0.01$, coincides with that obtained in CO-free solutions. A clear increase in θ_{CO} was only observed when $P_{CO} \geq 0.1\%$, although FTIR spectroscopy suggests that a coverage higher than 0.68 is already reached at $P_{CO} = 0.01\%$ (see Fig. 10.12). The CO coverage increases monotonically with increasing P_{CO} and already at $P_{CO} = 0.6\%$ reaches $\theta_{CO} = 0.75 \pm 0.01$, which corresponds to the (2×2)–3CO structure observed in CO-saturated solutions [41–43]. The higher coverage determined for $P_{CO} = 0.8\%$ and the higher standard deviation observed ($\theta_{CO} = 0.76 \pm 0.02$) suggest that at this CO concentration a nonnegligible amount of dissolved CO is reaching the CO surface and is oxidized in the positive sweep of the CO-stripping voltammogram.

The spectrum in CO-free solution (Fig. 10.12(a)), is characterized by a band at 1828 cm^{-1}, corresponding to CO$_B$ and characteristic of the $(\sqrt{19} \times \sqrt{19})R23.4°$–13CO structure ($\theta_{CO} = 0.685$) [41]. For $P_{CO} = 0.01\%$, and although, according to the data in Table 10.3, the coverage should still be 0.68 ± 0.01, a small shoulder at 1780 cm^{-1} corresponding to CO$_M$ emerges in the spectrum (Fig. 10.12(b)). This band can be considered as a fingerprint of the (2×2)–3CO structure [41], and its appearance is an indication of a slight increase in the CO coverage. This suggests that, although it is much more difficult to obtain accurate quantitative information regarding the CO

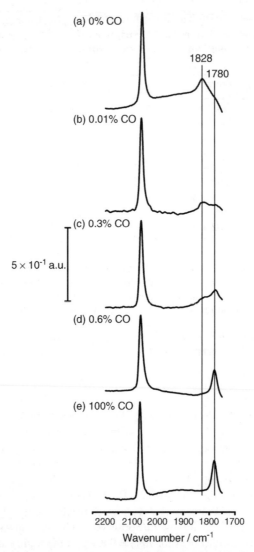

FIGURE 10.12 FTIR spectra at 0.05 V vs. RHE of CO adlayers on Pt(111) electrode in 0.1 M H_2SO_4 solutions in equilibrium with $N_2 + CO$ gas mixtures with $P_{CO} = 0\%$ (a), $P_{CO} = 0.01\%$ (b), $P_{CO} = 0.3\%$ (c), $P_{CO} = 0.6\%$ (d), and $P_{CO} = 100\%$ (e). The reference spectrum was taken at $E_{ref} = 0.80$ V vs. RHE. (From ref. 36. Copyright Wiley-VCH Verlag GmbH & Co. KGaA. Reproduced with permission.)

coverage from FTIR spectra than from CO-stripping voltammograms, the sensitivity of FTIR spectroscopy is high enough as to reflect changes in the CO coverage which cannot be detected by CO-stripping voltammetry. The increase of the CO coverage to $\theta_{CO} = 0.72 \pm 0.01$ when P_{CO} is increased to 0.3% provokes an increase in the intensity of the band of CO_M at 1780 cm^{-1} (Fig. 10.12(c)), and at $P_{CO} = 0.6\%$ the band of

CO_B has practically disappeared (Fig. 10.12(d)), the spectrum being nearly identical with that obtained in CO-saturated solutions (Fig. 10.12(e)), thus confirming the results shown in Table 10.3, according to which the CO coverage attains its maximum value of 0.75 ML already at $P_{CO} = 0.6\%$. Finally, we noted elsewhere [36] that the presence in the FTIR spectra of CO adlayers on Pt(111) electrodes with a coverage intermediate between 0.68 and 0.75 of bands for both CO_B and CO_M suggests that, at these intermediate coverages, the (2 x 2)–3CO and the $(\sqrt{19} \times \sqrt{19})R23.4°$–13CO structures coexist on the Pt(111) surface.

A similar dependence of θ_{CO} on P_{CO} is to be expected on Pt(100), Pt(110), or any other Pt surface. The measurements reported in the literature [36] showed unequivocally that in the coverage range between 0.68 and 0.75 ML all the adsorbed molecules are in equilibrium with the CO in solution. Accordingly, the Clausius–Clapeyron equation

$$\left(\frac{\partial \ln P_{CO}}{\partial (1/T)} \right)_{\theta_{CO}} = -\frac{\Delta H_{ads}}{R}$$

could be applied to obtain the heat of adsorption, $-\Delta H_{ads}$, from measurements of the equilibrium CO coverage as a function of P_{CO} at different temperatures.

10.3.4 Determination of Activation Energies for CO Oxidation on Pt Electrodes: Dependence on Dosing Potential

An important part of our work in the last years has been dedicated, as shown above, to find differences in the properties of CO adlayers on platinum electrodes as a function of E_d. This has been stimulated by the differences observed in CO-stripping voltammograms when E_d is increased, namely an increase in the peak potential and the disappearance of the pre peak observed when CO is adsorbed at potentials within the hydrogen adsorption region. In addition to the clear correlation between E_d and θ_{CO} reported in Section 10.3, we will show here that the activation energy of the electrooxidation of adsorbed CO also shows a clear dependence on the potential at which the CO adlayer is formed.

Herrero et al. [44] have shown that the apparent activation energy for the electrooxidation of adsorbed CO, E_{act}, can be obtained with the equation

$$E_p - E^0 = \frac{RT}{F} \ln \frac{j_p A_{-1} a_{H^+}}{Q_p A_1 A_2 \theta_{CO,p} \theta_{H_2O,p}} + E_{act} \qquad (10.5)$$

where E_p, j_p and Q_p are the peak potential, peak current density, and uncorrected charge, respectively, of the CO stripping peak, E^0 is the standard potential for the electrooxidation of chemisorbed CO, a_{H^+} is the activity of the protons, A_1, A_2, and A_{-1} are preexponential factors, and $\theta_{CO,p}$ and $\theta_{H_2O,p}$ are the coverages of CO and H_2O, respectively, at the potential of the CO stripping peak, with all the potentials referred to the SHE at 298 K. Obviously, E_{act} for the electrooxidation of adsorbed CO

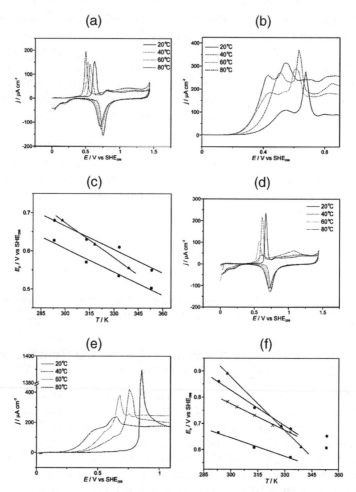

FIGURE 10.13 (a) Voltammograms at 20 mV s^{-1} of Pt wire electrode in 0.1 M HClO$_4$ at 20, 40, 60, and 80°C obtained with CO admission potential of 0.03 V vs. SHE at 298 K in solution first saturated with CO, after which CO in solution was eliminated by N$_2$ bubbling. (b) Voltammograms at 20 mV s^{-1} of Pt wire electrode in 0.1 M HClO$_4$ at 20, 40, 60, and 80°C obtained with CO admission potential of 0.03 V vs. SHE at 298 K in CO-saturated solution. (c) Plot of peak potential vs. SHE at 298 K of adsorbed CO electrooxidation on Pt wire in 0.1 M HClO$_4$ vs. absolute temperature obtained with CO admission potential of 0.03 V vs. SHE in solution first saturated with CO, after which the CO in solution was eliminated by N$_2$ bubbling (squares) and in CO-saturated solution (circles). The triangles are the peak potentials (estimated by us from the CVs in Figure 10.8 of ref. 46) for the electrooxidation of CO on Pt electrodeposited on glassy carbon in CO-saturated 0.5 M H$_2$SO$_4$ with an admission potential of 0.10 V vs. RHE at the cell temperature. (d) Voltammograms at 20 mV s^{-1} of Pt wire electrode in 0.1 M HClO$_4$ at 20, 40, 60, and 80°C obtained with CO admission potential of 0.35 V vs. SHE at 298 K in solution first saturated with CO, after which the CO in solution was eliminated by N$_2$ bubbling. (e) Voltammograms at 20 mV s^{-1} of Pt wire electrode in 0.1 M HClO$_4$ at 20, 40, 60, and 80°C obtained with CO admission potential of 0.35 V vs. SHE at 298 K in CO-saturated

(as that of any other electrochemical reaction) is potential dependent, and the value obtained with Eq. 10.5 corresponds to E_{act} at E^0.

According to Eq. 10.5, the intercept of a plot of E_p versus the absolute temperature should yield E_{act} minus E^0 (the latter was taken to be about 0.0 V vs. SHE), provided that the uncorrected charge under the CO-stripping peak, the CO peak current density, the CO coverage at the CO peak potential, and the water coverage at the CO peak potential remain constant. However, as noted elsewhere [45], this limitation is far from stringent, and, as a matter of fact, at the lower E_d the same value of the slope is obtained irrespective of the presence of CO in solution, although this would lead to sizable changes in some or all of the above parameters.

We have used Eq. 10.5 and CO-stripping voltammograms for determining the activation energy for the electrooxidation of a CO adlayer formed on a smooth Pt wire at $E_d = 0.03$ V versus SHE. However, at a higher E_d of 0.35 V versus SHE there is some electrooxidation of adsorbed CO at the higher temperatures, which makes impossible to maintain the required condition of constant CO coverage. We circumvented this limitation by using the peak potentials obtained in CO-saturated solution [45]. This strategy was based on two facts: (i) even in CO-saturated solution the peak potential is mainly determined by the electrooxidation of adsorbed CO; (ii) as said above, even sizable changes in the four parameters appearing in the first term in Eq. 10.5 should affect little, if at all, the value of the slope. The validity of this approach was confirmed by testing it at $E_d = 0.03$ V, at which potential valid measurements can be obtained both in the absence and presence of CO in solution.

Figure 10.13 shows the CO-stripping voltammograms in 0.1 M HClO$_4$ at 20, 40, 60, and 80 °C and the corresponding plots of the peak potential versus the absolute temperature for CO adlayers formed at $E_d = 0.03$ V versus SHE (Fig. 10.13(a), (b), and (c)) and for CO adlayers formed at $E_d = 0.35$ V versus SHE (Fig. 10.13(d), (e), and (f)). The triangles in Figures 10.13(e) and (f) are the peak potentials (estimated by us from the CVs in Figure 10.8 of ref. 46) for the electrooxidation of CO on Pt electrodeposited on glassy carbon in CO-saturated 0.5 M H$_2$SO$_4$ with an admission potential of 0.10 V versus RHE at the cell temperature. The crosses in Figure 10.13(f) are the peak potentials (estimated by us from the CVs in Figure 10.1 of ref. 47) of adsorbed CO electrooxidation on carbon-supported Pt nanoparticles in CO-free 0.5 M H$_2$SO$_4$ with a CO admission potential of 0.30 V versus RHE at the cell temperature. In the case of CO adlayers formed at 0.03 V,

←———————————————————————————————————————

FIGURE 10.13 (*Continued*) solution. (f) Plot of peak potential vs. SHE of adsorbed CO electrooxidation on Pt wire in CO-saturated (circles) and in CO-free (squares) 0.1 M HClO$_4$ vs. absolute temperature obtained with CO admission potential of 0.35 V vs. SHE. The triangles are the peak potentials (estimated by us from the CVs in Fig. 7 of ref. 46) for the electrooxidation of CO on Pt electrodeposited on glassy carbon in CO-saturated 0.5 M H$_2$SO$_4$ with a CO adsorption potential of 0.45 V vs. RHE. The crosses are the peak potentials (estimated by us from the CVs in Fig. 1 of ref. 47) of adsorbed CO electrooxidation on carbon-supported Pt nanoparticles in CO-free 0.5 M H$_2$SO$_4$ with CO admission potential of 0.30 V vs. RHE at the cell temperature. (Adapted from ref. 45. Copyright 2005, with permission from Elsevier.)

the peak potential decreases linearly with increasing temperature, with slopes of - 2.04 $\pm$ 0.20 and -2.05 $\pm$ 0.26 mV K^{-1} in the absence and presence of CO in the solution, respectively. These slopes are the same within experimental error, which fully confirms the above estimation of the negligible effect on the slope of the peak potential–temperature plot of large changes in the four parameters appearing in Eq. 10.5. The activation energy of CO electrooxidation, as given by the intercept of the plot, is 1.22 $\pm$ 0.06 V (118 $\pm$ 6 kJ mol^{-1}) and 1.28 $\pm$ 0.08 V (123 $\pm$ 8 kJ mol^{-1}) for CO-free and CO-saturated solutions, respectively. The 60-mV difference is due to the fact that the peak potentials are 60 mV higher in the latter case, since CO in solution will tend to adsorb on those Pt sites left free by the first adsorbed CO molecules oxidized, healing the CO adlayer and hindering further electrooxidation.

In the case of CO adlayers formed at $E_d = 0.35$ V, in CO-free solution the CO peak decreases already at 60°C and very much more at 80°C [see Fig. 10.13(d)]. Concomitantly there is an increase of the hydrogen charge in the first positive sweep, which is negligible at 20 and 40°C, while at 60 and 80°C it is 11 and 20% of the hydrogen charge for a completely CO-free surface. The obvious conclusion is that already at 60°C there is some electrooxidation of adsorbed CO. Consequently, the activation energy of 1.35 V obtained from the intercept of the plot of the peak potential–absolute temperature (Fig. 10.13(f), squares) is largely in error, because the conditions necessary for Eq. 10.5 to apply do not obtain. In CO-saturated solution, the activation energy at $E_d = 0.35$ V is 2.17 V (209 kJ mol^{-1}), which is 70% higher than that in CO-saturated solution with $E_d = 0.03$ V. This increase of the activation energy with increasing admission potential, although in apparent contradiction with the behavior of the CO coverage, whose dependence on E_d has been discussed in the previous section, is in agreement with the fact that a prepeak in the CO-stripping voltammogram appears only when $E_d < 0.30$ V, suggesting an intimate relationship between the activation energy and the presence or absence of a pre-peak.

10.3.5 Prepeak

As said in Section 10.2, it is now firmly established that the potential at which CO is first admitted to the electrolytic cell crucially determines the electrocatalytic properties of Pt for dissolved CO electrooxidation: If the admission potential is lower than 0.3 V versus RHE, a prepeak of dissolved (plus some adsorbed) CO electrooxidation appears at potentials as low as 0.5 V, followed by the usual peak at about 0.9 V, while if the Pt electrode is held at higher potentials (or at open circuit, as was the usual practice) during CO admission only the peak at 0.9 V appears. Actually, advantage has been taken of this dependence of the Pt activity on the CO admission potential to show in a simple, inexpensive way, without the need of any UHV spectroscopy and a transfer chamber, but only with conventional electrochemical equipment, that the double layer on a Pt electrode remains intact upon emersion under potential control [48]. However, the nature of the processes taking place in the prepeak remains so obscure that often CO electrooxidation on Pt is purposely studied using conditions under which the prepeak does not appear, despite it being much more interesting, from the point of view of electrocatalysis, than the main-peak region.

Initially, the prepeak was assigned, with basis on early infrared studies using electrochemically-modulated infrared spectroscopy (EMIRS), to the oxidation of bridge-bonded adsorbed CO [49]. However, the use of single-crystal electrodes in combination with STM and FTIR spectroscopy (see, e.g., the seminal paper by Villegas and Weaver [41]) led to the conclusion that the presence and relative coverage of different kinds of CO on the Pt surface (on-top, twofold, three or fourfold hollow) depend on the surface structure and the CO coverage, rapid interconversion between different kinds of adsorbed CO being produced by minor changes in CO coverage, and that it is impossible to relate a given voltammetric peak to the oxidation of a given kind of adsorbed CO.

Villegas and Weaver [41] were the first to unequivocally show by STM that in CO-saturated solution the oxidation of a fraction of the adsorbed CO which occurs in the prepeak produces a change in the structure of the CO adlayer on Pt(111), from the (2 x 2)–3CO ($\theta_{CO} = 0.75$) structure at potentials lower than 0.3 V versus RHE to the (v19 x v19)23.4°–13CO ($\theta_{CO} = 0.684$) structure at higher potentials, and that these structures had different IR spectra, with a peak of threefold-hollow CO and a peak of bridge CO, respectively, besides a common peak of linear CO. This change of structure has been confirmed by second harmonic generation (SHG) [12, 13, 50–54] and by in-situ surface X-ray scattering (SXS) [43, 55].

Typically, when a prepeak of dissolved plus some adsorbed CO electrooxidation appears in the first CV of a polycrystalline or single-crystal electrode, as a rule this prepeak does not appear or is much smaller in the second CV since then the adsorption of CO has taken place at high potentials during the first negative sweep. In experiments with Pt(111) in CO-saturated solutions in which the intensity of the SHG signal of a CO adlayer initially formed at 0.10 V was recorded simultaneously with the current [12, 13], it was found that in the first positive sweep the intensity of the SHG signal increased at around 0.60 V, together with the oxidation of a fraction of the adsorbed CO in the prepeak. Although in the first negative sweep the SHG signal recovered the level characteristic of the (2 x 2)–3CO structure at 0.60 V and retained this level in the second positive sweep up to 0.75 V, no prepeak appeared in the second CV. Similarly, formation of the (2 x 2)–3CO structure (as reflected by the corresponding X-ray diffraction intensity) in the first negative sweep (although with an intensity ~10% lower than that measured in the same region in the first positive sweep) has been observed by in situ surface X-ray diffraction [43, 55]. However, in the subsequent second positive sweep, and although the CO coverage and the structure of the CO adlayer, as indicated by the X-ray diffraction intensity, were the same as in the first positive sweep, the prepeak in the CV had disappeared. In addition to the failure of SHG and in situ surface X-ray diffraction to detect differences in the CO adlayer that determine the presence and absence of the prepeak of CO oxidation in the first and second positive sweeps, respectively, it has been found that cyclic voltammetry does not allow at all to predict, with basis on the amount of defects or the long-range order of the surface, if a Pt(111) electrode will show a prepeak or not [37]. Therefore, the differences in the structure of the substrate or of the CO adlayer responsible for the prepeak remain undetectable.

We decided to look for differences between the CO adlayers for which a prepeak can be observed (those adsorbed at E <0.30 V), and the CO adlayers for which the prepeak is absent (those formed at $E > 0.30$ V) in terms of coverage and activation energy, as we have described in Sections 10.3.3. and 10.3.4. The differences observed led us to suggest [37–39] that the oxidation of a fraction of the adsorbed CO molecules in the prepeak requires their surface diffusion to especially reactive sites, postulated to be step sites, and therefore should be completely absent for an ideal Pt(111) surface. Consequently, the rate at which the CO molecules diffuse to the reactive sites should decrease with decreasing CO coverage (and, most likely, should also depend on the long-range order of the CO adlayer), this then being the reason why a prepeak appears only for CO adlayers with the highest possible coverage.

We also suggested [37–39] that the main peak in the stripping CVs of adsorbed CO corresponding to the complete oxidation of the CO adlayer occurs once adsorbed OH can also form on the terraces. On the contrary, Koper and co-workers have provided evidence that, in acidic media, Pt step sites possess a unique activity for CO oxidation, all CO reacting to CO_2 at the step sites only, even in the main peak [56, 57]. However, the same group has explained the observation of a prepeak and up to two peaks in the electrooxidation of adsorbed CO on Rh single-crystal electrodes in sulfuric acid [58], and on Pt single-crystal electrodes in 0.1 M NaOH [59], as being due to the oxidation at different sites, terraces included, the deconvolution of the different contributions being possible due to the low mobility of CO adsorbed on the terraces in these cases.

Taking all these data into account, the prepeak must be due to the presence on the platinum surface of especially reactive sites, whose nature is at this moment unknown. When high-coverage CO adlayers are formed at $E_d < 0.3$ V, in the potential region corresponding to the prepeak up to 15% of the adsorbed CO can diffuse to these sites and react with adsorbed OH to CO_2. When lower coverage adlayers are formed at $E_d > 0.3$ V, in the region of the prepeak a negligible amount of adsorbed CO can diffuse to these sites, and consequently CO oxidation occurs nearly exclusively in the main-peak region.

Probably the most striking fact in relation with the prepeak is the finding that, in CO-containing solutions (and provided that CO has been adsorbed at $E_d < 0.3$ V), in the prepeak region *diffusion-limited* CO electrooxidation occurs in the presence of a high-coverage CO adlayer. This behavior was independently and nearly simultaneously explained by Wieckowski et al. [8] and by Kita et al. [9], who suggested and presented evidence that oxidation of $\sim$15% of adsorbed CO in the prepeak produces CO-free islands of Pt atoms that act as microelectrodes whose hemispherical diffusion layers would overlap, creating a planar diffusion front and behaving, under quiescent conditions, as a wholly active Pt surface. Not surprisingly, it has been found that these CO-free islands are also active sites for the electrooxidation in the lower potential range of H_2 on a Pt/Vulcan carbon fuel cell catalyst in 0.5 M H_2SO_4 at 60°C in the presence of 250 ppm CO [60].

Finally, we would like to note that, very recently, Kucernak and Offer have proposed an alternative interpretation of the electrooxidation of adsorbed CO [61] which is the exact opposite of the widely accepted model of adsorbed CO molecules diffusing to OH groups adsorbed in fixed step sites. In their model, it is the diffusion of OH groups that would control the rate at which the electrooxidation of adsorbed CO occurs.

10.4 CO POISONING OF Pt-BASED ELECTRODES IN RELATION WITH FUEL CELL TECHNOLOGIES

Arguably, a main, if not the critical, reason for the extensive work done during the last decades on the adsorption of CO on Pt electrodes is the ubiquitous presence of adsorbed carbon monoxide as a poison in fuel cell anodes. In some cases, CO is present in the fuel as an impurity, as in the hydrogen produced by methanol reforming, and in other cases, CO is a byproduct of the fuel electrooxidation, as in fuel cells burning formic acid or methanol.

The most usual strategy to decrease the extent of poisoning of the electrocatalyst by carbon monoxide is to alloy platinum (or, alternatively, to decorate the platinum surface) with one or several different metals, like Ru [62–67], Os [68, 69], Sn [6, 70, 71], Bi [72–74], Sb [75], and Mo [76]. The addition of a second and, sometimes, a third metal can decrease the strength of the bond between Pt and CO (electronic effect), increase the activity toward the oxidation of adsorbed CO through a Langmuir–Hinshelwood mechanism by facilitating the formation of adsorbed OH (bifunctional mechanism), or both.

In the case of formation of adsorbed CO from formic acid or methanol, a different strategy can be conceived. In order to form adsorbed CO, these molecules need to interact simultaneously with several atoms of the electrode surface in order to break their carbon–hydrogen bonds and, in the case of formic acid, one of the carbon–oxygen bonds, this being the reason why the electrooxidation of formic acid and methanol is sensitive to the surface structure. Accordingly, there must be a minimum number of surface atoms, with a given geometric distribution, necessary for a given reaction or reaction step to proceed. This is the critical atomic ensemble, a concept developed in heterogeneous gas-phase catalysis [77], from where it has been borrowed by electrocatalysis [78].

In our group, we have recently worked on the identification of the smallest atomic ensemble necessary for the formation of adsorbed CO during the electrooxidation of methanol on platinum using cyanide-modified Pt(111) electrodes [79]. Cyanide adsorbs on Pt(111) electrodes adopting a $(2\sqrt{3} \times 2\sqrt{3})$–$R30°$ structure, first detected using low-energy electron diffraction (LEED) by Stickney et al. [80] and whose atomic, real-space structure could be elucidated using in situ STM by Stuhlmann et al. [81] and later by Kim et al. [82]. The structure (Fig. 10.14) consists of hexagonally packed arrays, each containing six CN groups adsorbed on top of a hexagon of Pt atoms surrounding a free Pt atom. The cyanide coverage corresponding to the $(2\sqrt{3} \times 2\sqrt{3})$–$R30°$ structure on Pt(111) is $\theta_{CN} = 0.5$. As shown by Huerta et al. [83–87], the cyanide adlayer on Pt(111) is remarkably stable, no change being observed in the CVs of the cyanide-covered electrode upon repetitive cycling between 0.06 and 1.10 V versus RHE. Accordingly, as pointed out by Huerta et al. [83], the cyanide-covered Pt(111) electrode can be considered as a chemically modified electrode, with CN groups acting as a third body, blocking those Pt atoms onto which they are adsorbed but leaving unaffected the CN-free Pt atoms onto which H, OH, CO [87, 88], or NO [89] can adsorb.

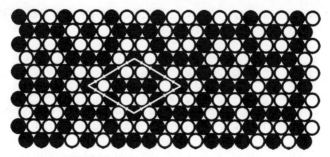

FIGURE 10.14 Ball model of $(2\sqrt{3} \times 2\sqrt{3})R30°$ structure adopted by cyanide adlayer on Pt(111) electrodes according to ref. 74. Black balls correspond to Pt atoms, and white balls to linearly chemisorbed CN groups. (Reprinted with permission from ref. 79. Copyright 2006 American Chemical Society.)

As can be seen in Figure 10.15, methanol can be oxidized on the cyanide-modified Pt(111) surface. The hydrogen adsorption region remains unaffected, and, contrary to what is observed in unmodified Pt(111) electrodes, there is nearly no hysteresis between the positive- and the negative-going sweeps. These two observations suggest that no poison intermediate (adsorbed CO) is being formed on the surface of the cyanide-modified Pt(111) electrode, as confirmed by the FTIR spectra shown in Figure 10.16 (the band with frequencies increasing with potential from 2086 to $2135\,cm^{-1}$ corresponds to CN adsorbed on-top [81–84, 90, 91]).

There are four possible kinds of reactive sites on the cyanide-modified Pt(111) surface (see Fig. 10.14): (i) a single Pt atom, (ii) two adjacent Pt atoms, (iii) three Pt atoms arranged linearly, and (iv) three Pt atoms arranged forming a chevron with a

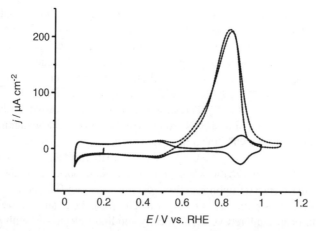

FIGURE 10.15 Cyclic voltammogram at $50\,mV\,s^{-1}$ of cyanide-modified Pt(111) electrode in 0.1 M HClO$_4$ +0.2 M CH$_3$OH. The scan starts at 0.20 V in the negative direction. The solid line is the cyclic voltammogram of cyanide-modified Pt(111) in the blank electrolyte. (Reprinted with permission from ref. 79. Copyright 2006 American Chemical Society.)

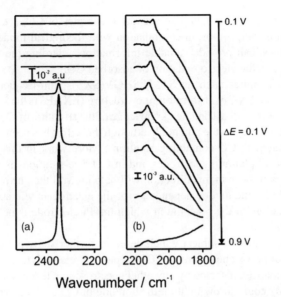

FIGURE 10.16 FTIR spectra at increasing potentials of a cyanide-modified Pt(111) electrode in 0.1 M HClO$_4$ + 0.2 M CH$_3$OH. The spectra in the frequency region between 2500 and 2200 cm^{-1} (a) were calculated using the spectrum at 0.05 V as reference, while the spectra in the frequency regions between 2200 and 1800 cm^{-1} (b) were calculated using the spectrum at 1.30 V as reference. (Reprinted with permission from ref. 79. Copyright 2006 American Chemical Society.)

120° angle. Accordingly, we reached [79] the conclusion that none of these reactive sites suffices for the dehydrogenation reaction of methanol to CO to occur, at least three contiguous Pt atoms being necessary. On the contrary, and since simultaneous interaction of the methanol molecule with three Pt atoms on the cyanide-modified Pt(111) electrode seems unlikely, we suggested that the minimum atomic ensemble necessary for the oxidation of methanol to CO$_2$ through the non-CO pathway consists of two adjacent platinum atoms. Identification of the critical atomic ensemble for formation of adsorbed CO during the electrooxidation of methanol could help design an electrocatalyst with enhanced activity, in which platinum atoms would be forced to form groups on the electrocatalyst surface smaller than the critical atomic ensemble for CO formation from methanol.

10.5 SUMMARY AND CONCLUSIONS

It is unquestionable that the effort devoted in the last decades to the study of the CO–Pt system has been fruitful. At the macroscopic level, the CO charge displacement technique is now duly recognized as able to accurately measure the CO coverage, which was a subject of heated controversy not so long ago. Careful use of other electrochemical techniques, combined with a thermodynamic analysis of the data can provide us with important quantitative data, like activation energies of CO_{ads} electrooxidation or CO adsorption energies, or, in combination with theoretical modelling and simulation techniques, can provide valuable information regarding the role played by surface diffusion in the electrooxidation of adsorbed CO.

Advanced techniques such as STM, SHG, (SXS), sum-frequency generation (SFG), and extended X-ray absorption fine structure (EXAFS) (the last two not discussed here) have provided a detailed knowledge of the structure of CO adlayers on Pt and of their dependence on the potential, although this effort has been mainly focused on Pt(111) surfaces in CO-containing solutions. More work in this direction with other single-crystal platinum electrodes and in CO-free solutions is still necessary. Hopefully, new advances in this direction will help unravel the elusive nature of the sites responsible for the unique dependence of the electrocatalytic activity of Pt for CO electrooxidation on the potential at which the Pt electrode is held while CO is introduced in the cell.

In our opinion, another subject which deserves further study is the nature of the electronic transition of chemisorbed CO detected by electrolyte electroreflectance and which has not been detected by any other spectroscopic technique. We think that this kind of study could throw light on the bonding of CO to Pt, as we have recently shown for the case of CO adsorption on Ag and Au electrodes [92].

REFERENCES

1. L. Stryer, *Biochemistry*, W. H. Freeman, New York, 1988.
2. E. Abel, *Ed. Chem.*, 29 (1992) 46.
3. S. Gilman, *J. Phys. Chem.*, 68 (1964) 70.
4. M. W. Breiter, *J. Phys. Chem.*, 72 (1968) 1305.
5. M. W. Breiter, *J. Electroanal. Chem.*, 101 (1979) 329.
6. H. Kita, K. Shimazu, and K. Kunimatsu, *J. Electroanal. Chem.*, 241 (1988) 163.
7. J. A. Caram and C. Gutiérrez, *J. Electroanal. Chem.*, 305 (1991) 259.
8. A. Wieckowski, M. Rubel, and C. Gutiérrez, *J. Electroanal. Chem.*, 382 (1995) 97.
9. H. Kita, H. Naohara, T. Nakato, S. Taguchi, and A. Aramata, *J. Electroanal. Chem.*, 386 (1995) 197.
10. A. Couto, A. Rincon, M.C. Pérez, and C. Gutiérrez, *Electrochim. Acta*, 46 (2001) 1285.
11. J. A. Caram and C. Gutiérrez, *J. Electroanal. Chem.*, 305 (1991) 275.
12. W. Akemann, K. A. Friedrich, U. Linke, and U. Stimming, *Surf. Sci.*, 402–404 (1998) 571.
13. W. Akemann, K. A. Friedrich, and U. Stimming, *J. Chem. Phys.*, 113 (2000) 6864.
14. *Symp. Faraday, Soc.*, 4 (1970).
15. *Faraday Discuss. Chem. Soc.*, 56 (1973).
16. K. L. Shaklee, F. H. Pollak, and M. Cardona, *Phys. Rev. Lett.*, 15 (1965) 883.
17. A. Bewick, J. M. Mellor, and B. S. Pons, *Electrochim. Acta*, 25 (1980) 931.
18. D. M. Kolb, *Ber. Bunsenges. Phys. Chem.*, 92 (1988) 1175.
19. D. M. Kolb, in *Spectroelectrochemistry: Theory and Practice*, R. J. Gale (Ed.), Plenum, New York, 1988, pp. 87–188.
20. A. Cuesta and C. Gutiérrez, *J. Electroanal. Chem.*, 383 (1995) 195.

21. V. Climent, R. Gómez, J. M. Orts, A. Rodes, A. Aldaz, and J. M. Feliu,in *Interfacial Electrochemistry*, A. Wieckowski (Ed.), Marcel Dekker, New York, 1999, pp. 463–475.

22. M. J. Weaver, *Langmuir*, 14 (1998) 3932.

23. A. Cuesta, *Surf. Sci.*, 572 (2004) 11.

24. U. W. Hamm, D. Kramer, R. S. Zhai, and D. M. Kolb, *J. Electroanal. Chem.*, 414 (1996) 85.

25. P. Ramírez, R. Andreu, A. Cuesta, C. J. Calzado, and J. J. Calvent, *Anal. Chem.*, 79 (2007) 6473.

26. N. Kizhakevariam, I. Villegas, and M. J. Weaver, *J. Phys. Chem.*, 99 (1995) 7677.

27. S. Trasatti, in *Advances in Electrochemistry and Electrochemical Engineering*, Vol. 10, H. Gerischer and C. W. Tobias (Eds.), Wiley, New York, 1977, pp. 213–321.

28. F. T. Wagner, in *Structure of Electrified Interfaces*, J. Lipkowski and P. N. Ross (Eds.), *Frontiers in Electrochemistry*, Vol. 2, VCH, New York, 1993, p. 310.

29. W. Schmickler, *Interfacial Electrochemistry*, Oxford University Press, *New York*, 1996.

30. H. H. Rotermund, S. Jakubith, S. Kubala, A. von Oertzen, and G. Ertl, *J. Electr. Spec. Relat. Phen.*, 52 (1990) 811.

31. A. Miki, S. Ye, and M. Osawa, *Chem. Commun.*, (2002) 1500.

32. J. M. Orts, A. Fernández-Vega, J. M. Feliu, A. Aldaz, and J. Clavilier, *J. Electroanal. Chem.*, 327 (1992) 261.

33. M. J. Weaver, S.-C. Chang, L.-W. H. Leung, X. Jiang, M. Rubel, M. Szklarczyk, D. Zurawski, and A. Wieckowski, *J. Electroanal. Chem.*, 327 (1992) 247.

34. O. Wolter and J. Heitbaum, *Ber. Bunsenges. Phys. Chem.*, 88 (1984) 6.

35. A. López-Cudero, A. Cuesta, and C. Gutiérrez, *J. Electroanal. Chem.*, 548 (2003) 109.

36. A. Cuesta, M. d. C. Pérez, A. Rincón, and C. Gutiérrez, *ChemPhysChem*, 7 (2006) 2346.

37. A. López-Cudero, A. Cuesta, and C. Gutiérrez, *J. Electroanal. Chem.*, 579 (2005) 1.

38. A. López-Cudero, A. Cuesta, and C. Gutiérrez, *J. Electroanal. Chem.*, 586 (2006) 204.

39. A. Cuesta, A. Couto, A. Rincón, M. C. Pérez, A. López-Cudero, and C. Gutiérrez, *J. Electroanal. Chem.*, 586 (2006) 184.

40. N. P. Lebedeva, M. T. M. Koper, J. M. Feliu, and R. A. van Santen, *J. Electroanal. Chem.*, 524–525 (2002) 242.

41. I. Villegas and M. J. Weaver, *J. Chem. Phys.*, 101 (1994) 1648.

42. N. M. Markovic, B. N. Grgur, C. A. Lucas, and P. N. Ross, *J. Phys. Chem. B*, 103 (1999) 487.

43. C. A. Lucas, N. M. Markovic, and P. N. Ross, *Surf. Sci.*, 425 (1999) L381.

44. E. Herrero, J. M. Feliu, S. Blais, Z. Radovic-Hrapovic, and G. Jerkiewicz, *Langmuir*, 16 (2000) 4779.

45. A. Rincón, M. C. Pérez, A. Cuesta, and C. Gutiérrez, *Electrochem. Commun.*, 7 (2005) 1027.

46. R. J. Bellows, E. P. Marucchi-Soos, and D. T. Buckley, *Ind. Eng. Chem. Res.*, 35 (1996) 1235.

47. T. Kawaguchi, W. Sugimoto, Y. Murakami, and Y. Takasu, *Electrochem. Commun.*, 6 (2004) 480.

48. A. Couto, M. C. Pérez, A. Rincón, and C. Gutiérrez, *Langmuir*, 13 (1997) 2572.

49. K. Kunimatsu, H. Seki, W. G. Golden, J. G. GordonII, and M. R. Philpott, *Langmuir*, 2 (1986) 464.

50. B. Pozniak, Y. Mo, I. C. Stefan, K. Mantey, M. Hartmann, and D. A. Scherson, *J. Phys. Chem. B*, 105 (2001) 7874.

51. B. Pozniak, Y. Mo, and D. A. Scherson, *Faraday Discuss.*, 121 (2002) 313.

52. B. Pozniak and D. A. Scherson, *J. Am. Chem. Soc.*, 125 (2003) 7488.

53. B. Pozniak and D. A. Scherson, *J. Am. Chem. Soc.*, 126 (2004) 14696.

54. I. Fromondi and D. A. Scherson, *J. Phys. Chem. B*, 110 (2006) 20749.

55. D. S. Strmcnik, P. Rebec, M. Gaberscek, D. Tripkovic, V. Stamenkovic, C. Lucas, and N. M. Markovic, *J. Phys. Chem. C*, 111 (2007) 18672.

56. N. P. Lebedeva, M. T. M. Koper, J. M. Feliu, and R. A. van Santen, *J. Phys. Chem. B*, 106 (2002) 12938.

57. M. T. M. Koper, N. P. Lebedeva, and C. G. M. Hermse, *Faraday Discuss.*, 121 (2002) 00.

58. T. H. M. Housmans and M. T. M. Koper, *Electrochem. Commun.*, 7 (2005) 581.

59. G. García and M. T. M. Koper, *Phys. Chem. Chem. Phys.*, 10 (2008) 3802.

60. T. J. Schmidt, H. A. Gasteiger, and R. J. Behm, *J. Electrochem. Soc.*, 146 (1999) 1296.

61. A. R. Kucernak and G. J. Offer, *Phys. Chem. Chem. Phys.*, 10 (2008) 3699.

62. O. A. Petry, B. I. Podlovchenko, and A. N. Frumkin, *J. Electroanal. Chem.*, 10 (1965) 253.

63. S. Watanabe and S. Motoo, *J. Electroanal. Chem.*, 60 (1975) 267.

64. J. B. Goodenough, A. Hamnett, B. J. Kennedy, R. Manoharan, and S.A. Weeks, *J. Electroanal. Chem.*, 240 (1988) 133.

65. H. A. Gasteiger, N. M. Markovic, P. N. RossJr., and E. J. Cairns, *J. Phys. Chem.*, 97 (1993) 12020.

66. W. Chrzanowski and A. Wieckowski, *Langmuir*, 13 (1997) 5974.

67. C. Roth, N. Benker, R. Theissmann, R. J. Nichols, and D. J. Schiffrin, *Langmuir*, 24 (2008) 2191.

68. K. L. Ley, R. Liu, C. Pu, Q. Fan, N. Leyarovska, C. Segres, and E. S. Smotkin, *J. Electrochem. Soc.*, 144 (1997) 1543.

69. C. M. Johnston, S. Strbac, A. Lewera, E. Sibert, and A. Wieckowski, *Langmuir*, 22 (2006) 8229.

70. C. T. Hable and M. S. Wrighton, *Langmuir*, 9 (1993) 3284.

71. V. Stamenkovic, M. Arenz, B. B. Blizanac, K. J. J. Mayrhofer, P. N. Ross, and N. M. Markovic, *Surf. Sci.*, 576 (2005) 145.

72. J. Clavilier, A. Fernández-Vega, J. M. Feliu, and A. Aldaz, *J. Electroanal. Chem.*, 258 (1989) 89.

73. E. Casado-Rivera, Z. Gál, A. C. D. Angelo, C. Lind, F. J. DiSalvo, and H. D. Abruña, *ChemPhysChem*, 4 (2003) 193.

74. A. V. Tripkovic, K. D. Popovic, R. M. Stevanovic, R. Socha, and A. Kowal, *Electrochem. Commun.*, 8 (2006) 1492.

75. A. Fernández-Vega, J. M. Feliu, A. Aldaz, and J. Clavilier, *J. Electroanal. Chem.*, 258 (1989) 101.

76. S. L. Gojgovic, A. V. Tripkovic, R. M. Stevanovic, and N. V. Krstajic, *Langmuir*, 23 (2007) 12760.

77. C. T. Campbell, in *Handbook of Heterogeneous Catalysis*, Vol. 2, G. Ertl, H. Knözinger, and J. Weitkamp (Eds.), Wiley, Weinheim, 1997, p. 814.

78. P. N. Ross, in *Electrocatalysis*, J. Lipkowski and P. N. Ross (Eds.), Wiley, New York, 1998, p. 43.

79. A. Cuesta, *J. Am. Chem. Soc.*, 128 (2006) 13332.

80. J. L. Stickney, S. D. Rosasco, G. N. Salaita, and A. T. Hubbard, *Langmuir*, 1 (1985) 66.

81. C. Stuhlmann, I. Villegas, and M. J. Weaver, *Chem. Phys. Lett.*, 219 (1994) 319.

82. Y.-G. Kim, S.-L. Yau, and K. Itaya, *J. Am. Chem. Soc.*, 118 (1996) 393.

83. F. Huerta, E. Morallón, C. Quijada, J. L. Vázquez, and A. Aldaz, *Electrochim. Acta*, 44 (1998) 943.

84. F. J. Huerta, E. Morallon, J. L. Vazquez, and A. Aldaz, *Surf. Sci.*, 396 (1998) 400.

85. F. Huerta, E. Morallon, C. Quijada, J. L. Vazquez, and L. E. A. Berlouis, *J. Electroanal. Chem.*, 463 (1999) 109.
86. F. Huerta, E. Morallón, and J. L. Vázquez, *Surf. Sci.*, 431 (1999) L577.
87. F. Huerta, E. Morallón, and J. L. Vázquez, *Electrochem. Commun.*, 4 (2002) 251.
88. I. Morales-Moreno, A. Cuesta, and C. Gutiérrez, *J. Electroanal. Chem.*, 560 (2003) 135.
89. A. Cuesta and M. Escudero, *Phys. Chem. Chem. Phys.*, 10 (2008) 3628.
90. C. S. Kim and C. Korzeniewski, *J. Phys. Chem.*, 97 (1993) 9784.
91. W. Daum, K. A. Friedrich, C. Klünker, D. Knabben, U. Stimming, and H. Ibach, *Appl. Phys., A* 59 (1994) 553.
92. A. Cuesta, N. López, and C. Gutiérrez, *Electrochim. Acta*, 48 (2003) 2949.

CHAPTER 11

Exploring Metal Oxides:
A Theoretical Approach

MÓNICA CALATAYUD[1,2] and FREDERIK TIELENS[3,4]

[1]UPMC Univ Paris 06, UMR 7616, Laboratoire de Chimie Théorique, F-75005, Paris, France
[2]CNRS, UMR 7616, Laboratoire de Chimie Théorique, F-75005, Paris, France
[3]UPMC Univ Paris 06, UMR 7609, Laboratoire de Réactivité de Surface, F-75005, Paris, France
[4]CNRS, UMR 7609, Laboratoire de Réactivité de Surface, F-75005, Paris, France

11.1 INTRODUCTION

As there is ever concern over the provision of future energy, there is significant interest in the development of alternative energy sources. One of the more promising possibilities for future energy generation is the solid-oxide fuel cell. Such fuel cells represent an all-ceramic electrochemical power generation device that relies on the transport of oxide ions or protons across a ceramic membrane operating at temperatures in the range of 500–1000°C. Different families of fuel cells are the proton exchange fuel cell, the oxygen ion exchange fuel cell (solid-oxide fuel cell), and the proton exchange membrane (PEM) fuel cell. Metal oxides come into play at different stages of the fuel cell development.

Hydrogen as an energy source can be obtained primarily from water, coal, petroleum, and natural gas. The true renewable energy follows the following cycle with hydrogen as the energy carrier:

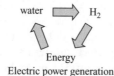

water ⟹ H₂

Energy
Electric power generation

Production of hydrogen from water is achieved with the aid of solar energy in indirect and direct approaches. The ideal process is the direct approach based on the photocatalytic splitting of water without the assistance of external bias potential.

Catalysis in Electrochemistry: From Fundamentals to Strategies for Fuel Cell Development,
First Edition. Edited by Elizabeth Santos and Wolfgang Schmickler.

However, this direct way of hydrogen production from water by solar energy in a commercially and technologically viable manner appears to be unrealistic in the near future. Many other strategies have been followed in hydrogen energy nevertheless based on the extraction of hydrogen from fossil fuel. This strategy will not solve the energy problem but will increase the efficiency of energy utilization and minimization of the negative impact of consuming fossil energy resources on the environment and human being.

The current biggest hurdle for hydrogen fuel cell technology is lack of efficient, stable, and abundant electrocatalyst for oxygen reduction and evolution. Some oxides have been shown to possess outstanding catalytic activities for both oxygen reduction and evolution, in some instances, as competitive as platinum. However, due to their complicated electronic and surface structures, as mentioned above, researchers have not been able to understand or reliably obtain its excellent performance yet. Recently, many electronic and structural studies have been explicated in perovskite-type oxides to study its magnetoresistive and multiferroic properties. Ongoing effort attempts to relate the better understanding of these oxides today to the past observed excellent catalytic activities of these materials.

Despite their technological and scientific importance, our understanding of the basic physics and chemistry of metal oxide surfaces is far behind that of metals and semiconductors. This is due in part to the great complexity of the crystal structures; one of the first goals of a surface science study is to locate the exact position of the atoms on the surface. Despite the apparent simplicity, it is a formidable task. In fact, the number of properly determined oxide surfaces is very small. Along with the complexity of the crystal structure comes a similar complex combination of chemical and physical properties. Many elements, especially transition metals, display a range of possible oxidation states and hence a series of oxides with different compositions. Vanadium oxides, for example, include not only VO, V_2O_3, VO_2, and V_2O_5 but also a wide range if intermediate phases. In addition, another consequence of chemical complexity is the wide range of chemical interactions possible with chemisorbed molecules.

The electronic structure of metal oxides is also much more complex than that of most metals or semiconductors. For example the bulk electronic structure of the late 3D transition metal oxides lies somewhere between itinerant and localized, and neither theoretical approach, the one for metals nor semiconductors or the one used for molecules, is entirely appropriate.

Experimentally, the preparation of a surface is central to the investigation, and herein lies what is possibly the greatest difficulty of all. Compared with elemental solids, the preparation of nearly perfect surfaces of any compound is difficult, and for oxides it is a research program on itself. Hopefully in the future it will be possible to prepare oxide surfaces that we know correspond to the surfaces modeled theoretically.

Today theoretical surface science starts to be well developed due to an increasing demand of collaboration with experimental groups. The boost from the last decade in computational surface science is mainly due to the development of powerful density functional theory (DFT) codes in the 1990s and this, of course, in parallel with the technological improvement of the computational power, making the simulation of complex and large systems, impossible some years ago, affordable.

In the present chapter a general overview is presented on the application of computational chemistry on metal oxides. After a short horizon screening of the role of oxides in electrocatalysis, a description of the use of model systems in the modeling of metal oxides is presented as well as an overview of their acid–base and redox properties. A series of case studies are shown to illustrate with concrete examples how metal oxides are studied using quantum-chemical computational techniques. Oxides such as TiO_2, V_2O_5, ZrO_2, and MgO are discussed in a variety of situations. In some cases the surface is considered, used as a support or not, in others the bulk in order to study diffusion properties. The effects of solvation, hydroxylation, and surface stability are shown. A quite complete collection of case studies are presented in order to lead the reader through the field of metal oxide simulations.

11.2 DESCRIBING OXIDES BY QUANTUM CHEMISTRY

The structure of metal oxides is basically of regular nature and the ions are distributed following crystallographic rules. The phenomena taking place in/on metal oxides may be tackled from different approaches and the model used should be carefully chosen to ensure a correct description. Thus, local phenomena are properly described with finite models: The solid is cut in a way that only selected atoms are explicitly treated. In other words, the whole solid is treated as a molecule or cluster. On the other side, the long-range order found in solids, and especially in metal oxides, needs to be taken into account for certain systems. In this context, models considering periodicity are indicated. We will briefly present the main advantages and drawbacks of these models and the most common programs available.

11.2.1 Cluster Models

The cluster model is built by selecting a portion of the surface to represent the whole system. Therefore it is of finite size, and it is possible to apply the conventional quantum-chemical techniques available in standard codes with low computational cost. A wide variety of local phenomena may be properly described by this model: adsorption, defects, vibrational frequency, polarizability. Despite its simplicity many factors must be carefully considered to select a proper model. The main drawback is that long-range effects are omitted. Let us present some specific problems encountered when using finite models and the ways to overcome these deficiencies.

The cluster is truncated with respect to the crystalline structure: Chemical bonds are cleaved, leaving "dangling bonds" by a decrease in the coordination. Even in the case that stoichiometry is respected, the atoms located on the border of the cluster are artificially undercoordinated. For instance, MgO rocksalt structure presents a coordination of 6 in the bulk crystal. Figure 11.1 shows two finite clusters built from the bulk structure: Despite the right stoichiometry, the atoms on the border are undercoordinated and therefore more reactive. This problem may be solved by saturating the border atoms with other atoms. The long-range effects may be incorporated in the model by embedding the cluster into a point charge net so the Madelung field is

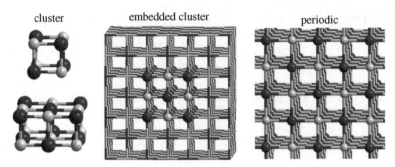

cluster embedded cluster periodic

FIGURE 11.1 Most widely used models for description of solids. A rocksalt NaCl structure is modeled by cluster, embedded cluster, or periodic models. Finite clusters, although stoichiometric, lead to undercoordinated atoms on the border. Long-range effects may be introduced by embedding a finite cluster into a point charge net. Finally, periodic boundary conditions allow considering translational symmetry in the crystal.

considered. This is schematically shown in Figure 11.1. Embedding techniques [1, 2] are also solutions that have been widely used [3]. It is thus important that the local structure (chemical environment, symmetry) is representative of the real system in a finite model, but the treatment of the border atoms should not be neglected.

11.2.2 Periodic Models

Periodic models involve translational symmetry by imposing periodic boundary conditions (PBCs). Although this approach has been used traditionally in molecular dynamics simulation, it has become nowadays reliable in ab initio treatment of crystals, surfaces, or polymers. Its application to the field of surface science and catalysis started in the 1980s and has considerably spread in the last years [4]. The implementation in computational programs is based on the definition of functions adapted to the crystal symmetry following the Bloch theorem (for more details on the formulation see ref. [5]). The band theory can be naturally applied in this frame; the periodic approach needs a fine sampling of k points in the reciprocal space. The explicit consideration of translational symmetry is both an advantage and a disadvantage: Long-range effects are taken into account; however, this imposes a perfect ordered system. The presence of imperfections in crystals or surfaces usually takes place in low concentration. The way to include them in a periodic model is thus to increase the size of the unit cell, the model becoming computationally more demanding. Figure 11.2 displays a two-dimensional real system and the periodic models derived depending on the choice of the unit cell. It can be observed that the smaller the unit cell, the more regular (ordered) the model. The ideal systems to be simulated in periodic boundary conditions are thus those involving high coverage and ordered patterns. Low-coverage adsorption on surfaces or low-concentration defects are examples of the need of large unit cells in periodic models.

The development of periodic programs lags behind that of standard molecular codes. A great effort has been devoted to apply periodic boundary conditions to the

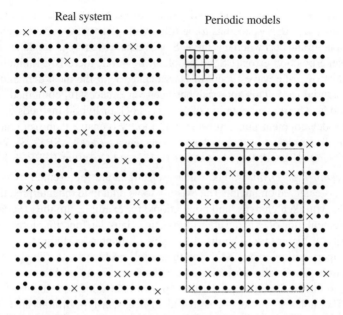

FIGURE 11.2 Schematic representation of two-dimensional system. The real system (left) may be modeled by periodic models. The unit cell chosen determines the pattern to be repeated. The smaller the unit cell, the more ordered the model. For low concentration of defects large unit cells are needed.

existing quantum chemistry in order to describe condensed matter from a physico-chemical point of view. A wide variety of properties can be calculated: ground-state and excites-state energy, vibrational spectrum, band structure, mechanical response, kinetic barriers, and so on. Nowadays a number of periodic user-friendly codes are available for carrying out such analyses [6]. The efficient parallelization will contribute to their wide use as a rich technique to characterize solids and surfaces. We will focus here on the two main implementations of the PBC based on the representation of the one-electron wavefunctions: plane waves and Gaussian-type orbitals. The representative programs available will be mentioned.

- *Plane Waves.* It constitutes a universal basis set since it does not need to be defined for different species. The mathematical manipulations are simple and easy to implement, but the number of plane waves necessary to describe correctly the wave function is huge. Two main approximations are done to decrease the computational demands: first, the core electrons are replaced by pseudopotentials and, second, only the plane waves below a threshold value are considered. Due to the expansion of the plane waves in the whole space, three dimensions need to be calculated. The trick to model surfaces is to take a large unit cell vector in the direction perpendicular to the surface, so that the surface model is indeed an infinite repetition of slabs separated by a vacuum region. The vacuum region has to be large enough to avoid interaction between successive layers. Tests can be performed on the potential to verify that it drops and remains flat in the vacuum region. The empty space has to be filled with plane waves and therefore has a computational cost when it is large. Since plane waves are used, there is no basis set superposition error (BSSE) problem; however, there is a limitation in the number of plane waves; a cutoff is associated with the most demanding pseudopotential used. This approach has been very successful combined with DFT methods. The implementation of Hartree–Fock or hybrid methods is currently under development but is substantially more expensive than using atomic orbitals. Examples of programs based on plane waves are VASP [7, 8], WIEN [9], DACAPO [10], CPMD [11], CASTEP [12], and Quantum-ESPRESSO [13].

- *Gaussian-Type Orbitals.* The valence electrons are represented by contracted orbitals centered on nuclei. The existing tools related to the interpretation of orbitals, Mulliken charges, and populations, for example, are thus available. The implementation of existing algorithms using Gaussian functions into a PBC code is not a difficult task. The disadvantages are their nonorthogonality, the basis set superposition error, and the mathematical problems associated with diffuse functions, especially for small unit cells. There is no problem to represent a system in two-dimensions. Hartree–Fock or hybrid methods are easily accessible. Examples of periodic codes based on Gaussian-type orbitals are CRYSTAL [14], SIESTA [15], and GAUSSIAN [16].

11.3 METAL OXIDE SURFACES

Reactivity of metal oxide surfaces is different from that of pure metals by its very nature. The presence of two ions with different electronegativity provides acid–base properties. Thus, cations (metals) are deprived of electrons and behave like acidic sites, while anions (oxygen) are rich in electrons and behave like basic sites. Such acid–base properties are not present in pure metallic systems. However, the chemical behavior of metal oxides is not only governed by their acid–base character but also depends on their ability to exchange electrons with a chemical partner. The redox behavior is related to the reducible character of the cationic site. For metal oxide surfaces, the electron transfer to or from the substrate finds a wide variety of applications such as gas sensing, electronic devices, or electrocatalysis. In this section we will present the adsorption mechanisms based on the electron count [17, 18]: The best adsorption modes contribute to restore the ideal electron count of clean and stoichiometric surfaces. We will present the mechanisms of adsorption as acid–base or redox processes between substrate and adsorbate and we will illustrate this point with selected case systems.

Metal oxides present electronic structure going from insulators to semiconductors, metals, and superconductors [19]. We will focus on a typical insulator structure where an energy gap lies between the valence (occupied) and the conduction (virtual) bands. The valence band (VB) is mainly composed of oxygen states. It is usually filled, and since it is rich in electrons, it presents a Lewis basic character. The conduction band

TABLE 11.1 Electronic Band Gaps for Selected Bulk Metal Oxides (eV)

Material	Band Gap
MgO	7.8
SiO_2	9
Al_2O_3	8.8
ZrO_2	7
TiO_2	3.0
SnO_2	3.6
V_2O_5	2.3
CeO_2	6.0

Note: The larger the gap, the more stable the system.

(CB) is mainly formed by metallic states. It is empty and possesses Lewis acidic character. A large gap indicates a high stabilization of the oxygen levels and/or a destabilization of the metallic levels. On the contrary, a small gap is associated with high reactivity: The oxygen levels (VB) are destabilized and thus increase in energy, basicity is enhanced, while the metallic levels (CB) are stabilized, decreasing in energy and becoming more accessible, and acidity is enhanced. Therefore, a general rule in (surface) metal oxides is the following: The larger the gap, the more stable the system. Reactivity is thus associated with small band gaps, the same way that reactive molecules present low highest–lowest occupied molecular orbital (HOMO–LUMO) gaps. Table 11.1 collects band gaps for selected bulk metal oxides. It can be seen that large gaps correspond to traditionally inert materials (MgO, SiO_2, Al_2O_3) while low band gaps are found in more reactive systems (TiO_2, SnO_2, V_2O_5). The electronic band gap can be tuned for obtaining desired properties following band theory principles. The presence of defects, such as impurities or oxygen vacancies, has a strong impact on the electronic structure of a metal oxide since it may alter the position of the bands or induce specific levels in the gap. Some of the levels may become accessible for an electron transfer leading to redox reactivity. Finally, we would like to stress that both substrate and adsorbate are determinants in the adsorption process.

Figure 11.3 shows schematically the adsorption mechanisms considered in this chapter, acid–base and redox, which are discussed below.

11.3.1 Acid–Base Properties

It has been mentioned above that metal oxides present both acidic (metallic) and basic (oxygen) sites and an acid–base reactivity is expected. In this section we will discuss different situations where the acidic–basic properties rule out the interaction between substrate and adsorbate on metal oxide surfaces: first, the basicity of alkaline earth oxides; second, the balance between molecular and dissociative adsorption of small molecules; and third, the adsorption of radicals coupled to cationic and anionic species.

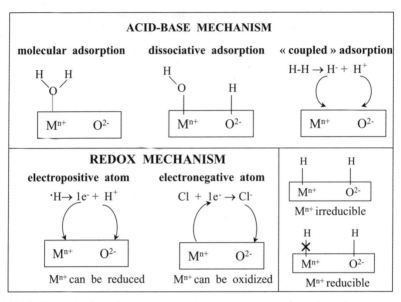

FIGURE 11.3 Schematic chart showing acid–base and redox mechanisms occurring on metal oxide surfaces. The stability of the surface species depends on the properties of both the adsorbate and the substrate.

Metal oxides have been traditionally classified into acid or basic materials. A typical example of basic oxides is the series of alkaline earth oxides MgO, CaO, SrO, BaO. While the former do not exhibit strong basic character, the basicity increases when increasing the cation size, so BaO is found to be very basic. Let us analyze in detail the reason for this behavior taking the example of regular nondefective (001) planes for the series MO (M: Mg, Ca, Sr, Ba). The structure of such a surface is regular, cations M^{2+} and anions O^{2-} alternate filling the space as in the bulk rocksalt crystal structure; see Figure 11.4. This surface is nonpolar and stable, and relaxation is poor. The increase in the cation size when going from MgO to BaO corresponds to an increase in basicity; the values of cell parameters are shown in Figure 11.4. The reason for such a strong difference in reactivity is intimately connected to the structure and the cation size. Pacchioni et al. [20] have shown that the larger the unit cell (or the cation size), the smaller the Madelung potential and the higher the basicity. Indeed, the oxygen ion is stabilized by the electrostatic potential generated by the charge distribution of the crystal (the Madelung potential). As a consequence, the charge distribution is more diffuse on oxygen sites in the case of CaO than for MgO, resulting in an ease of the charge transfer from the donor orbital of the surface anion.

More recently, the Madelung potential of alkaline earth metal oxides was studied in relation to the adsorption of CO on defect-free surfaces of MgO, CaO, and SrO. Halim [21] found that the strength of adsorption of CO on nondefective surfaces of MgO, CaO, and SrO decreases in the following order: SrO > CaO > MgO. This corresponds to cationic site distances $d(O–M)$ which are shortest for MgO and longest for SrO.

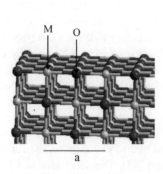
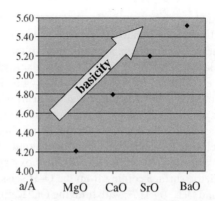

FIGURE 11.4 Left: schematic representation of regular (001) plane of MO materials. The bulk crystalline structure is rocksalt. Right: Value of the bulk cell parameter for different materials. The basicity increases with the cell parameter value.

This is in line with the cation size, which changes in the same order. Density functional calculations have also been used to study the interaction of MgO, CaO and SrO with water and methanol [22, 23]. In the case of MgO, it was found that water maintains its molecular character and dissociation was not observed at room temperature on the (001) surface. This is due to the large barriers involved in the separation of proton–hydroxyl structure [24]. At the same time it was found that defects can increase the activity and dissociation of water can occur on those places [25].

The interaction between a closed-shell molecule and a metal oxide surface is usually of electrostatic nature. The adsorption of a small molecule occurs on one or several metallic sites, the molecule behaving like a base in interaction with the acidic metallic site. This is the general behavior of adsorption processes. The adsorption energy increases with the basicity of the gas-phase molecule. Thus, NH_3, which is the usual probe for the acidic sites [26–29], adsorbs on the metal cation of the surface with the lone pair oriented toward the metal in MgO [30, 31]; the deviation from normality to favor an H bond as secondary interaction is slightly less favorable [32]. Ammonia adsorption on (110)-rutile TiO_2 takes place on the undercoordinated titanium sites [33, 34], its heat of adsorption being higher than for MgO due to the more acidic character of Ti^{4+} with respect to Mg^{2+}.

Water, less basic than NH_3, is also adsorbed without dissociation on flat terraces of MgO(100); it is oriented roughly parallel to the surface and forms hydrogen bonds with the surface basic sites at low coverage. The dissociative mode is not stabilized and the proton recombines with the adsorbed hydroxyl to form water. Note that the observed dissociation of water on MgO(100) surfaces is associated to the presence of defects (see refs. 35 and 36 and references therein). On stoichiometric rutile $TiO_2(110)$ the adsorption mode of water has been controversial. Recent experiments show a stable adsorbed phase of molecular water at low temperatures (see, for e.g., refs. 37), and dissociation at higher temperatures is only mediated by surface defects. Most theoretical calculations suggest facile dissociation on the perfect surface [17, 38–40] or mixed dissociative and molecular situations [41]. The adsorption mode is indeed very sensitive to the theoretical setup, in particular the slab thickness and the adsorbate orientation in periodic calculations. A careful choice of the theoretical parameters results in a stabilization of the molecular mode for a coverage of one monolayer [42]: 1.090 eV for the molecular mode versus 0.911 eV for the dissociated, 1.051 eV for the mixed. The Ti_{5f}–O_{water} bond length is generally overestimated in DFT methods, and this may lie at the heart of the problem. It is however clear that the oxygen basicity in the stoichiometric rutile surface is higher than that in MgO, the dissociative modes in the former being closer in energy to the molecular ones.

Methanol is found to dissociate on the rutile $TiO_2(110)$. Recent scanning tunneling microscopy (STM) studies identify methoxy and hydroxyl groups on the rutile stoichiometric surface coming from the O–H scission [43]. The adsorption as CH_3^+ on O^{2-} and OH^- on Ti^{4+}, with bridging OCH_3 groups and terminal OH groups is found from ab initio calculations [44–45]; the C–O bond is the weakest in the gas phase. The same has been found on rutile $SnO_2(110)$ [46]. Carboxylic acids are also found to dissociate on metal oxides, and the RCO_2^- groups bind by one or two bonds to the metal cations [47–52]. The acidic cleavage, less favorable in the gas phase, leads to RCO_2^- binding to surface cations (mono or bidentate) and H^+ binding to a surface anion. The basic cleavage that is the best in the gas phase would lead to an OH^- adsorbed on the surface cation (terminal OH in case of the rutile structure) and to a RCO^+ that binds to the surface anion, O^{2-}, forming a singly coordinated RCO_2^- group. The geometry of this species is not very favorable, even if a surface metal site is available close by, since it involves a lattice oxygen as well as the carboxylic oxygen. Thus the singly coordinated RCO_2^- is less favorable than the dicoordinated species. The interaction of simple amino acids has been recently studied by ab initio and molecular dynamics simulations [53], showing that carboxylic acid binds to two Ti_{5f} sites as a stable bidentate species [54].

The adsorption of radicals on irreducible metal oxides results on the formation of A^+/A^- species that interact with basic/acidic sites, respectively. This is the case for H_2 adsorption that forms H^-/Mg^{2+} and H^+/O^{2-} pairs. This mechanism implies an acid–base relation and avoids the reduction of the substrate. In a way, a redox process has taken place before adsorption:

$$2A\cdot \rightarrow A^+ + A^-$$

The presence of surface hydride species has been observed for irreducible metal oxides such as Ga_2O_3 [55], ZnO [56], or reduced TiO_2 surfaces (see ref. 19 and references therein), proving that this mechanism applies. Note that although M–H species are not stabilized on a reducible metal oxide, they are observed in TiO_2 after reducing conditions. Similar behavior might be expected for diatomic molecules like Cl_2 or dimers such as $(NO)_2$.

11.3.2   Redox Properties

A redox mechanism involves the exchange of electrons between substrate and adsorbate. For this process to occur, the electron transfer *must* be possible, so the adsorbate and the surface behave as an oxidation–reduction couple. Thus, the electron transfer occurs from the substrate to the adsorbate for an electronegative group only if the surface can be oxidized:

$$A \cdot + e^- \rightarrow A^-$$
$$(\text{Surface}) \rightarrow (\text{oxidized surface}) + e^-$$
$$(\text{Surface}) + A \cdot \rightarrow (\text{oxidized surface}) + A^-$$

The anionic A^- moiety would then adsorb on the metallic acidic sites through an acid–base mechanism. For instance, Cl adsorbs better on reduced TiO_2 than on the stoichiometric oxide, since the former presents available electrons at the Ti^{3+} sites. The electron transfer occurs from the adsorbate to the surface for the adsorption of an electropositive radical, provided that the surface can be reduced:

$$A \cdot \rightarrow A^+ + e^-$$
$$(\text{Surface}) + e^- \rightarrow (\text{reduced surface})$$
$$(\text{Surface}) + A \cdot \rightarrow (\text{reduced surface}) + A^+$$

The cationic A^+ moiety will then adsorb on basic O^{2-} surface sites via electrostatic interactions. This is the case of H adsorption on TiO_2, where OH groups are formed with reduction of the Ti^{4+} sites to Ti^{3+}. The surface metal site must be reducible to take the electron coming from H. Otherwise, the electron transfer to the surface is not possible and another way is preferred such as coupling H^+/H^- as in the acid–base case. This is the reason why, upon adsorption of atomic H on ZnO or MgO, two species M–H and O–H can be detected, whereas only OH species can be detected on V_2O_5 or TiO_2.

It is important to note that both the adsorbate and the surface properties need to be taken into account. In particular, the oxidation–reduction properties of the surface are determining in the final adsorption process. A significant change in the surface conductivity occurs upon a redox adsorption, so it can only take place if the electron transfer is possible. A consequence is that it applies not only to open-shell adsorbates but also to defective substrates (surfaces deviating from stoichiometry or doped) since the electron count is different from that on the perfect ones. The most common defect in metal oxides is the neutral oxygen vacancy. The cost for removing a neutral oxygen from a condensed phase (bulk or surface) depends on the nature and the structure and is usually higher in irreducible metal oxides. During the process two electrons are left in the surface: $O^{2-} \rightarrow O + 2e^-$.

If the metal is not reducible, as for MgO, an F center is formed: The electrons are trapped in the cavity left by the oxygen atom. The presence of an F center is associated with a state in the gap filled by the two electrons. If the metal is reducible, like TiO_2, the electrons are transferred to the metal site that is reduced. A state in the gap appears 0.6–1.1 eV below the conduction band, mainly formed of Ti $3d$ states. The localized nature of such a state is not well described by pure DFT methods but necessitates hybrid methods as recently pointed out by Di Valentin et al. [57]. The impact of oxygen vacancies in TiO_2, ZrO_2, V_2O_5, and CeO_2 is reviewed elsewhere [58]. Metallic systems can also be found by reduction of the metal oxide surface, as found for hydrogenated ZnO (10–10) [56].

In cases of oxidation (O adatoms), the valence band is not completely filled. The most favorable adsorption scheme is the redox mechanism which restores the situation of an insulator and the highest oxidation states for all the atoms. Note that the presence of defects is crucial for reactivity. However, a defective surface may lead to less reactive systems due to the electron count, as pointed out by Barteau [59].

We will illustrate the above concepts in detail for the following cases: H_2/TiO_2, $H_2/TiO_2/V_2O_5$ surfaces focusing on the electron count that controls the stability of the clean and defective systems. The effect of alkali doping will be commented on for the latter.

11.4 CASE STUDIES

11.4.1 Hydrogen on Reducible Metal Oxides

11.4.1.1 *TiO₂*

Titanium dioxide has received much attention in the last decades for technological applications in catalysis, photocatalysis, microelectronics, gas sensing, ceramics, or pigments [60]. Rutile is the most stable TiO_2 polymorph; anatase crystals are predominant in small nanoparticles; the brookite phase is less stable. The rutile $TiO_2(110)$ plane is the most exposed and has been largely used as a surface model in both experimental and theoretical studies (see ref. 61 and therein). Its structure is made of alternative horizontal and vertical polymers [31], and is classified of type II according to Tasker [61]. This surface is at the same time stable and reactive [62, 63]. The surface is reducible and the most predominant defect is the bridging oxygen vacancy. From the experimental point of view, it is easy to control the degree of reduction of these materials, as for rutile SnO_2 (110) [64]. The reconstruction is weak and restricted to a small relaxation and rumpling of the bridging oxygen atoms [65, 66]. The structure of the (110) TiO_2 plane is generally well described by ab initio techniques; however, it has been difficult to reproduce the experimental reconstruction of the outmost layers. The origin of such divergence seems related to numerical thresholds in the calculation setup [67] and the choice of the model [68] rather than on the method itself. Regarding the electronic structure, pure DFT methods are known to give an incorrect delocalized picture of the reduced surfaces [57], and the correct description needs to take into account an exchange contribution in the functional.

Despite the large number of works devoted to rutile (110), many fundamental aspects are still a matter of debate. In particular, the stability of hydrogenated phases is controversial. It is well known that rutile (110) reacts toward molecular H_2 only at high temperatures. However, it is reactive toward atomic hydrogen forming hydroxyl groups. Another procedure to obtain hydrogenated surfaces is the dissociation of water in oxygen vacancies. Both mechanisms are equivalent, as can be seen in Figure 11.5.

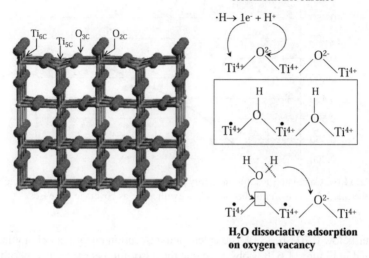

FIGURE 11.5   Left: slab model for rutile (110) surface. Right: schematic representation of two mechanisms leading to hydrogenated surface.

Hydrogenated (hydroxylated) surfaces have been characterized by different techniques: atomic force microscopy (AFM) [69], STM [70, 71], helium atom scattering (HAS) [72], electron energy loss spectroscopy (EELS) [73], and DFT [57, 74, 75]. We will briefly discuss the main conclusions raised from these studies.

Coverage   Despite the simplicity of the system, little information is available on hydrogenated rutile surfaces. STM [70, 71] measurements on low-coverage hydroxylated surfaces focus on the isolated nature of such species. They come mainly from water dissociation in oxygen vacancies. Other STM studies show the formation of a disordered layer of 0.25 monolayer (ML) [76]. Preparation of surfaces with higher content in hydrogen needs strong reducing conditions (high-atomic hydrogen pressure). Exposure to $150 \, L^{*}$ of H_2 leads to the formation of a phase close to 0.50 ML. The typical STM image of a stoichiometric surface shows dark rows in the position of titanium atoms; after hydrogenation a bright spot appears in the dark row and is associated with the formation of a hydroxyl group and a reduced titanium site [60]. The formation of hydoxyl species is accompanied by the electron transfer to the titanium $3d$ orbitals in the conduction band (see electronic structure in Figure 11.7), and this occupied level is visible in STM. Going beyond this coverage requires huge amounts of atomic hydrogen, 1200 L. In these conditions STM images show an increase in the number of bright spots aligned in the (001) direction. A pattern (1×1) appears in agreement with diffraction experiments [72]. The coverage remains

---

*1 L = 1 Langmuir = 10^{-6} Torr·s.

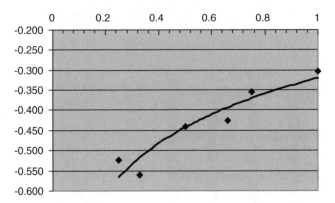

FIGURE 11.6 Calculated adsorption energy per H_2 molecule as function of coverage. Values are exothermic (eV). Adsorption becomes less exothermic as coverage increases.

however below 0.70 ML [73]. The calculated adsorption energy per H_2 molecule is displayed in Figure 11.6. It is observed that the adsorption energy is exothermic with respect to gas-phase H_2, indicating a favorable process. However, the values between -0.20 eV and -0.55 eV are small because of the high stability of the bare TiO_2 surface. For comparison, the calculated adsorption of H_2 on a ZnO surface is exothermic by around 3 eV. Note that in order to properly analyze the stability of hydrogenated slabs at different coverage, atomistic thermodynamics is required.

Electronic Structure   Titanium dioxide is a semiconductor with a band gap of 3.2 eV. The valence band (VB) is composed of oxygen $2p$ levels, and the conduction band (CB) is composed of titanium $3d$ orbitals. Figure 11.7 shows the calculated density of state (DOS) for bulk rutile TiO_2. The Fermi level indicates the level occupied by electrons; in the figure it corresponds to zero energy. The calculated band gap is 1.8 eV, far below the experimental value; this is a typical feature of pure DFT methods and can be corrected by using other functionals local density approximation with U (LDA+Uparameter); Becke, three-parameter, Lee-Yang-Parr (B3LYP). Decomposition of the DOS into the two different surface Ti sites, fivefold and sixfold, shows that the former mainly contributes to the lower part of the CB. It is thus expected that upon reduction this site will be the first populated. This picture is consistent with a higher reactivity of the undercoordinated Ti_{5C}. Hydrogenated slabs, where all the bridging sites are bonded to hydrogen (coverage $\theta = 1$), have been calculated and the corresponding DOS is displayed in Figure 11.7. It is observed that the Fermi level has shifted towards CB levels. In the VB region a peak appears associated with hydroxyl groups. The VB is decomposed into the contributions of the surface Ti sites. According to this analysis, the fivefold coordinated site is occupied while the sixfold site is empty. This is expected from the analysis of the bare system and is consistent with STM observations. Indeed, the bright line observed in STM lies in the (001) direction, as the Ti_{5C} sites are. Note that the STM experiment consists of filling CB levels by applying a potential; the final system is equivalent to hydrogenation since unoccupied states are also filled. Note that Figure 11.7 shows the DOS calculated

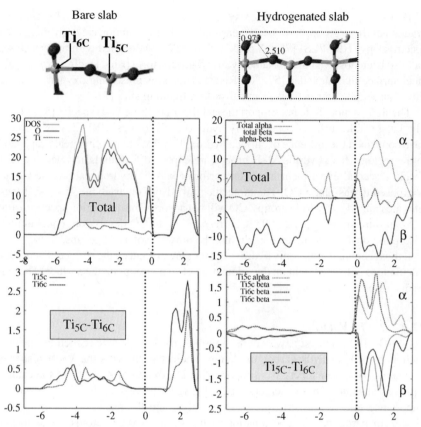

FIGURE 11.7 PBE calculated DOS for bare and hydrogenated rutile (110) TiO$_2$. The vertical dotted line indicates the Fermi level. The bare slab shows a VB composed of oxygen states, and a CB composed of titanium sites separated by an energy gap of 1.8 eV. Decomposition of the VB states into the titanium contributions shows that Ti$_{5C}$ will be the first populated sites upon hydrogenation. The hydrogenated slab shows alpha and beta contributions due to the spin-polarized character. The Fermi level is shifted toward the CB, and the Ti$_{5C}$ states are filled.

with a generalized gradient approximation (GGA) functional giving rise to a delocalized picture of the electron density (metallic state). Actually, hydrogenation induces a state in the electronic gap keeping the gap, as for an oxygen vacancy [58]. In order to properly describe this situation the exchange energy needs to be taken into account as in the hybrid B3LYP method or using a LDA+U-type functional [57].

Surface Species   The formation of surface hydroxyl groups is clear when the oxide surface is exposed to atomic hydrogen. The process is associated with reduction of the surface titanium sites:

$$H \rightarrow H^+ + 1e^- \tag{11.1}$$

$$Ti^{4+} + 1e^- \rightarrow Ti^{3+} \tag{11.2}$$

Protons are bonded to bridging oxygen sites leading to hydroxyl OH groups. Vibrations of these groups have been measured by High resolution electron energy loss spectroscopy (HREELS) [73]: A peak at 456 meV is associated with this species. It has also been observed that isolated hydroxyl groups form pairs, leading to nonequivalent surface titanium sites [57, 70]. Reduced surfaces presenting oxygen vacancies show important contributions from subsurface titanium sites [77].

Theoretical models do not support the stability of hydride Ti–H species [57, 74, 75]. The most stable surfaces contain only hydroxyl OH groups. HREELS experiments have confirmed the presence of surface hydroxyl groups but no other peak, especially in the region where hydride M–H groups are expected, is found. In order to double check, a second experiment has been performed: After hydrogenation, the surface is exposed to gas-phase CO. The molecule will adsorb on available Ti sites; if these sites are capped by hydride groups, CO will not adsorb. The desorption spectrum after CO exposure shows a peak at 95 K corresponding to CO desorption, concluding to adsorption on Ti sites and then excluding the presence of Ti–H groups.

It thus seems clear that only hydroxyl groups are present on the surface. Theoretical calculations help shed light on the orientation of such groups on the surface. This will be discussed in following sections.

11.4.1.2 H–V₂O₅/TiO₂

11.4.1.2 $H–V_2O_5/TiO_2$

The V_2O_5/TiO_2 catalyst is widely used in industry for selective reduction reactions such as $DeNO_x$ or hydrocarbon oxidation. The active phase is the V_2O_5 dispersed on anatase TiO_2 which acts as a support [78]. The extreme efficiency of the catalyst has been attributed to a synergetic effect between the two oxides [79]. Despite its importance, the atomic structure of the catalyst and the reactive sites is not well known. In particular, the coordination of the surface oxygen atoms seems to play a key role in reactivity: Mono-, di-, and three-coordinated oxygen atoms behave differently in catalytic conditions, although there is no unanimity in establishing the most reactive one for the oxidation process. The effect of the support is also of great importance since it is not always possible to correlate the results from unsupported crystals to those for supported catalysts. Supported materials possess reactive sites at the interface that do not exit in unsupported ones and could be directly involved in the key catalytic steps.

Active Site As mentioned above, the nature of the active surface oxygen sites is still uncertain. For pure vanadia, Tepper et al. [80] have conducted HREELS, angle-resolved ultraviolet photoemission spectroscopy (ARUPS), and X-ray photon spectroscopy (XPS) experiments to determine the degree of interaction of the V_2O_5 (001) surface with molecular and atomic hydrogen [note that in the literature the two surfaces (001) and (010) of V_2O_5 are equivalent]. While molecular hydrogen does not interact with the surface, atomic hydrogen induces significant changes in the spectra, indicating a strong interaction. According to their results, the most reactive oxygen atom is the bridging one, while terminal and three fold coordinated sites are stable with respect to the interaction with atomic hydrogen. Another important feature is that no OH groups were observed in the spectra even after exposure to high doses of

**TABLE 11.2 Adsorption energy (E_{ads}, eV) for Interaction of H with
V_2O_5/TiO_2 Slab Model**

Site	E_{ads}	OH Orientation
1	2.37	Towards O1, [010] direction
1bis	2.60	Towards O1
2	2.95	[100] direction
2bis	2.81	Towards O1
3	2.17	
s1	2.03	Towards s3
s2	2.58	Towards O3
s3	2.52	Towards 1bis

Note: Only values higher than 2.29 eV can be considered exothermic. The interface
oxygen sites V–O–Ti (sites 2 and 2bis, see Fig. 11.8) present the highest values.

atomic hydrogen. The hydroxyl groups would recombine to form water and oxygen
vacancies in the surface. Photoemission measures and theoretical cluster studies also
reveal the major reactivity of bridging oxygen atoms in the same surface [81, 82], and
in a paper by Kolczewski and Hermann the assignment of the experimental near edge
X-ray absorption fine structure (NEXAFS) peaks on V_2O_5 systems to the different
surface oxygen centers is proposed [83]. Ozkan et al. [84] have studied the reducibility
of V_2O_5 samples with H_2 and NH_3 by temperature programmed desorption (TPD) and
spectroscopic catalyst characterization, finding that ammonia is more effective in the
reduction of the substrate. Reduction takes place at the surfaces in a major extent; only
after high dosages of hydrogen is the bulk fully reduced. Reduction of the unsupported
catalysts has also been associated with the interaction of the reducing agent with
terminal oxygen atoms (vanadyl V=O groups). Upon reaction, vacancies are formed
in the surface that are replenished by gas-phase oxygen or lattice oxygen diffusion
as for V_2O_5/TiO_2 selective catalytic reduction (SCR) catalysts. They associate the
vanadyl sites with NH_3 adsorption and subsequent oxidation and the bridging oxygen
with the NO reduction in the SCR reaction. MSINDO cluster [85] and periodic DFT
[86] calculations on the hydrogen adsorption on the unsupported (010) V_2O_5 surface
conclude that terminal oxygen atoms of the vanadyl groups present the highest activity
toward hydrogen adsorption, that is, the adsorption energy on such sites is the most
exothermic (see Table 11.2). Even the threefold coordinated oxygen atoms in the
V_2O_5 (010) surface have been proposed to be responsible for the catalytic activity
of V_2O_5 [87]. Periodic semiempirical Hartree–Fock calculations indicate that, upon
creation of an oxygen vacancy in the perfect V_2O_5 (001) surface, the most stable
system corresponds to the uptake of the threefold coordination oxygen. This result is
confirmed by IR studies on the oxidation of dimethyl sulfoxide (DMSO) catalyzed
by V_2O_5 carried out in the same work. A cluster model consisting of a V_2O_7 unit
supported on the (100) anatase surface has been proposed by Jug and co-workers [88]
together with a reaction mechanism for the SCR process [89].

Support Effect Despite the research conducted on hydrogen interaction with
V_2O_5, it is not possible to correlate the results obtained from unsupported crystals to those obtained for supported catalysts. Besselman et al. [90] conclude from

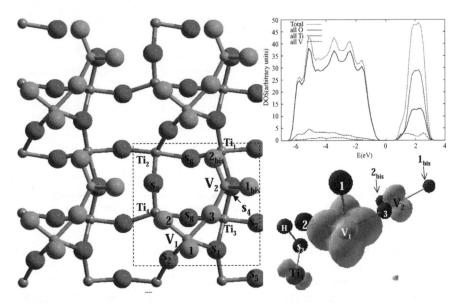

FIGURE 11.8 Left: top view of model used to represent V_2O_5 supported on anatase (001) (only uppermost atoms are shown for clarity). Oxygen sites are numbered 1–5 if they belong to the vanadia unit, s_1–s_6 if they belong to the support. Right: Total and projected density of states for bare model (top), and spin density isosurface (bottom) for same system hydrogenated on support site s_3. The spin density is localised on the V_1 atom despite the adsorption of H on a surface Ti–O–Ti site.

temperature-programmed reduction (TPR) experiments that supported vanadia/titania samples are easier to reduce than unsupported V_2O_5. Tops⊘e et al. [91] also find that the V–OH and V=O bands disappear while Ti–OH bands appear in H_2 atmosphere. The vanadia species would agglomerate upon reduction and disperse upon oxidizing conditions in a reversible process. More recent spectroscopic experiments carried out by Bulushev et al. [92] also report the disappearance of the V=O signals and the formation of new hydroxyl groups probably associated to titania; however, they did not study the region characteristic for the polymeric species because of a strong adsorbance by the titania support. They conclude that "the nature of hydroxyl groups associated with supported vanadia species and the structure of reduced vanadia species need further clarification." Wachs and co-workers have probed the redox properties of the vanadia-supported catalysts with TPR in a hydrogen environment [93, 94]. They conclude that the specific support does have an effect on the redox properties of surface vanadia species for the SCR process. In situ Raman spectra in SCR conditions show that the terminal V=O bond is perturbed by the reduction but the attribution of such bonds to the rate-determining step during SCR is doubtful; instead, the bridging V–O support bond could be responsible for it. This is supported by ^{18}O exchange experiments showing that the time required for exchanging the terminal V $=^{18}O$ to V $=^{16}O$ during the SCR is extremely long and it can be considered as too stable to react. The implication of the V–O support comes from the correlation of the catalytic activity with the electropositivity of the metal of the support ($ZrO_2 > TiO_2 \gg Al_2O_3 > SiO_2$). Went et al. [95] recorded the Raman spectra for vanadia-supported catalysts exposed to water and ethanol and found that the V=O band shifted to lower energies but the band associated with polymeric species (bridging oxygens) did not change, indicating that the vanadyl bond would be the only active site involved in such adsorption processes. For titania-supported samples, the adsorbates would coordinate to the oxygen sites but would not induce any hydrolysis of the V–O support bond which is strong.

11.4.2 Diffusion of Ions in Bulk

11.4.2.1 H/TiO₂

The presence of hydrogen defects in bulk metal oxides has been known since the 1960s. The ability of protons to act as minority charge carriers in metal oxides opens new possibilities in the field of fuel cells, in particular for solid-oxide fuel cells (SOFCs) [96]. The materials employed are based on conducting metal oxides such as ZrO_2, CeO_2, and $LaGaO_3$. Hydrogen may be stabilized in interstitial positions of the oxide host structure as neutral H·, proton H^+, or hydride H^- species [97]. Their stability depends on the acid–base and redox properties of the metal site, and the rules derived above also apply in the case of bulk materials. We will illustrate the case of hydrogen on the TiO_2 (110) surface, showing that the penetration of hydrogen from the surface to the bulk is a favorable process. This might have important consequences in technological applications related to ion mobility as well as on (photo) catalysis since diffusion could take place even at room temperature. Few works have dealt with the study of hydrogen interaction with TiO_2 [73, 98, 100].

We have discussed in Section 11.4.1.1 that hydrogen adsorbs on the surface (110) of rutile TiO_2 forming hydroxyl groups with a coverage always below 0.7 ML. High pressures of atomic hydrogen were needed to increase the surface coverage. In order to study the effect of temperature, a temperature desorption experiment TDS was conducted on the same system. The sample was heated and the desorption products were analyzed by mass spectrometry. Unexpectedly, even at temperatures of 650 K nothing was observed to desorb. However, HREELS, HAS, and STM experiments clearly show that the surface hydroxyl groups disappear from the surface [72, 73]. A plausible explanation is that hydrogen diffuses to the bulk. This might happen only if the kinetic barriers to desorb as H_2 or H_2O are larger than those for penetration in the bulk. Periodic DFT calculations have permitted to estimate the relative value for the desorption/penetration barriers on a hydroxylated (hydrogenated) rutile slab. A thick slab containing eight TiO_2 layers is used, see Figure 11.9. The bridging oxygen sites are all hydrogenated. A hydrogen atom is located on a threefold oxygen site simulating a hydrogen-rich atmosphere. All the atoms of the slab are able to relax during the calculations. The transition state structures leading to hydrogen insertion in subsurface positions have been searched with the nudged elastic band (NEB) method. An analysis of the results displayed in Figure 11.9 shows that the surface hydrogen is stabilized in subsurface positions with a gain in energy ranging from 0.31 to 0.61 eV. Indeed, hydrogen binds to oxygen sites forming hydroxyl groups as found for the surface, indicating high stability of the O–H groups. Moreover, bulk hydroxyl groups

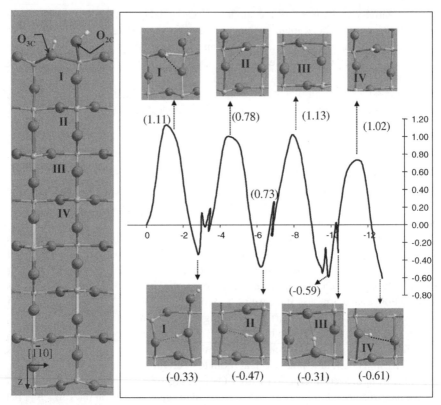

FIGURE 11.9 Left: side view of slab used for study of hydrogen diffusion on bulk rutile TiO_2 structure. The surface bridging sites O_{2C} are hydrogenated, and a hydrogen atom is located on one threefold oxygen site O_{3C}. Right: energy profile for diffusion of hydrogen in O_{3C} to bulk (energies in eV).

are further stabilized than on the surface because of the formation of hydrogen bonds. The cavity containing the hydroxyl group is found to distort locally to accommodate it, although the bulk structure is not destroyed.

The calculated barrier to penetrate to the subsurface is 1.11 eV. Once in the subsurface, hydrogen is found to easily move from one oxygen to another inside a cavity. This process requires little energy; our calculated barriers 0.20–0.50 eV lie far below the barrier for penetration into the subsurface position. This is confirmed from molecular dynamic runs showing high mobility of the hydrogen atom in the cage. Diffusion toward deeper positions takes place through barriers between 0.78 and 1.13 eV. From these energy values, the diffusion process is thus likely to occur.

Finally, let us comment that the electronic state of hydrogen in both the surface and bulk position is H^+, as corresponds to the redox process seen above. As a result, the titanium sites are reduced. The question of the localization of the electrons coming from the reduction arises: Is the electron of the bulk hydrogen in the titanium sites close to it or is it mainly located on surface titanium atoms? Is this extra electron located on one or several titanium sites? Calculations with GGA-type functionals show a delocalized picture of the electron density, all the surface titanium atoms possessing a part of the electron coming from the bulk hydrogen, irrespective of its position. As mentioned above, in order to properly describe the electronic features of this system a part of the exchange energy should be taken into account by using LDA+U or B3LYP methods. Preliminary calculations with the LDA+U approach show localization of the electron coming from the bulk hydrogen on one neighboring titanium site.

The diffusion of hydrogen from surface to bulk positions is thus a feasible process. For comparison, we have calculated the barrier to desorb two surface hydroxyl groups as H_2: 2.6 eV, which is far above the energy needed to diffuse into the bulk.

In conclusion, hydrogen is found to diffuse from surface to bulk in rutile TiO_2. This is observed indirectly in surface science experiments and is confirmed from calculations. Since the diffusion process may occur at room temperature, it would have important implications for ion-mobility-based technologies.

11.4.2.2 LiTiO₂–Anatase/Titanate Used for Renewable Energy Storage

Lithium ion secondary batteries are widely applied to electronic devices used in our everyday life, because of their high energy density and rechargeability [101–104]. TiO_2 is a lightweight, inexpensive, and environmentally friendly material and a potential material for energy storage, for example, the power source in electric vehicles. TiO_2 batteries have shown their efficiency in photoelectrochemical solar cells [105] and are frequently studied for their electrochromicity [106, 107], which can be used in display applications.

The electrochemical insertion reaction can be written as $xLi^+ + TiO_2 + xe^- \leftrightarrow Li_xTiO_2$, where x is the mole fraction of lithium present in the titanium dioxide framework. The reversibility of the above-mentioned reaction is the basis of the main property for this type of batteries. Insertion of lithium ions into the anatase framework results in a phase transition from the original tetragonal anatase structure toward the orthorhombic titanate phase for concentrations $x > 0.3$–0.5 [108, 109]. This transition represents a loss of symmetry referring to TiO_2 anatase as a central crystal structure. Indeed, $LiTiO_2$ crystals are often mixtures between both structures due to inhomogeneous Li distribution [110].

The topology of the anatase–TiO_2 structure has been investigated theoretically [111–115] and experimentally [110, 116–119] and also for TiO_2 structures with a wide range of Li concentrations [120, 121].

It is generally assumed that Li is located in sixfold coordinated intersticial sites, which are strongly elongated along the c direction so that the coordination becomes equal to 4. Moreover, the Li ion does not occupy the exact a center of the octahedron and is displaced along the c direction. This displacement results in a fivefold-coordinated environment for the Li ion. Several studies have been recently carried out to measure the extent of this displacement [116, 119–122].

The electrochemical performance or the rate at which the battery can be charged and discharged is directly related to the ion mobility. The latter is a quantity directly derived from the diffusion coefficient and can be approximated by the "hopping" rate Γ of the lithium ions between two vacant sites [123]. Activation barriers are used to approximate the diffusion coefficient D of the lithium ions in the lattice [123]. The "hopping" rates (Γ) [124], which on their turn can be used as a measure for the ion mobility of the lithium ions between the different interstitial spaces available, can be calculated using transition state theory by the relation

$$\Gamma = v^* \exp\left(\frac{-\Delta E_a}{kT}\right) \qquad (11.3)$$

where v^* is an effective vibrational frequency and ΔE_a the activation barrier to hop from one minimum to another. The effective vibrational frequency is typically of the order of 10^{13} s^{-1}. The activation energy for the hop to vacant neighboring sites, ΔE_a, is defined as the difference in energy at the transition state and the energy at the initial equilibrium state.

Previous experimental and computational [115, 125] studies determine a barrier for this hopping process around 0.5 eV [120, 126] in the special case of Li anatase. However, the dependence of the barrier energy with respect to the concentration in Li is not well understood since other theoretical results [114] do not agree with the experimental ones. The topology of anatase after Li insertion has been carried out by different groups experimentally [117, 118] and theoretically [127]. TiO$_2$ crystal structures anatase (tetragonal) and titanate (orthorhombic) were investigated in detail on their respective stability, geometry, and ion mobility for different concentrations of Li. The geometries of the crystals used are fully optimized in order to study the geometric transition (tetragonal–orthorhombic), and the position of the Li ion in the interstitial sites is discussed for the different concentrations considered. The latter has only been investigated for the Li$_{0.5}$TiO$_2$ structure [120, 128].

The anatase TiO$_2$ lattice structure belongs to the I4$_1$/amd space group [125]. The crystallographic unit cell of TiO$_2$ anatase is shown in Figure 11.10. It consists of four octahedrically coordinated titanium atoms and eight threefold coordinated oxygen atoms. The anatase structure can also be seen as a chain of TiO$_6$ octahedrons sharing edges [112]. The Ti atoms occupy the 4e special positions of the spatial group and the oxygen atoms the 8e positions. The rest of the volume is filled with empty "distorted" octahedrons centered at 4b special positions and "distorted" tetrahedrons at 16f special positions. The empty octahedrons also share edges. Both empty sites, octahedral and tetrahedral, are in principle candidates to host a Li atom in the anatase structure. The so-called octahedral empty voids are distorted and larger in size than the occupied ones (i.e., containing Ti atoms). Inserting impurities into these voids leads to fourfold-coordinated or when they are shifted up or down the c axis to fivefold-coordinated atoms. The lattice structure of TiO$_2$ anatase has been extensively studied both theoretically and experimentally.

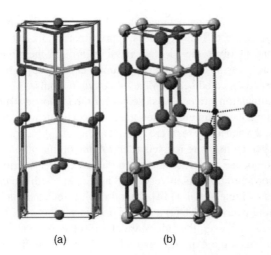

(a)   (b)

FIGURE 11.10 Unit cell of anatase structure. (a) Small spheres indicate center of empty octahedral sites in $1 \times 1 \times 1$ unit cell that could be occupied by Li ions. (b) Detail of octahedral environment around one interstitial site (dotted lines).

On the other hand, the titanate structure belongs to the space group Imma [108, 109]. The difference between the tetragonal anatase and the orthorhombic titanate structures mainly concerns the cell parameters and not the octahedron distribution in the space: Lattice parameters a and b are the same for anatase but differ in titanate, rendering a loss of symmetry. The small distortion associated with the transformation anatase–titanate generates more regularly shaped TiO_6 octahedrons.

The different TiO_2 anatase supercells considered above are doped with Li atoms, which are inserted in the interstitial cavities. Two different sites are formally possible: octahedral and tetrahedral. However, since the optimizations of the tetrahedral sites doped with Li ions led to serious problems of convergence and once converged were less stable than their octahedral counterparts, only octahedral sites are considered. The position of the empty octahedral sites available for Li intercalation in the anatase unit cell and the detail of the octahedral environment are displayed in Figure 11.10. The empty octahedral void is, as can be seen from Figure 11.10, highly distorted. When the Li is positioned in the center, one can speak about a formally fourfold-coordinated environment, since the apical oxygen atoms are quite far away. When the Li atom is displaced in the c direction toward an apical oxygen, it becomes fivefold coordinated [120, 128].

The effect of Li insertion, position, and concentration on the anatase structure is discussed in the paper on which this chapter is based [125].

Investigation of Mobility of Lithium Ions   The anatase structure contains two types of interstitial sites for the Li atoms, octahedral and tetrahedral, as described above. From the topology of the lattice and since the tetrahedral positions are high in energy, we only consider the path between the octahedral interstitial voids. The chain structure of edge-sharing octahedrons allows a migration through successive hops. The length of the diffusion path is approximately 3.1 Å (depending on the concentration) and involves the Li atom passing through the edge shared by two octahedrons, which was also suggested by Lunell et al. [122]. The transition state would then be exactly located between the two oxygen atoms.

The same optimization strategy was used as for the Li-doped structures; however, some difficulties were encountered when the atom positions were optimized. As the structure collapsed when all atoms were relaxed at the same time, only the two closest neighboring oxygen atoms were relaxed after volume optimization.

The magnitude of the ab initio calculated values, converging to 0.6 eV for high (up to $x = 0.5$) concentrations, is in very good agreement with the experimental and calculated values of Lunell et al. [114]. Lunell et al. calculated via intermediate neglect of differential overlap (INDO) an activation barrier between 0.66 and 0.51 eV for $x = 0.5$–0.0625. However, the values are close to the experimental ones, and the trend is not clear since the experimental values do not show the trend in the low-concentration area. Since no more experimental data were found for the activation barriers, the trend obtained in our calculations and those by Lunell et al. are in contradiction. In the latter study an increase of activation barrier was found, in contrast with a decrease in our predictions (see Table 11.3). Koudriachova et al. [129] calculated the open-circuit voltage (OCV), showing the experimental trend, that is decreasing hopping barrier with increasing Li concentration, as is found in our calculations as well (see Fig. 11.10 and Table 11.3). This trend is similar to the one in solution, since the ionic mobility also increases at high concentrations. Interesting to note is the kink present in Fig. 11.11. which might be associated with the change geometry from a low Li phase to a high Li phase (vide infra).

These results also show that there is a large difference in the mobility between anatase and titanate; the barriers differ by 0.6 eV. Van De Krol et al. [130] suggested that this difference would be the origin for the difference found between the insertion rate and the extraction rate of Li atoms in both coexisting phases.

From these barriers it is straightforward to calculate the hopping rate using Eq. (11.4) (see Table 11.3 and Fig. 11.11).

TABLE 11.3 Energy Barriers and Hopping Rate Γ for Different Li Concentrations in $Li_x TiO_2$ (Anatase–Titanate) between Two Neighboring Octahedrons

X	Energy Barrier (eV)	Hopping Rate Γ (s^{-1})
0.5	0.67	2.06×10^{15}
0.25	0.77	1.47×10^{16}
0.125	1.20	6.77×10^{19}
0.083	1.24	1.48×10^{20}
0.063	1.31	5.86×10^{20}

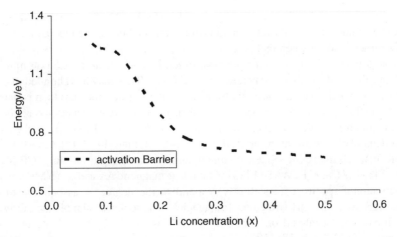

FIGURE 11.11 Hopping barrier for different Li concentrations between two octahedral cavities.

11.4.3 Water Interaction with Oxide Surfaces

11.4.3.1 Hydroxylated Surfaces

Ab Initio Study of Hydroxylated Surface of Amorphous Silica: Realistic Model
Silica, the most common mineral on Earth, is used for applications in catalysis and chromatography and is a key component in electronic devices, solar cells, and optical fibers [131]. Its chemical properties have been widely studied and reviewed [132, 133]. Silica polymorphs are composed of SiO_4 tetrahedra which polymerize forming different structures. The flexibility of the Si–O–Si bond explains the great number of existing polymorphs, either natural or synthetic: several crystalline forms such as quartz, cristobalite, tridymite, diatomite, and edingtonite, a number of noncrystalline glasses or sol–gel phases, and micro/mesoporous materials.

When created by cleavage, the silica surface exhibits undercoordinated atoms such as three coordinated Si atoms, terminal (nonbridging) oxygens, and strained Si–O–Si bridges in small size rings. The silica defects readily react with water in ambient conditions to form surface hydroxyl groups named silanols [133–136] which can make the surface hydrophilic. In contrast, a nondefective surface is hydrophobic [137–140].

Due to the noncrystalline nature of silica, the classical diffraction techniques cannot be used to give structural information. Yet its surface seems to exhibit a quite rich diversity of chemical groups.

The first important distinction that must be made here is between the terminal and geminal silanols. Terminal silanols are bound to a Si atom involved in three Si–O–Si siloxane groups, whereas geminal silanols complete the coordination sphere of a Si atom involved in two siloxane groups. Sometimes, longer range interactions are taken into account: Two silanols on Si atoms connected by a siloxane bridge are vicinal;

they are adjacent if the Si atoms are separated by a O–Si–O bridge. A terminal silanol could be engaged in either vicinal or adjacent relations, both, or neither, and the same holds for a silanol in a geminal pair.

In addition to this distinction based on through-bond interactions, silanols may differ according to through-space neighboring relations: They may be H bonded to other silanols (in which case they are said to be associated) or not; they are then isolated.

The methods available to characterize silica surfaces give information on either of these distinctions. The silanols on the surface of amorphous silica have been characterized mainly by means of infrared (IR) [141, 142], Raman [143, 144], as well as ^{29}Si magic angle spinning (MAS) and charge polarized magic angle spinning (CP-MAS) [138, 145–148] and ^{1}H MAS [149–156] nuclear magnetic resonance (NMR).

Vibrational spectroscopies mostly respond to the degree of association of the silanol groups. The OH-stretching frequencies of associated silanols are observed in a broad band centered on $3550\,cm^{-1}$, while isolated silanols vibrate at 3747–$3750\,cm^{-1}$ [132, 133, 157, 158].

^{29}Si NMR provides information on the state of covalent binding (geminal or terminal) of the silanol groups: Geminal silanols correspond to "Q^{2}" environments for the silicon atom, whose nucleus resonates at about $-90\,ppm$, and terminal silanols to "Q^{3}" environments for the silicon, resonating around $-100\,ppm$. Recently, methods were proposed to perform the quantification of protons by ^{1}H MAS NMR in silicas [159, 160].

Besides providing an independent estimate of the total density of silanols, they allowed to discriminate between associated, accessible isolated, and inaccessible isolated silanols. The insulator nature of silica compromises the use of classical surface tools due to sample charging effects. XPS has been scarcely used [161, 162] for surface site characterization, although thin SiO_2 films may be synthesized to overcome this technical limit and some studies on the reactivity of amorphous thin films of SiO_2 on Si(100) and Si(111) were performed [136, 163]. Freund's group synthesized a flat and homogeneous monocrystalline SiO_2 layer on Mo(112) [164, 165], which was used later by Goodman et al. and others [140, 166, 167].

A realistic model of the silica surface should reproduce what is known about the silanol density and the distribution among the various kinds of silanols. Some precipitated silicas can expose as many as $7\,OH/nm^{2}$ [157], but in most studies the average reported density is $4.9 \pm 1\,OH/nm^{2}$ [135], as compared to a total of 8 surface Si atoms/nm^{2} [168].

The measured proportion of geminal silanols is more uncertain and depends on the type of silica as well as on the degree of humidity of the atmosphere. Indeed, geminal silanols are expected to be stabilized in the presence of water (vapor or liquid). Various authors have reported that geminal silanols amount to 9–30% of the total silanol population on amorphous silica [148, 169, 170]. A quite typical value of 14% geminal silanols was obtained for an Aerosil-type silica by ^{29}Si NMR [171].

Other claims are quite different. Using the technique of evanescent wave cavity ringdown spectroscopy, Fisk et al. [172] have reported that the surface of a quartz prism in contact with a water solution exhibited two silanols with different acidities (pK_a) in the relative proportions of 27% (more acidic) to 73% (less acidic). They assigned the first group to terminal and the second to geminal silanols; that is, the geminal Si–OH would then constitute a $\frac{3}{4}$ majority. However, the basis for this assignment is unclear. Even if it is justified, the substrate under study was quite different from amorphous silica.

In an SFG (sum frequency generation) study of a silica prism in contact with solutions of variable pH, Ong et al. also distinguished two surface groups of different acidities with 15–20% of the more acidic one and 80–85% of the other, but they refrained from a precise assignment of both groups [173]. Thus, based on currently available evidence, we believe that a realistic model of silica should include a minority of geminal silanols in the 10–30% range.

The silanols are responsible for the silica reactivity. Among other studied probes, adsorption of water on the silanols has been the subject of systematic studies [174]. The experimental heat of adsorption of water on silanols was measured with microcalorimetry to be between 90 and 44 kJ/mol, the latter value being close to the latent heat of water liquefaction [174]; it was concluded that silanols are hydrophilic.

H-bonded silanols in silanol nests have been suggested to be more reactive to water than isolated ones and to allow water clustering [175, 176]. Many other adsorption or grafting reactions depend on the existence of several Si–OH in close vicinity. For instance, in a rather old study, Kol'tsov et al. proposed to use $VOCl_3$ grafting from the gas phase to quantify the number of groups of two or three neighboring silanols on an amorphous silica surface as a function of the pretreatment temperature [177].

Because of the multiple sites exhibited on silica surface, it might be expected that adsorbed molecules would give rise to intricate spectra corresponding to the superposition of the responses of a variety of adsorbed species. However, some instances are known where rather sharp spectra are observed (e.g., UV spectra of $[Ni^{II}L_x(H_2O)_{6-x}]^{2+}$ [178], ^{13}C NMR of amino acids [179]) suggesting that adsorbed molecules have a strong preference to interact with specific groups.

Many theoretical studies have been performed on silica in order to complement the lack of information due to its amorphous nature. Different approaches have addressed different topics. On the one hand, force field calculations have allowed to propose large size models (several thousands of atoms) of vitreous/amorphous silicas which have been compared with bulk data such as Si–O–Si angles, Si–O bond lengths, and ring size distributions [180–185]. The bulk, surfaces, and their reactivity toward water to generate a hydroxylated surface have been much less studied: After the seminal work of Feuston and Garofalini on water reactivity with a silica surface [180, 181] a few works have been devoted to the water/silica interaction [186–188] Ab initio calculations have been combined with classical molecular dynamics in order to get an insight on the strength of the physisorption sites and the chemical reactivity of the anhydrous surface [188]. Only very recently a new model for the hydroxylated amorphous silica surface has been proposed [189]. The high interest of this new model is to propose a hydroxylated surface with hydrophilic zones coexisting with hydrophobic ones.

On the other hand, first-principles calculations on small-size clusters (10–50 atoms) have focused on the spectroscopic properties and reactivity of the silanols with small adsorbates [190]. In particular, the interaction of water with the surface was modeled

[191–193]. The adsorption of NH_3 or alcohols [194], isobutene [195], benzyl- and phenyl-containing probes [196], metallic complexes [197–204], and small-size amino acids [179, 205, 206] onto the silica surface was also investigated. It is well known that cluster studies allow the use of high-accuracy quantum methods which describe energetics, reaction paths, and intermediates but fail to reproduce the long-range constraints and forces.

From the large-size models obtained with classical methods, it is possible to extract small systems and to use them as a starting point for elaborating a new model with ab initio methods [207, 208, 209]. Alternatively, it is possible to start from molten silica models and to cool them down slowly [210, 211]. Bulk amorphous silica may also be generated by cooling a preheated β-cristobalite via classical molecular dynamics simulation and perform a further ab initio molecular dynamics optimization [168].

Very few models [212, 213] of hydroxylated amorphous silica surfaces have been modeled using ab initio periodic calculations trying to reproduce what is empirically known of the density, nature, and distribution of surface silanols and compare predicted and observed spectroscopic properties. In this chapter we report results from first-principle density functional calculations on a realistic model for an amorphous hydroxylated silica surface.

The amorphous silica surface model is inspired from the model of Garofalini [181] used in molecular mechanics calculations on pure SiO_2 glasses. Since this model is much too large to be used in periodic ab initio calculations, it was truncated to form a slab with dimensions of about $13\,\text{Å} \times 17.5\,\text{Å} \times 25\,\text{Å}$ (of which $15\,\text{Å}$ is vacuum) containing 26 SiO_2 units. To this structure, 13 water molecules were added as terminating OH groups. The total number of atoms of the new hydrated slab is 120 ($Si_{26}O_{65}H_{27}$). The slab has a density of approximately $1.7\,\text{g/cm}^3$. The number of Si atoms belonging to tetrahedra exposed on the surface is approximately $8/\text{nm}^2$.

The atom positions and cell dimensions were relaxed using ab initio MD at constant temperature (400 K). The time step was set at 3 fs and the geometries were sampled up to 5 ps to have a reliable image of the equilibrium geometry at 400 K using a microcanonical ensemble. Higher temperatures show dehydroxylation and ring closing, making the surface less hydrophilic, which is in agreement with experimental observations [214]. In a second step the equilibrated slab was completely (i.e., without geometric constraints) optimized at 0 K. Because of the large cell size, the Γ point was used for the Brillouin zone integration and a plane-wave energy cutoff of 230 eV, followed by an optimization performed at a $3 \times 2 \times 1$ k-point mesh for the Brillouin zone integration with an energy cutoff of 400 eV. After the geometry optimization the obtained surface was $12.8\,\text{Å}$ long $\times 17.6\,\text{Å}$ wide with a slab thickness varying from $5.2\,\text{Å}$ to $8.2\,\text{Å}$, corresponding to a three-layer slab (see Fig. 11.12).

The gas-phase acidity of the hydroxyls was evaluated using the deprotonation energy ΔE, for the reaction $AH \rightarrow A^- + H^+$. The ΔE was calculated as the difference between the energy of the optimized protonated structure and the energy of the optimized conjugated base. An increase in deprotonation energy indicated a decrease in acidity.

Top view

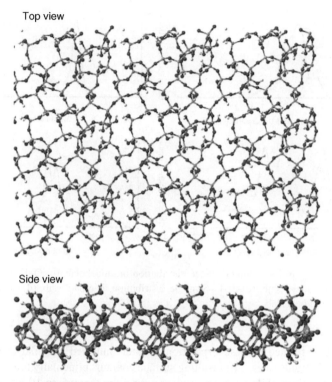

Side view

FIGURE 11.12   Model slab seen from top and side.

To calculate the Hessian matrix, finite differences are used, that is, each ion is displaced in the direction of each Cartesian coordinate, and the Hessian matrix is determined from the forces. All atoms are displaced in all three Cartesian directions. The frequency calculations were performed considering one k point.

The first-principles NMR calculations were performed within Kohn–Sham DFT using the PARATEC code [215]. The PBE generalized gradient approximation [216–219] was used and the valence electrons were described by norm-conserving pseudopotentials [220] in the Kleinman–Bylander [221] form. The core definition for O is $1s^2$ and $1s^2 2s^2 2p^6$ for Si. The core radii are 1.2 a.u. for H, 1.5 a.u. for O, and 2.0 a.u. for Si. The wavefunctions are expanded on a plane-wave basis set with a kinetic energy cutoff of 1088 eV.

The integrals over the first Brillouin zone are performed using a Monkhorst–Pack $2 \times 2 \times 1$ k-point grid [222] for the charge density and chemical shift tensor calculation. The shielding tensor is computed using the gauge including projector augmented waves (GIPAW) [223] approach, which permits the reproduction of the results of a fully converged all-electron calculation. The isotropic chemical shift δ_{iso} is defined as $\delta_{iso} = -(\sigma - \sigma^{ref})$, where σ is the isotropic shielding and σ^{ref} is the isotropic shielding of the same nucleus in a reference system as previously described [224, 225].

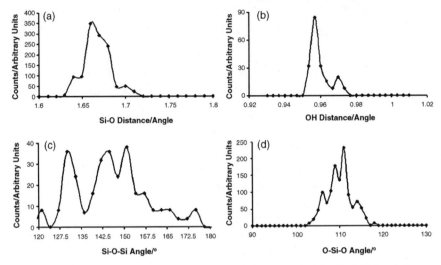

FIGURE 11.13 (a) Si–O bond distance distribution in model slab. (b) O–H bond distance distribution in model slab. (c) Si–O–Si angle distribution in model slab. (d) O–Si–O angle distribution in model slab.

The slab is formed by a superposition of interconnected irregular $(SiO_2)_n$ ring structures of various sizes. The resulting surface contains principally four-, five-, and six-membered rings. There are also some larger rings (from 7 to 10 tetrahedra). In our model slab the distribution is 4, 3, 4, 1, 1, 1, and 1 for the 4-, 5-, 6-, 7-, 8-, 9-, and 10-membered rings, respectively. Smaller rings (two and three rings) were absent, as intuitively expected since they are more strained and rigid and are known to open after hydration of the surface.

The Si–O distance analysis shows the distribution being centered on 1.66 Å (ranging from 1.64 to 1.72 Å). The value of 1.66 Å is in very good agreement with the high-level calculations on the disiloxane molecule [226]. The average Si–O bond length in amorphous silica has been reported to be 1.62 Å [227].

The Si–O–Si angles range from 130° to almost 180° (see Fig. 11.13), having a maximum around 142°–150°. Small values for the Si–O–Si angle generally correspond to small rings. However, it was shown by Athanasopoulos et al. [228] that the decrease in the Si–O–Si angle is related not directly to the presence of small rings but rather to the deformation of five- and six-membered rings. Indeed, since no strongly strained two- or three-membered rings are present in our model, the mean value agrees with the presence of deformed five and six membered rings. It is also known that the Si–O–Si angle has a very low bending barrier (the barrier to linearity is only 1.80 kJ/mol) [226], which is confirmed by the broad distribution of this parameter as can be seen in Figure 11.13.

The distribution of the O–Si–O angle (see Fig. 11.13) shows a maximum around 111° (very close to the tetrahedral angle). This distribution is in agreement with observations in α-quartz where a small shift to higher values is noticed as well.

TABLE 11.4   Overview of Different Number/Nature of Silanols on Slab Compared with Experiment

Density and Nature of Silanols on Two Faces of Slab	Top Surface	Bottom Surface	Experimental Values
Silanol density (OH/nm²)	5.8	5.8	4.9
Geminal/(geminal + terminal) ratio	0.23	0.46	0.1–0.3
Vicinal/(vicinal + adjacent) total ratio	0.92	1	Not reported
Silanols involved in H bonds	0.38	0.31	Not reported

The correct description of the Si–O–Si angle is very important as it provides an extremely useful test for theoretical calculation methods. The geometric parameters may be directly compared with high-level DFT results obtained for the elementary building block of silicas, silanol, and disiloxane studied by one of us in the past [226]. In contrast, the distribution of the O–Si–O angles is much narrower, in agreement with the intuitive idea that the structure of silica polymorphs is made of rigid (SiO₄) tetrahedra connected by Si–O–Si hinges.

As exposed in Section 11.1, the main purpose of this chapter is to propose a model of amorphous silica taking into account the density and nature of silanols and H bond network present on a real amorphous surface, as those silanols are responsible for silica reactivity. Table 11.4 summarizes the characteristics of our model concerning those SiOH groups: Two opposite surfaces were created when generating the slab. We refer to them (arbitrarily) as top and bottom surfaces, as shown in the Figure 11.12. The concentration of surface hydroxyls is 5.8/nm² on each surface, which is quite close to the experimental estimates on a hydroxylated surface (see Section 11.1). On the top surface, 23% silanols are geminal ones, in good agreement with experimental values; in contrast, the bottom surface presents a higher geminal density. Most of the silanols are vicinal ones, due to the relatively high silanol surface density. Only one silanol is surrounded by Q_4 Si atoms (Si atoms bearing no OH group), but even in this case it has adjacent silanols in the neighborhood. Finally, 38% (respectively 31%) of the silanols are involved in H bonds. To summarize, the slab proposed here exhibits a silanol density and distribution (geminal/terminal proportion) compatible with experiment; the large majority of silanols are vicinal ones, from which one-third are H bonded and two-thirds remain isolated. A summary of the different types of silanols on our model surface compared with experiment is shown in Table 11.4.

Various abinitio calculations have been carried out previously [170, 229–235] on silicate clusters in order to determine possible correlations between local structure and NMR parameters, with special attention to ^{17}O NMR. These results have been compared to experimental ones to translate a distribution of ^{17}O quadrupolar parameters into a distribution of structural parameters such as the Si–O–Si angles and Si–O distances [236, 237]. Alternative approaches involve the use of periodic boundary conditions: Pickard and Mauri [238] introduced a plane-wave pseudopotential method for calculating NMR parameters. Their calculated values for a variety of silica polymorphs were shown to be in excellent agreement with experimental data [239].

The relevance of our amorphous silica surface model was checked by comparing the calculated NMR parameters (using the periodic approach) with experimental values concerning Si–O–Si bonds in similar systems. The obtained results were also confronted with the relationships previously proposed between NMR properties and local environments. Obviously, it would be particularly important to understand the ^{1}H and ^{17}O NMR characteristics of the atoms included in silanol groups since the latter determine the adsorptive behavior. It would for instance be interesting to establish correlations between the NMR parameters and the hydrogen bonding state of silanols. So far, however, very few experimental ^{17}O NMR data [240, 241] have been reported for Si–OH groups.

The vibrational frequency analysis and comparison with experimental spectra are discussed elsewhere [213].

ADSORPTION OF WATER ON SILANOLS   Water can be adsorbed onto the surface silanols via hydrogen bonds with an adsorption energy higher than 44 kJ/mol (which is the latent heat of liquefaction of water) [174]. This result identifies the silica surface as hydrophilic. The interaction of a single water molecule was studied with a pair of geminal silanols and an isolated terminal silanol (Figure 11.14). It was found that the energy of interaction varies only slightly with the nature of the adsorption site, with an adsorption energy between 46 and 50 kJ/mol, slightly higher (4 kJ/mol) with the geminal than with the isolated silanol. In the latter case, two configurations were investigated, one with the water H bonded to one silanol only (Figure 11.14) and one with water H bonded to the silanol and a bridging O (Fig. 11.14).

Very similar heats of adsorption (51 kJ/mol) were obtained by other theoretical works for water on vicinal silanols [168].

A slightly higher heat of adsorption of water on geminal (46.5 kJ/mol) than an isolated silanols (37.5 kJ/mol) has been calculated by Ugliengo et al. [192] Other works conclude to an identical heat of adsorption of water of 38 and 39 kJ/mol on the isolated and geminal silanols, respectively [193].

ZrO2   Zirconia is an important material for applications ranging from catalysis to microelectronic gate dielectrics and to ceramic engineering. One of the main benefits of zirconia for many applications is its high thermal stability [242].

The versatile reactivity of zirconia originates among others from the amphoteric character of its hydroxyl groups [243]. In the methanol synthesis the hydroxyl species of zirconia are responsible for the formation of formate or carbonate species that are important intermediates of the reaction. In the atomic layer deposition (ALD) technique, which can be used to prepare thin films or catalysts, the hydroxyl groups take part in grafting the gaseous precursors to the substrate surface [244, 245]. It has been suspected that the hydroxyl groups influence the film growth rate [245]. In addition to the hydroxyl species, undercoordinated Zr^{4+} and O^{2-} species and defects such as oxygen vacancies exist on the zirconia surface. The relative concentration

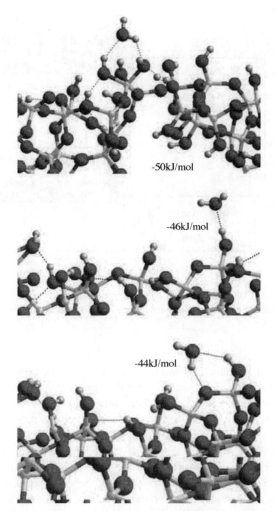

-50kJ/mol

-46kJ/mol

-44kJ/mol

FIGURE 11.14   Insertion of single water molecule with pair geminal silanols and isolated terminal silanol in interaction with and without lattice oxygen.

of the surface species depends on the phase of zirconia, the particle size, and the preparation route (e.g., impurities and dopants).

The hydroxylated surfaces of monoclinic zirconia, which is the thermodynamic stable phase of ZrO_2, have been studied from experimental [246–248] and theoretical methods [249–251]. On the monoclinic zirconia the hydroxyl groups are mainly terminal and threefold coordinated [246–248]. The interaction with water at low coverage leads to dissociative adsorption of the molecule and the presence of hydroxyl pairs. On a regular nondefective monoclinic ($\bar{1}11$) slab, this adsorption mode takes place on undercoordinated Zr sites and surface oxygen atoms in an acid–base mechanism described above. The process is exothermic by 1.2 eV, showing moderate affinity

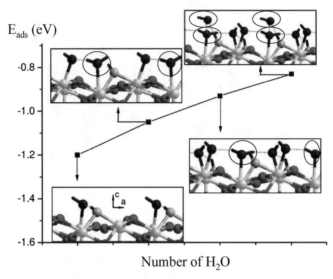

FIGURE 11.15   The Adsorption energy per water molecule as function of coverage for ($\bar{1}$11) surface of monoclinic zirconia, (eV). The structures of the most stable hydrated surfaces are displayed.

between water and zirconia. Figure 11.15 displays the structures for the calculated slabs. Increasing the coverage leads to a stabilization of the molecular form, so both hydroxyl and molecular water coexist. The average energy per water molecule becomes less exothermic with the increase in coverage due to the repulsion between adsorbates and despite the hydrogen bonds created.

The role of surface hydroxyl groups is crucial for technological applications. For instance, the interaction of CO with hydroxyl groups leads to the formation of a formate group. Such groups are key intermediates in the synthesis of methanol, among others. It has been shown in infrared measurements that the signal associated with hydroxyl groups decreases while the formate signal increases. It is concluded that CO inserts in the hydroxyl group, as in Figure 11.16. This reaction does not take place at room temperature but needs activation. Different formate species may exist on the surface, and theoretical calculations bring valuable information to elucidate their role as intermediates. In a combined IR-DFT work [252], we have shown that the most stable formate species on monoclinic zirconia ($\bar{1}$11) is bidentate (see structure I in Fig. 11.16). The high stability probably indicates poor reactivity. In the oxidation to CO_2, the bidentate formate is most likely a spectator. Species II and III show the weakest adsorption energy. Species IV, the monodentate formate, presents intermediate adsorption energy—It is stable enough to be formed and reactive enough to evolve to products:

A similar study carried out on Ga_2O_3 material [253] shows that the bidentate species accumulates due to its high thermodynamic stability, while monodentate formate is the key intermediate in the further oxidation to CO_2. The calculated energetic barriers reported therein support this mechanism. It is thus reasonable to suggest this

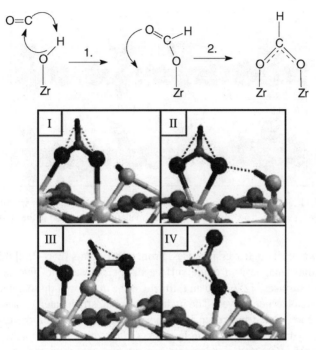

FIGURE 11.16 Top: mechanism of formation of formates on surface hydroxyl groups Bottom: four models for surface formates. The most stable is species I.

mechanism also for ZrO_2. The presence of surface hydroxyl groups is thus of prime importance in the chemical reactivity of interface systems.

TiO_2   As mentioned above, the (110) surface of rutile is hydroxylated upon interaction with hydrogen. In this section we will briefly comment on the molecular orientation of such hydroxyl groups for the hydrogenated conditions, which is formally equivalent to the result of water dissociation in an oxygen vacancy. For a general view on the electronic structure or coverage effects see Section 11.2.

Several works in the past have considered hydroxyl groups as being perpendicular to the surface [74, 75], while others found that such groups are tilted around $50°$ from the vertical [254]. Periodic PBE calculations carried out by us indicate that OH groups are tilted $54°$ from the vertical position, with a gain of 257 meV per OH [255]. Moreover, they are found to alternate in the row of bridging oxygen atoms; see Figure 11.17. The energy gain for this orientation is weak, 71 meV, but is found for all the coverage tested. Since the difference in energy is weak, it is likely that this arrangement is not visible even at low temperature.

11.4.3.2   Hydrated Surfaces: DFT Study of Water/MgO(100) Interface in Acidic and Basic Media

Pure water at the contact of MgO has been supposed to remain molecular by some authors and to dissociate by others. According to Henderson, MgO(100) represents

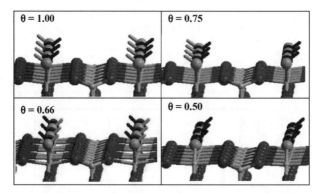

FIGURE 11.17 Hydroxyl groups on rutile (110) TiO_2 surface. They are aligned on the bridging oxygen row, tilted 54° from the vertical and alternate right–left in the (1–10) direction.

one of the two most controversial cases. From experiments [256–259] it is generally considered that "magnesium oxide surface dissociate water to give rise to highly hydroxylated surfaces" [256] and an estimate of the number of dissociated molecules per surface area has been given. The hydroxyls are mostly located on adsorbing sites that have 5-coordinated ions even if dissociation might occur on the corners being followed by hydroxyl migration on terraces. On the other hand, quantum-chemical calculations [25, 260–269] mostly agree in predicting that the perfect (100) faces do not dissociate water. Therefore it is supposed that dissociation occurs on defects [254, 270–273], steps [274], when surfaces are reconstructed [275, 276], under pressure [277], or when traces of acids are present [278]. The lack of dissociation on a perfect surface is a consequence of the poor basicity of the surface O^{2-} anions, which are fivefold-coordinated and stabilized by a strong Madelung field; MgO is an insulator with a valence band low in energy. Most of the adsorbates bind to the surface cations, and in anhydrous conditions the metal surface is predominantly acidic. The presence of water can modify the surface properties and reveal basic properties.

HCl dissociates in liquid water and partially in bulk ice [279–281]. Interaction between ice and HCl is responsible for pollution phenomena in the polar stratosphere [282–284]. It is less clear whether HCl dissociates on ice surfaces. Two mechanisms have been proposed: Either the HCl molecule adsorbs on the surface (forming eventually dihydrates or dissociating [285–291]) or it penetrates inside the ice structure at least in the upmost layers (see refs. 12–15 in ref. 292 and refs. 293–295). Experiments have been made to investigate the HCl adsorption on five bilayers of ice supported on MgO [296] showing that HCl-dihydrate coexists with icosahedral rotational symmetry with reflection symmetry (Ih) ice at 1 HCl monolayer ($T = 220$ K).

There are few studies of coadsorption of HCl with H_2O on solid surfaces. On Pt(111), the presence of hydronium was suggested [297]. On metal oxides (titania) Bourgeois et al. [278] have found that the hydroxylation is greatly enhanced by traces of fluorine resulting form the pollution of the UHV system by HF used for chemical etching. NaOH has been chosen to study the basic media because it is the most popular representative. Na^+ has not been observed on defect-free surfaces; it is believed to adsorb only in the vicinity of step edges or kink sites [298]. On NaCl surfaces, the adsorption energy decreases from 0.5 to 0.1 eV when water is present at the interface [299].

Modeling the interface between acid and base solution and metal oxide surface represents a complex system; indeed, it involves three kinds of interactions: those between water and surface, those between the solvated molecule and the surface, and finally those between the solvated molecule and the water. The acid and base easily dissociate into ionic fragments. The interaction which is of interest here, and which must be investigated, is the one of the two ions with the surface and the water media.

This complex system has been modeled through periodic quantum-chemical methods [300]. Since temperature is mostly not taken into account in such methods, water is modeled by ice. Ice is always the structure representing water and results have to be extrapolated to represent a liquid media. Molecular dynamics or Monte Carlo simulations can provide valuable information (see references in refs. 301 and 302), however with the known difficulties and drawbacks such as the needed computational power [301, 302]. In embedded techniques, the region of the interface is privileged and the liquid perturbs (reduces) mostly the energy without changing much of the structures [302]. QM calculations remain a useful tool of analysis, treating the different interactions on equal footing.

MgO is often chosen as representative for the metal oxides due to its simple structure; the bulk possesses a rock-salt structure with an alternation of ions with opposite charge [19, 303].

The (100) surface is stable, nonpolar, and classified of type I according to Tasker [304]. The clean surface is easy to prepare with well-defined stoichiometry [270], and it does not undergo large relaxations. We have modeled a three-layer-thick slab of MgO using a $\sqrt{2} \times \sqrt{2}$ unit cell with a lattice a ($= b$) parameter of 4.211 Å taken from the bulk value [305, 306].

This value also represents the parameter calculated by vienna ab initio simulation package (VASP) for the bulk optimization, 4.2118 Å. The slab is periodically repeated to generate a three-dimensional (3D) calculation. A fixed c value of 13 Å was chosen, allowing the accommodation of four layers of water molecules without dissociation [277]. For the study of the adsorption of individual molecules this value corresponds to an empty gap of 9 Å.

A slab with the (001) orientation has a square lattice periodicity, allowing a structure in epitaxy over MgO by a small contraction of the lattice parameter of the square unit cell, 4.272 Å [307]. This is compensated for by a vertical expansion of the water slab. The epitaxy gives a formal coverage of ½ for the first layer of water on the surface. It should be noted that this "sandwich" model with two surfaces is very suited to model the water oxide interface resulting effectively in a larger surface area.

The space between the successive MgO slabs was filled by a slab of proton-ordered water. The cubic ice (water) structure with $P4_12_12$ symmetry, where the sublattice of the O atoms has a face-centered-cubic (fcc) structure instead of a hexagonal close-packed (HCP) one, has been used; it is an antiferromagnetic structure [307].

In these systems, we have replaced a water molecule by an X molecule ($X = HCl$ or NaOH). The interspace between the MgO layers then contains one X molecule and three water molecules per unit cell. The energies, E_S, for these new systems have been defined relative to that of the unsubstituted system (four layers of water molecule in the interspace) and to those of H_2O and X molecules according to the expression

$$\Delta E_S = E(4\,H_2O \text{ layers inserted}) + [E(X) - E(H_2O)] - E(\text{system}) \qquad (11.4)$$

The adsorption energies ΔE_{ads} are defined as:

$$\Delta E_{ads} = [E(\text{slab}) + \sum E(Y_i)] - E(\text{system}) \qquad (11.5)$$

with $Y_i = H_2O$, HCl, or NaOH, yielding positive adsorption energies for exothermic interactions.

The calculations simulate UHV conditions. All the positions of the atoms were optimized, that is, the MgO layers as well as the water and/or HCl and/or NaOH. The calculated energies of the isolated molecules are $E(H_2O) = -14.248$ cV, $E(HCl) = -6.057$ eV, and $E(NaOH) = -11.411$ eV. For water, the distance and angle are $d_{OH} = 0.972$ Å and $106.4°$, respectively; for HCl the distance $d_{HCl} = 1.282$ Å, and for NaOH, the distances and angle are $d_{NaO} = 1.995$ Å, $d_{HO} = 0.971$ Å, and $Na\hat{O}H = 128°1$, respectively. These values have been calculated using a $10\,\text{Å} \times 10\,\text{Å} \times 10\,\text{Å}$ box.

(a) Water on MgO(100)   As a single molecule ($\theta = 0.5$), water adsorbs very weakly on the MgO surface with an energy slightly better than the energy of a single hydrogen bond, that is, 0.47 eV [308]. In this study we have introduced four layers of water with the geometry of $P4_12_12$ ice in the space between three layers of MgO. After complete optimization for a constant volume, it was found [277] that water does not dissociate and preserves its structure (see Fig. 11.18). Water binds to the two opposite surfaces of the MgO slab. On the bottom surface, water is adsorbed through H to a $O_{lattice}$ atom while on the upper surface water is adsorbed through an electron pair to a $Mg_{lattice}$ atom. The surfaces of MgO are slightly buckled. On the bottom surface, the $O_{lattice}$ atoms are external and the Mg atoms are 0.04–0.09 Å below. On the upper surface, a pair of Mg and O are shifted outward relatively to the other (by 0.07–012 Å).

(b) HCl in Water between MgO Slabs   The unit cell now contains three water molecules and one HCl molecule in the interspace between the MgO layers (see Fig. 11.18). The energy of the system is defined by the exchange of an HCl molecule with a water molecule between the system and the gas phase; it is referred relatively to the interface containing pure water and an HCl molecule in the gas phase as explained above [see Eq. (11.5)]. In pure water [279], HCl dissociates and the proton migrates to form an H_3O^+ with the molecules second neighbor relative to the Cl^- ion.

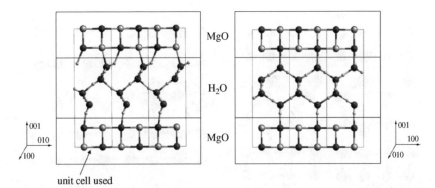

FIGURE 11.18 Four layers of proton ordered water [seen along (100) direction] of bulk unit cell epitaxially inserted in three-layer MgO slabs.

When HCl is placed within the water system of the MgO interspace, it also dissociates. In the most stable topologies, the Cl^- ions form a monolayer with water molecules, while the H^+ ions bind to the lattice oxygen and form OH^- ions. It is again a neutralization of HCl by the MgO surface. All of the protons are consumed by the surface O^{2-} ions and none is left in the solution that consists of water molecules and Cl^- ions. In our model, there is one HCl molecule per four surface oxygen atoms (two surfaces and a $\sqrt{2} \times \sqrt{2}$ unit cell) and most of the surface oxygens remain uncovered. We are therefore far below the situation where all of them would be turned into hydroxyl, which represents a passivated surface. Over this limit, the H^+ ions would still not stay in solution but would bind to the surface hydroxyls forming water; this would erode the surface and the Mg^{2+} surface ions remaining alone on the surface could trap the chloride ions forming $MgCl_2$ and inducing strong corrosion.

Energies for the different adsorption modes that we have calculated are gathered in Table 11.5. When the H^+ ions bind to the same surface as the Cl^- (Fig. 11.19), the process is exothermic by $\Delta E_S = 0.45\,eV$; when they bind to the opposite surface,

TABLE 11.5 Energy ΔE_S (eV) as Defined in Equation (11.5) for Substitution of HCl for H_2O in Interspace Filled with Water

Adsorption Site	Main Interaction	d_{MgCl} (Å)	ΔE_S (eV)
Cl in ice, H adsorbed on same surface	Cl^-/H_2O and H^+/O_{surf}	2.94	0.45
Cl in ice, H adsorbed on opposite surface	Cl^-/H_2O and H^+/O_{surf}	2.89	0.40
H and Cl adsorbed on opposite surfaces	Cl^-/H_2O and H^+/O_{surf}	2.40	0.15
Cl solvated + H_3O^+	—	—	−1.21
H and Cl adsorbed on same surface	Cl^-/Mg_{surf} and H^+/O_{surf}	2.44	−0.41
Cl in surface plane + 2 OH	—	—	−2.03

Note: Positive values for exothermic processes. The distances of the chlorine atom to the $Mg_{lattice}$ is indicated; the Mg–Cl distance in $MgCl_2$ is 2.18 Å; above 2.5 Å, the interaction between Mg and Cl is weak.

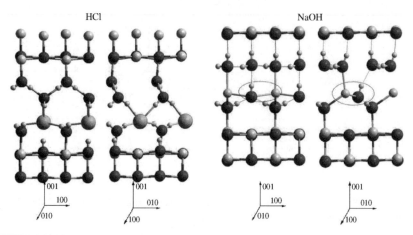

FIGURE 11.19 Side views along (100) and (010) direction of lowest energy structure for HCl, Na OH insertion. Left: Cl$^-$ ion is at 2.94 Å of surface Mg atom. Right: Na$^+$ ion is coordinated to 2 OH$^-$ and one H$_2$O.

the energy gain is slightly reduced, $\Delta E_S = 0.40$ eV. The MgO surfaces are buckled, all the sites bound to adsorbates being attracted outward: For the surface below, the O$_{lattice}$ of the hydroxyl groups and the Mg atoms bound to the water molecules are shifted up. The binding of the chloride ions to Mg is weaker than that of the H$^+$ to the O atoms of the surface. So, the surface Mg atoms remain below the average surface plane. For the surface above, the Mg atoms bound to the water molecules or the protonated oxygen atoms (on the upper surface) are shifted down.

We have also obtained a stable configuration exchanging the Cl$^-$ layer with the water layer beneath, bringing the Cl$^-$ close to the surface and moving the water molecules outward. The buckling of the surface layer is reversed to enhance the interaction with chloride ions closer to the surface. The proton is adsorbed on the opposite surface than the Cl. The energy of the system is slightly weaker, $\Delta E_S = 0.15$ eV.

The system where H$^+$ is bound to a water molecule and forms a hydronium cation is a secondary minimum high in energy and endothermic [$\Delta E_S = -1.21$ eV using the definition from Eq. (11.5)]. The O^{2-} atoms of the surface are much more basic than the O atoms from water, and thus H$^+$ naturally reacts better with them to form OH$^-$. This leaves the surface layer charged and with an unbalanced stoichiometry (MgOH$^+$), which should be a step toward the formation of Mg^{2+}(H$_2$O) with another HCl molecule that could react with Cl$^-$, forming magnesium chloride and thus peeling the first layer. Corrosion would then easily happen layer by layer.

Finally, the penetration of the Cl$^-$ has been investigated, which represents another mechanism for corrosion. First, we have adsorbed the two ions resulting from the HCl dissociation on the same surface: H$^+$/O^{2-} and Cl$^-$/Mg^{2+}. The three H$_2$O molecules form a homogeneous structure above; this structure is endothermic ($\Delta E_S = -0.41$ eV). The distance between the H and the Cl is 1.906 Å, which corresponds to HCl dissociation into adjacent ion pairs. Next, referring to the structure described in Figure 11.19, an O from the surface has been substituted by a Cl atom

TABLE 11.6 $d_{\text{Na-O}}$ in NaOH (Å) for NaOH Insertion into Different Systems Considered

System Number	Gas Phase	Pure Water	Adsorption on MgO	Coadsorption with H_2O on MgO	Coadsorption with H_2O on MgO	Water between MgO Slabs
1	1.95 (OH)	2.18 (OH)	2.17 (OH)	2.28 (OH)	2.51 (OH)	2.39 (OH)
2	—	2.29 (H_2O)	2.17 (OH)	2.35 (OH)	2.55 (OH)	2.41 (OH)
3	—	2.36 (H_2O)	2.50 (O_{lattice})	2.47 (H_2O)	2.07 (H_2O)	2.23 (H_2O)
4	—	—	—	—	2.15 (H_2O)	2.31 (H_2O)

Note: (Na–O bond distance in gas phase: 1.950 Å, both in this work and experiment 309). In parentheses is indicated whether the O from the Na–O bond pertains to a hydroxyl or a water molecule.

and one proton has been removed from the water layers for charge neutrality. The new system contains the same atoms per unit cell as the previous ones and corresponds to initiation of the penetration of Cl^-; indeed, the optimized structure represents the initial step for an exchange between a Cl^- anion and an OH^- hydroxyl from the surface, leading to corrosion of the oxide. The system is strongly endothermic $\Delta E_S = -2.03$ eV; the solution splits into two parts. A water molecule is dissociatively adsorbed on the unsubstituted surface and stabilized by a molecular water layer. The substituted surface that has a mixed composition between a chloride and an oxide is stabilized by a single water molecule. This mechanism for corrosion is therefore unlikely.

(c) NaOH in Water between MgO Slabs   NaOH dissociates within the water system of the interspace between the MgO slabs (see Fig. 11.19 and Table 11.6). The energy of the system referred to the molecules, $\Delta E_S = 1.94$ eV, corresponds to a highly exothermic process. The Na^+ cations are fourfold coordinated (Table 11.7); indeed, they form chains of ions with the OH^- (two neighbors) and are bound to two water molecules. The O atoms from OH^- are stabilized by two H bonds in addition to the two Na^+. This situation is close to that described in Figure 11.4(c) for the coadsorption. The NaOH thus interacts with the water and not directly with the surface. However, MgO contributes to modify the insertion mode of NaOH that differs from that in pure ice; this is because it stabilizes the ionic contributions in the ice. Note that the water media is also necessary to allow NaOH dissociation that did not occur with the same extent for the coadsorption [see Section (e) above]. The NaOH dissociation is also revealed by the Na–O distances; these are longer for the hydroxyls than for the neighboring water molecule (Table 11.6).

The situations where NaOH is closer to the surface are energetically less favorable, $\Delta E_S = 1.1$–1.3 eV (Table 11.7).

The substitution of an Mg atom of the surface by a Na is relatively unfavorable. In this case Na^+ alternates with hydroxyl groups in the "surface" layer while Mg^{2+} in the solution is bound to two OH^-. Cations Mg^{2+} and Na^+ remain in the interaction

TABLE 11.7 Energy ΔE_S (eV) as Defined in Eq. (11.5) for Substitution of NaOH for H$_2$O in Interspace Filled with Water

Adsorption Mode	ΔE_S (eV)
Ion pairs in ice forming chains	1.94
NaOH and H$_2$O without dissociation, Na from NaOH adsorbed on Mg$_{surf}$	1.34
NaOH without dissociation and H$_2$O with dissociation, two atoms from NaOH adsorbed	1.33
NaOH with dissociation and H$_2$O without dissociation, Na bound to 3 H$_2$O and OH adsorbed on Mg$_{surf}$	1.12
Exchange Na–Mg	0.43
Insertion between MgO layers	−3.5

Note: Positive values for exothermic processes.

with OH$^-$ ions. Finally, the insertion between the surface and the sublayer is strongly endothermic ($\Delta E_S = -3.5$ eV).

The main conclusion from these calculations is that the largest interaction concerns the ion pairs forming chains within the water phase. These ions do not interact directly with the surface; however, the presence of the MgO contributes to stabilize the charged species and favors the dissociation.

11.5 CONCLUSIONS AND PERSPECTIVES

Theoretical understanding of chemisorbed bonding on transition metals and metal oxides has been advanced notably by density-functional theory using metal clusters and slabs as model surfaces. This approach has been shown to yield binding-site energetics and vibrational frequencies for physi- and chemisorbed species on surfaces that are in reasonable accordance with UHV-based experimental results. Significantly the DFT results provide a detailed quantum-chemical description of the chemisorbate bonding. Given these developments, it is of considerable interest to develop also a quantum-chemical picture of the electrostatic field effects on surface bonding of relevance to electrochemical systems. There have been several previous studies along these general lines using semiempirical as well as ab initio theoretical approaches. An interesting yet largely unexplored issue is understanding how the electrostatic field-dependent chemisorbate bonding varies across the periodic table.

Most of the theoretical studies are undertaken on perfect slabs, corresponding to single crystals carefully prepared in ultrahigh vacuum. However, even freshly cleaved metal oxides show many defects such as edges, steps, corners, and kinks, which cannot be easily controlled in the actual experimental conditions. Experimentally their characterization has been approached by low-energy electron diffraction and by nondestructive adsorption of probe molecules, as reviewed elsewhere. It is clear that these sites have a distinct reactivity compared with the perfect sites found on the terraces.

Steps have often been considered in the model surfaces, but as is the case for the effect of an external electric field on the adsorption process, a lot of work has still to be undertaken. Many oxide surfaces have been let aside for the moment. Beside structural defects, oxygen vacancies, Schotky defects, F centers, adatoms, impurities, and so on, have been investigated.

Another perturbation one has to mention, and one seldom used in theoretical surface chemistry, is the effect of a solvent on the adsorption process. If solute–solvent interactions are strong, they may have a large impact on the electronic structure of a system and then on its excitation spectrum, reactivity, and properties. For these reasons, numerous models have been developed to deal with solute–solvent interactions in ab initio quantum-chemical calculations. A microscopic description of solvation effects can be obtained by a supermolecule approach or by combining statistical mechanical simulation techniques with quantum-chemical methods. Such methods, however, demand expensive computations. By contrast, at the phenomenological level, the solvent can be regarded as a dielectric continuum, and there are a number of approaches based on the classical reaction field concept. The reaction field formalism is based on a sharp partition of the system: The solute molecule (possibly supplemented by some explicit solvent molecules) is placed in a cavity surrounded by a polarizable dielectric. The surrounding is characterized mainly by its dielectric constant and density: An important parameter of the method is the size of the cavity; the dielectric medium is polarized by the solute, and this polarization creates a reaction field which perturbs the solute itself.

The most used model is based on the Kirkwood model and uses only spherical cavities; the other is called PCM (polarizable continuum model) and can use cavities of general shape modeled on the actual solute molecule. In the former case, the reaction field is computed as a truncated multipolar expansion and added as a perturbation to the one-electron Hamiltonian; in the latter case the reaction field is expressed in terms of a collection of apparent charges (solvation charges) spread on the cavity surface.

In conclusion one can say that theoretical metal oxide surface chemistry has known a dynamic start since the beginning of the 1990s due to the development of powerful algorithms for the DFT codes but also technological improvement of the calculators. Today most surface science problems can be investigated using these types of codes in their most simple states. As mentioned above, the effect of an external electrostatic field and inclusion of surface defects and solvent effects are still underrepresented in the investigations. It is clear that the second decade of the twenty-first century will be the start for a new kind of modeling techniques including more features in order to model more diverse surfaces in more diverse external conditions.

ACKNOWLEDGMENTS

The examples presented in this chapter are the result of fruitful research collaborations during the last five years with different people form different institutions: Professor J. Andrés (UJI, Spain), Dr. D. Costa (ENSCP, France), Professor J.F. Lambert (LRS, France), Dr. C. Gervais

(LCMC, France), Dr. A. Beltran (UJI, Spain), Professor C. Minot (LCT, France), and Professor C. Wöll (Bochum, Germany).

REFERENCES

1. C. Pisani, *Phys. Revi. B*, 17 (1978) 3143.
2. R. A. van Santen and L. H. Toneman, *Int. J. Quant. Chem.*, 12 Suppl. 2 (1977) 83.
3. F. Illas, A. Lorda, J. Rubio, J. B. Torrance, and P. S. Bagus, *J. Chem. Phys.*, 99 (1993) 389.
4. A. G. Pelmenschikov, G. Morosi, A. Gamba, S. Collucia, G. Martra, and E. A. Paukshtis, *J. Phys. Chem.* 100 (1996) 5011.
5. C. Pisani (Ed.), *Quantum-Mechanical Ab-Initio Calculation of the Properties of Crystalline Materials*, Springer, Berlin, 1996.
6. C. Pisani, *J. Molec. Struct. (Theochem.)*, 463 (1999) 125.
7. G. Kresse and J. Hafner, *Phys. Rev. B*, 47 (1993) 558.
8. G. Kresse and J. Hafner, *Phys. Rev. B*, 49 (1994) 14251.
9. P. Blaha, K. Schwarz, and J. Luitzand, WIEN97, Full-potential, linearized augmented plane wave package for calculating crystal properties, Institute of Technical Electrochemistry, Vienna University of Technology, Vienna, 1999.
10. D. Ricci, G. Pacchioni, P. V. Sushko, and A. L. Shluger, *J. Chem. Phys.*, 117 (2002) 2844.
11. 1990–2006, C. I. C. CPMD, MPI für Festkörperforschung, Stuttgart, 1997–2001.
12. M. C. Payne, M. P. Teter, D. C. Allan, and T. A. Arias, J. D. Joannopoulos, *Rev. Mod. Phys.*, 64 (1992) 1045.
13. Quantum-ESPRESSO; http://www.quantum-espresso.org/2000.
14. R. Dovesi et al. *CRYSTAL06 User's Manual*, University of Torino, Torino, Italy, 2006.
15. J. M. Soler et al. *J. Phys. Condens. Matter*, 14 (2002) 2745.
16. M. J. Frisch, et al. *Gaussian 03 (Revision A.1)*, Gaussian, Inc., Pittsburg, PA, 2003.
17. M. Calatayud, A. Markovits, A. Menetrey, B. Mguig, and C. Minot, *Catal. Today*, 85 (2003) 125.
18. M. Calatayud, A. Markovits, and C. Minot, *Catal. Today*, 89 (2004) 269.
19. V. E. Henrich and P. A. Cox. *The Surface Science of Metal Oxides*, Cambridge University Press, Cambridge, 1994.
20. G. Pacchioni, J. M. Ricart, and F. Illas, *J. Am. Chem. Soc.*, 116 (1994) 10152.
21. W. S. A. Halim and A. S. Shalabi, *Appl. Surf. Sci.*, 221 (2004) 53.
22. I. D. Gay and N. M. Harrison, *Surf. Sci.*, 591 (2005) 13.
23. W. S. A. Halim, *Appl. Surf. Sci.*, 253 (2007) 8974.
24. J. Carrasco, F. Illas, and N. Lopez, *Phys. Rev. Lett.*, 100 (2008) 016101.
25. W. Langel and M. Parrinello, *J. Chem. Phys.*, 103 (1995) 2340.
26. J. Ahdjoudj, A. Fahmi, and C. Minot, Periodic HF calculations of the adsorption of small molecules on TiO2, in *The Synergy Between Dynamics and Reactivity at Clusters and Surfaces*, L. J. Ferrugia (Ed.), Drymen, Scotland, 1995.
27. A. Auroux and A. Gervasini, *J. Phys. Chem.*, 94 (1990) 6371.
28. A. Gervasini and A. Auroux, *J. Therm. Anal.*, 37 (1991) 1737.
29. J. B. Peri, *J. Phys. Chem.*, 69 (1965) 231.
30. A. Markovits, J. Ahdjoudj, and C. Minot, *Il Nuovo Cimento*, 19D (1997) 1719.
31. J. Ahdjoudj, A. Markovits, and C. Minot, *Catal. Today*, 50 (1999) 541.
32. S. Pugh and M. J. Gillan, *Surf. Sci.*, 320 (1994) 331.

33. C. L. Pang, A. Sasahara, and H. Onishi, *Nanotechnology*, 18 (2007) 044003.
34. A. Markovits, J. Ahdjoudj, and C. Minot, *Surf. Sci.*, 365 (1996) 649.
35. C. Chizallet, G. Costentin, M. Che, F. Delbecq, and P. Sautet, *J. Phys. Chem. B*, 110 (2006) 15878.
36. D. Costa, C. Chizallet, E. Ealet, J. Goniakowski, and F. Finocchi, *J. Chem. Phys.*, 125 (2006) 054702.
37. F. Allegretti, S. O'Brien, M. Polcik, D. I. Sayago, and D. P. Woodruff, *Phys. Rev. Lett.*, 95 (2005) 226104.
38. A. V. Bandura, D. G. Sykes, V. Shapovalov, T. N. Troung, J. D. Kubicki, and R. A. Evarestov, *J. Phys. Chem. B*, 108 (2004) 7844.
39. J. Goniakowki and M. J. Gillan, *Surf. Sci.*, 350 (1996) 145.
40. A. Fahmi and C. Minot, *Surf. Sci.*, 304 (1994) 343.
41. P. J. D. Lindan, N. M. Harrison, and M. J. Gillan, *Phys. Rev. Lett.*, 80 (1998) 762.
42. L. A. Harris and A. A. Quong, *Phys. Rev. Lett.*, 93 (2004) 086105.
43. Z. Zhang, O. Bondarchuk, J. M. White, B. D. Kay, and Z. Dohnálek, *J. Am. Chem. Soc.*, 128 (2006) 4198.
44. A. Markovits, J. Ahdjoudj, and C. Minot, *Mol. Eng.*, 7 (1997) 245.
45. S. P. Bates, G. Kresse, and M. J. Gillan, *J. Phys. Chem. B*, 102 (1998) 2018.
46. M. Calatayud, J. Andres, and A. Beltran, *Surf. Sci.*, 430 (1999) 213.
47. R. E. Tanner, Y. Liang, and E. I. Altman, *Surf. Sci.*, 506 (2002) 251.
48. J. Ahdjoudj and C. Minot, *Catal. Lett.*, 46 (1997) 83.
49. H. Onishi, T. Aruga, and Iwasawa. Y., *J. Catal.*, 146 (1994) 557.
50. K. I. Fukui, H. Onishi, and Y. Iwasawa, *Phys. Lett.*, 280 (1997) 296.
51. S. A. Chambers, *Surf. Sci. Rep.*, 39 (2000) 105.
52. K. S. Kim and M. A. Barteau, *Langmuir*, 6 (1990) 1485.
53. V. Carravetta and S. Monti, *J. Phys. Chem. B*, 110 (2006) 6160.
54. M. Nilsing, S. Lunell, P. Persson, and L. Ojamäe, *Surf. Sci.*, 582 (2005) 49.
55. S. E. Collins, M. A. Baltanas, and A. L. Bonivardi, *Langmuir*, 21 (2005) 962.
56. Y. Wang, et al., *Phys. Rev. Lett.*, 95 (2005) 266104.
57. C. Di Valentin, G. Pacchioni, and A. Selloni, *Phys. Rev. Lett.*, 97 (2006) 166803.
58. M. V. Ganduglia-Pirovano, A. Hofmann, and J. Sauer, *Surf. Sci. Repts.*, 62 (2007) 542.
59. M. A. Barteau, *Chem. Rev.*, 96 (1996) 1413.
60. U. Diebold, *Surf. Sci. Repts.*, 48 (2003) 53.
61. P. W. Tasker, *J. Phys. C*, 12 (1979) 4977.
62. A. Markovits, M. K. Skalli, C. Minot, G. Pacchioni, N. Lopez, and F. Illas, *J. Phys. Chem. B*, 115 (2001) 8172.
63. A. Linsebigler, G. Lu, and J. T. Yates Jr., *Chem. Rev.*, 95 (1995) 735.
64. V. A. Gercher, D. F. Cox, and J.-M. Themlin, *Surf. Sci.*, 306 (1994) 279.
65. G. Charlton, et al., *Phys. Rev. Lett.*, 78 (1997) 495.
66. M. Ramamoorthy, D. Vanderbilt, and R. D. King-Smith, *Phys. Rev. B*, 49 (1994) 16721.
67. S. J. Thompson and S. P. Lewis, *Phys. Rev. B*, 73 (2006) 073403.
68. K. J. Hameeuw, G. Cantele, D. Ninno, F. Trani, and G. Iadonisi, *J. Chem. Phys.*, 124 (2006) 024708.
69. C. L. Pang, A. Sasahara, H. Onishi, Q. Chen, and G. Thornton, *Phys. Rev. B*, 74 (2006) 073411.
70. S. Wendt et al., *Phys. Rev. Lett.*, 96 (2006) 066107.
71. Z. Zhang, O. Bondarchuk, B. D. Kay, J. M. White, and Z. Dohnálek, *J. Phys. Chem. B.*, 110 (2006) 21840.
72. M. Kunat, U. Burghaus, and C. Wöll, *Phys. Chem. Chem. Phys.*, 6 (2004) 4203.

73. X.-L. Yin et al., *Chem. Phys. Chem.*, 9 (2008) 253.
74. A. Bouzoubaa, A. Markovits, M. Calatayud, and C. Minot, *Surf. Sci.*, 583 (2005) 107.
75. J. Leconte, A. Markovits, M. K. Skalli, C. Minot, and A. Belmajdoub, *Surf. Sci.*, 497 (2002) 194.
76. S. Suzuki, K.-i. Fukui, H. Onishi, and Y. Iwasawa, *Phys. Rev. Lett.*, 84 (2000) 2156.
77. P. Kruger et al. *Phys. Rev. Lett.*, 100 (2008) p. 055501.
78. Y. Cai and U. Ozkan, *Appl. Catal.*, 78 (1991) 241.
79. K. Devriendt, H. Poelman, and L. Fiermans, *Surf. Interf. Anal.*, 29 (2000) 139.
80. B. Tepper et al. *Surf. Sci.*, 496 (2002) 64.
81. K. Hermann et al. *J. Electr. Spectr. Rel. Phen.*, 98–99 (2004) 245.
82. A. Chakrabarti, K. Hermann, R. Druzinic, M. Witko, F. Wagner, and M. Petersen, *Phys. Rev. B*, 59 (1999) 10583.
83. C. Kolczewski and K. Hermann, *Surf. Sci.*, 552 (2004) 98.
84. U. S. Ozkan, Y. Cai, M. W. Kumthekar, and L. Zhang, *J. Catal.*, 142 (1993) 182.
85. T. Homann, T. Bredow, and K. Jug, *Surf. Sci.*, 515 (2002) 205.
86. X. Yin et al. *J. Phys. Chem. B*, 103 (1999) 4701.
87. R. Ramirez, B. Casal, L. Utrera, and E. Ruiz-Hitzky, *J. Phys. Chem.*, 94 (1990) 8960.
88. T. Bredow, T. Homann, and K. Jug, *Res. Chem. Intermed.*, 30 (2004) 65.
89. K. Jug, T. Homann, and T. Bredow, *J. Phys. Chem. A*, 108 (2004) 2966.
90. S. Besselmann, C. Freitag, O. Hinrichsen, and M. Muhler, *Phys. Chem. Chem. Phys.*, 3 (2001) 4633.
91. N.-Y. Topsøe, J. A. Dumesic, and H. Topsøe, *J. Catal.*, 151 (1995) 241.
92. D. A. Bulushev, L. Kiwi-Minsker, F. Rainone, and A. Renken, *J. Catal.*, 205 (2002) 115.
93. I. E. Wachs et al. *J. Catal.*, 161 (1996) 211.
94. M. A. Bañares and I. E. Wachs, *J. Raman Spectrosc.*, 33 (2002) 359.
95. G. T. Went, S. T. Oyama, and A. T. Bell, *J. Phys. Chem.*, 94 (1990) 4240.
96. K. D. Kreuer, *Annu. Rev. Mater. Res.*, 33 (2003) 333.
97. J. Robertson and P. W. Peacock, *Thin Solid Films*, 445 (2003) 155.
98. J. B. Bates and R. A. Perkins, *Phys. Rev. B*, 16 (1977) 3713.
99. G. D. Bromiley and A. A. Shiyaev, *Phys. Chem. Miner.*, 33 (2006) 426.
100. M. V. Koudriachova, S. W. de Leew, and N. M. Harrison, *Phys. Rev. B*, 70 (2004) 165421.
101. F. Bonino, L. Busani, M. Manstretta, B. Rivolta, and V. Scrosati, *J. Power Sources*, 6 (1981) 261.
102. S. Y. Huang, L. Kavan, I. Exnar, and M. Grätzel, *J. Electrochem. Soc.*, 142 (1995) L142.
103. T. Ohzuku, Z. Takehara, and S. Yoshizawa, *Electrochem. Acta*, 24 (1979) 219.
104. M. S. Whittingham, *Chem. Rev.*, 104 (2004) 4271.
105. B. O'Regan and M. Grätel, *Nature*, 353 (1991) 737.
106. C. Bechinger, S. Ferrere, A. Zaban, J. Sprague, and B. Gregg, *Nature*, 383 (1996) 608.
107. T. Ohzuku, and T. Hirai, *Electrochem. Acta*, 27 (1982) 1263.
108. R. J. Cava, D. W. Murphy, S. Zahurak, A. Santoro, and R. S. Roth, *J. Solid State Chem.*, 53 (1984) 64.
109. M. Wagemaker, D. Lützenkirchen-Hecht, A. A. van Well, and R. Frahm, *J. Phys. Chem. B*, 108 (2004) 12456.
110. M. Wagemaker, R. van de Krol, P. M. Kentgens, A. A. van Well, and F. M. Mulder, *J. Am. Chem. Soc.*, 123 (2001) 11454.
111. W. C. Mackrodt, *J. Solid State Chem.*, 142 (1999) 428.
112. M. Calatayud, P. Mori-Sanchez, A. Beltrán, and A. Martín Pendás, *Phys. Rev. B*, 64 (2001) 184118.
113. M. Calatayud and C. Minot, *Surf. Sci.*, 552 (2004) 169.

114. S. Lunell, A. Stashans, L. Ojamäe, H. Lindström, and A. Hagfeldt, *J. Am. Chem. Soc.*, 119 (1997) 7374.

115. M. V. Koudriachova, N. M. Harrison, and S. W. de Leeuw, *Phys. Rev. Lett.*, 86 (2001) 1275.

116. S. Södergren, H. Siegbahn, H. Rensmo, H. Lindström, A. Hagfeldt, and S.-E. Lindquist, *J. Phys Chem. B*, 101 (1997) 3087.

117. V. Luca, T. L. Hanley, N. K. Roberts, and R. F. Howe, *Chem. Mater.*, 11 (1999) 2089.

118. V. Luca, B. Hunter, B. Moubaraki, and K. S. Murray, *Chem. Mater.*, 13 (2001) 796.

119. M. Wagemaker, G. J. Kearly, A. A. van Well, H. Mutka, and F. M. Mulder, *J. Am. Chem. Soc.*, 125 (2003) 840.

120. P. Persson, R. Bergström, L. Ojamäe, and S. Lunell, *Adv. Quant. Chem.*, 41 (2002) 203.

121. F. Tielens, M. Calatayud, A. Beltrán, C. Minot, and J. Andrés, *J. Electroanal. Chem.*, 581 (2005) 988.

122. A. Stashans, S. Lunell, R. Bergström, A. Hagfeldt, and S.-E. Lindquist, *Phys. Rev. B*, 53 (1996) 159.

123. G. Garcia-Belmonte, V. S. Vikhrenko, J. Garcia-Cañadas, and J. Bisquert, *Solid State Ionics*, 170 (2004) 123.

124. A. Van der Ven and G. Ceder, *Electrochem. Solid-State Lett.*, 3 (2000) 301.

125. C. J. Howard, T. M. Sabine, and F. Dickson, *Acta Crystallogr. Sect. B Struct. Sci.*, 47 (1991) 462.

126. R. Van De Krol, A. Goossens, and E. Meulenkamp, *J. Electrochem. Soc.*, 146 (1999) 3150.

127. T. Ebina et al., *J. Power Sources*, 81–82 (1999) 393.

128. M. V. Koudriachova, S. W. de Leeuw, and N. M. Harrison, *Phys. Rev. B*, 69 (2004) 054106.

129. M. V. Koudriachova, N. M. Harrison, and S. W. de Leeuw, *Phys. Rev. Lett.*, 86 (2001) 1275.

130. R. Van De Krol, A. Goossens, and J. Schoonman, *J. Phys. Chem. B*, 103 (1999) 7151.

131. T. Bakos, S. N. Rashkeev, and S. T. Pantelides, *Phys. Rev. Lett.*, 88 (2002) 055508.

132. A. P. Legrand, *The Surface Properties of Silicas*, Wiley, New York, 1998.

133. R. K. Iler, *The Chemistry of Silicas*, Wiley, New York, 1979.

134. H. Knözinger, in *The Hydrogen Bond* vol. 3, P. Schuster, G. Zundel and C. Sandorfy (Eds.), North Holland Publishing Co, Amsterdam (1976), p. 1265.

135. L. T. Zhuravlev, *Langmuir*, 3 (1987) 316.

136. M. Nishijima, K. Edamoto, Y. Kubota, S. Tanaka, and M. Onchi, *J. Phys. Chem.*, 84 (1986) 6458.

137. V. Bolis, B. Fubini, L. Marchese, G. Martra, and D. Costa, *J. Chem. Soc., Faraday Trans.*, 87 (1991) 497.

138. D. W. Sindorf and G. E. Maciel, *J. Am. Chem. Soc.*, 105 (1983) 1487.

139. A. S. D'Souza, and C. G. Pantano, *J. Am. Ceram. Soc.*, 85 (2002) 1499.

140. S. Wendt, M. Frerichs, T. Wei, M. S. Chen, V. Kempter, and D. W. Goodman, *Surf. Sci.*, 565 (2004) 107.

141. B. A. Morrow and A. J. McFarlan, *J. Phys. Chem.*, 96 (1992) 1395.

142. P. Vandervoort, I. Gillisdhamers, K. C. Vranken, and E. F. Vansant, *J. Chem. Soc. Faraday Trans.*, 87 (1991) 3899.

143. B. Riegel, I. Hartmann, W. Kiefer, J. Gross, and J. Fricke, *J. Non-Crystal. Sol.*, 211 (1997) 294.

144. A. Pasquarello and R. Car, *Phys. Rev. Lett.*, 80 (1988) 5145.

145. G. E. Maciel and D. W. Sindorf, *J. Am. Chem. Soc.*, 102 (1980) 7606.

146. S. Léonardelli, L. Facchini, C. Fretigny, P. Tougne, and A. P. Legrand, *J. Am. Chem. Soc.*, 114 (1992) 6412.
147. A. Tuel, H. Hommel, A. P. Legrand, and E. S. Kovats, *Langmuir*, 6 (1990) 770.
148. C. C. Liu, and G. E. Maciel, *J. Am. Chem. Soc.*, 118 (1996) 5103.
149. H. Eckert, J. P. Yesinowski, L. A. Silver, and E. M. Stolper, *J. Phys. Chem.*, 92 (1988) 2055.
150. C. Bronnimann, R. C. Zeigler, and G. E. Maciel, *J. Am. Chem. Soc.*, 110 (1988) 2023.
151. I. S. Chuang and G. E. Maciel, *J. Am. Chem. Soc.*, 118 (1996) 401.
152. D. R. Kinney, I. S. Chuang, and G. E. Maciel, *J. Am. Chem. Soc.*, 115 (1993) 6786.
153. D. Freude, M. Hunger, and H. Pfeifer, *Chem. Phys. Lett.*, 91 (1982) 307.
154. S. Haukka, E. L. Lakomaa, and A. Root, *J. Phys. Chem.*, 97 (1993) 5085.
155. D. F. Shantz, J. S. auf der Gunne, H. Koller, and R. F. Lobo, *J. Am. Chem. Soc.*, 122 (2000) 6659.
156. J. Trebosc, J. W. Wiench, S. Huh, V. S. Y. Lin, and M. Pruski, *J. Am. Chem. Soc.*, 127 (2005) 3057.
157. A. Burneau, B. Humbert, O. Barres, J. P. Gallas, and J. C. Lavalley, *Coll. Chem. Silica Adv. Chem. Ser.*, 234 (1994) 199.
158. B. A. Morrow and A. J. McFarlan, *J. Phys. Chem.*, 96 (1992) 1395.
159. G. J. Kennedy, M. Afeworki, D. C. Calabro, C. E. Chase, and R. J. Smiley, *Appl. Spectr.*, 58 (2004) 698.
160. G. Hartmeyer, C. Marichal, B. Lebeau, S. Rigolet, P. Caullet, and J. Hernandez, *J. Phys. Chem. C*, 111 (2007) 9066.
161. A. Shchukarev, J. Rosenqvist, and S. Sjöberg, *J. Elect. Spect. Rel. Phen.*, 137–140 (2004) 171.
162. Y. Duval, J. A. Mielczarski, O. S. Pokrovsky, E. Mielczarski, and J. J. Ehrhardt, *J. Phys. Chem. B*, 106 (2002) 2937.
163. Y. J. Chabal and S. B. Christman, *Phys. Rev. B*, 29 (1984) 6974.
164. T. Schroeder, J. B. Giorgi, M. Bäumer, and H. J. Freund, *Phys. Rev. B*, 66 (2002) 165422.
165. T. Schroeder, M. Adelt, B. Richter, M. Naschitzki, M. Bäumer, and H. J. Freund, *Surf. Rev. Lett.*, 7 (2000) 7.
166. E. Ozensoy, B. K. Min, A. K. Santra, and D. W. Goodman, *J. Phys. Chem. B*, 108 (2004) 4351.
167. M. S. Chen, A. K. Santra, and D. W. Goodman, *Phys. Rev. B*, 69 (2004) 155404.
168. M. H. Du, A. Kolchin, and H. P. Cheng, *J. Phys. Chem.*, 120 (2004) 1044.
169. D. W. Sindorf, and G. E. Maciel, *J. Phys. Chem.*, 86 (1982) 5208.
170. X. Xue, and M. Kanzaki, *Phys. Chem. Miner.*, 26 (1998) 14.
171. S. Boujday, J.-F. Lambert, and M. Che, *Chem. Phys. Chem.*, 5 (2004) 1003.
172. J. D. Fisk, R. Batten, G. Jones, O'Reilly J.P., and A. M. Shaw, *J. Phys. Chem. B*, 109 (2005) 14475.
173. S. Ong, X. Zhao, and K. B. Eisenthal, *Chem. Phys. Lett.*, 191 (1992) 327.
174. V. Bolis, B. Fubini, L. G. M. Marchese, and D. Costa, *J. Chem. Soc., Faraday Trans.*, 87 (1991) 497.
175. T. Takei and M. J. Chikazawa, *J. Colloid. Interf. Sci.*, 208 (1998) 570.
176. B. Fubini, V. Bolis, A. Cavengo, E. Garrone, and P. Ugliengo, *Langmuir*, 9 (1993) 2712.
177. S. I. Kol'tsov, A. A. Malygin, A. V. Volkova, and V. B. Aleskovskii, *Zh. Fiz. Khim.*, 47 (1973) 988.
178. S. Boujday, J.-F. Lambert, and M. Che, *J. Phys. Chem. B*, 107 (2003) 651.
179. L. Stievano, L.-Y. Piao, I. Lopes, M. Meng, D. Cota, and J.-F. Lambert, *Eur. J. Miner.*, 19 (2007) 321.

180. B. P. Feuston and S. H. Garofalini, *J. Appl. Phys.*, 68 (1990) 4830.

181. S. H. Garofalini, *Non-Cryst. Solids*, 120 (1990) 1.

182. L. V. Woodcock, C. A. Angell, and P. Cheeseman, *J. Chem. Phys.*, 65 (1976) 1565.

183. T. F. Soules, *J. Chem. Phys.*, 71 (1979) 4570.

184. S. K. Mitra, M. Amini, D. Fincham, and R. W. Hockney, *Phil. Mag. B Phys. Cond. Matt. Stat. Mech. Elect. Opt. Magn. Prop.*, 43 (1981) 365.

185. J. D. Kubicki and A. C. Lasaga, *Am. Miner.*, 73 (1988) 941.

186. V. A. Bakaev and W. A. Steele, *J. Chem. Phys.*, 111 (1999) 9803.

187. E. A. Leed and C. G. Pantano, *J. Non-Crystal. Sol.*, 325 (2003) 48.

188. E. A. Leed, J. O. Sofo, and C. G. Pantano, *Phys. Rev. B*, 72 (2005) 155427.

189. A. A. Hassanali and S. J. Singer, *J. Phys. Chem. B*, 111 (2007) 11181.

190. J. Sauer, P. Ugliengo, E. Garonne, and V. R. Saunders, *Chem. Rev.*, 94 (1994) 2095.

191. T. R. Walsh, M. Wilson, and A. P. Sutton, *J. Chem. Phys.*, 113 (2000) 9191.

192. B. Civalleri, E. Garrone, and P. Ugliengo, *Chem. Phys. Lett.*, 299 (1999) 443.

193. A. G. Pelmenschikov, G. Morosi, and A. Gamba, *J. Phys. Chem. A*, 101 (1997) 1178.

194. M. A. Natal-Santiago and J. A. Dumesic, *J. Catal.*, 175 (1998) 252.

195. M. A. Natal-Santiago, J. J. dePablo, and J. A. Dumesic, *Catal. Lett.*, 47 (1997) 119.

196. B. Granqvist, T. Sandberg, and M. Hotokka, *J. Coll. Inter. Sci.*, 310 (2007) 369.

197. P. Pietrzyk, *J. Phys. Chem. B*, 108 (2005) 10291.

198. N. Lopez, F. Illas, and G. Pacchioni, *J. Phys. Chem. B*, 103 (1999) 8552.

199. D. Costa, G. Martra, M. Che, L. Manceron, and M. Kermarec, *J. Am. Chem. Soc.*, 124 (2002) 7210.

200. G. S. C. Martra, M. Che, L. Manceron, M. Kermarec, and D. Costa, *J. Phys. Chem. B*, 107 (2003) 6096.

201. S. T. Bromley, et al., *Chem. Phys. Lett.*, 340 (2001) 524.

202. L. Y. Ustynyuk, Y. A. Ustynyuk, D. N. Laikov, and V. V. Lunin, *Russ. Chem. Bull.*, 50 (2001) 2050.

203. M. Tada, T. Sasaki, T. Shido, and Y. Iwasawa, *Phys. Chem. Chem. Phys.*, 4 (2002) 5899.

204. M. Tada, T. Sasaki, and Y. Iwasawa, *J. Phys. Chem. B*, 108 (2004) 2918.

205. D. Costa, C. Lomenech, M. Meng, L. Stievano, and J.-F. Lambert, *J. Mol. Struct. (Theochem.)*, 806 (2007) 253.

206. C. Lomenech, G. Bery, D. Costa, L. Stievano, and J.-F. Lambert, *Phys. Chem. Phys.*, 6 (2005) 1061.

207. M. Benoit, S. Ispas, and M. E. Tuckerman, *Phys. Rev. B*, 64 (2001) 224205.

208. D. Ceresoli, M. Bernasconi, S. Iarlori, M. Parrinello, and E. Tosatti, *Phys. Rev. Lett.*, 84 (2000) 3787.

209. P. Masini and M. Bernasconi, *J. Phys. Condens. Matter*, 14 (2002) 4133.

210. J. Sarnthein, A. Pasquarello, and R. Car, *Phys. Rev. B*, 52 (1995) 12690.

211. R. M. Van Ginhoven, H. Jonsson, and L. R. Corrales, *Phys. Rev. B*, 71 (2005) 024208.

212. P. Ugliengo, M. Sodupe, F. Musso, I. J. Bush, R. Orlando, and R. Dovesi, *Adv. Mater.*, 20 (2008) 4579.

213. F. Tielens, C. Gervais, J. F. Lambert, F. Mauri, and D. Costa, *Chem. Mater.*, 20 (2008) 3336.

214. A. S. D'Souza and C. G. Pantano, *J. Am. Ceram. Soc.*, 28 (1999) 1289.

215. F. Mauri, M. Cote, Y. Yoon, C. Pickard, and P. Heynes, in *PARATEC (PARAllel Total Energy Code)*, B. Pfrommer, D. Raczkowski, A. Canning, and S. G. Louie (Eds.), Lawrence Berkeley National Laboratory, www.nersc.gov/projects/paratec.

216. J. P. Perdew, K. Burke, and M. Ernzerhof, *Phys. Rev. Lett.*, 77 (1996) 3865.

217. J. P. Perdew, K. Burke, and M. Ernzerhof, *Phys. Rev. Lett.*, 78 (1997) 1396.

218. H. Zhang and J. F. Banfield, *J. Mater. Chem.*, 8 (1998) 2073.
219. J. P. Perdew and K. M. E. Burke., *Phys. Rev. Lett.*, 77 (1996) 3865.
220. N. Troullier and J. L. Martins, *Phys. Rev. B*, 43 (1991) 1993.
221. L. Kleinman and D. Bylander, *Phys. Rev. Lett.*, 48 (1982) 1425.
222. H. J. Monkhorst and J. D. Pack, *Phys. Rev. B*, 13 (1976) 5188.
223. C. J. Pickard and F. Mauri, *Phys. Rev. B*, 63 (2001) 245101.
224. C. Gervais et al., *Magn. Reson. Chem.*, 42 (2004) 445.
225. C. Gervais, et al., *J. Phys. Chem. A*, 109 (2005) 6960.
226. F. Tielens, F. De Proft, and P. Geerlings, *J. Mol. Struct. (Theochem.)*, 542 (2001) 227.
227. R. L. Mozzi and B. E. Warren, *J. Appl. Cryst.*, 2 (1969) 164.
228. D. C. Athanasopoulos and S. H. Garofalini, *J. Chem. Phys.*, 97 (1992) 3775.
229. J. Casanovas, G. Pacchioni, and F. Illas, *Mater. Sci. Ing.*, 68 (1999) 16.
230. J. A. Tossell and P. Lazzeretti, *Chem. Phys.*, 112 (1987) 205.
231. J. A. Tossell and P. Lazzeretti, *Phys. Chem. Miner.*, 15 (1988) 564.
232. C. G. Lindsay and J. A. Tossell, *Phys. Chem. Miner.*, 18 (1991) 191.
233. T. M. Clark and P. J. Grandinetti, *Solid State Nucl. Magn. Reson.*, 16 (2000) 55.
234. X. Xue and M. Kanzaki, *Solid State Nucl. Magn. Reson.*, 16 (2000) 245.
235. X. Xue and M. Kanzaki, *J. Phys. Chem. B*, 205 (2001) 3422.
236. I. Farnan, et al., *Nature*, 358 (1992) 31.
237. T. M. Clark and P. J. Grandinetti, *J. Phys. Condens. Matter.*, 15 (2003) 2387.
238. C. J. Pickard and F. Mauri, in *Calculation of NMR and EPR Parameters: Theory and Applications*, M. Knaupp (Ed.), Wiley-VCH, Weinheim, 2004.
239. M. Profeta, F. Mauri, and C. J. Pickard, *J. Am. Chem. Soc.*, 125 (2003) 541.
240. X. Cong and R. J. Kirkpatrick, *J. Am. Ceram. Soc.*, 79 (1996) 1585.
241. E. R. H. van Eck, M. E. Smith, and S. C. Kohn, *Solid State Nucl. Magn. Reson.*, 15 (1999) 181.
242. S. Damyanova, L. Petrov, M. A. Centeno, and P. Grange, *Appl. Catal. A gen.*, 224 (2002) 271.
243. K. T. Jung and A. T. Bell, *J. Mol. Catal. A*, 163 (2000) 27.
244. R. L. Puurunen, *Chem. Vap. Depos.*, 11 (2005) 79.
245. A. B. Muckhopadhyay, J. Fdez. Sanz, and C. B. Musgrave, *Chem. Mater.*, 18 (2006) 3397.
246. B. Bachiller-Baeza, I. Rodriguez-Ramos, and A. Guerrero-Ruiz, *Langmuir*, 14 (1998) 3556.
247. G. Cerrato, S. Bordiga, S. Barbera, and C. Morterra, *Appl. Surf. Sci.*, 115 (1997) 53.
248. F. Ouyang, J. N. Kondo, K. Maruya, and K. Domen, *J. Chem. Soc. Faraday Trans.*, 93 (1997) 169.
249. A. Ignatchenko, D. G. Nealon, R. Dushane, and K. Humphries, *J. Mol. Catal. A*, 256 (2006) 57.
250. I. M. Iskandarova, A. A. Knizhnik, E. A. Rykova, A. A. Bagatur'yants, B. V. Potapkin, and A. A. Korkin, *Microelectron. Eng.*, 69 (2003) 587.
251. S. T. Korhonen, M. Calatayud, and A. O. I. Krause, *J. Phys. Chem. C*, 112 (2008) 6469.
252. S. T. Korhonen, M. Calatayud, and A. O. I. Krause, *J. Phys. Chem. C*, 112 (2008) 16096.
253. M. Calatayud, S. Collins, and M. A. B. Baltanás, *Phys. Chem. Chem. Phys.*, 11 (2009) 1397.
254. R. Schaub, et al., *Phys. Rev. Lett.*, 87 (2001) 266104.
255. M. Calatayud, Modelling metal oxide surfaces: From single crystals to supported catalysts. Habilitation (in French), Université Pierre et Marie Curie (UPMC), Paris, 2008.
256. D. Abriou and J. Jupille, *Surface Sci.*, 430 (1999) L527.
257. S. Coluccia, S. Lavagnino, and L. Marchese, *Mater. Chem. Phys.*, 18 (1988) 445.

258. V. Coustet and J. Jupille, *Nuovo Cimento Della Societa Italiana Di Fisica D-Condensed Matter Atomic Molecular and Chemical Physics Fluids Plasmas Biophysics*, 19 (1997) 1657.

259. E. Knozinger, K. H. Jacob, S. Singh, and P. Hofmann, *Surface Sci.*, 290 (1993) 388.

260. C. A. Scamehorn, A. C. Hess, and M. I. McCarthy, *J. Chem. Phys.*, 99 (1993) 2786.

261. C. A. Scamehorn, N. M. Harrison, and M. I. McCarthy, *J. Chem. Phys.*, 101 (1994) 1547.

262. J. Goniakowski and M. J. Gillan, *Surface Sci.*, 350 (1996) 145.

263. J. Goniakowski and C. Noguera, *Surface Sci.*, 330 (1995) 337.

264. J. Goniakowski, S. Bouetterusso, and C. Noguera, *Surface Sci.*, 284 (1993) 315.

265. S. Russo and C. Noguera, *Surface Sci.*, 262 (1992) 245.

266. W. Langel and M. Parrinello, *Phys. Rev. Lett.*, 73 (1994) 504.

267. W. Langel, *Surface Sci.*, 496 (2002) 141.

268. J. Ahdjoudj, A. Markovits, and C. Minot, *Catal. Today*, 50 (1999) 541.

269. M. Calatayud, A. Markovits, M. Menetrey, B. Mguig, and C. Minot, *Catal. Today*, 85 (2003) 125.

270. C. Duriez, C. Chapon, C. R. Henry, and J. Rickard, *Surf. Sci.*, 230 (1990) 123.

271. C. A. Scamehorn, N. M. Harrison, and M. I. McCarthy, *J. Chem. Phys.*, 101 (1994) 1547.

272. M. J. Stirniman, C. Huang, R. C. Smith, S. A. Joyce, and B. D. Kay, *J. Chem. Phys.*, 105 (1996) 1295.

273. M. Ménétrey, A. Markovits, and C. Minot, *Surf. Sci.*, 524 (2003) 49.

274. W. Langel and M. Parrinello, *Phys. Rev. Lett.*, 73 (1994) 504.

275. K. Refson, R. A. Wogelius, D. G. Eraser, M. C. Payne, M. H. Lee, and V. Milman, *Phys. Rev. B*, 52 (1995) 10823.

276. N. H. Deleeuw, G. W. Watson, and S. C. Parker, *J. Phys. Chem.*, 99 (1995) 17219.

277. C. Minot, *Surf. Sci.*, 562 (2004) 237.

278. S. Bourgeois, L. Gitton, and M. Perdereau, *J. Chim. Phys. Phys.-Chim. Biol.*, 85 (1988) 413.

279. M. Calatayud, D. Courmier, and C. Minot, *Chem. Phys. Lett.*, 369 (2003) 287.

280. A. B. Horn, M. A. Chesters, and M. R. S. McCoustra, *J. Chem. Soc. Faraday Trans.*, 88 (1992) 1077.

281. L. Delzeit, B. Rowland, and J. P. Devlin, *J. Phys. Chem.*, 97 (1993) 10312.

282. C. Girardet and C. Toubin, *Surf. Sci. Repts.*, 44 (2001) 163.

283. S. Solomon, *Nature*, 347 (1990) 347.

284. S. Solomon, R. R. Garcia, F. S. Rowland, and D. J. Wuebbles, *Nature*, 321 (1986) 755.

285. K. Bolton and J. B. C. Pettersson, *J. Am. Chem. Soc.*, 123 (2001) 7360.

286. G. Bussolin, S. Casassa, C. Pisani, and P. Ugliengo, *J. Chem. Phys.*, 108 (1998) 9516.

287. B. Demirdjian et al., *J. Chem. Phys.*, 116 (2002) 5143.

288. M. J. Isakson and G. O. Sitz, *J. Phys. Chem. A*, 103 (1999) 2044.

289. C. J. Pursell, M. Zaidi, A. Thompson, C. Fraser-Gaston, and E. Vela, *J. Phys. Chem. A*, 104 (2000) 552.

290. M. Svanberg, J. B. C. Pettersson, and K. Bolton, *J. Phys. Chem. A*, 104 (2000) 5787.

291. C. Toubin et al., *J. Chem. Phys.*, 116 (2002) 5150.

292. M. Calatayud, D. Courmier, and C. Minot, *Chem. Phys. Lett.*, 369 (2003) 287.

293. I. Xueref and F. Dominé, *Atmos. Chem. Phys.*, 3 (2003) 1779.

294. F. Dominé and I. Xueref, *Anal. Chem.*, 73 (2001) 4348.

295. F. E. Livingston, J. A. Smith, and S. M. George, *Anal. Chem.*, 72 (2000) 5590.

296. B. Demirdjian et al., *J. Chem. Phys.*, 116 (2002) 5143.

297. F. T. Wagner, and T. E. Moylan, *Surface Sci.*, 216 (1989) 361.

298. M. I. McCarthy, G. K. Schenter, M. R. Chacon-Taylor, J. J. Rehr, and G. E. Brown, Jr., *Phys. Rev. B*, 56 (1997) 9925.
299. E. V. Stefanovich, and T. N. Truong, *Chem. Phys. Lett.*, 299 (1999) 623.
300. F. Tielens, and C. Minot, *Surface Sci.*, 600 (2006) 357.
301. M. R. Chacon-Taylor, and M. I. McCarthy, *J. Phys. Chem.*, 100 (1996) 7610.
302. E. V. Stefanovich, and T. N. Truong, *J. Chem. Phys.*, 106 (1997) 7700.
303. M. Causà, R. Dovesi, C. Pisani, and C. Roetti, *Surf. Sci.*, 175 (1986) 551.
304. P. W. Tasker, *J. Phys. C*, 12 (1979) 4977.
305. Z. Feng and M. S. Seehra, *Phys. Rev. B*, 45 (1992) 2184.
306. A. Kuzmin, and N. Mironova, *J. Phys. Condens. Matter*, 10 (1998) 7937.
307. S. Casassa, M. Calatayud, J. Doll, C. Minot, and C. Pisani, *Chem Phys Lett.*, 409 (2005) 110.
308. J. Ahdjoudj and C. Minot, *Surf. Sci.*, 402–404 (1998) 104.
309. K. Kuchitsu, *Structure of Free Polyatomic Molecules — Basic Data*, Springer, Berlin, 1998.

CHAPTER 12

Electrocatalysis at Liquid–Liquid Interfaces

BIN SU[1], HUBERT H. GIRAULT[1], and ZDENĚK SAMEC[2]

[1]Laboratoire d'Electrochimie Physique et Analytique, Ecole Polytechnique Fédérale de Lausanne, CH-1015 Lausanne, Switzerland
[2]J. Heyrovsky Institute of Physical Chemistry, Academy of Sciences of Czech Republic, Prague 8, Czech Republic

12.1   INTRODUCTION TO ELECTROCHEMISTRY AT LIQUID–LIQUID INTERFACES

When considering electrocatalysis, one often thinks of a metallic electrode, the surface of which has been tailored to favor a reaction pathway involving an electron transfer reaction. Major applications as listed in this book include hydrogen evolution, oxygen reduction, and organic redox reactions in general.

Another type of interface that can be externally polarized is the liquid–liquid interface, or the so-called the interface between two electrolyte solutions (ITIES) [1–3]. Indeed, when an interface is formed between an aqueous solution containing mainly hydrophilic ions and an organic solution containing mainly lipophilic ions, it is possible to polarize the interface. In other words, we can form a capacitor with an excess of cations in one phase and of anions in the other, as schematically illustrated in Figure 12.1.

When considering the distribution of the electrical potential across the two phases, we can see in Figure 12.2 that the potential drop occurs mainly in the two back-to-back diffuse layers. The back-to-back layers can be treated in a first approximation by the classical Gouy–Chapman theory [4].

The interface polarization can be controlled externally using a four-electrode potentiostat together with two reference electrodes and two counterelectrodes. Alternatively, the polarization of the interface can be controlled either by the distribution of the ions or by redox equilibria or more generally by a combination of both.

Catalysis in Electrochemistry: From Fundamentals to Strategies for Fuel Cell Development,
First Edition. Edited by Elizabeth Santos and Wolfgang Schmickler.
© 2011 John Wiley & Sons, Inc. Published 2011 by John Wiley & Sons, Inc.

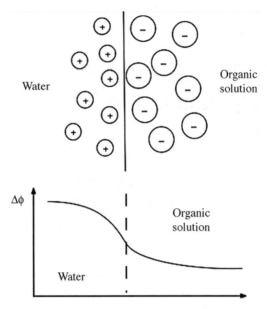

FIGURE 12.1   Liquid–liquid interface polarization (top) and potential distribution (bottom).

The simplest case comprises two salts, one in each phase but with a common ion. If the respective counterions are hydrophilic for one and lipophilic for the other, then the Galvani potential difference is given by the Nernst equation for the distribution of the common ion, i:

$$\Delta_o^w \phi = \phi^w - \phi^o = \Delta_o^w \phi_i^\ominus + \frac{RT}{z_i F} \ln \left(\frac{a_i^o}{a_i^w} \right) \tag{12.1}$$

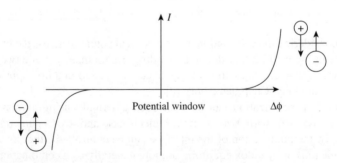

FIGURE 12.2   Potential window for a polarizable liquid–liquid interface limited at negative potentials by the transfer of a lipophilic cation or a hydrophilic anion and at positive potentials by the transfer of a hydrophilic cation or a lipophilic anion.

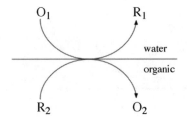

FIGURE 12.3 Heterogeneous electron transfer reaction at a liquid–liquid interface.

with $\Delta_o^w \phi_i^{\ominus}$ the standard transfer potential equal to the Gibbs energy of transfer expressed in a voltage scale,

$$\Delta_o^w \phi_i^{\ominus} = \frac{\Delta G_{tr,i}^{\ominus,w\rightarrow o}}{z_i F} = \frac{\mu_i^{\ominus,o} - \mu_i^{\ominus,w}}{z_i F} \tag{12.2}$$

Coming back to the case of a polarizable interface containing a hydrophilic salt, such as LiCl, in water and a lipophilic organic electrolyte, say C^+A^-, we can polarize the interface up to a point where one of the ions of the supporting electrolytes will cross the interface. Indeed, according to Eq. (12.1) when the potential difference approaches the standard value of the transferring ion, the concentration ratio of this ion between the two phases will adjust according to the Nernst equation. The transferring ions thus define the potential window limited in this case by ion transfer reactions. For example, on the positive side, if the organic anion (A^-) is very lipophilic, Li^+ will transfer as soon as the potential difference is sufficient to provide Li^+ energy enough to overcome its standard energy of transfer. The same on the negative side, if the organic cation (C^+) is very lipophilic as well, Cl^- will transfer as soon as the potential difference is sufficient to overcome its standard energy of transfer.

Now let us consider the case when we add to these respective supporting electrolytes a hydrophilic redox couple O_1^w/R_1^w to the aqueous phase and a lipophilic organic redox couple O_2^o/R_2^o to the organic phase such that a heterogeneous electron transfer reaction between them occurs as illustrated in Figure 12.3.

The Nernst equation for this electron transfer reaction at the interface is given by

$$\Delta_o^w \phi = \Delta_o^w \phi_{ET}^{\ominus} + \frac{RT}{nF} \ln \left(\frac{a_{R_1}^w a_{O_2}^o}{a_{O_1}^w a_{R_2}^o} \right) \tag{12.3}$$

with $\Delta_o^w \phi_{ET}^{\ominus}$ the standard redox potential for the interfacial redox reaction

$$\Delta_o^w \phi_{ET}^{\ominus} = \left[E_{O_2^o/R_2^o}^{\ominus} \right]_{SHE}^o - \left[E_{O_1^w/R_1^w}^{\ominus} \right]_{SHE}^w \tag{12.4}$$

where $[E_{O_1^w/R_1^w}^{\ominus}]_{SHE}^w$ represents the classical standard redox potential in water on the standard hydrogen electrode (SHE) scale. The standard redox potential of the redox couple in an organic solvent referred to an aqueous standard hydrogen electrode

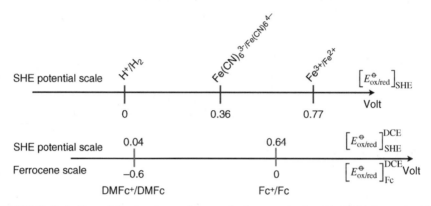

FIGURE 12.4 Potentials of various redox couples in water (top) and in DCE (bottom) with respect to the SHE scale. In DCE, the Fc^+/Fc scale is also illustrated for comparison.

$[E^{\ominus}_{O_2^{\ominus}/R_2^{\ominus}}]^{O}_{SHE}$ is a term that cannot be formally measured, because it is impossible to assemble directly such a cell without having a liquid junction. Indeed, such a measurement would involve an estimation of the liquid junction potential. In the case of the water–1,2-dichloroethane (DCE) system, the standard redox potential for the ferrocenium/ferrocene (Fc^+/Fc) couple has been estimated to be $[E^{\ominus}_{Fc^+/Fc}]^{DCE}_{SHE} = 0.64$ V [5]. In this way, all the other organic standard redox potentials can be measured experimentally with respect to the Fc^+/Fc couple.

As observed in Figure 12.4, it is easy to compare the redox scale between the two phases when the interface is not polarized. For example, the standard redox Galvani potential difference between ferri/ferrocyanide [$Fe(CN)_6^{3-}/Fe(CN)_6^{4-}$] in water and Fc^+/Fc in DCE is

$$\Delta^w_o \phi^{\ominus}_{ET} = \left[E^{\ominus}_{Fc^+/Fc}\right]^{DCE}_{SHE} - \left[E^{\ominus}_{Fe(CN)_6^{3\check{S}}/Fe(CN)_6^{4\check{S}}}\right]^{H_2O}_{SHE} = 0.64 - 0.36 = 0.28 \text{ V}$$
(12.5)

This means that we have to polarize the interface to positive values to allow the oxidation of Fc by $Fe(CN)_6^{3-}$ to occur.

From an experimental viewpoint, all the classical electrochemical methodologies used to study electrode reactions can be transposed to study charge transfer reactions at polarized liquid–liquid interfaces, the most ubiquitous being of course cyclic voltammetry. For example, Figure 12.5 shows a cyclic voltammogram exhibiting two waves. The one on the left corresponds to the transfer of tetrapropylammonium ion ($TPrA^+$) used here as an internal calibrator for the Galvani potential scale. Indeed, the half-wave potential for this ion transfer reaction is associated with the Gibbs energy of transfer as given by Eq. (12.2). The second wave corresponds to the heterogeneous electron transfer between $Fe(CN)_6^{3-}/Fe(CN)_6^{4-}$ in water and Fc^+/Fc in DCE.

In addition to the classical electrochemical methodologies, many spectroelectro-chemical techniques have been developed lately. One can cite voltabsorptometry and chronoabsorptometry [6–8], voltfluorometry and chronofluorometry [9, 10], potential

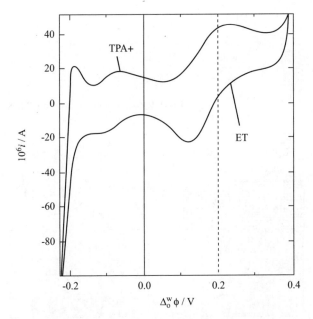

FIGURE 12.5 Cyclic voltammogram showing the transfer of TPrA$^+$ (left wave) and the heterogeneous electron transfer between Fe(CN)$_6^{3-}$/Fe(CN)$_6^{4-}$ in water and Fc$^+$/Fc in DCE (right wave).

modulated techniques such potential-modulated absorption and fluorescence [11, 12], and time-resolved techniques such as fluorescence lifetime measurements [13, 14]. More adapted to the study of buried interfaces, surface second-harmonic generation (SSHG) [15] and sum frequency generation (SFG) [16] have also been widely used. These techniques have been used to monitor both ion transfer and electron transfer reactions and to probe the dynamics of the interface.

From a structural point of view, a liquid–liquid interface is, by definition, a molecular interface between two fluids where capillary waves take place. The roughness of these molecular interfaces at the molecular level has been largely debated over the last decades, and of course the concept of roughness is time scale dependent. On the time scale of molecular motions, that is, nanoseconds, the interface can be seen as rough. So far, our vision of the interface stems mainly from molecular dynamics calculations that can provide a pictorial description [17, 18]. However, contrary to solid electrodes, it is here easier to measure the surface tension of the interface by classical techniques such as pendant drop [19] or drop time methods [20] or optically by quasi-elastic light scattering (QELS) [21]. In this way, a thermodynamic approach can be used to determine from electrocapillary data the different surface excesses, therefore providing a quantitative description.

Recently, the organic electrolyte phase has been replaced with ionic liquids [22]. In particular, Kakiuchi et al. have synthesised room temperature molten salts made with very lipophilic ions with large Gibbs energy of transfer that result in a large potential window when used in contact with an aqueous solution [23].

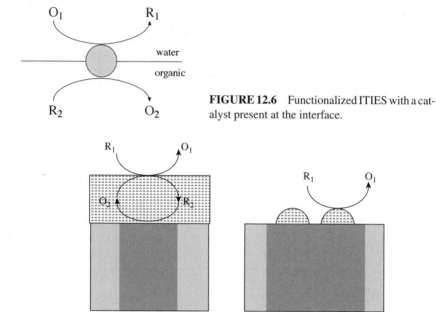

FIGURE 12.6 Functionalized ITIES with a catalyst present at the interface.

FIGURE 12.7 Supported ITIES using a thin liquid film (left) or microdroplets (right).

Electrocatalysis using ITIES can be performed by two different methods. The first one consists in tailoring the interface with an adsorbed catalyst being either a molecule or a nanoparticle as illustrated in Figure 12.6 to form what could be called *functionalized ITIES*. The second method, which can be called *supported ITIES electrocatalysis*, consists in covering a solid electrode with a thin layer either of water or of an organic solution. A redox couple is dissolved in the thin film, thereby acting as the electrocatalyst, as schematically illustrated in Figure 12.7. Alternatively, the thin layer can be replaced with droplets to form another type of supported ITIES called three-phase junctions (Fig. 12.7).

12.2 MOLECULAR ELECTROCATALYSIS AT FUNCTIONALIZED ITIES

The first work on electrocatalysis at functionalized liquid–liquid interfaces can be traced back to the work of Volkov et al. at the Frumkin Institute of Moscow [24]. They reported that chlorophyll adsorbed at the water–decane interface could catalyze the electron transfer reactions between lipophilic electron acceptors (such as 2*N*-methylamino-1,4-naphtoquinone or vitamin K_3) and hydrophilic electron donors (such as NADH or potassium ascorbate). Indeed, they realized early that an ITIES can display catalytic properties for interfacial charge transfer reactions and that redox potential scales can be tuned by changing the solvent composition, thereby permitting reactions at the interface that cannot proceed in a bulk phase [25, 26]. Also, they emphasized that by polarizing the interface we can drive redox reactions that cannot proceed if the Galvani potential difference is not controlled (see Fig. 12.4).

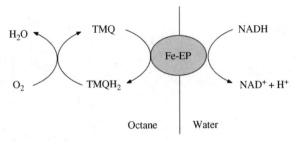

FIGURE 12.8   FeEP and TMQ-catalyzed oxygen reduction at the water–octane interface. (Adapted from ref. 29.)

In 1983, Volkov et al. studied the adsorption of the iron complex of coproporphyrin III tetramethyl ether at the octane–water interface and found that the adsorption process follows a Frumkin isotherm and that the angle between the porphyrin ring and the interface was $-65°$ [27]. Then, they monitored the heterogeneous electron transfer reaction between 2-N-methylamino-1,4-naphthoquinone in octane upon addition of a reducing agent such as NADH or ascorbate into the aqueous phase in the presence of coproporphyrin. The change of potential drop at the octane–water interface was found to be proportional to the amount of adsorbed porphyrin molecules, indicating that the coproporphyrin played an active catalytic role [27]. Following this work, they investigated the cobalt and ironcomplex of ethioporphyrin (EP) at the water–octane interface and found that CoEP and FeEP were active for the reduction of vitamin K_3 or 1,4-naphtoquinone present in octane by NADH in water but the free-base EP was not [28]. Furthermore, FeEP could also catalyze oxygen reduction at the octane–water interface [28]. By using FeEP and tetramethylquinone (TMQ) as the interfacial cocatalysts, NADH oxidation and oxygen reduction could be coupled through the reduced form TMQ ($TMQH_2$), as illustrated in Figure 12.8 [29].

Volkov et al. also investigated water photooxidation catalyzed by adsorbed chlorophyll a at the water–octane interface using potassium ferricyanide [$K_3Fe(CN)_6$], NAD^+, or NADP as the electron acceptor in water and dinitrophenol (DNP) or pentachlorophenol (PCP) as the proton acceptor in octane [30, 31]. The photopotential generated at the water–octane interface is proportional to the amount of adsorbed chlorophyll a and to the oxygen evolution rate. A four-electron transfer mechanism involving a hydrated chlorophyll oligomer adsorbed at the interface was then proposed [31].

These early works were indeed pioneering and have been summarized by Volkov in 1998 in a review published in *Analytical Sciences* [32], but the experimental methodology used did not give this approach a large visibility. The major drawback was to choose water–alkane interfaces (such as water–decane and water–octane interfaces) that are experimentally difficult to polarize, and as a consequence most of the experimental evidence had to be provided by surface potential measurements carried out using a Kelvin probe.

In Volkov's early works, all catalysts studied are porphyrins or contain porphyrin substructures. Porphyrins are of course well-known electrocatalysts for oxygen reduction. A landmark contribution was the work by Collman et al., who demonstrated in

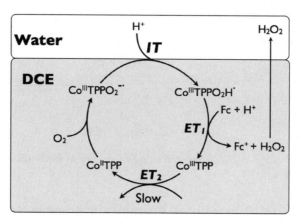

FIGURE 12.9 Catalytic cycle for oxygen reduction in biphasic system using CoTPP as catalyst. (Adapted from ref. 34.)

1979 that cofacial porphyrins deposited by solvent deposition on a graphite electrode could be used as the catalysts for oxygen reduction [33]. The importance of the oxygen reduction reaction (ORR) for fuel cell research has been a major driving force for electrocatalysis.

More recently, a very clear demonstration of electrocatalysis by porphyrins adsorbed at liquid–liquid interfaces was proposed by Samec et al. [35] for the reduction of oxygen by decamethylferrocene (DMFc) in DCE, and the work was extended by Partovi-Nia et al. for two ferrocene derivatives, ferrocene (Fc) and 1,1′-dimethylferrocene (DFc) [34]. The catalytic cycle for these systems is shown in Figure 12.9 and is similar to the cycle proposed by Fukuzumi et al. [36, 37] for homogeneous reactions. The difference is the supply of protons from the aqueous phase in the former case, which is of course potential dependent. In homogeneous reactions, the second electron transfer reaction for the reduction of Co(III) to Co(II) is the slowest of the ET steps. Here, the rate-determining step can also be the proton transfer reactions at low positive potentials.

To demonstrate the catalytic cycle, one can use cyclic voltammetry as shown elsewhere [35] or apply the Galvani potential difference by fixing the distribution potential between the two phases by dissolving respectively an hydrophilic salt and a lipophilic salt having a common ion. In this case, the distribution is given by the Nernst equation [Eq. (12.1)] for this common ion. Figure 12.10 illustrates the Galvani potential differences (shown by the dashed lines) obtained for different common ions [38]: bis(triphenylphosphoranylidene)ammonium (BTPPA$^+$), tetrabutylammonium (TBA$^+$), tetraethylammonium (TEA$^+$), tetramethylammonium (TMA$^+$), and tetrakis(pentafluorophenyl)borate (TPFB$^-$).

For the CoTPP catalytic cycle, the proton transfer reaction can be driven at very positive potentials by using lithium tetrakis(pentafluorophenyl)borate (LiTPFB) in the aqueous phase and bis(triphenylphosphoranylidene)ammonium tetrakis(pentafluorophenyl)borate (BTPPATPFB) in the organic phase. After 30 min

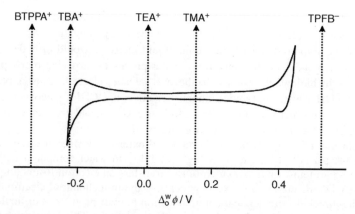

FIGURE 12.10   Galvani potential differences controlled by different common ions.

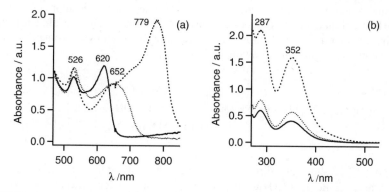

FIGURE 12.11   UV–VIS absorption spectra of the DCE phase (a) and the aqueous phase treated by excess NaI (b) after 30 min of shake flask experiments in presence of 50 μM CoTPP with 5 mM Fc (full line), DFc (dotted line), and DMFc (dashed line). The aqueous pH was controlled at 2 by HCl. (Adapted from ref. 34.)

of vigorous shaking, the two phases were separated and analyzed by UV–VIS spectroscopy. As demonstrated in Figure 12.11(a), oxidations of Fc, DFc, and DMFc were clearly illustrated with absorption bands of Fc$^+$, 1,1′-dimethylferrocenium (DFc$^+$), and decamethylferrocenium (DMFc$^+$) at 620, 652, and 779 nm, respectively. To show that hydrogen peroxide was produced and extracted to the aqueous phase, the latter was treated by addition of NaI, which is known to react with H_2O_2 to form I_3^-, which adsorbs in the UV range with two absorption bands at 287 and 352 nm [Fig. 12.11(b)] [34].

12.3   BIOCATALYSIS AT FUNCTIONALIZED ITIES

The liquid–liquid interface provides a simple approach to biomembrane mimicking. The bioassay applications of liquid–liquid interfaces have been recently reviewed by

Shao et al. [39]. In comparison to other systems such as vesicles, the major advantage of an ITIES is the fact that one can control the interfacial polarization. As a result, the electrochemical approach using the liquid–liquid interface would provide valuable information for understanding electron transfer processes involving redox proteins or enzymes in biological systems. In the past decade, electron transfer reactions between NADH in water and some quinone derivatives in DCE [40], between ascorbic acid/ascorbate in water and chloranil in nitrobenzene [41], and between cytochrome c in water and DFc in DCE [42] have been reported.

Cytochrome c is one of most important and extensively studied electron transfer proteins, partially because of its high solubility in water compared with other redox-active proteins. In vivo, cytochrome c transfers an electron from complex III to complex IV, membrane-bound components of the mitochondrial electrontransfer chain. The direct electrochemical communication to cytochrome c is difficult as the redox active heme center is buried well inside the protein. This difficulty has been circumvented using modified electrodes, with monolayers able to interact with both the heme center and the underlying electrode. Another interesting approach is based on the liquid–liquid interface electrochemistry as reported by Dryfe et al., with which they have shown a direct electron transfer reaction between cytochrome c in water and DFc in DCE using the following cell [42]:

$$\text{Ag}_{(s)} \left| \text{AgTPB}_{(s)} \right| 0.02\,\text{M BTPPATPB}_{(DCE)} \left\| \begin{array}{c} 0.1\,\text{M NaCl} \\ 0.088\,\text{M NaOH} \\ 0.0125\,\text{M Na}_2\text{B}_4\text{O}_{7\,(aq)} \end{array} \right| \text{AgCl}_{(s)} \left| \text{Ag}_{(s)} \right. \qquad \text{Cell 1}$$

Figure 12.12 shows the direct electron transfer between the reduced protein and DFc, while Figure 12.13 shows the direct electron transfer between the oxidized protein and DFc$^+$.

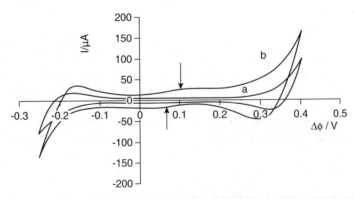

FIGURE 12.12   Cyclic voltammograms of (a) cell 1 with 0.02 M DFc in the organic phase and (b) cell 1 with 0.02 M DFc in the organic phase and 4×10^{-4} M ferricytochrome c in the aqueous phase. Voltage scan rate was 0.05 V s^{-1}. The arrows denote the electrontransfer process. (Reproduced from ref. 42 with permission from the American Chemical Society.)

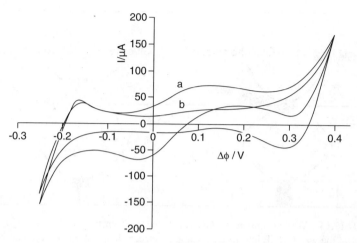

FIGURE 12.13 Cyclic voltammograms of cell 1 with 4×10^{-4} M ferricytochrome c in the aqueous phase in the presence (a) and absence (b) of 1×10^{-4} M DFc$^+$ in the organic phase. Voltage scan rate was 0.05 V s^{-1}. (Reproduced from ref. 42 with permission from the American Chemical Society.)

In this work, the initial oxidation state of the protein was confirmed by the UV–VIS spectrum, recording the absorption peak centered at 540 nm. In its oxidized state, cytochrome c shows a single peak, but a split peak is seen for the reduced protein. No charge transfer was observed when reduced cytochrome c was present in the aqueous phase of cell 1 and DFc in the organic phase. On replacement of the reduced cytochrome c with its oxidized form, a charge transfer was seen on both the forward and reverse scans, with the peak on the forward scan attributed to the heterogeneous transfer of an electron from DFc to cytochrome c.

In biological systems, all redox processes occur at the interfaces and in most cases are catalyzed by enzymes embedded inside the biomembranes. Therefore, electrochemistry at liquid–liquid interfaces provides an interesting platform for the investigations of the adsorption of enzymes and their electrocatalysis. Williams et al. first studied glucose oxidase (GO$_x$)–mediated electron transfer between β-D-glucose and DFc$^+$ at the polarized water–DCE interface by means of scanning electrochemical microscopy [43]. Two different reaction mechanisms were proposed in which the enzyme-catalyzed electron transfer occurs heterogeneously at the water–DCE interface or homogeneously in the water phase, as shown in Figure 12.14, though the reaction mechanism remained to be clarified.

The group of Osakai has also studied GO$_x$-catalyzed electron transfers between some oxidants in nitrobenzene (NB) and glucose in water by cyclic voltammetry [44]. When an electrically neutral compound, chloranil (CQ), was employed as the oxidant in NB, the enzymatic reaction could not be regulated because of the spontaneous transfer of CQ from NB to water. However, when an ionic oxidant, DFc$^+$, was employed as the oxidant, the electrochemical control of the enzymatic reaction was achieved by controlling the interfacial transfer of DFc$^+$ (as illustrated in Fig. 12.15). The

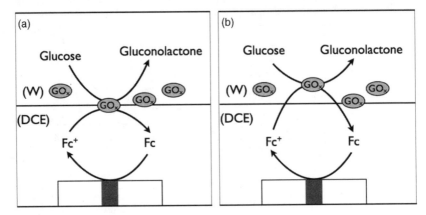

FIGURE 12.14 Mechanisms of the GO_x-mediated electron transfer between β-D-glucose and DFc^+ at the water–DCE interface by scanning electrochemical microscopy: (a) heterogenous electron transfer pathway; (b) ion transfer of DFc^+ followed by homogeneous electron transfer. (Adapted from ref. 43.)

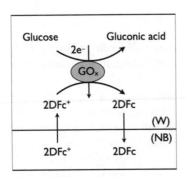

FIGURE 12.15 Mechanism of the GO_x-catalyzed oxidation of glucose by DFc^+ at the water–NB interface. (Adapted from ref. 44.)

voltammetric behaviors were successfully explained by a digital simulation based on the iontransfer mechanism which involves the interfacial transfer of DFc^+ and the succeeding GO_x-catalyzed electron transfer which occurs not heterogeneously at the interface but homogeneously in the water phase.

12.4 NANOPARTICLE ELECTROCATALYSIS AND PHOTOCATALYSIS AT ITIES

Electrodeposition of metal nanoparticles at the liquid–liquid interfaces has attracted a great deal of interest as a novel approach to prepare metal catalysts or photocatalyst. In 1857, Michael Faraday reported the electron transfer reactions at the ether–water or carbon disulfide–water interfaces to prepare colloidal metal particles by reducing metal salts [45], which has been recognized as the birth of modern nanotechnology. Guainazzi et al. demonstrated for the first time that the direct current applied to flow through the interface between Cu^{2+} cation in water and the $V(CO)_6^-$ anion in DCE leads to the formation of a copper layer at the liquid–liquid boundary [46]. A similar effect was seen when silver ions were present in the aqueous phase. In 1996 Cheng and Schiffrin [47] showed that the rate of deposition of Au particles formed by the electron transfer reaction between $AuCl_4^-$ in DCE and $Fe(CN)_6^{4-}$ in water can be controlled by the applied potential. Using UV–VIS spectroscopy and SEM, these authors demonstrated that the deposited Au particles have a diameter greater than $\sim 100\,nm$.

More recently, the polarization measurements have been used by Kontturi et al. [49–51], Dryfe et al. [48, 52, 53], and Unwin et al. [54] to study the deposition of Pd [48, 50], Pt [48, 53], Pd–Pt [48], and Ag [54] particles on the macroscopic [49–51], microscopic, or nanoscopic [54] and the membrane-supported [48, 52, 53] liquid–liquid interfaces. In the latter case, an array of liquid–liquid interfaces of micrometer or submicrometer size was formed, which made it possible to control the size of the grown particles and to manipulate with the metal deposit for the purpose of scanning electron microscopy (SEM), high-resolution transmission electron microscopy (HRTEM), or X-ray diffraction (XRD) analysis [48, 52, 53]. SEM image of a Pt-loaded porous membrane is demonstrated in Figure 12.16. HRTEM and SEM images indicated that the Pd and Pt deposits at the water–DCE interface consist of aggregates of discrete nanoparticles between 3 and 5 nm in diameter [53]. Metal nanoparticle can be stabilized at the liquid–liquid interface by simultaneous nucleation and polymerization of a monomer, as shown by Cunnane et al. [55]

Available experimental data on metal deposition apparently support the deterministic description that has been developed assuming that the basic concepts of nucleation and growth of metal particles at liquid–liquid interfaces are essentially similar to those at a solid electrode [56]. Although the liquid–liquid interface can be expected to be free of defects acting as permanent preferential nucleation sites, the metal deposition is likely to include the formation of nucleus, which can grow after reaching a critical size [50]. Consequently, the exclusionprogressive or the exponential nucleation rate law has been anticipated [50], which also describes the multiple nucleations at solid electrodes (see, e.g., ref. 57 for a review). The observation that may cast doubts on this picture is the morphology of the Pt layer deposited on the array of microscopic interfaces (Fig. 12.16) [48]. While all these interfaces should have equal conditions for the nucleation and growth, SEM of the deposited Pt layer clearly indicates that nucleation and growth take place only on some of these microscopic interfaces while on others do not occur at all [48]. Nonuniform morphology was ascribed to the autocatalytic mechanism, where the nucleation of a particle is triggered by the presence of an initial particle, leading to the particle distribution deviating from uniformity [48]. An alternative explanation has been proposed by Samec et al. [58], who have shown that the repeated potential-step experiments provide the initial rate of the Pt deposition on the newly prepared (bare) ITIES that attains a broad range of values even approaching zero, while the shape of current transients can be described by the available theory [50]. This surprising feature is likely to be due to the random rate of the formation of the Pt nuclei with a critical size that is necessary for stable growth to occur [58].

Catalysis by electrochemically prepared metal nanoparticles [59, 60] or by chemically prepared metal colloids [61] at the ITIES has been reported by Schiffrin and Cheng [59], Samec et al. [60], and Kontturi et al. [61]. In the pioneering study [59], electrochemically prepared interfacial Pd particles were shown to act as a catalyst for

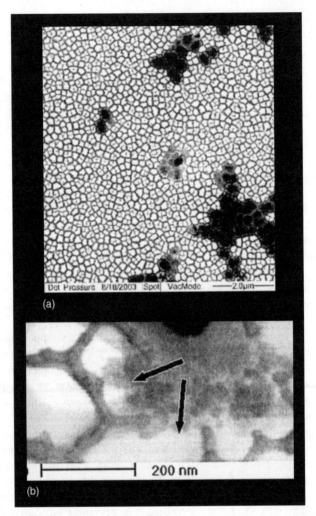

FIGURE 12.16 (a) SEM image of Pt-loaded aluminum membrane. The deposition time was 5 min, stepping from a potential of 0.2 to 0.45 V. (b) Higher resolution SEM image of platinum particles trapped within a membrane. The arrows show the spill-over from one filled pore to its neighbor. (Reproduced from ref. 48 with permission from the Royal Chemical Society.)

dehalogenation of 2-bromoacetophenone (BrAc) to acetophenone (Ac) at the water–DCE interface. The reaction has been investigated further in a two-phase system with tetramethylammonium cation as a phase transfer catalyst that controls the interfacial potential difference (distribution potential), DMFc in the DCE phase as the electron donor, and the aqueous Au or Pd colloid as electron transfer catalyst [61]. A mechanism has been proposed to consist of the sequence of two interfacial steps: (a) charging of the metal colloid through the electron uptake from DMFc and (b) dehalogenation of BrAc to Ac catalyzed by the negatively charged metal colloid [61]. Such a process can be termed two-phase electrocatalysis, defined as the interfacial catalytic reaction that requires electrical field across the liquid–liquid interface [61]. While this definition is broad enough to cover various catalytic phenomena at ITIES, the classical definition of electrocatalysis is worth comparing. The latter has been related to the catalytic effect of the electrode on the electrode reaction or, alternatively, to the effect of the nature of the electrode material on the rate of the electrode reaction, which is usually associated with the adsorption of reactants, intermediates, or products on the electrode surface. Since a metal particle at ITIES is likely to behave as a bipolar electrode [50], it should be possible to make use of the classical concept with an interesting exclusion. Electrochemical measurements at ITIES allow comparing the rate of the interfacial reaction in the presence and absence of metal. Such an approach is not applicable in classical electrocatalysis.

More recently, evidence and the quantitative evaluation of electrocatalysis of the potential-dependent reduction of oxygen in water by DMFc in DCE by electrochemically prepared metal particles have been reported [60]. This electron transfer reaction has been previously found to proceed with a low yet measurable rate [62]. DMFc was found to be a suitable reducing agent also for the insitu deposition of the metal particles at ITIES [60]. Its standard potential is more negative than that of DFc [62], which was used for the preparation of the interfacial Pd or Pt nanoparticles by reducing Pd(II) or Pt(II) chloro-complexes in water in the previous studies [48, 53]. The nucleation and growth of the Pt particles at the polarized water–DCE interface can be described by the equation

$$2\,\text{DMFc (o)} + \text{PtCl}_x^{2-x}\,\text{(w)} + \text{Pt}_n\,(\sigma) \rightarrow 2\,\text{DMFc}^+\,\text{(o)} + \text{Pt}_{n+1}\,(\sigma) + x\,\text{Cl}^-\,\text{(w)} \tag{12.6}$$

where σ represents the interface. Figure 12.17 shows the cyclic voltammograms of the water–DCE interface obtained using the air-saturated solutions in the presence of DMFc in DCE (dashed line) and in the presence of both DMFc in DCE and the Pt(II) complex in water (full line). When only the Pt(II) complex is present, a single voltammetric peak can be seen at $E \approx 0.15\,\text{V}$ (Fig. 12.17, inset), which corresponds to the diffusion-controlled transfer of the divalent anion PtCl$_4{}^{2-}$ [58].

In the presence of DMFc and in the absence of the Pt(II) complex, an enhancement of the positive current can be seen at cell potentials $E > 0.2\,\text{V}$, while on the reverse sweep a voltammetric peak appears at $E \approx 0.1\,\text{V}$ (Fig. 12.17, dashed line). The increased positive current was shown to be associated with the interfacial reduction of the air oxygen present in the aqueous phase [62]:

$$4\,\text{DMFc (o)} + \text{O}_2\,\text{(w)} + 4\,\text{H}^+\,\text{(w)} \rightarrow 4\,\text{DMFc}^+\,\text{(o)} + 2\,\text{H}_2\text{O}\,\text{(w)} \tag{12.7}$$

The voltammetric peak at $E \approx 0.1\,\text{V}$ corresponds to the reversible transfer of DMFc$^+$,

$$\text{DMFc}^+\,\text{(w)} \rightarrow \text{DMFc}^+\,\text{(o)} \tag{12.8}$$

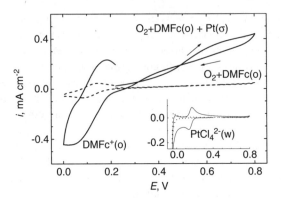

FIGURE 12.17 Cyclic voltammograms $(0.02\,V\,s^{-1})$ of water–DCE interface in the presence of 0.05 M DMFc in DCE (dashed line) or 0.05 M DMFc in DCE and 5 mM $PtCl_4^{2-}$ in water. Inset: cyclic voltammogram in the presence of 5 mM $PtCl_4^{2-}$ in water only. Air-saturated solutions containing background electrolytes, aqueous solution pH 4.9. (Adapted from ref. 60.)

which is produced in the electron transfer reaction, Eq. (12.7). In the presence of both DMFc and the Pt(II) complex, the positive current at potentials $E > 0.2$ V, as well as the voltammetric peak at $E \approx 0.1$ V, increases remarkably (Fig. 12.17, full line). The latter effect was attributed to the overlap of the ion transfer currents of $PtCl_4^{2-}$ and $DMFc^+$, reflecting also an enhanced production of $DMFc^+$ at more positive potentials [60]. The same experiment performed with the deaerated solutions clearly indicated that the substantial current increase at positive potentials is associated with the presence of oxygen in both solutions. A conclusion was made that this current represents an overlap of currents due to the interfacial reduction of the Pt(II) complex by DMFc, Eq. (12.6), and the interfacial reduction of oxygen as described by Eq. (12.7), which is catalyzed by the deposited Pt particles acting as a bipolar electrode (as illustrated in Fig. 12.18).

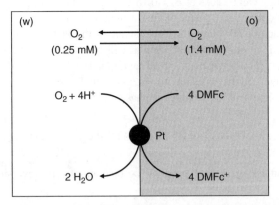

FIGURE 12.18 Mechanism of electrocatalysis of oxygen reduction at polarized water–DCE interface by simultaneously deposited Pt particles acting as a bipolar electrode.

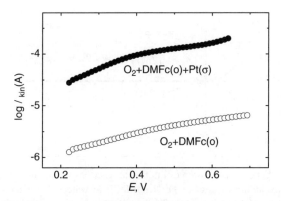

FIGURE 12.19 Tafel plots obtained by convolution analysis of voltammograms recorded using air saturated in presence of 0.05 M DMFc in DCE and in the presence (solid circles) or absence (empty circles) of 5 mM $PtCl_4^{2-}$ in water, pH 3.4. (Adapted from ref. 60.)

Convolution analysis of voltammograms obtained at various values of pH of the aqueous phase made it possible to consider the stoichiometry of the overall electron transfer reaction and to infer kinetic data (Tafel plots). Figure 12.19 compares the Tafel plots evaluated in the absence and presence of the Pt particles at the interface demonstrating the electrocatalytic effect. This analysis indicated that (a) the catalytic oxygen reduction proceeds as a four-electron transfer reaction; (b) the rate of the oxygen reduction by DMFc increases by more than one order of magnitude in the presence of the interfacial Pt particles; (c) the rate-determining step is the first electron uptake not including the concerted proton transfer; and (d) a transition from the normal ($\alpha = 0.26 - 0.39$) to the activationless ($\alpha < 0.1$) region occurs, which is probably associated with a high overpotential for the rate-determining step.

First evidence of photoinduced two-phase redox reaction catalyzed by electro-chemically generated interfacial metal particles at the polarized liquid–liquid interface was reported by Lahtinen et al. [63]. The particles were prepared by heterogeneous reduction of aqueous ammonium tetrachloropalladate ($PdCl_4^{2-}$) by the electron donor ferrocene (Fc) in DCE, that is, Eq. (12.6). It was observed that as-grown Pd particles act as electron transfer mediators in the photoreduction of tetracyanoquinodimethane (TCNQ) by the photoexcited water-soluble zinc tetrakis(carboxyphenyl)porphyrin ($ZnTPPC^{4-}$) adsorbed at the interface. Interfacial reactions taking place in this system are depicted schematically in Figure 12.20.

A conclusion was made that this effect could open new perspectives in the field of photocatalysis and photosynthesis [63]. In particular, the hydrogenation properties of Pd particles combined with the interfacial properties of porphyrins can provide new photosynthetic routes. For instance, the photoreduction of benzoquinone by por-phyrin species at the water–DCE interface depends strongly on the lifetime of the photoinduced charges [64], which can be increased by storing electrons in the metal particles [63].

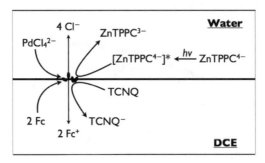

FIGURE 12.20 Schematic diagram of interfacial reactions taking place at polarized water–DCE interface (excluding transfer of Fc^+ from DCE to water). The generation and growth of Pd particles effectively mediate the electron transfer from the porphyrin excited state to TCNQ. The predominant reaction is the photooxidation of Fc by the porphyrin excited state $[ZnTPPC^{4-}]^*$ in the absence of the palladate salt. (Adapted from ref. 63.)

Study of the electrocatalysis of the electron transfer reactions by nanoparticles adsorbed at the polarized ITIES has been expanded by considering the photoelectrochemical properties of TiO_2 anatase nanoparticles. Fermin et al. [65] were first to show that the photoreactivity of these nanoparticles in solutions can be studied by photocurrent measurements at the polarized ITIES under potentiostatic conditions. The photoelectrochemical properties of TiO_2 nanoparticles were studied at the water–DCE interface. The interfacial concentration of the electrostatically stabilized particles can be effectively tuned by the Galvani potential difference and the pH of the aqueous phase. At pH values lower than that corresponding to the point of zero zeta potential, TiO_2 particles are positively charged as a result of the protonation of the surface oxide and they assemble in the liquid–liquid boundary region when the Galvani potential difference $\Delta_o^w\phi$ is made positive. Upon the band gap illumination, photocurrent responses are observed in the presence of Fc in DCE, as shown in Figure 12.21.

The potential and wavelength dependencies of the photocurrent unambiguously reveal that the photoresponses arise from the transfer of valence band holes to the redox couple in DCE. The hole transfer is mediated via the generation of OH_s^g radicals at the particle surface. The overall mechanism can be represented by

$$X_s + h\upsilon \rightarrow h_{VB}^+ + e_{CB}^- \tag{12.9}$$

$$h_{VB}^+ + OH_s^- \rightarrow OH_s^g \tag{12.10}$$

$$e_{CB}^- + OH_s^g \rightarrow OH_s^- \tag{12.11}$$

$$Fc\,(o) + OH_s^g \rightarrow Fc^+\,(o) + OH_s^- \tag{12.12}$$

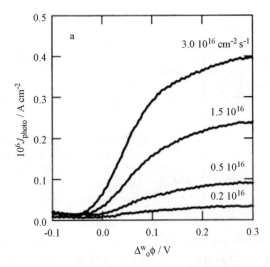

FIGURE 12.21 Photocurrent–potential curves obtained under 12-Hz chopped illumination (325 nm) and lock-in detection at various photon fluxes obtained in the presence of 0.1 g dm^{-3} TiO$_2$ in water and 1 mM Fc in DCE at pH 3.5. (Reproduced from ref. 65 with permission from the American Chemical Society.)

where X$_s$ represents a site in the nanocrystalline lattice which is photoinduced under the band-gap illumination.

On the other hand, photocurrents of the opposite sign were observed in the presence of TCNQ at pH higher than 10. In this pH range, the particles are negatively charged, and the formation of the interfacial assembly takes place when the Galvani potential difference $\Delta_o^w \phi$ is made negative. The effect of the pH and the Galvani potential difference on the photocurrent suggests that heterogeneous charge transfer is in effective competition with recombination via OH$_s^g$ radicals as well as the oxygen evolution at the particle surface.

Photoelectrochemical responses of the TiO$_2$ nanoparticles electrostatically assembled at the polarized water–DCE interface were extended into the visible region by the dye sensitization, which was accomplished by the homogeneous complexation in the presence of chlorin e-6, catechol, or alizarin or by the interfacial complexation involving alizarin in the DCE phase [66]. Photocurrent responses associated with the photooxidation of Fc in DCE were observed upon illumination of the system by polychromatic light from an arc Xe lamp in conjunction with a grating monochromator or by laser illumination at 442 nm. For the TiO$_2$/alizarin and TiO$_2$/catechol systems, the photocurrent spectra showed the characteristic metal-to-ligand charge transfer absorption, indicating that the photoelectrochemical process is initiated by an ultrafast electron injection from the dye into the conduction band of the particle.

The origin of the photoelectrochemical response for the TiO$_2$/alizarin system can be rationalized in terms of the mechanism shown schematically in Figure 12.22. Essentially, the ultrafast injection from the alizarin excited state into the particle is followed by interfacial electron transfer from Fc to the oxidized dye. Consequently, the

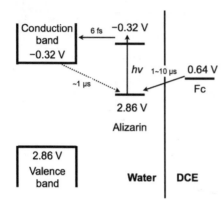

FIGURE 12.22 Mechanism of photoresponses observed for alizarin-sensitised TiO_2 particles at water–DCE interface. Characteristic time constant for the main processes involved is indicated. (Adapted from ref. 66.)

quantum yield of the photoelectrochemical process is determined by the competition between the back electron transfer from the particle to the oxidized dye and the interfacial regeneration of the dye by Fc. The lifetime of the charge-separated state is comparable to the transient time for electron tunnelling from Fc across the water–DCE boundary. The steady-state condition is then determined by matching the electron flux across the interface with the diffusion of redox species in DCE and the capture of injected electrons from the particle surface by species in the aqueous phase.

12.5 SUPPORTED ITIES ELECTROCATALYSIS

Anson et al. [67, 68] have developed an ingenious cell design to support thin organic film on an edge plane graphite electrode (EPG), which is shown schematically in Figure 12.6 (left). Those supported ITIES as discussed below have interesting mass transport characteristics as they display two polarizable interfaces in series that are bound by the condition of current continuity. The method was applied first to measure the electron transfer rate across the water–nitrobenzene (NB) interface [67]. Redox reactants dissolved in NB included DMFc and zinc tetraphenylporphyrin. Reactants in the adjoining aqueous phase were multiply charged anions, including $Fe(CN)_6^{3-/4-}$, $Ru(CN)_6^{4-}$, $Mo(CN)_8^{4-}$, and $IrCl_6^{2-}$. These pioneering measurements have been extended further to study the effects of the driving force and reaction reversibility on the electron transfer rate [69, 70] and to monitor electrochemically the proton transfer across the water–NB and water–benzonitrile (BN) interfaces [71]. The method of evaluation of electron transfer rates [67] has been criticized from the fundamental point of view [72].

More recently Anson et al. [73, 74] have investigated the electrocatalytic reduction of dioxygen to water via cobalt porphyrins dissolved in thin layers of NB (or BN) on the surface of EPG, which was immersed in the aqueous acidic solution. In their seminal work [73] these authors have examined first the electrochemical behavior of Co(II) 5,10,15,20-tetrakis(α, α, α, α-2-pivalamidophenyl)porphyrin (CoTpivPP) in BN by conventional cyclic voltammetry on EPG. The electrochemical responses were found

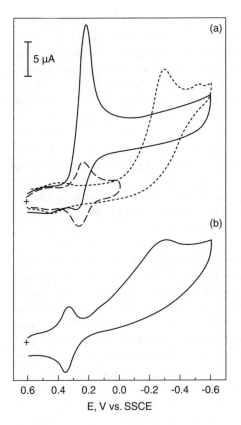

FIGURE 12.23 Cyclic voltammograms of CoTPP and CoTpivPP dissolved in thin layers of acidified BN in presence of O_2. (a) 1.8 mM CoTPP in thin layer of BN placed on EPG electrode immersed in aqueous solution containing 2 M $HClO_4$. The concentration of $HClO_4$ in the equilibrated thin layer of BN was ~ 25 mM. Dashed curve: in absence of O_2. Solid curve: after aqueous phase (and equilibrated thin layer of BN) was saturated with air. The dotted curve is the response obtained for the reduction of O_2 when the BN layer contained no porphyrin. (b) As for the solid curve in panel (a) except that 1.9 mM CoTpivPP was used in place of CoTPP. (Reproduced from ref. 73 with permission from the American Chemical Society.)

to change significantly by the addition of 1-methylimidazole, which coordinates once to the Co(II) and twice to the Co(III) porphyrin. The formation of the O_2 adduct with the Co(II) porphyrin resulted in only small changes in the cyclic voltammetric pattern [73]. Further insight into electrocatalysis of the O_2 reduction was obtained by dissolving the porphyrin catalysts in a thin layer of BN supported on EPG. A comparison of CoTpivPP and Co(II) tetraphenylporphyrin (CoTPP) as catalysts for the electroreduction of O_2 indicated that the latter is the superior catalyst both when adsorbed on graphite electrodes and when dissolved in thin layers of acidified BN [73].

The difference between the catalytic activity with respect to the O_2 reduction of CoTpivPP and CoTPP is demonstrated in Figure 12.23, which shows cyclic voltammograms obtained with an EPG electrode on which was placed a 30-μm layer of BN containing CoTPP [panel (a)] or CoTpivPP [panel (b)]. The dashed curve in panel (a) represents the response from the reversible Co(III/II) couple in the absence of O_2. After the aqueous solution was saturated with air, the solid curve was obtained indicating the remarkable catalytic enhancement of the reduction current. On the other hand, the catalytic effect of CoTpivPP is much less pronounced [solid curve in panel (b)].

The study of electrocatalysis of the O_2 reduction by CoTPP dissolved in a thin layer of BN or NB placed on the surface of EPG electrode has been extended to make a comparison in catalytic activity between CoTPP dissolved and that adsorbed on the bare electrode surface [74]. Essentially, the same quantity of CoTPP catalyst that was present in the BN layers was deposited on an EPG electrode by applying an aliquot of a solution of CoTPP in CH_2Cl_2 to the electrode surface and allowing the solvent to evaporate. In either case, the electrode was equilibrated with aqueous solutions of $HClO_4$ saturated with air. The presence of the bare organic layers enhanced the rate of the electroreduction and affected its stoichiometry compared with results obtained at uncoated electrodes. Upon dissolving CoTPP in the thin layers of BN produced further increases in reaction rates and greater fractions of the O_2 were reduced to H_2O. Importantly, CoTTP dissolved in acidified thin layers of BN was found to be a much more potent catalyst for the four-electron electroreduction of O_2 than is the same quantity of CoTPP adsorbed directly on the surface of EPG electrodes. A possible catalytic mechanism involved the formation of an O_2–CoTPP complex [74].

Several methods have been used to modify the surface of a solid electrode with a single redox droplet of an organic solvent immiscible with water or with an array of such droplets, as shown schematically in Figure 12.7 (right). These methods, as well as various aspects of electrochemistry of redox systems immobilized within the organic solvent or aqueous droplet, have been recently reviewed [75–77]. A related system is the solid electrode the surface of which is modified with an ionic liquid (room temperature molten salt) [78]. Typically, the electrochemical oxidation (or reduction) of the redox-active molecule present in a droplet unsupported with an electrolyte is coupled to an ion transfer across the organic solvent droplet–water or aqueous droplet–organic solvent interface so as to maintain the electroneutrality of the droplet. For this reason, the main application of the droplet-modified electrodes has been so far in the determination of Gibbs transfer energies of inert ions and in the development of highly selective sensors for both ions and molecules [75]. On the other hand, the reactions between the products of the coupled electrochemical oxidation (or reduction) and the ion transfer have led to voltammetric effects that are characteristic for electrocatalysis, for example, copper deposition from the aqueous droplets on a graphite electrode [76]. Since, however, these processes cannot be strictly classified as electrocatalysis at a liquid–liquid interface and are out of scope of this review.

12.6 OUTLOOK

The interface between two immiscible electrolyte solutions (ITIES) is a soft interface with molecular and heterogeneous nature. It offers a spatial separation of the reactants/products according to their lipophilicity as well as a versatile location for various molecular and nanoparticle catalysts. Furthermore, the ITIES can be polarized like the solid–liquid interface, thereby providing an electrochemical control and allowing investigations of not only electron transfer reactions but also ion transfer reactions. With these properties, the ITIES presents a challenge of both fundamental and technological interest for electrocatalysis research.

In nature life is sustained by bioenergetics occurring at the soft liquid interfaces, that is, biomembranes. The ITIES presents a simple model system mimicking biological interfacial processes, in particular those coupled charge transfer reactions (e.g., proton-coupled electron transfer reactions) and those catalyzed by embedded enzymes. For example, biomimetic respiratory oxygen reduction has been performed at the water–1,2-dichloroethane (DCE) interface with protons located in water, electron donors in the organic phase, and amphiphilic catalysts at the interface. This oxygen reduction process can be controlled by driving the potential dependent proton transfer reaction, which is however not accessible by the classical electrochemical methodologies because of their inertness to ionic proton transfer processes.

The electrocatalysis at the ITIES, called electrocatalysis at soft interfaces, also presents industrial perspectives relevant to the energy conversion and storage, that is, artificial photosynthesis, hydrogen generation, and oxygen and carbon dioxide reduction. In this context, design and synthesis of catalysts for interfacial catalysis will represent one of the major aspects. Another aspect will be introduction of new model interfaces, such as supported water–organic interfaces on porous solid materials with open cell structures and water–ionic liquid interfaces. Lipophilic ionic liquids represent an eco-friendly substitute of the organic electrolytes, due to their ability to wet solid surface so as to form a thin layer on an electrode, thereby forming supported soft interfaces, and their property to dissolve large amounts of gas such as CO_2.

In summary, electrocatalysis at liquid–liquid interface will be an alternative paradigm to classical electrochemical approaches, which all rely on the use of solid electrodes for applications such as electrolysis, fuel cells, and photoelectrochemical cells.

REFERENCES

1. H. H. Girault and D. J. Schifferin, in *Electroanalytical Chemistry*, Vol. 15, A. J. Bard (Ed.), Dekker, New York, 1989, p. 1.
2. F. Reymond, D. Fermin, H. J. Lee, and H. H. Girault, *Electrochim. Acta*, 45 (2000) 2647.
3. Z. Samec, *Pure Appl. Chem.*, 76 (2004) 2147.
4. C. Gavach, P. Seta, and B. D'Epenoux, *J. Electroanal. Chem.*, 83 (1977) 225.
5. N. Eugster, D. J. Fermin, and H. H. Girault, *J. Phys. Chem. B*, 106 (2002) 3428.
6. Z. Ding and P. F. Brevet, *Chem. Commun.*, (1997) 2059.
7. Z. Ding, R. G. Wellington, P.-F. Brevet, and H. H. Girault, *J. Electroanal. Chem.*, 420 (1997) 35.
8. Z. Ding, R. G. Wellington, P. F. Brevet, and H. H. Girault, *J. Phys. Chem.*, 100 (1996) 10658.
9. T. Kakiuchi and Y. Takasu, *Anal. Chem.*, 66 (1994) 1853.
10. T. Kakiuchi, Y. Takasu, and M. Senda, *Anal. Chem.*, 64 (1992) 3096.
11. D. J. Fermin, Z. Ding, P. F. Brevet, and H. H. Girault, *J. Electroanal. Chem.*, 447 (1998) 125.
12. H. Nagatani, R. A. Iglesias, D. J. Fermin, P.-F. Brevet, and H. H. Girault, *J. Phys. Chem. B*, 104 (2000) 6869.
13. R. A. W. Dryfe, Z. Ding, R. G. Wellington, P. F. Brevet, A. M. Kuznetzov, and H. H. Girault, *J. Phys. Chem. A*, 101 (1997) 2519.

14. H. D. Duong, P. F. Brevet, and H. H. Girault, *J. Photochem. Photobiol., A*, 117 (1998) 27.

15. D. A. Higgins and R. M. Corn, *J. Phys. Chem.*, 97 (1993) 489.

16. M. C. Messmer, J. C. Conboy, and G. L. Richmond, *J. Am. Chem. Soc.*, 117 (1995) 8039.

17. I. Benjamin, *Chem. Rev.*, 96 (1996) 1449.

18. I. Benjamin, *Annu. Rev. Phys. Chem.*, 48 (1997) 407.

19. H. H. J. Girault, D. J. Schiffrin, and B. D. V. Smith, *J. Colloid Inter. Sci.*, 101 (1984) 257.

20. T. Kakiuchi and M. Senda, *Bull. Chem. Soc. Jpn.*, 56 (1983) 1753.

21. A. Trojanek, P. Krtil, and Z. Samec, *J. Electroanal. Chem.*, 517 (2001) 77.

22. B. M. Quinn, Z. Ding, R. Moulton, and A. J. Bard, *Langmuir*, 18 (2002) 1734.

23. N. Nishi, S. Imakura, and T. Kakiuchi, *Anal. Chem.*, 78 (2006) 2726.

24. L. I. Boguslavskii, A. A. Kondrashin, I. A. Kozlov, S. T. Metel'skii, V. P. Skulachev, and A. G. Volkov, *FEBS Letters*, 50 (1975) 223.

25. Y. I. Kharkats and A. G. Volkov, *J. Electroanal. Chem.*, 184 (1985) 435.

26. Y. I. Kharkats and A. G. Volkov, *Biochim. Biophys. Acta*, 891 (1987) 56.

27. A. G. Volkov, M. A. Bibikova, A. F. Mironov, and L. I. Boguslavskii, *Bioelectrochem. Bioenerg.*, 10 (1983) 477.

28. A. G. Volkov, M. I. Gugeshashvili, A. F. Mironov, and L. I. Boguslavskii, *Bioelectrochem. Bioenerg.*, 10 (1983) 485.

29. A. G. Volkov, *Electrochim. Acta*, 44 (1998) 139.

30. M. D. Kandelaki, A. G. Volkov, A. L. Levin, and L. I. Boguslavskii, *Bioelectrochem. Bioenerg.*, 11 (1983) 167.

31. A. G. Volkov, *Bioelectrochem. Bioenerg.*, 12 (1984) 15.

32. A. G. Volkov, *Anal. Sci.*, 14 (1998) 19.

33. J. P. Collman, M. Marrocco, P. Denisevich, C. Koval, and F. C. Anson, *J. Electroanal. Chem.*, 101 (1979) 117.

34. R. Partovi-Nia, B. Su, F. Li, C. Gros, J.-M. Barbe, Z. Samec, and H. H. Girault, *Chem.-Eur. J.*, 15 (2009) 2335.

35. A. Trojanek, V. Marecek, H. Janchenova, and Z. Samec, *Electrochem. Commun.*, 9 (2007) 2185.

36. S. Fukuzumi, S. Mochizuki, and T. Tanaka, *Inorg. Chem.*, 28 (1989) 2459.

37. S. Fukuzumi, K. Okamoto, C. P. Gros, and R. Guilard, *J. Am. Chem. Soc.*, 126 (2004) 10441.

38. B. Su, R. Partovi-Nia, F. Li, M. Hojeij, M. Prudent, C. Corminboeuf, Z. Samec, and H. H. Girault, *Angew. Chem. Int. Ed.*, 47 (2008) 4675.

39. P. Jing, S. He, Z. Liang, and Y. Shao, *Anal. Bioanal. Chem.*, 385 (2006) 428.

40. H. Ohde, K. Maeda, Y. Yoshida, and S. Kihara, *Electrochim. Acta*, 44 (1998) 23.

41. M. Suzuki, S. Umetani, M. Matsui, and S. Kihara, *J. Electroanal. Chem.*, 420 (1997) 119.

42. G. C. Lillie, S. M. Holmes, and R. A. W. Dryfe, *J. Phys. Chem. B*, 106 (2002) 12101.

43. D. G. Georganopoulou, D. J. Caruana, J. Strutwolf, and D. E. Williams, *Faraday Discuss.*, 116 (2000) 109.

44. T. Sugihara, H. Hotta, and T. Osakai, *Phys. Chem. Chem. Phys.*, 6 (2004) 3563.

45. M. Faraday, *Philos. Trans. R. Soc. London.*, 147 (1857) 145.

46. M. Guainazzi, G. Silvestri, and G. Serravalle, *J. Chem. Soc., Chem. Commun.*, (1975) 200.

47. Y. Cheng and D. J. Schiffrin, *J. Chem. Soc., Faraday Trans.*, 92 (1996) 3865.

48. M. Platt and R. A. W. Dryfe, *Phys. Chem. Chem. Phys.*, 7 (2005) 1807.

49. C. Johans, K. Kontturi, and D. J. Schiffrin, *J. Electroanal. Chem.*, 526 (2002) 29.

50. C. Johans, R. Lahtinen, K. Kontturi, and D. J. Schiffrin, *J. Electroanal. Chem.*, 488 (2000) 99.

51. C. Johans, P. Liljeroth, and K. Kontturi, *Phys. Chem. Chem. Phys.*, 4 (2002) 1067.

52. M. Platt, R. A. W. Dryfe, and E. P. L. Roberts, *Chem. Commun. (Cambridge, U. K.)*, (2002) 2324.
53. M. Platt, R. A. W. Dryfe, and E. P. L. Roberts, *Electrochim. Acta*, 49 (2004) 3937.
54. J. Guo, T. Tokimoto, R. Othman, and P. R. Unwin, *Electrochem. Commun.*, 5 (2003) 1005.
55. R. Knake, A. W. Fahmi, S. A. M. Tofail, J. Clohessy, M. Mihov, and V. J. Cunnane, *Langmuir*, 21 (2005) 1001.
56. R. Greef, R. Peat, L. M. Peter, and D. Pletcher, *Instrumental Methods in Electrochemistry*, Ellis Horwood, Chichester, 1985.
57. M. E. Hyde and R. G. Compton, *J. Electroanal. Chem.*, 549 (2003) 1.
58. A. Trojanek, J. Langmaier, and Z. Samec, *J. Electroanal. Chem.*, 599 (2007) 160.
59. Y. Cheng and D. J. Schifferin, in *International Seminar on Charge Transfer at Liquid-Liquid and Liquid-Membrane Interfaces*, Kyoto, 1996.
60. A. Trojanek, J. Langmaier, and Z. Samec, *Electrochem. Commun.*, 8 (2006) 475.
61. R. Lahtinen, C. Johans, S. Hakkarainen, D. Coleman, and K. Kontturi, *Electrochem. Commun.*, 4 (2002) 479.
62. V. J. Cunnane, G. Geblewicz, and D. J. Schiffrin, *Electrochim. Acta*, 40 (1995) 3005.
63. R. M. Lahtinen, D. J. Fermin, H. Jensen, K. Kontturi, and H. H. Girault, *Electrochem. Commun.*, 2 (2000) 230.
64. R. Lahtinen, D. J. Fermin, K. Kontturi, and H. H. Girault, *J. Electroanal. Chem.*, 483 (2000) 81.
65. H. Jensen, D. J. Fermin, J. E. Moser, and H. H. Girault, *J. Phys. Chem. B*, 106 (2002) 10908.
66. D. J. Fermin, H. Jensen, J. E. Moser, and H. H. Girault, *ChemPhysChem*, 4 (2003) 85.
67. C. Shi and F. C. Anson, *J. Phys. Chem. B*, 102 (1998) 9850.
68. C. Shi and F. C. Anson, *Anal. Chem.*, 70 (1998) 3114.
69. C. Shi and F. C. Anson, *J. Phys. Chem. B*, 103 (1999) 6283.
70. C. Shi and F. C. Anson, *J. Phys. Chem. B*, 105 (2001) 8963.
71. T. D. Chung and F. C. Anson, *Anal. Chem.*, 73 (2001) 337.
72. A. L. Barker and P. R. Unwin, *J. Phys. Chem. B*, 104 (2000) 2330.
73. B. Steiger and F. C. Anson, *Inorg. Chem.*, 39 (2000) 4579.
74. T. D. Chung and F. C. Anson, *J. Electroanal. Chem.*, 508 (2001) 115.
75. C. E. Banks et al., *Phys. Chem. Chem. Phys.*, 5 (2003) 4053.
76. T. J. Davies, S. J. Wilkins, and R. G. Compton, *J. Electroanal. Chem.*, 586 (2006) 260.
77. F. Scholz, U. Schroder, and R. Gulaboski, *The Electrochemistry of Immobilized Particles and Droplets*, Springer-Verlag, Berlin, 2004.
78. M. Opallo, A. Lesniewski, J. Niedziolka, E. Rozniecka, and G. Shul, *Rev. Polarogr.*, 54 (2008) 21.

CHAPTER 13

Platinum-Based Supported Nanocatalysts for Oxidation of Methanol and Ethanol

ERNESTO R. GONZÁLEZ[1], EDSON A. TICIANELLI[1], and ERMETE ANTOLINI[2]

[1] Instituto de Chimica de São Carlos, Universidade de São Paulo, 13560-970 São Carlos, Brazil
[2] Scuola Scienza Materiali, 16016 Cogoleto (Genova), Italy

13.1 PREPARATION METHODS OF DISPERSED ELECTROCATALYSTS

13.1.1 Preparation and Pretreatment of Carbon Support

Carbon blacks are widely used as catalyst support in low-temperature fuel cells. The carbon blacks are produced by the oil furnace and acetylene processes. The most important is the furnace black process in which the starting material is fed to a furnace and burned with a limited supply of air at about 1400°C. Due to its low cost and high availability, oil furnace carbon black (e.g., Vulcan XC-72) has been widely used as the support for platinum catalyst in low-temperature fuel cells. Generally, carbon blacks have high specific surface area but are constituted mostly by micropores of less than 1 nm, which are more difficult to be fully accessible. This is a disadvantage when the carbon is used as catalyst support. Indeed, when the average diameter of the pores is less than 2 nm, supply of a reactant may not occur smoothly and the activity of the catalyst may be limited. Moreover, it is known that the micropores of these type of amorphous carbon are poorly connected. Compared with carbon blacks, mesopore carbons generally present higher surface area and a lower amount or absence of micropores. A large mesopore surface area promotes the dispersion of Pt particles, which results in a large effective surface area of Pt with a high catalytic activity. The mesoporous structure facilitates smooth mass transport and leads to high limiting currents. On this basis, novel nonconventional carbon materials have attracted much interest as electrocatalyst support because of their good

Catalysis in Electrochemistry: From Fundamentals to Strategies for Fuel Cell Development,
First Edition. Edited by Elizabeth Santos and Wolfgang Schmickler.
© 2011 John Wiley & Sons, Inc. Published 2011 by John Wiley & Sons, Inc.

electrical and mechanical properties and their versatility in pore size and pore distribution tailoring. These materials present a different morphology than carbon blacks
both at the nanoscopic level in terms of their pore texture (e.g., mesopore carbon)
and at the macroscopic level in terms of the shape of the particles (e.g., micro-
spheres). The examples are supports produced from ordered mesoporous carbons,
carbon aerogels, carbon nanotubes, carbon nanohorns, carbon nanocoils, and carbon
nanofibers.

Generally, before their use as catalyst support, carbon materials are activated to
increase metal dispersion and the catalytic activity. There are two ways to activate the
carbon materials: chemical activation and physical activation. Derbyshire et al. [1]
found that the surface chemistry of carbon (surface functional groups) resulting from
a pretreatment is of critical importance in determining the catalytic activity of carbon-
supported metal catalysts. The functionalities present on the carbon surface containing
oxygen (e.g., carboxylic groups, phenolic groups, lactone groups, ether groups) are
responsible both for the acid–base and the redox properties of the carbon. The oxidative treatment of the carbon surface gives rise to the formation of surface acidic sites
and to the destruction of surface basic sites. The treatment can be performed with
different oxidants: HNO_3, H_2O_2, O_2, or O_3. The effect of oxidative pretreatment of
the carbon on platinum dispersion has produced contradictory results. According to
some authors [2, 3], the dispersion increases for increasing number of oxygen surface
groups in the support. Carbons functionalized with weak oxidants, which develop
acidic sites with moderate strength and show strong interaction with H_2PtCl_6 during
impregnation, would favor the Pt dispersion on the carbon surface. Other authors
[4–6], instead, reported that the presence of oxygen surface groups on carbon
decreases the metal dispersion. The dependence of the Pt dispersion on the total
surface oxygen content of the support is reported in Figure 13.1 [5]. According to
the authors, the decrease in dispersion with the increase in the total surface oxygen
is due to the reduction of the number of surface basic sites, which are centers for the
strong adsorption of $PtCl_6^{2-}$. The platinum content in the catalyst also depends on
the oxidative treatment of the carbon and decreases for increasing CO_2 groups in the
support.

The physical activation consists in a thermal treatment of the carbon performed
under inert atmosphere at 800–1100°C or in air/steam at 400–500 °C, with the aim of
removing the impurities present on the carbon surface. Pinheiro et al. [7] investigated
the preparation of carbon-supported Pt using three types of carbon substrates: Vulcan
XC-72 powder, Shawinigan black, and a fullerene soot consisting of the residue after
C60/C70 fullerene extractions. Heat treatments of the carbons were carried out under
two conditions: (i) argon atmosphere at 850 °C for 5 h; (ii) argon atmosphere at 850 °C
for 5 h followed by water vapor at 500 °C for 2.5 h. Following both heat treatments,
from cyclic voltammetry (CV) measurements the three carbons showed an increase
of the capacitive current, due to the elimination of surface impurities. With the heat
treatments, Pt catalysts present an increase of approximately 50% in the active surface
area for Shawinigan and Vulcan carbons. After thermal treatments of the carbons, Pt
supported on the Shawinigan and fullerene substrates showed similar active areas,
somewhat smaller than that of Pt supported on heat-treated Vulcan. Recently Yu and

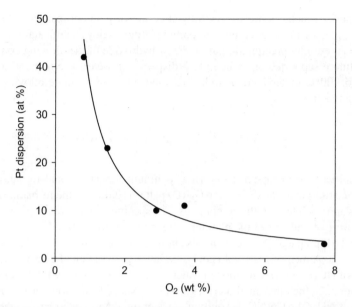

FIGURE 13.1 Dependence of platinum dispersion in Pt/C catalysts on total surface oxygen content of support. [Reproduced with permission from M. A. Fraga et al., *J. Catal.,* 209:355–364 (2002).]

Ye [8] reviewed new advances related to the physicochemical and electronic interactions at the catalyst–support interface and the catalyst activity enhancement through improved Pt–C interaction, especially focusing on the surface modification of the carbon support to form proper functional groups and chemical links at the Pt–C interface.

13.1.2   Preparation of Carbon Supported Alloys

The key parameters in the preparation of carbon-supported alloys are:

1. The actual Pt–M atomic ratio obtained, which can be different than the nominal one
2. The amount of M alloyed, which can be lower than the total M content in the catalyst
3. The metal particle size, as the electrocatalytic activity depends on the metal surface area

The preparation methods can be divided in high- and low-temperature methods.

13.1.2.1   High-Temperature Methods: Alloying Method

The alloying method of preparation of PtM/C consists in the formation of carbon-supported platinum followed by the deposition of the second metal on Pt/C and alloying at high temperatures. This thermal treatment at high temperatures gives rise to an undesired metal particle growth by sintering of platinum particles. Jalan and Taylor [9] prepared various carbon-supported platinum alloy electrocatalytsts (Pt_3Cr, Pt_3V, Pt_3Ti, etc.) by precipitating an oxide or hydroxide of the alloying component onto platinum supported on carbon black dispersed in water. The material was then treated to 900°C under inert or reducing atmosphere to promote alloying by the following reaction:

$$Pt + MO_x + (\tfrac{1}{2}x)C \rightarrow Pt\text{---}M + (\tfrac{1}{2}x)\,CO_2 \qquad (13.1)$$

The shift of face-centered-cubic (fcc) platinum reflections and the absence of superlattice lines in X-ray diffraction (XRD) patterns indicated the formation of solid solutions and not intermetallic compounds. The specific surface area of pure platinum was $135\,m^2\,g^{-1}$ and that of the platinum alloys ranged from 70 to $90\,m^2\,g^{-1}$. Beard and Ross [10] prepared PtCo/C catalysts starting from commercial Pt/C (10% Pt) by two methods. A method (series A) consisted in the preparation of an acidic (pH 2 by addition of HCl) $Co(OH)_2$ solution in water–methanol followed by Pt/C addition into this solution. In the other method (series B) Pt/C was added into a basic (pH 11 by addition of NH_4OH) $Co(NO_3)_2$ solution in water–methanol. In both cases the Pt : Co atomic ratio was 3 : 1. Thermal treatments at 700, 900, and 1200°C were performed under inert atmosphere on each catalyst. In both series and in the absence of thermal treatment the catalyst has a lattice parameter, obtained from XRD, indicative of pure Pt. Following thermal treatments in series A the lattice parameter decreased with increasing heating temperature, indicative of alloy formation. In series B the lattice parameter decreased after heating, but to a less extent than in series A. No clear trend appeared in the lattice parameter change due to thermal treatment temperature. The particle size for series A at each thermal treatment temperature was larger than the corresponding catalyst in series B. The final particle size of the series A 1200°C sample (12 nm) was about four times larger than that of the original Pt catalyst.

More recently, Min et al. [11] prepared three carbon-supported alloy catalysts (PtCo/C, PtNi/C, and PtCr/C) starting from commercial Pt/C (10%) catalyst. Appropriate amounts of $CoCl_2$, $NiCl_2$, or $Cr(NO_3)_3$ solutions were added to Pt/C. The atomic ratio Pt to M was 3 : 1. These catalysts were subjected to thermal treatment at 700, 900, or 1100°C in a reducing atmosphere. XRD measurements indicated a decrease of the lattice parameter with increasing heating temperature, that is, the degree of alloying increased with temperature. The particle size, obtained from both XRD and transmission electron microscopy (TEM) measurements, increased with increasing thermal treatment temperature.

13.1.2.2   Low-Temperature Methods

In view of the fact that the sintering process of Pt particles by thermal treatment takes place appreciably for temperatures higher than 600°C, a way to tailor carbon-supported Pt-based alloy with small particle size is the simultaneous impregnation on the carbon support of Pt and M precursors followed by reduction at low (<100°C) or intermediate (200–500°C) temperature.

Xiong and Manthiram [12] synthesised highly dispersed Pt–M (M = Fe and Co) alloy catalysts on a carbon support by the microemulsion method using sodium bis(2-ethylhexyl)sulfosuccinate as the surfactant, heptane as the oil phase, and NaBH$_4$ as the reducing agent. The synthesis was done at room temperature and, for the sake of comparison, the same catalysts were also prepared by a high-temperature method (alloying at 900°C). The nominal Pt–M atomic ratio was 80 : 20. By XRD analysis the samples prepared by the microemulsion method showed broad reflections compared to those obtained by the high-temperature route, indicating a smaller particle size for the former. Also, the reflections of the Pt–M samples were shifted to higher angles compared to those of Pt, indicating a contraction of the lattice and alloy formation. However, the shift is more significant for the samples prepared by the high-temperature route compared to those prepared by the microemulsion method, suggesting a greater alloy formation in the former case. It is possible that in the samples prepared by the microemulsion method some M may not be incorporated into the Pt lattice and it may be present as free M or its oxides. The Pt–M/C catalysts prepared by the microemulsion method had smaller particle size than those prepared by the high-temperature route. With the microemulsion method, the Pt–Co/C sample prepared by a modified procedure (three microemulsion solutions) showed a slightly smaller particle size than the sample prepared using two microemulsion solutions, which could be due to differences in the nucleation and growth rates in the two procedures. Regarding the EDX analysis, while the Pt content in the samples prepared by the high-temperature procedure was close to the expected nominal value, it resulted lower than the nominal value for the samples prepared by the microemulsion method. The discrepancy is particularly large in the case of the modified microemulsion procedure. According to the authors, the lower Pt content values could be due to the difficulty of effectively reducing the metal ions in the microemulsions since the nucleation and growth processes in the microemulsions are very different from those in the bulk, and there is loss of metal ions or smaller particles in the filtrate.

The polyol synthesis method is a very efficient way to prepare nanosized noble metals or noble metal/transition metal bimetallic clusters. Zhou et al. [13, 14] synthesized carbon-supported Pt, Pt–Ru, Pt–Pd, Pt–W and Pt–Sn and Pt–Fe binary electrocatalysts with a sharp particle size distribution. Generally, this preparation method uses H$_2$PtCl$_6$ and chloride salts as metal sources and ethylene glycol as solvent and reducing agent. The reaction occurs at 130–160°C. Liang et al. prepared carbon-supported ternary Pt–Ru–Ir (1 : 1 : 1) [15] and Pt–Ru–Ni (1 : 1 : 1) [16] catalysts having 40 wt % metal by using a microwave-irradiated polyol plus annealing (MIPA) synthesis method. The small particle size and the homogeneous size distribution of these catalysts are ascribed to the rapid reduction of the metal salts and easy nucleation of the metal particles in ethylene glycol facilitated by microwave irradiation.

A way to prepare unsupported nanosized Pt-based alloys is the use of organometallic compounds as precursors. By thermal decomposition or reduction of organometallic precursors, small nanoparticles of metals or alloys with narrow size distribution can be obtained. Among the various organometallic precursors used, metal–carbonyl complexes have been employed to obtain carbon-supported metal or alloy catalysts. By the reduction of Pt–Ru carbonyl molecular precursors, Nasher et al. [17] and Hills et al. [18] prepared carbon-supported bimetallic Pt–Ru alloys, while by decomposition of Pt and Ru carbonyls in organic solvents Dickinson et al. [19] obtained carbon-supported Pt–Ru alloys. In all cases, the Pt–Ru catalysts presented small particle size and narrow size distribution. Recently, Manzo-Robledo et al. [20] prepared carbon-supported Pt–Sn catalysts through the carbonyl chemical route. Yang et al. used this method to prepare carbon-supported Pt–Cr [21] and Pt–Ni [22]. After the synthesis of Pt–Ni carbonyl complexes, Vulcan XC-72 carbon was added to the mixture under N_2 gas flow. The alloying temperature under hydrogen ranged from 200 to 500°C. In the case of Pt–Cr/C, instead, Pt carbonyl and $CrCl_3$ were mixed with the carbon powder, then thermally treated at 500°C The total metal loading of all the carbon-supported catalysts was about 20 wt %. The energy-dispersive X-ray spectroscopy (EDX) composition of all the catalysts was very close to the nominal value. According to the authors, the nearly linear relationship between the lattice parameter and EDX composition again attested that Cr and Ni are completely alloyed with Pt. The dependence of the lattice parameters of Pt–Cr and Pt–Ni on EDX composition is shown in Figure 13.2 [23]. In the same way, Figure 13.3 [23] shows the metal particle size from transmission electron microscopy (TEM) versus EDX composition plot. For both Pt–Cr and Pt–Ni the metal particle size decreases with increasing content of nonprecious metal in the alloy. The low synthesis temperature could be favorable for the formation of carbon-supported alloy nanoparticles with small particle size and a narrow size distribution and thus with a good dispersion.

Another way to prepare carbon-supported Pt–M alloy catalysts is based on the reduction of the precursors with formic acid at room temperature. The method consists in the treatment of carbon powder with formic acid before the Pt and M impregnation. The treated carbon reduces Pt and M precursors on its surface. Carbon supported

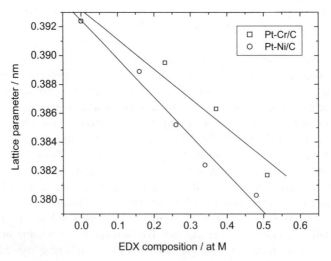

FIGURE 13.2 Dependence of lattice parameters of Pt–Cr/C and Pt–Ni/C electrocatalysts with metal content 20 wt % prepared by carbonyl route on EDX composition. [Reproduced with permission from E. Antolini et al., *Mater. Chem. Phys.*, 101:395 (2006).]

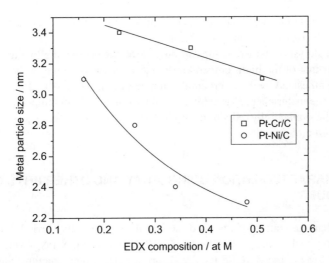

FIGURE 13.3 Metal particle size from TEM of Pt–Cr/C and Pt–Ni/C electrocatalysts with metal content 20 wt % prepared by carbonyl route vs. EDX composition. [Reproduced with permission from E. Antolini et al., *Mater. Chem. Phys.*, 101:395 (2006).]

Pt–Ru [24], Pt–Mo [7], and Pt–Sn [25] were successfully prepared using this method. Salgado et al. [26] prepared carbon-supported Pt–Co electrocatalysts in the Pt–Co atomic ratios 70 : 30 and 60 : 40 through the formic acid route. Instead of a supported alloy, EDX and XRD measurements indicated the formation of a supported bimetallic material, that is, two metals present on the same support but not alloyed. Indeed, the reducing power of the formic acid is not strong enough to allow the formation of Pt–Co crystallites. Following thermal treatment at 900°C of these electrocatalysts, the presence in the XRD pattern of secondary Pt reflections shifted to higher angles indicated partial alloy formation.

Carbon-supported Pt–Ru [27], Pt–Co [28], Pt–Ni [29], and Pt–Cr [30] alloy electrocatalysts were successfully prepared by impregnating high-surface-area carbon with Pt and M precursors followed by reduction of the precursors with $NaBH_4$. The metal particle size was in the range 3.8–4.8 nm. It has to be remarked that, independently of the EDX composition, the actual composition of the alloy was around 92 : 8. As a consequence, for low M content (10–15 at % M) a high degree of alloying was attained, while the degree of alloying was low for the catalyst with high M content (30 at % M).

Summarizing, Pt–M alloy catalysts have been synthesized by both impregnation of carbon support with Pt and M precursor solutions and adsorption of Pt and M in colloidal form onto the carbon surface. Platinum and base metal reduction following the impregnation or adsorption step has been carried out by different methods: heating in flowing hydrogen in the temperature range 300–500°C or chemical reduction with sodium borohydride, sodium formate, hydrazine, and formic acid at temperatures lower than 100°C. The most promising methods are the microemulsion method, the carbonyl route, and the impregnation followed by reduction with $NaBH_4$. The metal particle size of the catalysts prepared by these methods is in the range 2.3–4.8 nm, smaller than that of the catalysts prepared by alloying at 900°C, which results in Pt–M particles with sizes in the range 4.5–7.5 nm.

For the catalysts obtained by the carbonyl route, the degree of alloying was nearly the same as that of the high-temperature reduced catalysts, while in the case of the catalysts synthesized by the microemulsion method, the degree of alloy was slightly lower than the materials prepared by thermal treatment at high temperature. Finally, regarding the synthesis by reduction with NaBH$_4$, a high degree of alloy was obtained for low Pt–M atomic ratio.

13.2 CHARACTERIZATION USING X-RAY AND OTHER PHYSICAL TECHNIQUES

Physical characterization of Pt-based nanocatalysts is usually carried out by using several X-ray techniques, including EDX, XRD, and in situ X-ray absorption spectroscopy (XAS) in the XANES (X-ray absorption near edge structure) and EXAFS (extended X-ray absorption fine structure). Key techniques also employed are TEM and high-resolution TEM (HRTEM).

13.2.1 EDX and XRD Analyses

A common approach consists in analyzing first the catalyst samples by using the EDS and XRD techniques. In the first case the composition of the samples regarding metallic components can be estimated [23, 25, 29, 31–40]. Table 13.1 presents results of EDS analyses of Pt–Sn/C catalysts prepared by the so-called formic acid method [31]. These results show that all the EDX compositions of the prepared catalysts are very near the nominal values, confirming the efficiency of the preparation method [31].

XRD can be used for obtaining physical properties such as lattice parameters and the average crystallite sizes of the metallic components, which can be estimated from selected peaks of the Pt diffraction pattern by using Scherrer's equation [31–40]:

$$D = \frac{k\lambda}{B\cos\theta} \tag{13.2}$$

where D is the average crystallite size in angstrom, k a coefficient whose value is 0.9 for spherical particles [31], λ the wavelength of the X rays used (1.5406 Å), B the width of the diffraction peak at half height in radians, and θ the angle at the position of the peak maximum.

As an example, Figure 13.4 presents the XRD patterns for Pt–Sn/C materials with different Sn contents tested for ethanol oxidation [31]. As in many other examples [23, 25, 29, 32–40], here the XRD patterns indicate that all Pt–Sn/C electrocatalysts present the fcc structure of platinum but with shifted reflection angles due to differences in the lattice parameters. These results provide good evidence of the formation of true platinum–tin alloys, with no indication of significant segregation of Pt or Sn

TABLE 13.1 Physical Parameters of Catalysts Prepared and/or Used in this Work

Catalyst	EDX Composition (%)	Lattice Parameter (nm)	Nonalloyed Sn (at %)	Relative Pt$_3$Sn Content (%)	Crystallite Size from XRD (nm)	Metal Particle Size from TEM (nm)	Pt Active Area by CO Stripping (m^2 g^{-1} Pt)
Pt$_9$Sn/C	88:12	0.39431	2.5	13	3.3	4.2 ± 1.4	56.79
Pt$_3$Sn/C	77:23	0.39593	9.2	37	3.9	4.5 ± 1.4	38.53
Pt$_2$Sn/C	66:34	0.39756	15.9	62	4.9	5.4 ± 2.5	31.08
Pt/C E-TEK	(100)	0.39150	—	2.9	3.0 ± 1.2	69.05	
Pt$_3$Sn/C E-TEK	n.d.	0.40015	100% alloy assumed	100	4.1	4.2 ± 1.3	33.09

Source: From ref. 31.

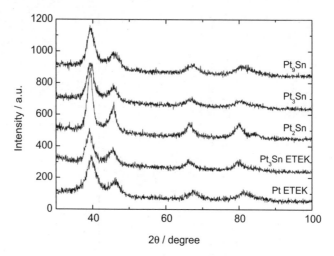

FIGURE 13.4   XRD diffractograms of carbon-supported 20 wt % Pt/C and Pt–Sn/C catalysts. [Reproduced with permission from *J. Electrochem. Soc.*, **154**:B39 (2007). Copyright 2007, The Electrochemical Society.]

as pure metallic or oxide phases. As mentioned above, the physical properties of electrocatalysts can be obtained from the analysis of the XRD results, and the lattice parameters and the average sizes of the Pt crystallites estimated by Scherrer's equation for some Pt alloy systems are shown in Table 13.1. It is seen that the lattice parameters for the Pt–Sn materials are all larger than for pure Pt and increase with the increase of the Sn content. For other samples, such as Pt–Ru/C [32], Pt–Ni/C [33], and Pt–Co/C [33–36], there is a shrinkage of the Pt crystallite, following a good correlation with the atomic size of the alloy elements. These parameters can be used to estimate the real content of the hosting element forming the true alloy with Pt by using Vergard's law [31]. In the case of Pt–Sn/C, these analyses evidenced the presence of a Pt_3Sn phase in all materials, whose amount increases with the Sn content. Extensive examples of XRD data for Pt-based catalysts can be found in the literature [e.g., 23, 25, 29, 31–40].

13.2.2   TEM and HRTEM Analyses

TEM and HRTEM analyses provide fundamental information about the particle size distribution and the morphology of the Pt-based catalysts. In Figures 13.5(a)–(c) there are typical HRTEM images at low and high magnification and the particle size distribution for a Pt–Sn–Ru/C $(1:1:0.3)$ catalyst, showing the high dispersion of metal particles with uniform distribution on the carbon surface [39]. The particle size distribution ranges from 1.5 to 8.0 nm, with a mean diameter of 3.2 nm. This value is different than the crystallite size obtained from XRD (2.7 nm) and this can be related to the fact that the broadening of the X-ray diffraction peak is not only related to the crystallite size but is also influenced by nonhomogeneities in the solid solution. The

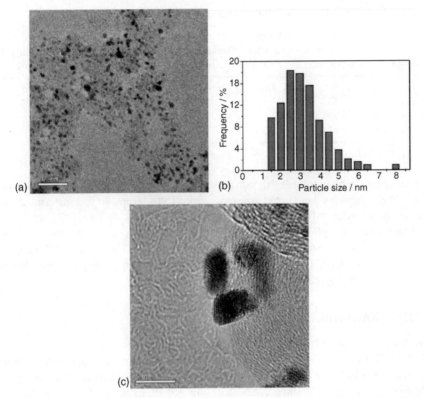

FIGURE 13.5   HRTEM image at low magnification (a), histogram of particle size distribution (b), and HRTEM image at high magnification (c) of Pt–Sn–Ru/C (1 : 1 : 0.3, atom proportions) catalyst. [Reproduced with permission from E. Antolini et al., *Electrochem. Commun.*, 9:398 (2007).]

high-magnification HRTEM image reveals the asymmetric faceted shape, typically cubooctrahedral, of the Pt–Sn–Ru particles, as usually seen for such nanosized Pt-based particles [41, 42].

Table 13.2 shows the diameter ranges and peak relative abundances obtained by TEM for Pt–Co/C catalysts prepared with different Pt–Co contents by the reduction of H_2PtCl_6 and $Co(NO_3)_2$ using $NaBH_4$ [36]. It is seen that the Pt/C catalyst has the largest particles (mean diameter 4.8 nm) and the broader size distribution, with 2.9% of the particles in the range of 15–46 nm. The presence of either large particles or the aggregation of small particles is evidenced by these TEM results for Pt/C. The measurements on the Pt–Co/C 1 : 1 catalyst indicated a mean particle diameter of 2.9 nm with only 0.5% of the particles presenting diameters in the range of 15–19 nm, indicating a good particle spatial dispersion without the formation of large particle aggregates. The Pt–Co/C 1 : 3 catalyst has even smaller particles (mean diameter 2.7 nm), and the size distribution histogram shows the existence of only 1.4% of the particles within the range of 15–40 nm. The Pt–Co/C 1 : 5 catalyst has the sharpest

TABLE 13.2 Morphological Informations on Catalysts Obtained by TEM

Catalyst	Nominal Atomic Co Content (%)	Mean Diameter (nm)	Standard Deviation (±) (nm)	Maximum Diameter (nm)	Diameter Range (nm) of Peak Relative Abundance (%)
Pt	0	4.8	4.2	45.6	2.5–3.0; 15
Pt-Co/C 1:1	50	2.9	2.0	18.4	1.5–2.0; 25
Pt-Co/C 1:3	75	2.7	3.0	39.2	1.5–2.0; 30
Pt-Co/C 1:5	83	2.0	1.4	15.2	1.0–1.5; 30

Source: From ref. 36.

size distribution histogram and the smallest particles (mean diameter 2.0 nm). Hence, 0.2% of the particles were in the size range from 15 to 16 nm, so this material shows the smallest tendency for particle aggregation.

Other examples of TEM and HRTEM results can be found in the literature, as, for example, for Pt–Sn/C [25, 31, 43], Pt–Co/C [23, 33–35, 40], Pt–Ni/C [23, 33, 34, 40], and Ag–Co/C [44], where representative analyses regarding the particle size distributions are carried out.

13.2.3 XANES and EXAFS Analyses

The electronic properties of the platinum alloy catalysts can be investigated by XANES. Figure 13.6 shows examples of these results obtained at the Pt L_3 edge for the Pt/C electrocatalyst in alkaline medium at several electrode potentials [45]. Figures 13.7(a) and (b) show XANES results for Pt–Co/C, Pt–Cr/C, and Pt–V/C

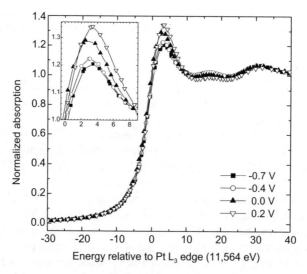

FIGURE 13.6 XANES spectra at Pt L_3 edge for Pt/C electrocatalysts at different electrode potentials (vs. Hg/HgO) in KOH 1.0 mol L^{-1}. [Reproduced with permission from F. H. B. Lima et al., *J. Brazil. Chem. Soc.,* 16:328 (2005).]

composites obtained at –0.7 and 0.2 versus Hg/HgO, respectively [45]. The absorption at the Pt L_3 edge (11.564 eV) corresponds to $2p_{3/2}$–$5d$ electronic transitions, and the magnitude of the absorption hump or white line located at ~5 eV is directly related to the occupancy of the $5d$ electronic states; the higher is the hump the lower is the occupancy and vice versa.

In Figure 13.6 it seen that the white-line magnitude increases with the increase of the electrode potential for the Pt/C catalyst. This phenomenon is attributed to the emptying of the Pt $5d$ band, in agreement with the presence of an electron-withdrawing effect of the oxygen present in a well-known Pt oxide layer formed above –0.1 V on the catalyst particle surface in alkaline media. The same trend is observed for Pt/C in acid media [32]. From Figure 13.7(a) it is seen that under reducing potential the

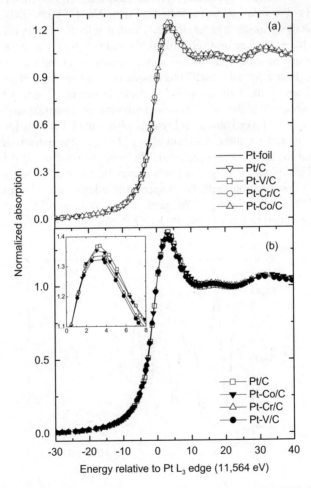

FIGURE 13.7   XANES spectra at Pt L_3 edge for Pt/C and as-received Pt–M/C electrocatalysts and for Pt-foil in KOH 1.0 mol L^{-1}: (a) –0.7 V and (b) 0.2 V vs. Hg/HgO. [Reproduced with permission from F. H. B. Lima et al., *J. Brazil. Chem. Soc.,* 16:328 (2005).]

magnitude of the white lines is essentially the same for the different metal alloys. On the other hand, an important aspect is observed in Figure 13.7(b), where it is seen that the increase of the Pt $5d$ band vacancy, caused by the oxide formation on Pt, is less pronounced for the alloys. These results indicate less Pt oxide formation in the alloys, following a consistent trend with respect to the electronegativity of the hosting element (V < Cr < Co). The invariance of the Pt $5d$ band occupancy for reduced catalysts is also observed from XANES studies of Pt–Co/C and Pt–Ni/C alloys [33, 36] and also for Pt–Sn/C in acid media [31, 44, 46]. However, in the case of Pt–Ru/C, Pt–Mo/C, and Pt–Fe/C, in acid media, a clear tendency of an emptying of the Pt $5d$ band is seen [32, 45, 47].

Important morphological informations of Pt alloy catalysts can be obtained in the EXAFS region of the XAS spectrum. Normalized EXAFS signals for a Pt foil and for the Pt/C and Pt–V/C catalysts are presented in Figure 13.8 [45]. The EXAFS functions for several catalysts present a great similarity with that for the Pt foil, evidencing that all materials have the same unit cell structure as pure Pt (fcc). However, the EXAFS signal represents the superimposition of contributions of several coordination shells and, thus, the Fourier transform (FT) technique is used to obtain information about the contributions of the individual shells. Peaks in the radial structure of the FT magnitude correspond to the contribution of individual coordination shells around the metallic atom under investigation. In Figure 13.9 are the FT results for the EXAFS oscillations obtained for these platinum alloys [45]. For comparison the results for the Pt foil and for Pt/C are also included. The peak centered at ∼2.6 Å in the FT–radial coordinate plots for these materials results from the contributions related to the first Pt–Pt coordination shell. For dispersed Pt alloys with $4d$ elements such as Ru and Ag [32, 37], a splitting of this peak is observed and this is assigned to the radiation backscattering from Pt and Pt–Ru or Pt–Ag neighbors in the first shell. In

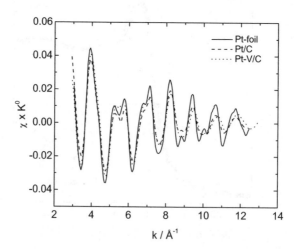

FIGURE 13.8 EXAFS signals at Pt L$_3$ edge for (solid line) Pt-foil; (dashed line) Pt/C; and (dotted line) Pt–V/C; E = –0.7 V vs. Hg/HgO. [Reproduced with permission from Lima F. H. B. et al., *J. Brazil. Chem. Soc.* 16:328 (2005).]

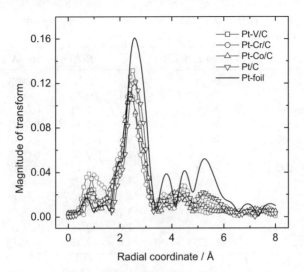

FIGURE 13.9 Fourier transform (k^3 weighted) of EXAFS oscillations for as-received Pt–M/C electrocatalyts at –0.7 V vs. Hg/HgO. [Reproduced with permission from F. H. B. Lima et al., *J. Brazil. Chem. Soc.* 16:328 (2005).]

Figure 13.6, results do not evidence a splitting of this first FT peak, and this is probably because the separation of both contributions is not large enough, as expect for these low-atomic-number Pt-coordinated backscatterers.

Further analyses of the Pt–M/C EXAFS results can be made from the Pt–Pt and Pt–M first coordination shell signals in the Fourier transform. Phase and amplitude data as a function of the radial coordinate (r) are fitted to those calculated for a Pt–M/C model using the FEFF program [48] to obtain the coordination number (N), the bond distances (R), the Debye–Waller factor (Δs^2), and the edge energy shift (ΔE_0). Table 13.3 presents a summary of the physical parameters for some Pt alloy samples [45]. Here a slight decrease in the Pt–Pt interatomic distance is noted in the Pt–M/C materials, compared to Pt/C, while the Pt–M bond distances are even smaller, particularly when M = Cr and Co. This indicates a shrinkage of the Pt crystalline structure, this being one of the effects of the incorporation of a second metal atom with smaller radius in the Pt fcc crystals. The ratio of N_{Pt-M}/N_{Pt-Pt} indicates real compositions of ∼1 : 0.15, 1 : 0.9, and 1 : 0.6 (Pt–M) for the as-received Pt–V/C, Pt–Cr/C, and Pt–Co/C materials, respectively. In the case of Pt–V/C (and to a lower extent for Pt–Co/C) it is seen that only a fraction of the non–noble metal atoms is alloyed with Pt, while the other atoms must be forming an amorphous oxide phase, as also proposed from XRD results [32–36]. The small incorporation of the V atoms explains the little effect on the Pt–Pt bond distance observed for the Pt–V/C material (Table 13.3). Heating treatments under hydrogen atmosphere conducted on the Pt–V/C catalyst lead to some incorporation of V atoms in the Pt lattice, but the final proportion does not exceed 1 : 0.2 (Pt–V).

TABLE 13.3   Results of In Situ EXAFS Analyses for Carbon-Supported Pt–M/C Electrocatalysts

Electrocatalyst	Coordination Shell	N	$R(\text{Å})$	$\Delta\sigma^2(\text{Å}^2)$	$\Delta E_0(\text{eV})$
Pt-foil	Pt–Pt	12	2.76	0.0043	9.15
Pt/C	Pt–Pt	10.1	2.74	0.0057	7.93
Pt–V/C as received	Pt–Pt	9.5	2.72	0.0053	6.55
	Pt–V	1.4	2.69	0.0057	6.55
Pt–V/C heat treated	Pt–Pt	9.3	2.72	0.0040	6.64
at 500°C	Pt–V	1.6	2.70	0.0066	6.64
Pt-V/C heat treated	Pt–Pt	8.8	2.72	0.0051	8.07
at 850°C	Pt–V	1.9	2.69	0.0036	8.07
Pt–Cr/C	Pt–Pt	6.3	2.70	0.0037	0.84
	Pt–Cr	5.7	2.63	0.0030	0.84
Pt–Co/C	Pt–Pt	7.0	2.71	0.0026	5.63
	Pt–Co	4.1	2.62	0.0162	5.63

Source: From ref. 45.

Note: Heat treatment of Pt–V/C was conducted under H_2 atmosphere for 2 h.

13.3   CHARACTERIZATION USING ELECTROCHEMICAL TECHNIQUES

Electrochemical techniques have been used for several decades to characterize the oxidation of low-molecular-weight alcohols on catalytic surfaces, mainly polycrystalline platinum and single-crystal platinum. On supported nanoparticles, this kind of work started much later, but progress on the understanding of methanol oxidation has been fast. The direct methanol fuel cell (DMFC) is already commercially available for some applications. On the other hand, and in spite of the fact that ethanol is a much more convenient fuel than methanol, the study of ethanol oxidation has been frustrated by the fact that the complete 12-electron oxidation to carbon dioxide is extremely difficult to achieve electrochemically.

Early investigations showed that the results obtained with electrochemical techniques on well-characterized metal surfaces were not directly applicable to supported metal nanoparticles, let alone actual fuel cell electrodes. This prompted many investigations on supported catalysts that allowed a better understanding of these complex systems.

One of the main problems related to electrochemical experiments on supported catalysts is to use the proper working electrodes. For stationary systems a diffusion electrode (GDE) of the type used in actual fuel cells can be employed in a proper electrochemical cell in an arrangement usually called a half-cell configuration. This is a usual three-electrode electrochemical cell with the active face of the working electrode facing the solution and the back in contact with the reagent (gaseous or in solution). The diffusion electrode is not convenient for use in a rotating-disk electrode (RDE) configuration, necessary to study the kinetics and separate mass transport effects in some reactions, typically the ORR. Early work with this configuration was carried out with the so-called thin porous coating electrode, in which the rotating shaft is provided with a recess, typically a few tenths of a millimeter, filled with the supported catalyst agglutinated with a dilute polytetrafluoroethylene (PTFE) dispersion. This system presented mass transport effects within the relatively thick porous layer, which demanded proper simulation and modeling in order to interpret correctly the results [49, 51]. More recently, Schmidt et al. [50, 51] introduced the so-called ultrathin-layer electrode in which a very thin layer of the supported catalyst is applied onto a suitable conducting substrate, like glassy carbon, and is held in place with a thin layer of Nafion® cast onto the catalyst powder. Provided the active layer has a thickness of about 1 μm or less no mass transport effects need to be considered. This type of electrode has become the standard to make electrochemical experiments with supported catalysts.

13.3.1 Potentiodynamic Techniques

Typical potentiodynamic techniques used to examine the oxidation of alcohols on dispersed catalysts are CV and linear sweep voltammetry (LSV).

The first aspect that characterizes these techniques in electrocatalysis is the necessity of using suitable electrodes. Bulk platinum or bulk alloys are classical surfaces in electrocatalysis and do not need further consideration. To work with supported catalysts thin-layer electrodes of the type described above are necessary.

The typical CV profile for the oxidation of alcohols on Pt-based dispersed catalysts is well known and, as a result of much research, it can be interpreted reasonably well [51, 53]. The profile shows an irreversible reaction, so great care should be exercised in extracting the limited information provided. Several problems in the use of CV can be detected in published material. Most of these problems can be traced to the use of CV to compare catalysts with different composition, different particle size, and so on. Often a comparison is made of the so-called *onset potential* for the reaction under study (a modern version of the discharge potential). This is an arbitrary, ill-defined magnitude which, nevertheless, can be used for qualitative purposes when results within the same work are compared. In this case, an arbitrary but sensible definition can be adopted, like the potential at which the current is, say, one-tenth of the peak current, to avoid leaving the reader of an article wondering how the onset potential was determined. The comparison of peak current densities is another common practice, sometimes extrapolated to the comparison of actual numerical values. This can only be considered acceptable to some extent if the current densities are normalized by the specific area (i.e., the electrochemical active area) of the catalysts under comparison.

An early way of correcting currents, particularly in LSV experiments, was to correct the oxidation currents in the presence of the alcohol by the current observed in the absence of the alcohol [51, 54]. This is indeed an approximate procedure. Presently, there are two forms of expressing the activity of a catalyst: mass activity and specific activity. A superior mass activity is relevant for economic/technological purposes because it implies that less expensive precious metal catalyst is necessary to have the same electrochemical activity. But this may be more related to catalyst utilization than to real catalytic activity. This last concept can only be used if specific activities are considered.

In conclusion, CV and LSV must be used with great care for qualitative purposes only and never as a single technique to report catalytic activity, particularly when comparisons are made.

Very early it was recognized that platinum was the best catalyst for the oxidation of low-molecular-weight alcohols. But, at the same time, it was realized that platinum was easily poisoned by the intermediates formed in the oxidation. This is a problem that heavily hampers the development of DAFCs and has prompted intensive research to overcome it. The essence of the problem is that in order to oxidize the intermediates it is necessary to have participation of oxygenated species formed at the surface. On platinum, these oxygenated species are formed at potentials that are too high and exceed those considered acceptable for fuel cell anodes. Because of this, one subject that has engaged much active research is platinum alloys. Commonly, these are formed by platinum and a transition metal that forms oxygenated species at lower potentials than platinum and promotes the so-called bifunctional mechanism [53, 55]. The first and most studied of these metals was ruthenium [3].

The oxidation of methanol on Pt–Ru catalysts was studied by CV and LSV by Valbuena et al. [51, 54]. In the absence of methanol, CV shows much larger charging currents for Pt–Ru than for Pt, which was attributed to the formation of oxygenated species at low potentials. On the other hand, in the presence of methanol, the charge corresponding to the anodic part of the voltammogram diminishes by more than 70% showing the presence of adsorbed methanol and species formed by dissociation [54, 56]. LSV proved to be an adequate technique to evaluate the effect of the composition of the catalyst and the concentration of methanol on the activity [51, 54]. The conclusion was that Pt75Ru25 was the best catalyst and the optimum concentration of methanol was 2 M.

On the other hand, in the case of small-size DMFCs, where methanol transport is done at room temperature by diffusion only, the best concentration of methanol was 5 M [55, 57]. This shows once again that results obtained by electrochemical techniques cannot be used directly to predict the behavior in DMFCs.

In the case of the oxidation of ethanol, recent work compared different catalysts by CV and LSV [56, 58] and showed the importance of normalizing the currents by the specific area determined through the oxidation of a CO monolayer. The results show larger currents at low potentials for Pt–Ru as a consequence of the facility with which Ru provides oxygenated species.

13.3.2 Specific Area Determinations

The comparison of activities when properties and/or parameters of the catalyst or the solution are changed requires the normalization of the current densities, because the use of the geometric area of the electrode is, in general, not acceptable. The two most used methods of normalization lead to the so-called mass activity (MA) and specific activity (SA). The mass activity is reported as the activity per unit mass of platinum. Sometimes, in the case of platinum alloys, it is reported per unit mass of metal. The mass activity is a practical parameter in the sense that a larger mass activity means that the particular catalyst/electrode requires less platinum. This may be related to a better platinum utilization or to an enhancement of the activity by other metals in the alloy but does not reveal actual catalytic effects. Using this concept, the oxidation of methanol was compared on Pt catalysts modified with Ru and Mo [7, 59]. Mass activities were larger for Pt–Ru–Mo than for the binary catalysts, probably due to better catalyst utilization.

On the other hand, the specific activity is directly related to the actual catalytic activity because it is given by the number of active surface sites in the catalyst. In the case of pure platinum, the SA is easily determined by CV, through the integration of the area corresponding to the hydrogen adsorption (or desorption) region. In the case of Pt alloys the situation is more complex, because not all metals used in these alloys adsorb hydrogen. In these cases it is more proper to use the area determined through the charge necessary for the oxidation of a monolayer of adsorbed CO. Usually this is obtained by LSV on a surface in which CO was previously adsorbed. A suitable use of the SA is exemplified by results on the oxidation of ethanol on Pt modified by Ru and Rh [56, 58].

13.3.3   Quasi-Steady State Techniques: Chronoamperometry

Potentiodynamic techniques are useful for the study of alcohol oxidation but quite often do not reveal the real activity of a catalyst. Therefore it is very risky to assume that a dynamic result will be valid for the steady-state situation. In some works, this problem is minimized with a pragmatic solution, carrying out the dynamic experiments at a very slow sweep rate ($\sim$1 mV s^{-1}) which, nevertheless, is an approximation. A quasi-steady-state technique that has proven to be very useful is chronoamperometry. But it should be pointed out here that sometimes, as in the oxidation of methanol on pure bulk Pt [57, 60], the buildup of intermediates is so slow that the current is not stabilized after many hours. Even if this is the case, chronoamperometry is very useful to compare activities, provided that proper normalization of the currents is carried out as described above.

A good example is the use of chronoamperometry to determine the activity of Pt–Ru catalysts for methanol oxidation as a function of the concentration of methanol [58, 61]. The measurements were carried out in a half cell and the results expressed as mass activities, but the currents were corrected by those observed in the absence of methanol. Independently of the Ru content and of thermal treatments applied to the catalysts, the optimum concentration of methanol was found to be 2 M.

For the oxidation of ethanol, a recent study used chronoamperometry to compare the activity of Pt skins prepared on Ru and Rh nuclei [59, 62]. The results show that the comparisons are more relevant than those done with results of CV and LSV.

13.3.4   Quasi-Steady State Techniques: Impedance Spectroscopy

Electrochemical impedance spectroscopy (EIS) is considered a steady-state technique. However, some adjustment has to be made for reactions like the oxidation of alcohols because the buildup of adsorbed impurities may be a very slow process. On pure bulk Pt in acid solutions, it is not uncommon to observe that after 40 h the steady-state current is still falling. Under these conditions, it is not possible theoretically to carry out impedance experiments that require a constant current. Thus, an alternative condition is to verify that the current decay is small and constant.

An aspect often neglected is to recognize that EIS is the only technique that allows probing the reaction, particularly from the mechanistic point of view. There is no doubt that Pt–Ru is more active than Pt for the oxidation of methanol in acid solutions at potentials that are of interest for the DMFC. However, at potentials more positive than 0.5 V versus the reference hydrogen electrode (RHE) there is an inversion and Pt becomes more active than Pt–Ru. This was elucidated by EIS and rationalized by comparing the rate-determining step on both catalysts, which is influenced by the adsorption energy of OH, which is higher on Pt–Ru than on Pt [54, 56].

An impedance model based on three state variables has been developed for the kinetics and mechanistic investigation of H_2/CO electrooxidation on Pt/C, PtRu/C, and PtSn/C catalysts [60]. The simulation results have shown that reversing the impedance pattern of the II and III quadrants is due to the change in the rate-determining step from CO oxidation to CO adsorption. The agreement that has been found between experiments and the simulation has given rise to the opinion that the adsorbed OH species is found to be responsible for the CO oxidation. The same model has also been used to differentiate the reaction mechanisms for the PtRu/C and PtSn/C systems. It has been concluded that the promoted OH generation is the primary reason for enhanced activity toward CO oxidation on PtRu/C. The high activity of the PtSn/C system toward H_2/CO oxidation is due to the combination of the promoted OH generation, exclusion of CO on Sn sites, and minimization of CO adsorption caused by the intermetallic bonding.

13.3.5 Steady-State Techniques

Real steady-state techniques involve building a potential–current plot using true steady-state current values. These can be obtained by performing chronoamperometry experiments at different potentials and recording the steady-state values of the current. Another way is to perform a true steady-state experiment. With supported catalysts it is usual to construct a fuel cell diffusion electrode and use it in a half-cell configuration (see above). This allows us to record point by point a true steady-state polarization curve.

Obviously, experiments carried out in a single fuel cell are true steady-state experiments and they will be described in Section 13.5.

A good example of the application of steady-state techniques is to determine the dependence of the current density, at a given potential, as a function of the concentration of alcohol. This is particularly important for the fuel cell operation because the larger the concentration of alcohol the more important will be the problem of crossover (see below). Results on Pt–Ru catalysts show that the mass activity is larger for a 10 at % Ru than for 20 at % Ru for concentrations up to 2 M methanol [55, 57]. This is for half-cell experiments, where there are no limitations by methanol diffusion.

Perhaps a more typical example is the building of steady-state current–potential plots. These plots helped to detrmine the true mass activities for the oxidation of methanol of some Pt catalysts containing Ru and Mo [7, 59].

13.3.6 Determination of Products of Reaction

When developing catalysts for the oxidation of alcohols, the first goal is to demonstrate high current densities for the process. However, this is far from being the whole story. The oxidation of an alcohol may not yield carbon dioxide and water as the only products of complete oxidation, but it can be partial, meaning that the corresponding aldehyde and acid can be found as products of the reaction. This is particularly true in the case of ethanol. To date, the situation is such that in acid media the main products formed in the oxidation of ethanol on platinum and platinum alloys are acetaldehyde and acetic acid. Therefore, it is essential to have suitable methods for the determination of product distribution.

Three methods have been generally used for the determination of products and intermediates: online Fourier transform infrared (FTIR), spectrometry online mass spectrometry (differential electrochemical mass spectrometry, DEMS), and chromatography. Obviously these are not electrochemical techniques, but FTIR and DEMS are usually performed simultaneously with CV, so the intermediate/product formation can be followed as a function of the potential. On the other hand, chromatography is more an ex situ technique. Samples are periodically taken from the cell and analyzed. These techniques have contributed much to our present understanding of the oxidation of alcohols on noble metal catalysts and alloys.

Methanol oxidation on a supported Pt fuel cell catalyst was investigated by online DEMS at continuous electrolyte flow and defined catalyst utilization, employing a thin-film electrode setup and a thin-layer flow-through cell [61]. The active surface of the Pt/Vulcan (E-TEK) high-surface-area catalyst was characterized quantitatively by hydrogen underpotential deposition (H_{upd}) and preadsorbed CO monolayer stripping. Methanol stripping DEMS experiments, oxidizing the adsorbed dehydrogenation products formed upon methanol adsorption at potentials in the hydrogen adsorption region, show that the coverage of these products and hence the methanol uptake depend on the electrode potential, in contrast to the potential-independent CO_{ad} coverage. The dehydrogenation products cannot be displaced by H_{upd}. The number of close to two electrons used per oxidation of one adsorbed dehydrogenation product identifies this as CO_{ad} species. Further methanol dehydrogenation is hindered when the CO adlayer reaches a density of one-third of a monolayer. Side reactions during bulk methanol oxidation were identified directly by DEMS, showing methylformate formation in addition to the main product, CO_2. The extent of formaldehyde and formic acid formation was estimated from mass spectrometry and faradaic currents to be between 25 and 50% per dehydrogenation step. The exclusive formation of fully deuterated methylformate upon oxidation of deuterated methanol underlines the irreversibility of methanol dehydrogenation and rules out H/D exchange. A rather low kinetic H/D isotope effect implies that the removal of poisoning CO_{ad} intermediates rather than C–H bond dissociation determines the methanol oxidation rate, although there is a contribution from the latter step. Reduction of an anodically preformed PtO monolayer by methanol under open-circuit conditions indicates that Pt oxy species are equally active for methanol oxidation.

The influence of catalyst loading ($7-35 \mu g \, Pt \, cm^{-2}$) on methanol oxidation product yields over Vulcan XC72 supported Pt catalyst by DEMS, both under potentiodynamic and potentiostatic (0.6V vs. RHE) conditions, was investigated by Jusys et al. [62]. The electrochemical efficiencies, product distribution, and turnover frequencies of partial reactions (methanol oxidation to formaldehyde, formic acid, and CO_2) during the methanol oxidation reaction (MOR) showed a pronounced dependence on the catalyst loading. With increasing Pt loading the current efficiency for methanol oxidation to formaldehyde decayed significantly (from 40 to almost 0%), while that for complete oxidation to CO_2 increased from 50 to 80%. The variation in current efficiency for methanol oxidation to formic acid was small ($\sim 10\%$). The product distribution varied accordingly; the absolute numbers, however, were different: low (5–20%) formic acid yields were accompanied by a decaying formaldehyde yield (60% to zero) and increasing CO_2 yield (30–80%) with increasing catalyst loading.

Determination of reaction products is particularly important in the oxidation of ethanol because on Pt- and Pt-based catalysts partial oxidation is observed and the main products of the reaction are acetaldehyde and acetic acid and the minor product is CO_2. Recently, in situ FTIR was used to follow the CO_2/acetic acid ratio formed on Pt, Pt–Ru, and Pt–Rh [59]. The results show that CO_2 and acetic acid are formed at lower potentials on Pr–Ru but the ratio decreases rapidly for increasing potentials. The opposite is observed on Pt, while the best ratio was observed on Pt–Rh.

Infrared techniques were also used to examine the products formed for the oxidation of ethanol on Pt–Sn [63]. Although the main products formed were acetaldehyde and acetic acid, the formation of some CO_2 was also followed as a function of potential, indicating that Pt–Sn has some activity for breaking the C–C bond and oxygenated species formed on tin participate in the oxidation of intermediate CO.

13.4 PERFORMANCE FOR METHANOL OXIDATION REACTION AND ETHANOL OXIDATION REACTION

13.4.1 MOR on Pt Binary and Ternary Catalysts: Performance and Product Distribution

The use of methanol as energy carrier and its direct electrochemical oxidation in DMFCs represent an important challenge for the polymer electrolyte fuel cell technology, since the complete system would be simpler without a reformer and reactant treatment steps. The use of methanol as fuel has several advantages in comparison to hydrogen: It is an inexpensive liquid fuel, easily handled, transported, and stored, and with a high theoretical energy density. Although a lot of progress has been made in the development of the DMFC, its performance is still limited by the poor kinetics of the anodic reaction [64]. Methanol oxidation is a slow reaction that requires active multiple sites for the adsorption of methanol and the sites that can donate OH species for the desorption of methanol residues. Methanol oxidation has been extensively investigated since the early 1970s with two man objectives: identification of the reaction intermediates, poisoning species and products, and modification of the Pt surface in order to achieve higher activity at lower potentials and better tolerance to poisoning. The results have been reviewed by several authors [65–67]. The main reaction product is CO_2, although significant amounts of formaldehyde, formic acid, and methyl formate were also detected. Most studies conclude that the reaction can proceed according to multiple mechanisms. However, it is widely accepted that the most significant reactions are the adsorption of methanol and the oxidation of CO according to this simplified reaction mechanism:

$$CH_3OH \rightarrow Pt - (CH_3OH)_{ads} \tag{13.3}$$

$$(CH_3OH)_{ads} \rightarrow (CO)_{ads} + 4H^+ + 4e^- \tag{13.4}$$

$$(CO)_{ads} + H_2O \rightarrow CO_2 + 2H^+ + 2e^- \tag{13.5}$$

Platinum is the most active metal for dissociative adsorption of methanol, but, as is well known, at room or moderate temperatures it is readily poisoned by carbon monoxide, a byproduct of methanol oxidation. To date, the remedy has been using binary or ternary eletrocatalysts based on platinum. The superior performance of these binary and ternary electrocatalysts for the methanol oxidation with respect to Pt alone was attributed to the bifunctional effect (promoted mechanism) [68, 69] and to the electronic interaction between Pt and alloyed metals (intrinsic mechanism) [67, 68]. According to the promoted mechanism, the oxidation of the strongly adsorbed oxygen-containing species is facilitated in the presence of M oxides, which supply oxygen atoms at an adjacent site at lower potentials than those accomplished with pure Pt. The intrinsic mechanism postulates that the presence of M modifies the electronic structure of Pt and, as a consequence, affects the adsorption of oxygen-containing species and even the dissociative adsorption of methanol. From the 1960s to the 1990s extensive studies on Pt-based alloy materials for methanol electrooxidation in acid electrolytes at ambient and elevated temperatures have been carried out [4, 70, 71]. These studies have shown that, among the binary Pt-based alloys, Pt–Ru alloys are the best candidate catalysts for methanol electrooxidation.

13.4.1.1 *Binary Pt–Ru/C*

Conversely to bulk Pt–Ru alloys, it has to be remarked that in carbon-supported catalysts the amount of Ru alloyed with Pt is generally lower than the nominal Ru content in the material [72] and depends on the preparation method of the supported catalyst. The composition and structure of supported PtRu nanoparticles affect the catalytic activity. It has been reported that for supported and unsupported nanoparticle Pt–Ru catalysts that are 40–60 at % Ru give the optimum catalytic activity for methanol oxidation [73–75]. The formation of a PtRu alloy, however, is not an essential requirement for a high-MOR activity. It remains an open question whether the bifunctional effect or the electronic effect is predominant. In recent years, Long et al. [76] strongly advocated avoiding the use of Pt–Ru alloys; they found that the catalytic activity would be larger by orders of magnitude if Ru existed as hydrous oxides in comparison with the alloyed form. The benefits of RuO_xH_y were attributed to the electron and proton conductivities and to the presence of surface OH groups. Ren et al. [77] also showed that the higher the RuO_xH_y content in nano-Pt–Ru catalysts, the better the DMFC performance. In addition, the presence of Nafion was less relevant in the anode because of the protonic conductivity of RuO_xH_y. However, they also pointed out that the presence of RuO_xH_y was not a prerequisite: indeed, a high DMFC performance was also achieved in their work using a completely alloyed Pt–Ru catalyst.

13.4.1.2 Ternary Pt–Ru-Based Catalysts

Regarding the anode catalysts, the question is whether the catalytic activity of a Pt–Ru alloy catalyst, which is the most active system for methanol oxidation, can be further improved by using ternary catalysts which include elements such as W, Mo, Co, Ni, Os, and Sn. These metals present a cocatalytic activity for the anodic oxidation of methanol if used either as platinum alloys or as adsorbate layers on platinum. Evaluations of the activity for methanol oxidation in half-cell and/or in DMFCs indicated that many ternary Pt–Ru–M catalysts (M = W in WO_x or W_2C form, Mo, Ir, Ni, Co, Os, V) perform better than commercial Pt–Ru and/or Pt–Ru prepared by the same method as the ternary catalysts. Experimental and theoretical combinatorial and high-throughput screening methods pointed to Co, Ni, Fe, Ir, Rh, and W as suitable elements to form ternary Pt–Ru-based catalysts with improved activity for CO and methanol oxidation. The improvement in the activity of Pt–Ru for methanol oxidation by the addition of a third metal has been explained in different ways. Liang et al. [16] assumed that in the Pt–Ru–Ir system a RuO_2–IrO_2 interaction promotes the formation of hydroxyl species by dissociating water at a lower potential with respect to the Pt–Ru system. Moreover, they inferred that this interaction could also weaken the bonding between the hydroxyl species and the catalyst surface as compared with the bonding on Pt–Ru nanoparticles. The more weakly adsorbed hydroxyl species further promote the electrooxidation of CO_{ads} on the active metal sites at lower potentials, thus improving the catalyst performance. Kim et al. [78] ascribed the enhancement of the activity for the MOR of Pt–Ru–Sn to the synergic effects of Ru as a water activator and Sn as an electronic modifier of Pt. On this basis, in ternary Pt–Ru–M catalysts with almost completely alloyed Pt–Ru, Ru acts as an electronic modifier of Pt, so M has to be a water activator and, as a consequence, the third metal has to be in the nonalloyed form. Conversely, if Ru is poorly alloyed, Ru acts as a water activator, so M has to be an electronic modifier of Pt and then has to form alloys with Pt. An effect of the amount of the third metal in these catalysts was observed. For some compositions [79, 80], and at fixed Pt–Ru atomic ratio, the MOR activity increased for increasing contents of the third element in the catalyst. On the other hand, above a certain amount of the third metal, its presence has a detrimental effect on the performance of the catalyst.

13.4.1.3 *Other Binary Pt-Based Catalysts*

When Pt–Ru or Pt–Ru-M is used as anode electrocatalyst the power density of a DMFC is about a factor of 10 lower than that of a proton exchange membrane fuel cell operated on hydrogen if the same Pt loading is used. Therefore, a number of Ru-alternative elements, showing a cocatalytic activity for the anodic oxidation of methanol, if used either as platinum alloys or as adsorbate layers on platinum, have been investigated. Among them the most interesting are Pt–Sn and Pt–W.

Pt–Sn/C Catalysts   Pt–Sn/carbon nanocomposites have been studied for decades as anode catalysts for the electrooxidation of methanol and other small carbohydrate fuels [81]. These nanocomposites have been prepared by a variety of electrochemical or chemical deposition methods, and discrepancies in catalyst performance have been reported. The activity for methanol oxidation of Pt–Sn catalysts varies significantly with the preparation method, which leads to different degrees of alloying, surface composition, and particle size. Wang et al. [82], using well-defined alloy surfaces, observed a significant enhancing effect of tin for CO_{ads} oxidation but no effect for methanol oxidation. They also presented a model in which CO_{ads} resulting from the dehydrogenation of methanol is in a different state than that directly adsorbed from gaseous CO. Generally, Pt–Sn alloy catalysts present a higher activity for CO oxidation but a lower activity for the MOR than Pt–Ru catalysts, probably due to a decreased methanol adsorption on Pt–Sn. Thus, Pt–Sn electrocatalysts can perform better than pure Pt under two conditions directed to minimize the negative effect of alloying on methanol adsorption/dehydrogenation: (i) with a moderate degree of alloying and (ii) with the cell operating at low current density.

Pt–W/C   The addition of WO_x to Pt has attracted some attention, and an improved methanol oxidation activity of WO_x containing Pt-based catalysts has been reported [83]. The question is whether the improved MOR activity on the WO_x-containing catalysts is due to true catalytic properties and/or whether the addition of WOx modifies physical properties such as particle, size and surface area. Indeed, the presence of W in the Pt/C catalyst gives rise to coalescence during the formation of the catalyst particle, which results in a particle size larger than that of pure Pt/C. Then, the presence of W in a Pt/C catalyst affects the morphological characterictics of the material. Moreover, dissolution of WO_x in the acid environment is also reported, causing an increase in the Pt surface area. The positive effect of W presence in ternary Pt–Ru-based catalysts was ascribed also to its intrinsic cocatalytic effect, which enhances CO oxidation. Strasser et al. [84], instead, adduced that the enhanced electrocatalytic activity in the presence of W is due to a corrosion effect, which increases the surface area of the catalyst. But, according to Yang et al. [85], the addition of tungsten to Pt–Ru results mainly in physical modifications of the catalytically active Pt and Ru surface components such as differences in electroactive surface area rather than a promotion of the CH_3OH oxidation reaction via a true catalytic mechanism.

13.4.2 EOR on Pt Binary and Ternary Catalysts: Problem of C–C Bond Breaking—Performance and Product Distribution

Several studies on the electrooxidation of ethanol have been devoted mainly to identifying the adsorbed intermediates on the electrode and elucidating the reaction mechanism by means of various techniques, as differential electrochemical mass spectrometry (DEMS), insitu Fourier transform infrared spectroscopy (FTIRS), and electrochemical thermal desorption mass spectroscopy (ECTDMS) [86–89]. Based on the foregoing work, the global oxidation mechanism of ethanol in acid solution may be summarized in the following scheme of parallel reactions:

$$CH_3CH_2OH \rightarrow [CH_3CH_2OH]_{ad} \rightarrow Cl_{ad}, C2_{ad} \rightarrow CO_2 \quad \text{(total oxidation)} \quad (13.6)$$

$$CH_3CH_2OH \rightarrow [CH_3CH_2OH]_{ad} \rightarrow CH_3CHO \rightarrow CH_3CHOO \quad \text{(partial oxidation)}$$
$$(13.7)$$

The formation of CO_2 goes through two adsorbed intermediates, Cl_{ad} and $C2_{ad}$, which represent fragments with one and two carbon atoms, respectively. In spite of many advances in the understanding of the mechanism of ethanol oxidation, there are still some unclear aspects. For instance, there is some controversy on whether acetic acid is formed in one step or through the aldehyde. Also, there is no agreement regarding the nature of the adsorbed species. According to some workers, the carbon–carbon bond is preserved, so a larger quantity of intermediates of type C2 are formed [86, 87], but others claim that the main intermediates contain only one carbon atom and are of type C1 [88, 89]. Breaking the C–C bond for a total oxidation to CO_2 is a major problem in ethanol electrocatalysis. Thus, high yields of partial oxidation products, CH_3CHO and CH_3COOH, are formed on Pt catalysts. These parallel reactions cause a considerable lowering of the fuel capacity to generate electricity and produce undesirable substances. This aspect also raises a question on the concept of activity of a catalyst because higher current densities are not necessarily associated with complete oxidation.

Carbon-supported platinum is commonly used as anode catalyst in low-temperature fuel cells. Pure Pt, however, is not the most efficient anodic catalyst for the direct ethanol fuel cell. Indeed, as discussed above, the electrooxidation of a partially oxygenated organic molecule, such as a primary alcohol, can only be performed with a multifunctional electrocatalyst. Platinum itself is known to be rapidly poisoned on its surface by strongly adsorbed species coming from the dissociative adsorption of ethanol. Efforts to mitigate the poisoning of Pt have concentrated on the addition of cocatalysts, particularly ruthenium and tin, to platinum. The more extensively investigated anode materials for DEFCs are the binary Pt–Ru and Pt–Sn and the correlated ternary Pt–Ru-based and Pt–Sn-based catalysts. As in the case of methanol oxidation, the superior performance of these binary and ternary electrocatalysts for the oxidation of ethanol with respect to Pt alone was attributed to the bifunctional effect (promoted mechanism) and to the electronic interaction between Pt and alloyed metals (intrinsic mechanism). Conversely to the case of methanol oxidation, the best binary catalyst for ethanol oxidation in an acid environment is not Pt–Ru but Pt–Sn. The optimum Sn content in the catalyst is not well established and depends on the ratio of alloyed and nonalloyed tin and the cell temperature. Controversial results regarding the effect of the degree of alloying of Sn in the Pt fcc structure on the EOR activity have been reported, depending on both the intrinsic characteristics of the material (surface composition, particle size and particle size distribution of the alloy and the oxide, alloy/oxide interactions) and the external conditions (current density, temperature, type of experiment). The addition of Sn to Pt catalysts, however, notwithstanding the enhancement of the activity for ethanol oxidation, inhibits the C–C bond cleavage reaction. The addition of Rh to Pt seems to promote the C–C bond cleavage [38, 90], but the overall EOR activity of these catalysts is lower than that of Pt–Sn.

Ternary Pt–Ru-based catalysts tested for the ethanol oxidation reaction always performed better than Pt–Ru, but contradictory results regarding their EOR activity with respect to that of Pt–Sn have been reported. Conversely, ternary Pt–Sn-based catalysts seem to perform better than Pt–Sn [29, 39, 91], so, in the light of these results, upcoming research on materials for ethanol oxidation should be focused on Pt–Sn-based instead of Pt–Ru-based ternary catalysts. Ternary Pt–Sn–Ru catalysts with nominal Ru–Sn atomic ratio <1 seem to be the most promising anode materials for direct ethanol fuel cells. DEFCs with these catalysts as anode materials perform better than those with binary Pt–Ru or Pt–Sn catalysts. A suitable nonalloyed Ru–nonalloyed Sn ratio seems to be ~0.4. And in these materials the RuO_2–SnO_2 interactions promote the formation of hydroxyl species by dissociating water at a lower potential with respect to the Pt–Ru systems. Moreover, this interaction could also weaken the bonding between the hydroxyl species and the catalyst surface as compared with the bonding on Pt–Ru nanoparticles. The more weakly adsorbed hydroxyl species further promote electrooxidation of adsorbed CO and/or acetaldehyde species on the active metal sites at a lower potential, thus improving the performance. On the other hand, for higher Ru contents, as in the case of Pt–Sn–Ru (1 : 1 : 1) [the RuO_2–SnO_2 ratio in Pt–Sn–Ru (1 : 1 : 1) is twice that in Pt–Sn–Ru (1 : 1 : 0.3)], RuO_2 substitutes SnO_2 in the interaction with Pt. Moreover, the presence of RuO_2 on the particle surface of the Pt–Sn–Ru catalyst decreases the active surface area of Pt and, as a consequence, part of the noble metal becomes inactive due to the blocking of the Pt surface by Ru oxide. Obviously, the active surface area loss by the presence of Ru oxides increases with increasing Ru content in the catalyst: These effects could explain the inferior performance of the Pt–Sn–Ru/C catalyst in the atomic ratio 1 : 1 : 1 than that of Pt–Sn/C (1 : 1) [29, 39]. It has to be pointed out, however, that the addition of Ru to Pt–Sn enhances the electrical performance of the DEFC, that is, the EOR activity of the catalyst, but does not modify the product distribution, the ternary Pt–Sn–Ru catalyst being unable to activate the C–C bond cleavage [91].

Fundamentally, the C–C bond scission remains the main problem in the ethanol oxidation reaction. Presently, to attain the performance of DMFCs, the DEFC catalyst development targets require a two- to threefold enhancement of the activity for ethanol oxidation, and this has not been achieved, but further development of Pt–Sn–M catalysts appears to be a suitable way of achieving those goals. As the C–C bond cleavage is the primary condition to achieve good results, particular efforts should be addressed to the formulation of ternary Pt–Sn–M catalysts, the third metal being able to promote the C–C bond scission. The same experimental and theoretical combinatorial and/or high-throughput screening methods applied to the development of new anode alloy catalysts with improved CO tolerance or enhanced activity for methanol oxidation could be used to achieve suitable ternary catalysts for the direct ethanol fuel cells. Apart from the control of particle size and particle size distribution, of fundamental importance is the optimization of the nominal composition, the surface composition, the degree of alloying, and the oxide content of these promising materials. Considering also that the chemical and physical characteristics of these catalysts depend on the synthesis method, the way of preparation becomes a key factor regarding their electrochemical activity.

13.5 SINGLE-CELL TESTS

13.5.1 Single Cells and Associated Equipment

The use of single fuel cells is becoming increasingly popular in many laboratories involved in fuel cell research and development. The first reason for this is that the ultimate test for the activity of a catalyst is its incorporation in a diffusion electrode and subsequent test in a single fuel cell. The other reason is that it is widely recognized that results obtained in an electrochemical research cell cannot be easily extrapolated to a fuel cell, mainly because there are too many differences between the two systems. The basic equipment for this activity involves the single cell and a suitable station to control the cell operation and for data acquisition. In the 1970s and 1980s the only way of having these equipments available was to build them in house. Today, several companies offer commercial single cells and control stations, which has contributed to popularize their use.

The development of the DAFC stems from the earlier work on proton exchange membrane fuel cells (PEMFCs) in which hydrogen is the fuel. Essentially, the only difference between the PEMFC and the DAFC is that in latter the anode is fed with an aqueous solution of the corresponding alcohol. Regarding the electrolyte, Nafion® membranes, which are commercially available in different thicknesses, are almost universally used, although they present the well-known problem of the crossover, discussed in Section 13.5.4.

When the equipment is available, the catalyst under study can be incorporated into a diffusion electrode, following well-known techniques, which in turn is used to build the membrane/electrodes assembly (MEA) that forms the heart of the single fuel cell.

13.5.2 Single-Cell Results for Oxidation of Methanol

The oxidation of methanol on Pt-based catalysts proceeds through the bifunctional mechanism that requires the presence of metals with the capability of supplying oxygenated species for the oxidation of intermediates. In the presence of these metals improvement in the performance and power density delivered by single cells is generally observed. So, not only traditional bimetallic catalysts such as Pt–Ru have been used but also materials that have been investigated for the reduction of oxygen rather than for the anodic reaction. In this category, Pt–Co and Pt–Ni have shown activity in DMFC provided that the catalyst contains an adequate amount of oxides [33, 92, 93].

Single-cell results are important to discuss the influence of several parameters on the performance of DMFCs. Due to the different conditions, the results are often in conflict with those observed in electrochemical cells. In particular, results on single DMFCs with the classical catalyst for methanol oxidation, Pt–Ru, showed that, within certain limits, the morphology of the catalyst is more important than the composition or the particle size [24, 51, 54, 58, 61, 94].

13.5.3   Single-Cell Results for Oxidation of Ethanol

The oxidation of ethanol is no doubt a difficult issue and requires the consideration of several aspects when single cells are considered. Globally, Pt–Ru and Pt–Sn as anode catalysts in DEFCs present a much larger activity than Pt [25]. And of the two catalysts Pt–Sn showed the larger power densities. Also, the effect of temperature in the interval 70–100°C is much more pronounced in the bimetallic catalysts, a phenomenon related to the adsorption characteristics of the ethanol molecule.

13.5.4   Problem of Crossover

Methanol, and to a lesser extent ethanol, presents a phenomenon known as crossover. The methanol that does not react at the anode of the fuel cell may cross the membrane to the cathode side, where it is oxidized at the high potential of the operational cathode. This generates a mixed potential at the cathode, lower than that in the absence of the alcohol, which reduces the efficiency of the fuel cell. This problem is actually more important than the obvious problem of fuel loss through the membrane. In principle, it may be difficult to see the connection between this problem and that of electrocatalysis, but this is not the case. Two of the more direct solutions to this problem involve the development of membranes with decreased crossover in comparison with Nafion®, and many efforts are directed to develop membranes that make the crossover difficult. Others are directed to develop membranes capable of working at higher temperatures. In this last case it has to be considered that an increase in temperature increases the crossover but also increases the amount of alcohol fuel reacted at the anode, and it seems that the trade-off is beneficial.

Another way of dealing with the problem of crossover, entirely related to electrocatalysis, is to develop cathode catalysts that are tolerant to the presence of alcohol. That is, while presenting a high activity for the oxygen reduction reaction their activity for the oxidation of alcohols would be very low. In this sense, when a tolerant catalyst is developed, the objective is to look for the lowest activity toward the oxidation of the alcohol.

Several results involving catalysts for the oxygen reduction reaction (ORR) with tolerance to methanol were reviewed recently [92]. Pt–Cr catalysts with Cr content as low as 9 at % proved to have low activity for methanol oxidation [30, 95]. Moreover, in the overall evaluation as cathode material it was concluded that the tolerance to methanol was more important than the intrinsic activity for the ORR.

Considering classic bimetallic materials for the ORR, a Pt70Ni30 catalyst showed a lower activity than Pt toward the ORR in methanol-free solutions but resulted in more activity in the presence of methanol, showing again the importance of methanol tolerance [96, 97]. In this catalyst the amount of alloyed Ni is too low, not enough to improve the activity of Pt for the ORR in the absence of methanol [97, 98]. But it seems to be enough to improve the tolerance of the catalyst in the presence of methanol to the point of making it a better catalyst for the cathode. These considerations apply also to Pt–Co catalysts [35].

For the case of ethanol studies of crossover through Nafion® are scarcer, but the behavior seems to be similar to the case of methanol, although the tendency to cross over is lower for ethanol.

REFERENCES

1. F. J. Derbyshire, V. H. J. de Beer, G. M. K. Abotsi, A. W. Scaroni, J. M. Solar, and D. J. Skrovanek, *Appl. Catal.*, 27 (1986) 117.
2. (a) P. Ehrburger, O. P. Majahan, and P. L. Jr. Walker, *J. Catal.*, 43 (1976) 61. (b) A. Guerrero-Ruiz, P. Badenes, and I. Rodriguez-Ramos, *Appl. Catal. A*, 173 (1998) 313.
3. M. Watanabe, M. Uchida, and S. Motoo, *J. Electroanal. Chem.*, 229 (1987) 395.
4. N. Giordano et al., *Electrochim. Acta*, 36 (1991) 1979.
5. M. A. Fraga, E. Jordao, M. J. Mendes, M. M. A. Freitas, J. L. Faria, and J. L. Figueredo, *J. Catal.*, 209 (2002) 355.
6. D. Duff, T. Mallat, M. Schneider, and A. Baiker, *Appl. Catal. A*, 133 (1995) 133.
7. A. L. N. Pinheiro et al., *J. New Mater. Electrochem. Syst.*, 6 (2003) 1.
8. X. Yu and S. Ye, *J. Power Sources*, 172 (2007) 133.
9. V. Jalan and E. J. Taylor, *J. Electrochem. Soc.*, 130 (1983) 2299.
10. B. C. Beard and P. N. Ross, *J. Electrochem. Soc.*, 133 (1986) 1839.
11. M. Min, J. Cho, K. Cho, and H. Kim, *Electrochim. Acta*, 45 (2000) 4211.
12. L. Xiong and A. Manthiram, *Electrochim. Acta*, 50 (2005) 2323.
13. W. J. Zhou et al., *Appl. Catal. B*, 46 (2003) 273.
14. W. Li et al., *Electrochim. Acta*, 49 (2004) 1045.
15. Y. Liang et al., *J. Catal.*, 238 (2006) 468.
16. Y. Liang, H. Zhang, Z. Tian, X. Zhu, X. Wang, and B. Yi, *J. Phys. Chem. B*, 110 (2006) 7828.
17. M. S. Nasher, A. I. Frenkel, D. Somerville, C. W. Hills, J. R. Shapley, and R. G. Nuzzo, *J. Am. Chem. Soc.*, 120 (1998) 8093.
18. C. W. Hills, M. S. Nasher, A. I. Frenkel, J. R. Shapley, and R. G. Nuzzo, *Langmuir*, 15 (1999) 690.
19. A. J. Dickinson, L. P. L. Carrette, J. A. Collins, K. A. Friedrich, and U. Stimming, *Electrochim. Acta*, 47 (2002) 3733.

20. A. Manzo-Robledo, A. C. Boucher, E. Pastor, and N. Alonso-Vante, *Fuel Cells*, 2 (2002) 109.
21. H. Yang, N. Alonso-Vante, J. M. Leger, and C. Lamy, *J. Phys. Chem. B*, 108 (2004) 1938.
22. H. Yang, W. Vogel, C. Lamy, and N. Alonso-Vante, *J. Phys Chem. B*, 108 (2004) 11024.
23. E. Antolini, J. R. C. Salgado, R. M. da Silva, and E. R. Gonzalez, *Materials Chemistry and Physics*, 101 (2007) 395.
24. W. H. Lizcano-Valbuena, D. Caldas de Azevedo and E. R. Gonzalez, *Electrochim. Acta*, 49 (2004) 1289.
25. F. Colmati, E Antolini, and E. R. Gonzalez, *Electrochim. Acta*, 50 (2005) 5496.
26. J. R. C. Salgado, E. Antolini, and E. R. Gonzalez, *J. Power Sources* 141 (2005) 13.
27. A. M. Castro Luna, G. A. Camara, V. A. Paganin, E. A. Ticianelli, and E. R. Gonzalez, *Electrochem. Commun.*, 2 (2000) 222.
28. J. R. C. Salgado, E. Antolini, and E. R. Gonzalez, *J. Electrochem. Soc.*, 151 (2004) A2143.
29. E. Antolini, J. R. C. Salgado, A. M. dos Santos, and E. R. Gonzalez, *Electrochem. Commun.*, 8 (2005) A226.
30. E. Antolini, J. R. C. Salgado, L. C. R. A. Santos, G, Garcia, E. A. Ticianelli, E. Pastor, and E. R. Gonzalez, *J. Appl. Electrochem.*, 36 (2006) 355.
31. F. Colmati, E. Antolini, and E. R. Gonzalez, *J. Electrochem. Soc.*, 154 (2007) B39.
32. G. A. Camara, M. J. Giz, V. A. Paganin, and E. A. Ticianelli, *J. Electroanal. Chem.*, 537 (2002) 21.
33. E. Antolini, J. R. C. Salgado, and E. R. Gonzalez, *J. Electroanal. Chem.*, 580 (2005) 145.
34. F. H. B. Lima, J. R. C. Salgado, E. R. Gonzalez, and E. A. Ticianelli, *J. Electrochem. Soc.*, 154 (2007) A369.
35. J. R. C. Salgado, E. Antolini, and E. R. Gonzalez, *Appl. Catal., Environ.*, 57 (2005) 283.
36. F. H. B. Lima, W. H. Lizcano-Valbuena, E. Teixeira-Neto, F. C. Nart, E. R. Gonzalez, and E. A. Ticianelli, *Electrochim. Acta*, 52 (2006) 385.
37. F. H. B. Lima, C. D. Sanches, and E. A. Ticianelli, *J. Electrochem. Soc.*, 152 (2005) A1466.
38. F. Colmati, E. Antolini, and E. R. Gonzalez, *J. Alloys Compounds*, 456 (2008) 254.
39. E. Antolini, F. Colmati, and E. R. Gonzalez, *Electrochem. Communi.*, 9 (2007) 398.
40. E. Antolini, J. R. C. Salgado, R. M. da Silva, and E. R. Gonzalez, *Mater. Chem. Phys.*, 101 (2007) 395.
41. A. R. West, *Solid State Chemistry and Its Applications*, Wiley, New York, 1984.
42. S. Mukerjee, *J. Appl. Electrochem.*, 20 (1990) 537.
43. F. Colmati, E. Antolini, and E. R. Gonzalez, *J. Solid State Electrochem.*, 12 (2008) 591.
44. F. H. B. Lima, J. F. R. de Castro, and E. A. Ticianelli, *J. Power Sources*, 161 (2006) 806.
45. F. H. B. Lima, M. J. Giz, and E. A. Ticianelli, *J. Braz. Chem. Soc.*, 16 (2005) 328.
46. R. Sousa, Jr., F. Colmati, E. G. Ciapina, and E. R. Gonzalez, *J. Solid State Electrochem.*, 11 (2007) 549.
47. L. G. S. Pereira, V. A. Paganin, and E. A. Ticianelli, *Electrochim. Acta*, submitted.
48. J. J. Rehr and R. C. Albers, *Phys. Rev. Sect. B*, 41 (1990) 139.
49. J. Perez, A. A. Tanaka, E. R. Gonzalez, and E. A. Ticianelli, *J. Electrochem. Soc.*, 141 (1994) 431.
50. T. J. Schmidt, H. A. Gasteiger, G. D. Stab, P. M. Urban, D. M. Kolb, and R. J. Behm, *J. Electrochem. Soc.*, 145 (1998) 2354.
51. K. Lasch, L. Joerissen, and J. Garche, *J. Power Sources*, 84 (1999) 225.
52. W. H. L. Valbuena, V. A. Paganin, and E. R. Gonzalez, *Electrochim. Acta*, 47, 3715 (2002).
53. S. Gilman, *J. Phys. Chem.*, 68 (1964) 70.
54. D. C. Azevedo, W. H. Lizcano-Valbuena, and E. R. Gonzalez, *J. New Mater. Electrochem. Syst.*, 7 (2004) 191.

55. F. Colmati, V. A. Paganin, and E. R. Gonzalez, *J. Appl. Electrochem.*, 36 (2006) 17.
56. F. H. B. Lima and E. R. Gonzalez, *Electrochim. Acta*, 53 (2007) 2963.
57. A. L. N. Pinheiro et al., *J. New Mat. Electrochem. Syst.*, 6 (2003) 1
58. W. H. L. Valbuena, A. Souza, V. A. Paganin, C. A. P. Leite, F. Galembeck, and E. R. Gonzalez, *Fuel Cells*, 2 (2003) 159.
59. F. H. B. Lima and E. R. Gonzalez, *Appl. Cat. B: Environ.*, 79 (2007) 341.
60. X. Wang and I.-M. Hsing, *J. Electroanal. Chem.*, 556 (2003) 117.
61. Z. Jusys and R. J. Behm, *J. Phys. Chem. B*, 105 (2001) 10874.
62. Z. Jusys, J. Kaiser, and R. J. Behm, *Langmuir*, 19 (2003) 6759.
63. F. C. Simões et al., *J. Power Sources*, 167 (2007) 1.
64. T. Iwasita and F. C. Nart, *J. Electroanal. Chem.*, 317 (1991) 291.
65. R. Parsons and T. VanderNoot, *J. Electroanal. Chem.*, 257 (1988) 9.
66. S. Wasmus and A. Kuver, *J. Electroanal. Chem.*, 461 (1999) 14.
67. T. Iwasita, *Electrochim. Acta*, 47 (2002) 3663.
68. N. M. Markovic, H. A. Gasteiger, P. N. Ross, X. Jiang, I. Villegas, and M. J. Weaver, *Electrochim. Acta*, 40 (1995) 91.
69. S. L. Goikovic, T. R. Vidakovic, and D. R. Durovic, *Electrochim. Acta*, 48 (2003) 3607.
70. K. Ota, Y. Nakagava, and M. Takahashi, *J. Electroanal. Chem.*, 179 (1984) 179.
71. P. A. Christensen, A. Hamnett, and G. L. Troughton, *J. Electroanal. Chem.*, 362 (1993) 207.
72. E. Antolini, *Mater. Chem. Phys.*, 78 (2003) 563.
73. J. B. Goodenough, A. Hamnett, B. J. Kennedy, R. Manoharan, and S. A. Weeks, *J. Electroanal. Chem.*, 240 (1988) 133.
74. D. Chu and S. Gilman, *J. Electroanal. Chem.*, 143 (1996) 1685.
75. Z. Jusys, J. Kaiser, and R. J. Behm, *Electrochim. Acta*, 47 (2002) 3693.
76. J. W. Long, R. M. Stroud, K. E. Swider-Lyons, and D. R. Rolison, *J. Phys. Chem. B*, 104 (2000) 9772.
77. X. M. Ren, M. S. Wilson, and S. Gottesfeld, *J. Electrochem. Soc.*, 143 (1996) L12.
78. T. Y. Kim, K. Kobayashi, M. Takahashi, and M. Nagai, *Chem. Lett.*, 34 (2005) 798.
79. T. Kessler and A. M. Castro Luna, *J. Solid State Electrochem.*, 7 (2003) 593.
80. J. Choi, K. Park, B. Kwon, and Y. Sung, *J. Electrochem. Soc.*, 150 (2003) A973.
81. M. J. Gonzalez, C. T. Hable, and M. S. Wrighton, *J. Phys. Chem. B*, 102 (1998) 9881.
82. K. Wang, H. A. Gasteiger, N. M. Markovic, and P. N. Ross, *Electrochim Acta*, 41 (1996) 2587.
83. J. Shim, C.-R. Lee, H.-K. Lee, J.-S. Lee, and E. J. Chairns, *J. Power Sources*, 102 (2001) 172.
84. P. Strasser, Q. Fan, M. Devenney, W. H. Weinberg, P. Liu, and J. K. Norskov, *J. Phys. Chem. B*, 107 (2003) 11013.
85. L. X. Yang, C. Bock, B. MacDougall, and J. Park, *J. Appl. Electrochem.*, 34 (2004) 427.
86. J. Willsau and J. Heitbaum, *J. Electroanal. Chem.*, 194 (1985) 27.
87. T. Iwasita and E. Pastor, *Electrochim. Acta*, 39 (1994) 531.
88. B. Bittins-Cattaneo, S. Wilhelm, E. Cattaneo, H. W. Buschmann, and W. Vielstich, *Ber. Bunsenges, Phys. Chem.*, 92 (1988) 1210.
89. J. F. E. Gootzen, W. Visscher, and J. A. R. Van Veen, *Langmuir*, 12 (1996) 5076.
90. J. P. I. de Souza, S. L. Queiroz, K. Bergamaski, E. R. Gonzalez, and F. C. Nart, *J. Phys. Chem. B*, 106 (2002) 9825.
91. S. Rousseau, C. Coutanceau, C. Lamy, and J.-M. Leger, *J. Power Sources*, 158 (2006) 18.
92. E. Antolini, J. R. C. Salgado, and E. R. Gonzalez, *Appl. Cat. B: Environ.*, 63 (2006) 137.

93. E. Antolini, J. R. C. Salgado, and E. R. Gonzalez, *Electrochem. Solid State Lett.*, 8 (2005) A226.

94. W. H. L. Valbuena, V. A. Paganin, C. A. P. Leite, F. Galembeck, and E. R. Gonzalez, *Electrochim. Acta*, 48 (2003) 3869.

95. E. Antolini et al., *J. Appl. Electrochem.*, 36, 355 (2006).

96. E. Antolini, J. R. C. Salgado, and E. R. Gonzalez, *J. Power Sources*, 155 (2006) 161.

97. E. Antolini, J. R. C. Salgado, M. J. Giz, and E. R. Gonzalez, *Int. J. Hydrogen Energy*, 30 (2005) 1213.

98. J. R. C. Salgado, E. Antolini, and E. R. Gonzalez, *J. Phys. Chem. B*, 108 (2004) 17767.

CHAPTER 14

Impact of Electrochemical Science on Energy Problems

ELIZABETH SANTOS[1,2] and WOLFGANG SCHMICKLER[2]

[1]Instituto de Física Enrique Gaviola (IFEG-CONICET), Facultad de Matemática, Astronomia y Física Universidad Nacional de Córdoba, Córdoba, Argentina
[2]Institute of Theoretical Chemistry, Ulm University, D-89069 Ulm, Germany

The way, by which the most important technological problems of all, the provision of cheap energy, can be solved, this way has to be found by electrochemistry. If we have a galvanic element, which generates electrical energy directly from coal and from the oxygen of the air, and with an efficiency close to the theoretical value, then we stand before a technological revolution, which will dwarf the one that was caused by the invention of the steam engine.

Just imagine how the incomparably comfortable and elastic distribution, which electrical energy permits, will change the appearance of our industrial cities! No more smoke, no soot, no boilers, no steam engines.

—*F. W. Ostwald, 1894*

Since the beginning of civilization, energy has been a central problem for mankind. The Greek origin of the word (*en+ergos*) means "at work" or "in action". It is connected with *power* (*pouair, povoir, podir, potere*: "to be able") and consequently with politics. There has been a continuous search for energy sources by mankind, from fire to nuclear energy. Unfortunately, nature has been seriously affected by the quest for this precious "untouchable essence".

In the nineteenth century coal transformed the economies of Europe, its cities, and its landscapes. The chemical energy released during the oxidation of coal was first converted to heat and then to mechanical energy by the steam engine. The literature of that century, from economic treaties like Karl Marx's *Kapital* to novels like Zola's

Catalysis in Electrochemistry: From Fundamentals to Strategies for Fuel Cell Development,
First Edition. Edited by Elizabeth Santos and Wolfgang Schmickler.
© 2011 John Wiley & Sons, Inc. Published 2011 by John Wiley & Sons, Inc.

Germinal, bears ample witness to the revolution which coal and steam caused in all aspects of human life. While the engineers developed more efficient heat engines, scientists discovered the laws that limit the conversion of heat to energy and pose insuperable barriers to their efficiency. Fuel cells had already been invented in the 1830s by Schönbein and Grove but were viewed as little more than a curiosity, until Ostwald pointed out that they are not limited by the second law of thermodynamics and have a theoretical efficiency of near 100% and painted the picture of an efficient economy based on the direct conversion of chemical into electric energy. Like the true visionary that he was, he foresaw the beneficial effects that this would have on the environment.

What has happened in the more than one hundred years that have passed since this seminal article by Ostwald? We have widened the energy basis of our economy, but it still overwhelmingly depends on fossil fuels. The industrial cities of Europe and North America have become cleaner, but we have polluted the whole atmosphere. The problem becomes worse as big countries like China, India, and Brazil develop rapidly and justly claim their share of the dwindling energy resources. Today, China has become the country with the highest absolute amount of carbon dioxide emission, yearly producing about 6000 millions tons. However, in terms of per-capita production the United States is the leader country. Statistically, a U.S. citizen produces five times more CO_2 than a Chinese citizen. Consequently, the search for alternative energies has become of paramount importance.

With wind, solar, hydro, geothermal, nuclear, tidal energy, with biodiesel and bio-gas, we have a number of alternatives to fossil fuels, so which are the best options? There are no objective answers, and every sector can argue in favor or against according to their interest. As stated above, energy is intrinsically tied to politics. The ecosystem is so complicated that, depending on the factors regarded, the methodologies employed, or the time periods considered, the various options can appear as very promising or totally absurd. As an example, we may consider biofuels. The production of the first generation of biofuels converting crops into ethanol or biodiesel has been claimed to supply clean and cheap energy. It is assumed that the emissions from the combustion of biofuels are canceled out by the carbon dioxide absorbed during the growth of the biomass. However, these calculations do not take into account that the total emissions from the production process are still higher than for the petroleum products they replace [1]. There are also other negative aspects such as the increase of food price due to the energy markets in competition with food markets, as pointed out in a recent policy research supported by the World Bank [2]. An article that has just appeared in *Science* [3] discusses another risk: An increased reliance on biofuel would strain water supplies and worsen water pollution. Effectively, water requirements for energy production from petroleum extraction require approximately 10–40, while the amount required for soybean biodiesel irrigation is about 13,900,000–27,900,000 $L Mwh^{-1}$. As the authors of this article mention: "it trades an oil problem for a water problem". Since converting the entire grain harvest of the United States would only produce 16% of its auto fuel needs, it seems that the improvements that the use of biofuels can achieve would not be dramatic. The situation is somewhat different in Brazil, where today the production of ethanol fuel from sugar cane provides 18% of the country's automotive fuel. Here we have analyzed only the economical implications and we have not considered the social problems. Especially in the third world, the conversion of large areas of virgin rainforest to the production of biofuels entails the violent displacement of people from their natural habitat.

Other renewable energy sources, in particular wind and sunlight, are increasingly being used, mainly in remote and isolated places. However, none of these resources shows sufficient efficiency to supply the whole energy required. They can be considered as sources complementary to others.

What has happened to fuel cells, which were meant to provide clean energy efficiently? They are used in niches, but on a global scale they play an entirely negligible role. It is not as if electrochemists had not listened to Ostwald's call. He sparked off a wave of research, and only two years later William Jacques claimed to have developed the first fuel cell based on coal. Of course he was mistaken, but since that time there has been a constant ebb and flow in fuel cell research. Periods of hectic and optimistic activity have alternated with virtual hibernation. The idea of using carbon has been given up as impractical; hydrogen is viewed instead as the energy vector of the future. This may be seen a change for the better, since the oxidation of hydrogen produces water and not CO_2. But there are no natural sources of hydrogen, so it must be produced from other energy sources. Also its transport is problematic, since it cannot be efficiently transmitted through pipelines. Generating hydrogen from reforming fuels like gasoline, diesel, or natural gas can produce larger amounts of CO_2 than the direct combustion of these fuels and therefore offers no solution. Thus, an important aspect is the development of new technologies to electrolyze water employing alternative energy sources.

The first fuel cell built by Grove had a voltage of 0.8 V and used platinum as electrode material; the PEM (polymer electrolyte membrane) fuel cells of today have the same voltage and use the same catalysts. So the principal problems of fuel cells have not changed: The efficiency is too low, and the electrode material too expensive. The crux of the matter is the oxygen electrode: At all known materials, oxygen evolution/reduction proceeds immeasurably slow near the theoretical equilibrium potential. Therefore, oxygen reduction requires an overpotential of a few hundred millivolts, which is lost in the overall energy balance. In contrast, hydrogen reaction on platinum, and on several other metals as well, proceeds so fast that it has been used to define the standard reference electrode.

However, in spite of the principal problem of oxygen reduction, much progress has been achieved. Even a superficial comparison of Grove's cell with a modern version (see Figure 14.1) shows that the design has little in common. The first cells could light a small bulb; modern cells provide enough energy and power to be used in cars or motorcycles. The scooter shown in Figure 14.2 (developed in the Environment Park in Torino, Italy) requires a small cylinder storing hydrogen and can then be used to cruise quietly and rapidly through the city, emitting no other fumes but water vapor. After 50–70 km it requires a fresh charge, but for local transport this is no problem. Obviously, cars require much more hydrogen and for long-distance travel would require a network of hydrogen-charging stations; nevertheless, several companies have built fuel cell cars in test or even small production series. A few countries

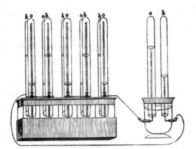

FIGURE 14.1 Schematic drawing of Grove's fuel cell (left) and picture of a modern, experimental PEM cell (right). The PEM cell was developed in the Environment Park, Torino, Italy. (Courtesy of Dr. Davide Damoso.)

have begun to experiment with transport systems based on hydrogen as fuel. Public hydrogen refueling stations have been opened in several European cities. In Latin America, Brazil has implemented a program called Brazilian Hydrogen Autobus, and the first prototype of a hydrogen fuel cell bus (see Figure 14.3) will begin to operate in Sao Paulo this year (2009).

The greatest progress in fuel cell technology has been in the design of the membrane that separates the anode and the cathode regions of the cell to prevent oxygen from reaching the anode and vice versa. While Grove's cell had the same voltage at zero load as the modern versions, this would rapidly break down when current was drawn, because the liquid junctions that join the two electrodes had a high resistance. Obviously, the separating membrane should be as thin and as conductive as possible. Let us take a brief look at how this is achieved in the two most promising types of fuel cells, the PEM cell that we already briefly mentioned and the solid-oxide fuel cell (SOFC). We shall not discuss methanol and ethanol cells, although much research is being spent on them and their theoretical energy density is high. However, they are inefficient at both electrodes. Effectively, methanol produces as intermediate CO which strongly interacts with the catalyst and poisons the available area

FIGURE 14.2 Scooter and car running on a PEM fuel cell.

FIGURE 14.3 Prototype of a hydrogen bus.

(see Chapter 10). In the case of ethanol, it is very difficult to break the C–C bond, and usually acetaldehyde and acetic acid appear as main end products (see Chapter 13).

The PEM cells use acid aqueous solutions, and reactions are the oxygen reduction and hydrogen oxidation in their familiar forms:

$$\tfrac{1}{2}O_2 + 2H^+ + 2e^- \rightarrow H_2O \qquad \text{cathode} \qquad (14.1)$$

$$H_2 \rightarrow 2H^+ + 2e^- \quad \text{anode} \qquad (14.2)$$

The membrane must conduct protons, which are generated at the anode and consumed at the cathode. The most common choice is nafion, which is based on polytetrafluoroethylene, but other perfluorocarbonsulfonic acid ionomers are also used. These membranes have a high conductivity, are chemically stable, and operate in an aqueous environment. The proton is the smallest and the most mobile of all ions, which is a big advantage. On the other hand, the presence of water limits the temperature range in which these cells can operate. Since the oxygen overpotential is lower at higher temperatures, this lowers the efficiency. However, operating near room temperature reduces risks in mobile applications.

In SOFCs, the mechanisms at the electrodes are different:

$$H_2 + O^{2-} \rightarrow H_2O + 2e^- \quad \text{anode} \qquad (14.3)$$

$$\tfrac{1}{2}O_2 + 2e^- \rightarrow O^{2-} \qquad \text{cathode} \qquad (14.4)$$

In this case, the O^{2-} migrates from the cathode to the anode, which are separated by a solid oxide such as yttrium-stabilized zirconia (YSZ). The conduction occurs via oxygen vacancies, which requires high temperatures. At present, their use is limited to stationary applications.

There are various other kinds of fuel cells discussed in countless articles and monographs. We limit ourselves to these prime examples to demonstrate in which

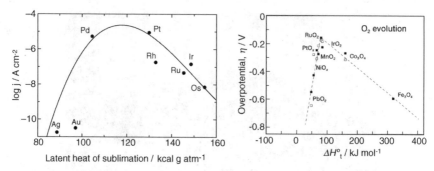

FIGURE 14.4 Examples of volcano plots for the oxygen reduction (left) [5] and evolution [6] (right). In the right figure, ΔH_f° is the standard enthalpy of transition from a lower to a higher oxide and η the overpotential at a current density of 1 A cm^{-2}.

ways effective ion transport between the two electrodes can be achieved. Although the present methods leave room for improvement—and much research and development effort are spent on this—this is definitely not the weak part of modern fuel cells. The problem is still the catalysis of the electrode reactions, and herein lies the challenge for electrochemists.

So what are the chances of finding better catalysts for the oxygen reaction? Although the overall reaction scheme is simple, the number of possible reaction steps for oxygen reduction is large. For example, in a recent density functional theory (DFT) study [4] reaction energies for more than 10 possible steps have been calculated, but not activation energies, because the latter are at present beyond the capabilities of DFT. This illustrates how complex the problem is.

Therefore, much of our knowledge is based on correlations which usually result in volcano curves (see Chapters 1 and 2). As an early example we show a plot of the exchange current density for oxygen reduction versus the latent heat of sublimation of the metal (see Figure 14.4). The choice of the abscissa is, of course, quite arbitrary. It only makes sense because the heat of sublimation correlates with more directly relevant parameters like the properties of the d bands. Note that the interpolation near the top of the curve is quite large and leaves sufficient room for better catalysts.

Near the theoretical equilibrium potential for oxygen reduction most metals are covered by an incipient oxide film, and the reverse direction, oxygen generation, which is important for electrolytic hydrogen production, usually takes place on oxides. For this reaction, a plot of the overpotential versus the standard enthalpy of transition from a lower to a higher oxide also produces a volcano curve. If one takes this plot as an indicator, there would be little hope to find a better oxide catalyst than RuO$_2$.

Like all volcano plots, those in Figure 14.4 are based on the idea that both a strong and a weak interaction of the reactant with the substrate is disadvantageous. In the former case it is difficult to desorb a species, in the latter it adsorbs too weakly; therefore, an intermediate interaction should be best. Obviously, such empirical relations are no more than a rough guide. With the advance of computational chemistry, key quantities like adsorption and formation energies can be calculated accurately and serve as a basis to estimate rate constants. The most comprehensive attempt in this direction has been made by Nørskov et al. [7]. This group calculated the O and OH binding energies on the most densely packed surface on a number of metals. They noted that the two adsorption energies are proportional and therefore took one of them, the O binding energy, as an indicator. They proposed a reaction mechanism, estimated the rate constants k_i of the various steps i from the associated free energy—activation energies not being available—and took as a measure of the activity A the rate constant of the slowest step:

$$A = k_B T \min_i \left[\ln \left(\frac{k_i}{k_0} \right) \right] \tag{14.5}$$

Once more, a plot of the calculated activity versus the O binding energy results in a volcano curve (see Figure 14.5). On the ascending branch the rate is limited by the removal of adsorbed O or OH; on Ag and Au it is the splitting of O_2. If this were indeed the correct plot for oxygen catalysis, there would be little hope to find better catalysts, since there is not much room at the top. However, this is just the result of a calculation, and though it represents the present state of the art, it need not be the final word. The descending branch is based on two points for silver and gold only, and the prediction that gold is so much worse than silver is at variance with Figure 14.4(a), which is based on experimental data. Also, in alkaline solutions gold is a good catalyst for oxygen, so this plot should be regarded as the result of a good theoretical calculation, but not as the ultimate word on oxygen catalysis.

In fact, the descending branch of volcano plots, where the rate decreases with decreasing, more exergonic, adsorption energy, is always problematic. Often there is more than one adsorption site, and the reaction may pass via an adsorbate that is less strongly bound or simply use another pathway. A case in point is hydrogen evolution

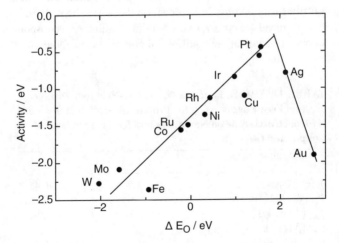

FIGURE 14.5 Calculated activity for oxygen reduction as a function of the oxygen binding energy [7].

on Pt(111), where the intermediate is not the strongly bound species adsorbed in the face-centered-cubic (fcc) threefold-hollow sites; instead, it passes via a weakly adsorbed species, which is probably adsorbed on top, and has a more favorable adsorption energy. A further complication for oxygen reduction is the fact that on many metals oxygen reduction competes with incipient surface oxidation, which may block reaction sites.

So, from a theoretical point of view our present understanding of oxygen reduction and evolution is limited and remains a challenge for the future. However, from a practical point of view a huge number of possible catalysts have been tried, since the potential gains are enormous, but so far to little avail. It would be surprising, if somebody found a bulk material that has much better catalytic properties than platinum. Therefore, in accord with current trends, hope has turned to nanostructured materials, which can have electronic properties quite different from those of bulk matter (see Chapters 3, 4, 6, 9, and 13).

As an illustration, we consider an idealized system, monoataomic wires of gold and copper. As bulk metals, both are poor catalysts for hydrogen evolution, and proton adsorption is endergonic at the equilibrium potential. As discussed in Chapter 6 on the catalysis of electron transfer, this is due to the fact that the d band lies well below the Fermi level and therefore does not contribute to the bonding of hydrogen. In the form of monoatomic nanowires, these metals experience not only an increase of the work function and a shortening of the bond distance (see Table 14.1), but also a shift of the d band. As shown in Figure 14.6, in the wires the d bands extend right to the Fermi level. Obviously, the Fermi level is the upper limit for the d band density of states, since also in the wires all d valence orbitals must be filled. This shift of the d band has far-reaching consequences for hydrogen adsorption: On the surface of the *bulk* metal, the hydrogen $1s$ orbital forms bonding and antibonding states with the d bands. The bonding state lies below the the the d band; the antibonding state lies above but below the Fermi level; therefore both bonding and antibonding states are filled and do not contribute to bonding. In contrast, on the wires the antibonding states lie above the Fermi level and are empty, so the bonding results. As a consequence, proton adsorption is now exergonic at the equilibrium potential (see Table 14.1).

TABLE 14.1 Work Function Φ, Nearest-Neighbor Distance d_{nn}, and Free-Energy ΔG_{ad} for Proton Adsorption at Standard Hydrogen Electrode on Nanowires and fcc(111) Surfaces of Copper and Gold

	Cu	Au
Φ [eV] (wire)	5.24	6.43
Φ [eV] (111) surface	4.50	5.261
d_{nn}/ [Å] (wire)	2.61	2.64
d_{nn}/ [Å] (bulk)	2.59	2.93
ΔG_{ad} [eV] (wire)	−0.31	−0.51
ΔG_{ad} [eV] (111) surface	0.10	0.41

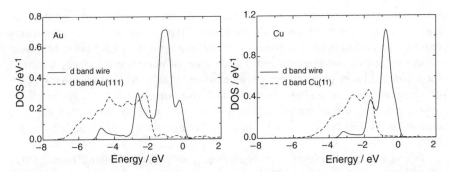

FIGURE 14.6 The *d* band of monoatomic gold and copper wires in comparison with the bulk metals [8]. The Fermi level has been taken as the energy zero.

Thus, gold and copper nanowires promise to be good catalysts for hydrogen evolution. Naturally, free-standing wires, for which the calculations have been performed, would not be a practical electrode material. However, preliminary results indicate that the good catalytic activities survive when the wires are adsorbed at the steps of graphite, which is a popular catalyst support. Gold is much cheaper and abundant than platinum; therefore a stable gold catalyst for hydrogen oxidation would greatly reduce the cost of fuel cells.

For oxygen reduction, where the greater challenge lies, there are presently two strands of promising approaches. We discuss them briefly, but with a word of warning: Since so much is at stake in this area, it is wise to remain somewhat skeptical until the results have been thoroughly examined and reproduced by other groups. One approach employs so-called skin catalysts, where the uppermost surface layer consists of platinum and the second layer of cheaper metal such as Fe, Co, Ni, Ti, and V. These can be prepared either in the form of flat surfaces or as nanoparticles [see Fig. 14.7(a)]. Preparation is aided by the fact that alloys of the form Pt_3X, X

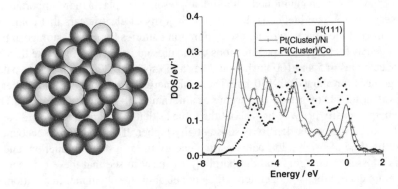

FIGURE 14.7 (a) Nanoparticle of a Pt_3Co skin catalyst (schematic); the blue (dark) spheres represent platinum, the yellow (light) cobalt. (b) *d* Bands for Pt(111) and Pt_3X clusters; our calculations.

being one of the above-mentioned metals, exhibit surface segregation with Pt forming the top layer and an excess of X in the second layer [9,10]. Such catalysts not only require less platinum, but the presence of the other metal actually has a beneficial effect on the catalytic properties of the top layer of platinum: It induces a shift of the d band to lower energies, which reduces the adsorption of nonreactive species. This shift must not be too large, because otherwise platinum would lose its reactivity toward oxygen. At the time of writing, Pt_3Co looks most promising, but this may change before this book is actually in print, because there is much activity in this field.

The other approach uses non–precious metal catalysts, which should be as good as platinum but much cheaper. In particular, several oxides have been shown to possess outstanding catalytic activities for both oxygen reduction and evolution, and in some instances they can compete with platinum (see Chapter 11). Iron complexes are also good candidates, not least because our own body uses an iron complex to bind oxygen reversibly. Very recently, a major breakthrough has been claimed: the synthesis on an iron-based catalyst supported on microporous carbon that can compete with platinum [11]. Although the nature of the catalytic sites in these systems is not known precisely, it seems that the presence of surface nitrogen and of carbon micropores is essential. Obviously, these results require verification, and it remains to be seen if these catalysts can be used in real cells.

Sometimes it is suggested that we should learn from nature how to develop catalysts—after all, nature had billions of years to develop and optimize energy consumption. However, nature can afford to be wasteful and compensate inefficiency by abundance. Thus a tree has an abundance of foliage to compensate for the low efficiency of photosynthesis (about 5%). If it were otherwise, a tree might look like a Magritte painting: a stem with a single large leaf. Also, biological systems are short-lived; therefore all living beings have repair and replacement systems, which would be difficult to implement in technological cells. So we should look at nature for inspiration, but not to find recipes for electrocatalysts.

Before we suggest scenarios for the future, a few words about batteries, which so far we have not been mentioned, although at present they play a more important role in everyday life than fuel cells. Indeed, the majority of electric cars that come onto the market at the time of writing use lithium ion batteries for energy storage and not hydrogen as fuel. Typical specifications of the present prototypes read like this: The car accelerates in 5 s to 100 $km\,h^{-1}$ and has a top speed of 200 $km\,h^{-1}$. One battery charge lasts for about 100–150 km. The total energy stored in an 180-kg battery is equivalent to 4 L of petrol; recharging at a normal 220-V plug takes 4–5 h. Thus, the energy density per weight is about the same as that of a normal lead–acid battery [12], though the power density is substantially higher. If one takes into account the finite lifetime of lithium ion batteries, the energy used in their production, and the energy losses in charging and discharging, it is easy to see that these cars actually use more energy than a moderately efficient petrol or diesel engine. In addition, our present power grid could not possibly supply the required electricity if the majority of cars used batteries as an energy source; a whole new network would have to be constructed.

So where does this leave us, what can we say about the impact of electrochemistry on energy problems? Since, as they say, predictions are difficult to make, especially if they are about the future, we consider two different scenarios. In the pessimistic version progress in fuel cell and battery technology remains incremental, as it has been in the past 20 years. Efficiency will slowly rise, and prices decrease, but there will be no breakthrough in catalysis. Nevertheless, electrochemical power will play an increasing role in transport. Hybrid cars combining lithium batteries with conventional motors already have an impact on the market. Their share will grow, and in a few years the first commercial all-electric cars using fuel cells or batteries will invade our cities. This development will not come about because it will save energy, but because there is a political will for it. It will reduce noise and emission in the cities and give a boost to the ailing car industries. Politicians will claim that they do everything possible to solve environmental and energy problems, and in the automobile shows they will stand smilingly next to the latest all-electric Porsche, taking the place of the lightly clad models of the past, which have been eliminated for reasons of political correctness. The electric cars will be light, sleek, and sexy and run out of energy when leaving the city limits, so that their owners will switch to trains for long-distance traveling. Industry will demand the foundation of more research institutes for applied electrochemistry, as they are doing in Germany now. All in all this will be a development for the good, but it will do little to save energy or reduce the emission of greenhouse gases, except in the case of the few countries that can produce electricity cheaply from natural resources. A prime example is Iceland, which has an abundance of geothermal power (see Figure 14.8), and using the generated electricity directly in cars would save energy and not pollute the environment. This should make Iceland a perfect testing ground for electric cars.

FIGURE 14.8   Geothermal power station in Iceland.

In the optimistic scenario we find a stable catalyst that is both considerably cheaper and more effective than platinum. This would make Ostwald's dream come true. Electrochemistry would have made a huge contribution to a future with clean energy, and we could leave the remaining problems, like storage and distribution of hydrogen, to our colleagues from the engineering department.

Finally, we like to mention an option which can strongly contribute to solve the problem of energy resources and which does not depend on scientific breakthroughs: the decrease of consumption. However, human nature being what it is, perhaps it is easier to find the elixir of life, the philosophers' stone, to transform lead into gold, to find the ideal catalyst, than to change social attitudes!

REFERENCES

1. T. D. Searchinger et al., *Science*, 326 (2009) 527.
2. D. Mitchell, A note on rising food crisis, The World Bank, Policy Research Working Paper No. 4682, 2008
3. R. F. Service, *Science*, 326 (2009) 516.
4. P. Vassilev and M. T. M. Koper, *J. Phys. Chem. C*, 111 (2007) 2607.
5. A. J. Appleby, *Catalysis Rev.*, 4 (1970) 221.
6. S. Trasatti, *J. Electroanal. Chem.*, 111 (1980) 125.
7. J. K. Nørskov et al., *J. Phys. Chem. B*, 108 (2004) 17886.
8. E. Santos, P. Quaino, G. Soldano, and W. Schmickler, *Electrochem. Commun.*, 11 (2009) 1764.
9. V. Stamenkovic et al., *Ang. Chemie Int. Ed.*, 45 (2006) 2897.
10. V. Stamenkovic et al., *Nature Mater.*, 6 (2007) 241.
11. M. Lefevre, E. Proietti, F. Jaouen, and J. P. Dodelet, *Science*, 324 (2009) 71.
12. L. Michels, fortu Research GmbH, private communication.

Catalysis in Electrochemistry: From Fundamentals to Strategies for Fuel Cell Development,
First Edition. Edited by Elizabeth Santos and Wolfgang Schmickler.
© 2011 John Wiley & Sons, Inc. Published 2011 by John Wiley & Sons, Inc.